Molecular Layer Deposition for Tailored Organic Thin-Film Materials

This book provides concepts and experimental demonstrations for various types of molecular layer deposition (MLD) and organic multiple quantum dots (organic MQDs), which are typical tailored organic thin-film materials produced by MLD. Possible applications of MLD to optical interconnects, energy conversion systems, molecular targeted drug delivery, and cancer therapy are also proposed. First, the author reviews various types of MLD processes including vapor-phase MLD, liquid-phase MLD, and selective MLD. Next, he introduces organic MQDs. The author then describes the design of light modulators/optical switches, predicts their performance, and discusses impacts of the organic MQDs on them. He then also discusses impacts of the organic MQDs on optical interconnects within computers and on optical switching systems. Finally, the author presents MLD applications to molecular targeted drug delivery, photodynamic therapy, and laser surgery for cancer therapy. This book is intended for researchers, engineers, and graduate students in optoelectronics, photonics, and any other field where organic thin-film materials can be applied.

Optics and Photonics

Series Editor: Le Nguyen Binh
Huawei Technologies, European Research Center, Munich, Germany

Molecular Layer Deposition for Tailored Organic Thin-Film Materials

Tetsuzo Yoshimura

CRC Press
Taylor & Francis Group
Boca Raton London New York

CRC Press is an imprint of the
Taylor & Francis Group, an **informa** business

Designed cover image: Naoko Yoshimura

First edition published 2023
by CRC Press
6000 Broken Sound Parkway NW, Suite 300, Boca Raton, FL 33487-2742

and by CRC Press
4 Park Square, Milton Park, Abingdon, Oxon, OX14 4RN

CRC Press is an imprint of Taylor & Francis Group, LLC

Library of Congress Cataloging-in-Publication Data
Names: Yoshimura, Tetsuzo, author.
Title: Molecular layer deposition for tailored organic thin-film materials / Tetsuzo Yoshimura.
Description: First edition. | Boca Raton : CRC Press, [2023] |
Series: Optics & photonics | Includes bibliographical references and index. |
Identifiers: LCCN 2022042617 (print) | LCCN 2022042618 (ebook) |
ISBN 9780367554743 (hardback) | ISBN 9780367555603 (paperback) | ISBN 9781003094012 (ebook)
Subjects: LCSH: Organic thin films. | Molecular electronics–Materials. |
Monomolecular films. | Quantum dots. | Atomic layer deposition.
Classification: LCC TK7872.T55 Y67 2023 (print) | LCC TK7872.T55 (ebook) |
DDC 621.3815/2–dc23/eng/20221206
LC record available at https://lccn.loc.gov/2022042617
LC ebook record available at https://lccn.loc.gov/2022042618

ISBN: 978-0-367-55474-3 (hbk)
ISBN: 978-0-367-55560-3 (pbk)
ISBN: 978-1-003-09401-2 (ebk)

DOI: 10.1201/9781003094012

Typeset in Times New Roman
by codeMantra

To the memory of my wife, Yoriko

Contents

Preface

Tailored organic thin-film materials, in which molecules are artificially assembled in designated arrangements, improve the performance of materials/devices/systems and generate new phenomena due to their capability to control the dimensionality and shapes of electron wavefunctions. Molecular layer deposition (MLD) is an ideal method to produce the tailored organic thin-film materials. MLD grows thin films with monomolecular steps by introducing different kinds of source molecules sequentially in designated order onto surfaces.

MLD was inspired by atomic layer deposition (ALD). While ALD enables layer-by-layer growth to form inorganic thin films, MLD enables dot-by-dot growth, namely, molecule-by-molecule growth of polymer wires and molecular wires, to form organic thin films. In MLD, since the wire locations and orientations can be controlled, three-dimensional growth is possible to build wire networks with designated molecular arrangements in principle. MLD is performed in the vapor phase or in the liquid phase.

The present book describes fundamentals of MLD and demonstrates the prospect of the tailored organic thin-film materials, especially featuring organic MQDs grown by MLD. Enhancement of the optical nonlinearity in the organic MQDs is theoretically predicted, and the impacts on light modulators/optical switches, optical interconnects within boxes of computers, and optical switching systems are discussed. Furthermore, applications of MLD to the energy conversion field like photovoltaic/photosynthesis devices and the biomedical/cancer therapy field like the molecular targeted drug delivery and the photodynamic therapy are proposed.

In 2011, a book entitled *Thin-Film Organic Photonics: Molecular Layer Deposition and Applications* was published by CRC Press/Taylor & Francis. In the present book, the structure is reorganized to make the story clear, newly-obtained results are added, and proposals related to the energy conversion and the cancer therapy are enriched. In the final chapter, achievements of forefront research groups for the microelectronics field, including resists for lithography, diffusion barriers, and flexible insulators/semiconductors/conductors, and the molecular sieving/catalysis field, including batteries, gas separation/water purification, dry reforming, and photocatalysis, are reviewed.

The author would like to thank Prof. Yoshio Nakai and Prof. Riso Kato of Kyoto University, and Prof. Masao Kamada of Saga University for initiating him into the solid-state physics and spectroscopy. The author would like to thank Prof. Kunihiko Asama of Tokyo University of Technology for supporting him to continue the research work and giving instruction and encouragement for a long time, and General Manager Kohei Kiyota of Fujitsu Laboratories for instructing him in the scientific research skill and in molecular nanotechnologies. The author would like to thank General Managers Masao Tanaka, Akira Ihaya, Hideo Takahashi, Masaaki Kobayashi, Hirofumi Okuyama, Osamu Takahashi, and Isao Fudemoto of Fujitsu Laboratories, and President Yoshio Honda of Fujitsu Computer Packaging Technologies (FCPT), San Jose, California, for their support and encouragement. The author would also like to thank Drs. Azuma Matsuura, Wataru Sotoyama, Tomoaki Hayano, and Satoshi Tatsuura of Fujitsu Laboratories for their great contribution to the research on organic nonlinear optical materials/devices and theoretical predictions of optical nonlinearity by the molecular orbital method, colleagues of FCPT for giving him exciting experiences to generate new research ideas, and students who joined Yoshimura Laboratory in Tokyo University of Technology for their collaboration in the research.

Finally, the author would like to express his sincere gratitude to Mr. Marc Gutierrez, Mr. Nick Mould, Ms. Kianna Delly and Ms. Kirsty Hardwick, Editors of CRC Press/ Taylor & Francis, Ms. Julia Hanney, Production Editorial Manager of CRC Press/ Taylor & Francis, and Ms. Sathya Devi, Project Manager of CodeMantra, for giving the opportunity to write the present book and instructing him to complete it.

Author

Tetsuzo Yoshimura was born in Tokyo, Japan in 1951, and graduated from Tokyo Metropolitan Aoyama High School in 1970. He received the B.Sc. degree in Physics from Tohoku University in 1974, and the M.Sc. and Ph.D. degrees in Physics from Kyoto University in 1976 and 1985, respectively.

In 1976, he joined Fujitsu Laboratories Ltd., where he was engaged in research on dye sensitization, electrochromic thin films, amorphous superlattices, organic nonlinear optical materials, and polymer optical circuits. He invented the molecular layer deposition (MLD) and the self-organized lightwave network (SOLNET).

From 1997 to 2000, he was with Fujitsu Computer Packaging Technologies (FCPT), Inc., San Jose, California, where he planned the strategy of 3-D integrated optical interconnects within boxes of computers.

From 2001 to 2017, he was a professor of Tokyo University of Technology, where he extended the research on MLD-based nanotechnologies and SOLNET-based optoelectronics for optical interconnects, solar energy conversion systems, and cancer therapy.

He is currently a professor emeritus of Tokyo University of Technology.

1 Introduction

Tailored organic thin-film materials, in which molecules are artificially assembled in designated arrangements, improve the performance of materials/devices/systems and generate new phenomena due to their capability to control the dimensionality and shapes of electron wavefunctions.

The dimensionality, which affects optical and electronic properties, is easily controlled in organic thin-film materials, especially between dimensionality of one and zero. Figure 1.1 depicts examples of dimensionality control in polymer wires. A conjugated polymer wire is regarded as a one-dimensional (1-D) system, that is, a quantum wire, in which π-electron wavefunctions are widely spread. By inserting molecular units that disconnect the spread of π-electron wavefunctions, the dimensionality is reduced to somewhere between one and zero to make a 0.5-dimensional system, which corresponds to a long quantum dot (QD). By increasing the density of the disconnecting molecular units, the spread of π-electron wavefunctions becomes narrow, creating a zero-dimensional (0-D) system, which corresponds to a short QD. Similar dimensionality control might be achieved in molecular wires consisting of molecules connected to each other by electrostatic force.

The electron wavefunction shapes in organic thin-film materials are controlled by placing constituent molecules artificially in designated arrangements. For example, as depicted in Figure 1.2, widely spread, confined, centrosymmetric, and noncentrosymmetric wavefunctions can be formed in polymer/molecular wires by controlling the molecular sequence, generating desired material properties.

In addition, organic materials will contribute to overcoming the resource issue because they are mainly made of carbon (C), hydrogen (H), nitrogen (N), and oxygen (O),

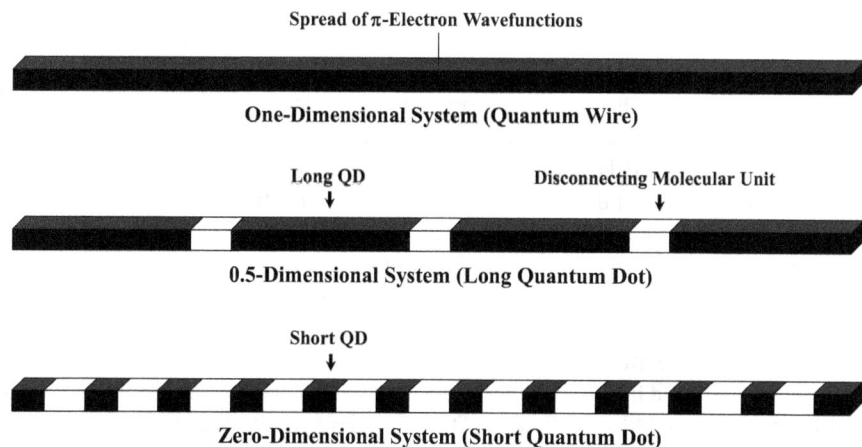

FIGURE 1.1 Control of dimensionality in polymer wires.

DOI: 10.1201/9781003094012-1

Electron Wavefunction

Molecule

Widely-Spread Electron Wavefunction

Confined Centrosymmetric Electron Wavefunction

Confined Non-Centrosymmetric Electron Wavefunction

Various Types of Electron Wavefunction

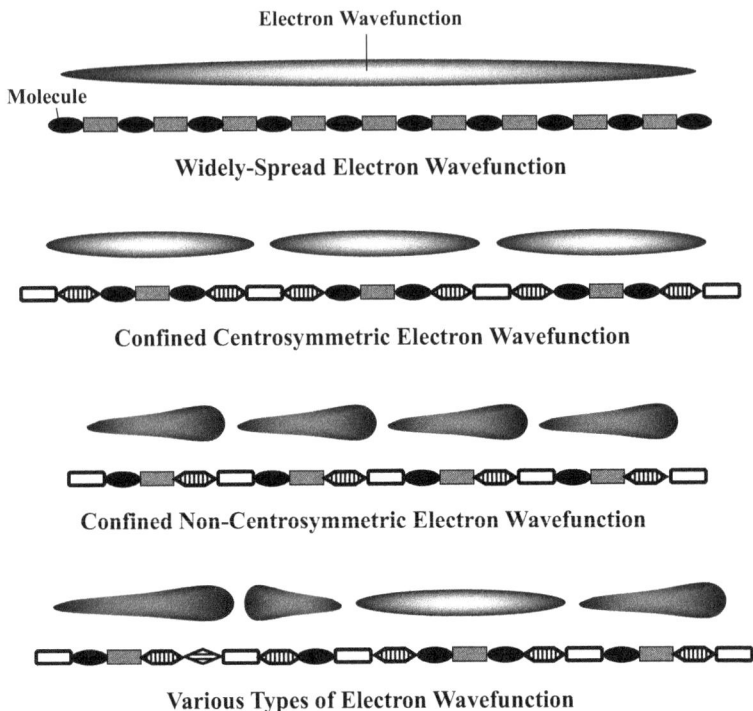

FIGURE 1.2 Control of electron wavefunction shapes by placing constituent molecules in designated arrangements.

which exist in abundance at the surface of the Earth. As summarized in Figure 1.3, currently, energy like electric power and fuel and materials like semiconductors, metals, and plastics are mostly obtained from resources in the Earth's interior. This destroys energy balance of the Earth and causes resource shortness. These problems will be solved if energy is obtained by photovoltaics and materials are obtained by photosynthesis from C, H, N, and O at the Earth's surface.

In order to produce the tailored organic thin-film materials, a method to assemble molecules in designated arrangements is essential. Molecular layer deposition (MLD) is ideal to do this. In MLD, different kinds of molecules are sequentially introduced onto a surface to connect the molecules one by one in the designated order, enabling us to grow tailored organic thin films consisting of polymer/molecular wires formed with monomolecular growth steps. MLD also enables us to grow ultrathin/conformal organic thin films not only on flat surfaces but also on three-dimensional (3-D) surfaces such as trenches, holes, particles, and various rough/porous/tortuous structures. MLD is performed both in the vapor phase and in the liquid phase.

MLD was inspired by atomic layer deposition (ALD), which was originally called atomic layer epitaxy (ALE). While ALD realizes layer-by-layer growth of inorganic thin films, MLD realizes dot-by-dot growth, namely, molecule-by-molecule growth, of organic thin films. MLD has a potentiality of 3-D growth to build polymer/

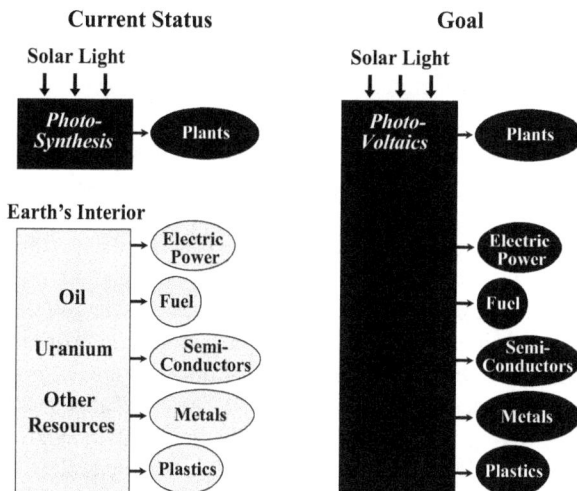

FIGURE 1.3 Solution of the energy balance destruction and resource shortness: energy production by photovoltaics and organic material production by photosynthesis from C, H, N, and O.

molecular wire networks because the wire locations and orientations can be controlled by selective MLD.

MLD is promising to produce tailored organic thin-film materials in the photonics/electronics field including nonlinear optical materials, semiconductors, and optical circuit materials. MLD is also promising in the energy conversion field including photovoltaics, photosynthesis, and sensitization. Liquid-phase MLD (LP-MLD), which is performed in the liquid phase, has the potentiality to be applied to the biomedical/cancer therapy field including molecular targeted drug delivery, photodynamic therapy, and laser surgery.

MLD has also been applied to the microelectronics field such as resists for lithography, diffusion barriers, and flexible insulators/semiconductors/conductors, and to the molecular sieving/catalysis field such as batteries, gas separation/water purification, dry reforming, and photocatalysis.

In the present book, the concept of MLD, experimental proof of concepts, and prospects of the tailored organic thin-film materials grown by MLD are described. As a typical example of the tailored organic thin-film materials, the organic multiple quantum dots (MQDs), namely, the polymer MQD and the molecular MQD, are proposed, and their impact on nonlinear optical materials, light modulators/optical switches, optical interconnects within boxes of computers, and optical switching systems are discussed. Applications of MLD to the energy conversion field and the biomedical/cancer therapy field are also proposed. To widen the scope of the book, a review of forefront research in the microelectronics and molecular sieving/catalysis fields is presented.

In Chapter 2, examples of inorganic tailored materials, in which atomic arrangements are artificially controlled, are presented, including wavefunction control in quantum corrals by scanning tunneling microscope (STM), wavefunction control in photonic/electronic devices by molecular beam epitaxy (MBE), amorphous superlattices fabricated by plasma chemical vapor deposition (plasma CVD), and wavefunction control in electrochromic thin films by sputtering.

In Chapters 3–5, fundamentals of MLD are described. Chapter 3 starts with a brief explanation of ALD and the beginning of MLD inspired by ALD. Concept, equipment, and growth mechanisms of MLD are explained. And then, the proof of concepts of the monomolecular-step growth in MLD is experimentally demonstrated for polymer wires and stacked structures of p-type and n-type dye molecules. Chapter 4 presents the selective MLD. After processes for selective organic thin-film growth are classified, area selective growth on hydrophilic/hydrophobic surfaces, horizontally-aligned selective growth on anisotropic atomic-scale structures, and electric-field-assisted selective growth are explained. And then, vertically-aligned selective MLD utilizing seed cores with self-assembled monolayers (SAMs) is described, and a possibility of location-/orientation-controlled selective MLD is suggested. Chapter 5 summarizes featured capabilities of MLD and discusses expected impacts of the tailored organic thin-film materials grown by MLD on the photonic/electronic field. The chapter gives possible examples of advanced organic photonic/electronic devices and MLD processes, and the molecular nano-duplication (MND) for mass production.

In Chapters 6–9, the polymer MQD and its applications to nonlinear optical materials and light modulators/optical switches are described. Chapter 6 presents the concept of the polymer MQD and the MLD process for the MQD growth. Quantum confinement of electrons in QDs is experimentally demonstrated to indicate the prospects of dimensionality control accomplished by MLD. In Chapter 7, the electro-optic (EO) effect, which is one of the nonlinear optical effects, in polymer MQDs is theoretically predicted using the molecular orbital (MO) method. It is revealed that drastic enhancement of the EO effect is expected by controlling electron wavefunctions in the polymer MQDs, where donor/acceptor groups are distributed in designated arrangements. In Chapter 8, some results of experimental work on currently-available EO molecular crystals and poled polymers are presented. Optical switches made of the EO poled polymers and optical waveguides fabricated by the selective growth are described. In Chapter 9, light modulators/optical switches are classified. Details of the design and the predicted performance are described for the waveguide prism deflector (WPD) optical switch, and the impacts of the polymer MQD on the performance are discussed.

Chapter 10 describes the self-organized lightwave network (SOLNET). SOLNETs enable us to form self-aligned coupling optical waveguides between misaligned devices, to construct 3-D optical wiring in free spaces, and to guide light waves onto specific objects. These capabilities are effective to build optical interconnects and optical switching systems described in Chapter 11, solar energy conversion systems described in Chapter 12, and photo-assisted cancer therapy described in Chapter 13.

In Chapters 11–13, expected applications of MLD are proposed and the impacts of the organic MQD on the m are discussed. Chapter 11 presents the 3-D integrated optical interconnects within boxes of computers and the 3-D micro-optical switching systems (3D-MOSSs). Chapter 12 presents organic-MQD sensitization, waveguide-type thin-film photovoltaic (WTP) devices, film-based integrated solar energy conversion systems, and thin-film artificial photosynthesis devices. Chapter 13 proposes applications of LP-MLD to cancer therapy, including *in-situ* synthesis of drugs at cancer cell sites for molecular targeted drug delivery, the two-photon photo-dynamic therapy (PDT), and the SOLNET-assisted laser surgery.

From Chapters 2 to 13, the content is on the lines of the author's interests based on the research work performed in the author's research groups. Meanwhile, in Chapter 14, to attempt a comprehensive survey related to MLD, accomplishments by leading research groups in the world are reviewed for basic growth mechanisms and applications to microelectronics, molecular sieving/catalysis, and other fields.

2 Examples of Inorganic Tailored Materials

Inorganic tailored materials, in which atomic arrangements are artificially controlled, have been studied for a long time. In the present chapter, after a brief summary of thin-film growth methods, four examples of inorganic tailored materials are given to show how they improve the material/device performance or generate new phenomena. The first example is electron wavefunction control by scanning tunneling microscope (STM), which realizes the quantum corral and the quantum mirage. The second is wavefunction control in photonic/electronic devices by molecular beam epitaxy (MBE), which realizes the high-electron-mobility transistor (HEMT) and the multiple quantum well (MQW) light modulator. The third is transfer-doping control in amorphous superlattices fabricated by plasma chemical vapor deposition (plasma CVD). Based on the findings in the superlattice study, influence of the gate insulator composition on the performance of thin-film transistors (TFTs) is discussed. The fourth is wavefunction control in WO_x electrochromic thin films by sputtering, which realizes enhancement of small-polaron absorption.

2.1 CLASSIFICATION OF THIN-FILM GROWTH METHODS

In Table 2.1, thin-film growth methods are summarized. Vacuum evaporation and sputtering are common methods. By selecting atoms provided onto a substrate, inorganic thin films with intended compositions are deposited. When molecules are used instead of atoms, organic thin films are obtained. Chemical vapor deposition (CVD) grows inorganic thin films utilizing chemical reactions between source molecules. Film compositions are varied by controlling source molecular gases. Organic CVD, which grows organic thin films, is the organic counter part of CVD. Molecular beam epitaxy (MBE) grows inorganic thin films utilizing chemical reactions between atoms (and sometimes molecules) in ultra-high vacuum. MBE can control film

TABLE 2.1

Various Inorganic and Organic Thin-Film Growth Methods

Inorganic Thin Films	Organic Thin Films
Vacuum Evaporation	
Sputtering	
Chemical Vapor Deposition (CVD)	Organic CVD
Molecular Beam Epitaxy (MBE)	Vacuum Deposition Polymerization (VDP)
Atomic Layer Deposition (ALD)	Molecular Layer Deposition (MLD)
Scanning Tunneling Microscope (STM)	

DOI: 10.1201/9781003094012-2

compositions along film thickness directions precisely. Vacuum deposition polymerization (VDP), which grows polymer thin films in vacuum, is the organic counter part of MBE.

Atomic layer deposition (ALD) is an ideal growth method of inorganic thin films. It enables layer-by-layer growth by switching source molecules alternately. ALD utilizes the self-limiting effect to achieve strict monolayer growth. Molecular layer deposition (MLD), which forms organic thin films with monomolecular growth steps, is the organic counterpart of ALD and is an ideal growth method of organic thin films.

Scanning tunneling microscope (STM) is a method to observe atomic-scale objects. STM also works as a method to position atoms and molecules in designated distribution.

In the present chapter, four examples of inorganic tailored materials produced by STM, MBE, plasma CVD, and sputtering are given. For ALD, since it is closely related to MLD, a detailed explanation is provided in Chapter 3 along with comparison to MLD. MLD and tailored organic materials grown by MLD are described in Chapters 3–14.

2.2 WAVEFUNCTION CONTROL IN QUANTUM CORRALS BY SCANNING TUNNELING MICROSCOPE

STM was invented by Binnig and Rohrer [1] originally for observing atomic-scale objects. STM also enables us to manipulate atoms and molecules on a substrate and detect electron wavefunctions. In the present section, examples of electron wavefunction control by STM are described.

Eigler et al. developed an atomic manipulation process, which utilizes atomic force between an atom on a substrate and atoms at the edge of an STM tip as depicted in Figure 2.1. Potential energy arising from Pauli's exclusion principle is positive and decreases rapidly with increasing the atom–atom distance while that arising from the van der Waals' force is negative and varies gradually. As a result, the total potential energy curve drawn by a solid line has a minimum. Attractive force works between the atoms for a distance longer than the minimum point, enabling the atomic manipulation as follows [2]:

(1)→(2): By moving the tip down to an atom on a substrate, attractive force works between the atom and the tip.

(2)→(3): By moving the tip horizontally, the atom slides on the surface to a definite position.

(3)→(4): The tip is moved up to leave the atom at the position.

They drew characters with atoms by the sliding method.

Figure 2.2 shows the tunneling process in STM. When a tip is placed on a sample surface with a small gap of 1-nm scale, electron wavefunctions penetrate through the gap by the tunneling effect. The tunneling probability of electrons is proportional to $e^{-d\sqrt{\frac{2m}{\hbar^2}(V-E)}}$, where, d, m, and E are, respectively, gap length, electron mass, and

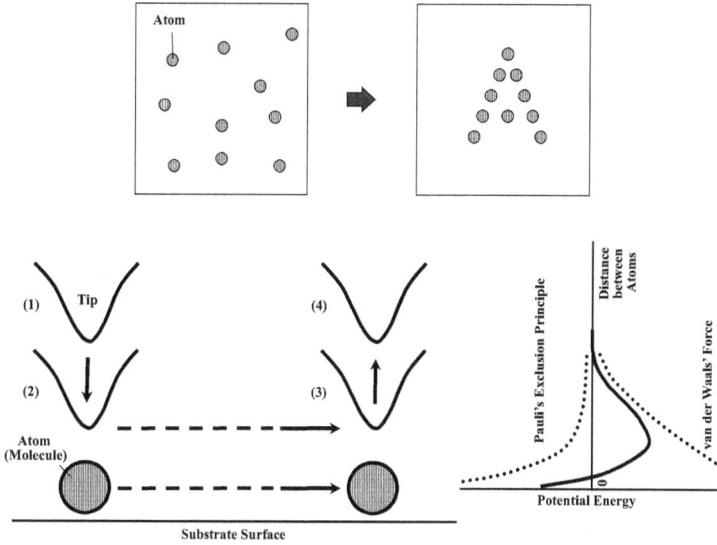

FIGURE 2.1 The sliding method, which is an atomic manipulation process to slide atoms across a substrate surface to definite positions utilizing the atomic force.

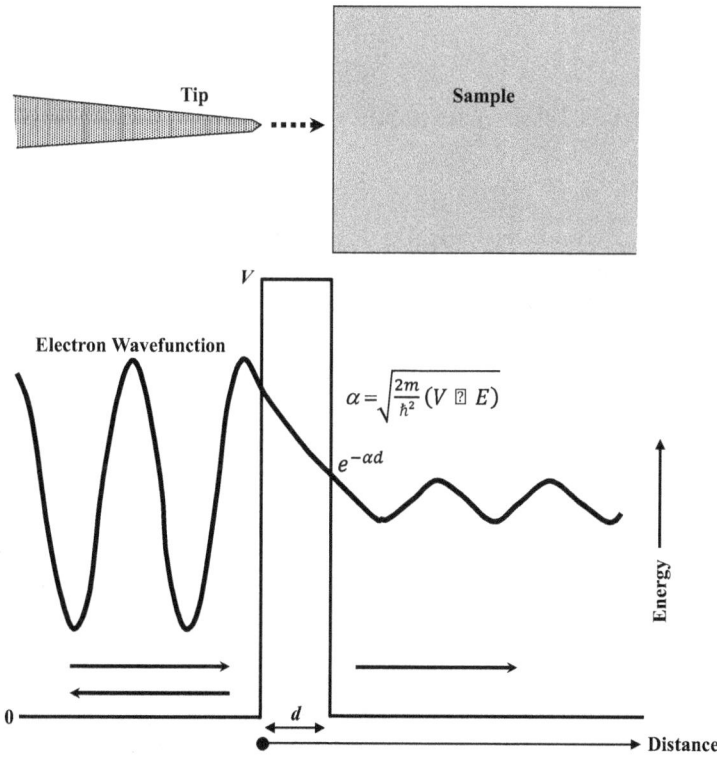

FIGURE 2.2 Tunneling process in scanning tunneling microscopy.

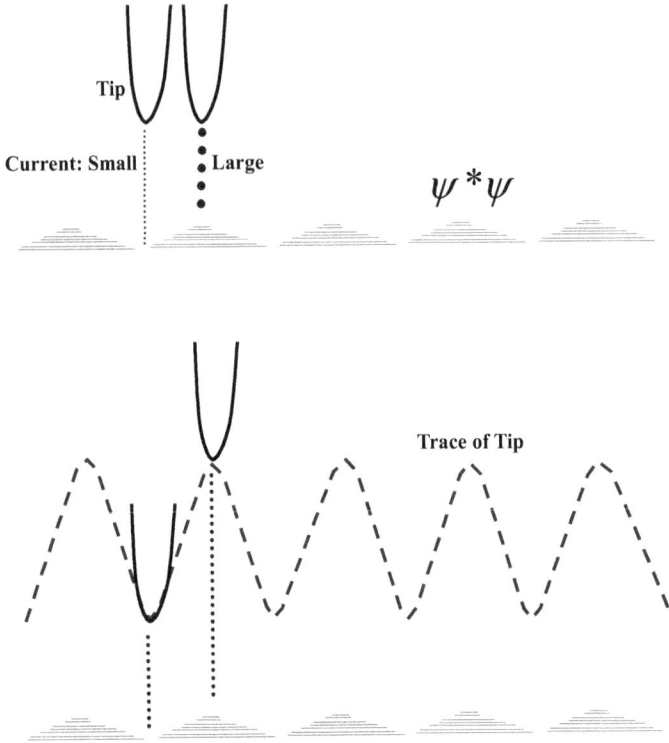

FIGURE 2.3 Detection of wavefunction shapes by scanning tunneling microscopy.

electron energy. V is the barrier height at the gap, and $\hbar$ is the Planck constant divided by 2π. The tunneling probability increases with decreasing d. Meanwhile, the probability of the electron transfer between the tip and the sample is proportional to $\psi^*\psi$, where ψ represents a wavefunction for an electronic state in the sample. Therefore, in the constant current mode, as Figure 2.3 shows, the tip goes away from the surface in the region with large $\psi^*\psi$, and gets close to the surface in the region with small $\psi^*\psi$ to keep the tunnel current constant. By tracing the up–down movement of the tip, the shape of $\psi^*\psi$ can be drawn [3,4].

Figure 2.4 shows an experiment of electron confinement in a quantum corral reported by Crommie et al. [3]. By STM, 48 Fe atoms were positioned into a ring with 7.13-nm radius on a Cu surface to construct a quantum corral. Due to the quantum confinement effect, a standing wave of ψ for an electron confined in the corral appears. By STM, the shape of $\psi^*\psi$ were observed.

Manoharan et al. succeeded in generating a quantum mirage in an elliptical quantum corral [4]. As Figure 2.5 shows, they placed a Co atom at f_1, which is one of the two focuses of the corral, and observed a quantum mirage projected around the other focus, f_2. Due to interference of electronic waves originated from the Co atom, the electronic structures surrounding the Co atom is regenerated near f_2 just like the mirage constructed by interference of light waves in an elliptical mirror.

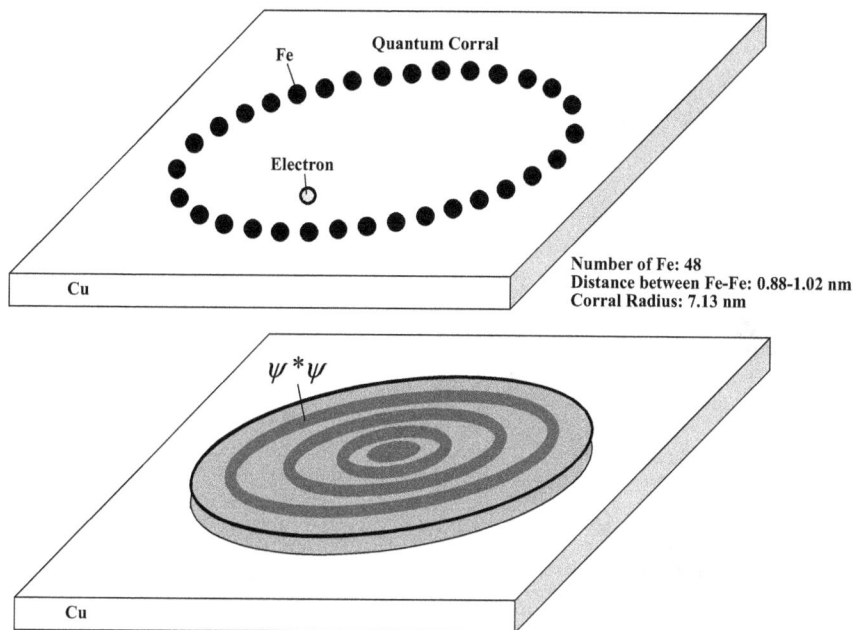

Quantum Corral

Fe

Electron

Cu

Number of Fe: 48
Distance between Fe-Fe: 0.88-1.02 nm
Corral Radius: 7.13 nm

$\psi^*\psi$

Cu

FIGURE 2.4 Electron confinement in a quantum corral.

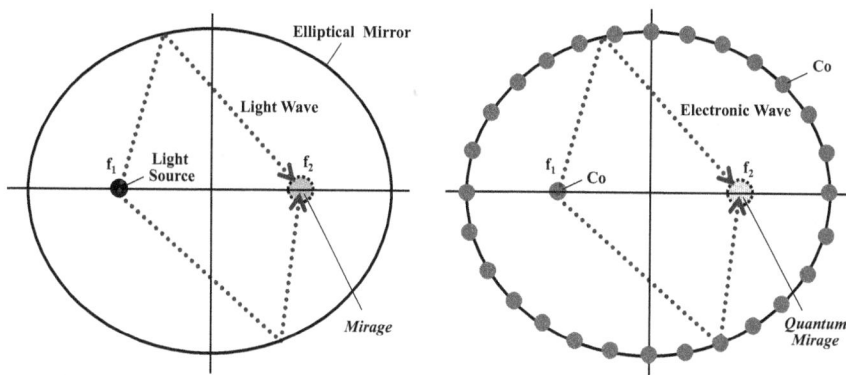

Elliptical Mirror

Light Wave

f_1 Light Source f_2

Mirage

Co

Electronic Wave

f_1 Co f_2

Quantum Mirage

FIGURE 2.5 Comparison between the mirage arising from the light wave interference and the quantum mirage arising from the electronic wave interference.

2.3 WAVEFUNCTION CONTROL IN PHOTONIC/ELECTRONIC DEVICES BY MOLECULAR BEAM EPITAXY

MBE is an important growth technique for inorganic thin films. Because MBE has monoatomic-layer thickness controllability achieved by strict thickness monitoring using the reflection high-energy electron diffraction (RHEED), it has greatly contributed to the photonic/electronic industry, realizing advanced transistors, laser diodes (LDs), light modulators, and so on.

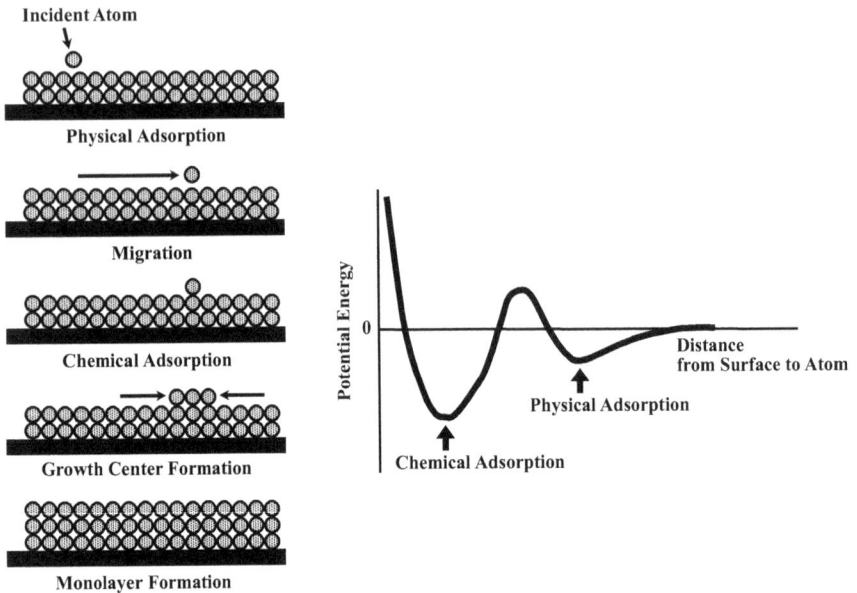

FIGURE 2.6 Growth mechanism of molecular beam epitaxy.

Figure 2.6 shows the growth mechanism of MBE. Incident atoms on the surface behave under a potential energy curve with two minimum points. The shallow minimum causes physical adsorption arising from weak force such as van der Waals' force, while the deep minimum causes chemical adsorption with chemical bonds to the surface. Incident atoms are weakly bound on the surface as physical adsorption initially. After the atoms migrate onto the surface for a while, the state changes from physical adsorption to chemical adsorption to fix the atoms at specific sites. Incident atoms that newly arrive and migrate on the surface reach the previously-fixed atoms to form growth centers. A monoatomic layer is grown from the growth centers to cover the entire surface. This procedure is repeated to grow thin films.

One problem of MBE is a lack of conformality. MBE cannot grow thin films on surfaces with three-dimensional (3-D) structures. Another problem is controllability. Even for flat surfaces, in order to control the growth thickness with monoatomic-layer accuracy, high-quality monitoring by RHEED is required. Another problem is mass productivity. Film growth can be performed only on a limited number of substrates. ALD solves these problems. The ALD is described in Section 3.1.

HEMT is a typical device realized by precise wavefunction control using MBE. Figure 2.7 shows a HEMT structure. In a wide-band-gap $Al_{1-x}Ga_xAs$ layer, Si atoms are doped as donors. Electrons from the dopants are transferred to an adjacent non-doped GaAs layer to generate a two-dimensional electron gas that acts as a channel. Thus, in HEMT, the doped layer is separated from the channel layer to reduce scattering of electrons, increasing the carrier mobility. This situation is similar to that of Silicon Valley, where there is little rain and water is supplied from other places for a comfortable life [5].

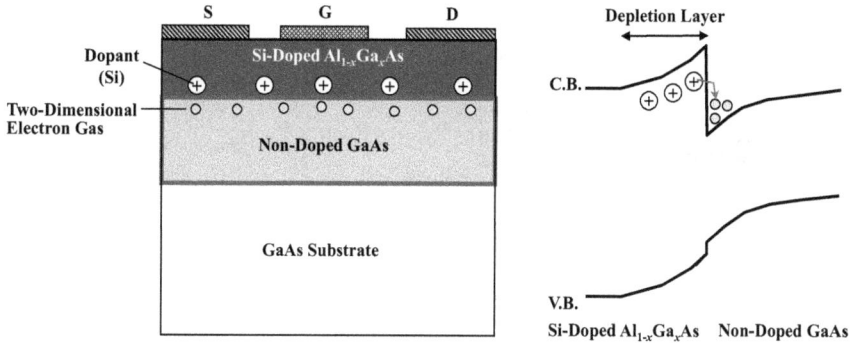

FIGURE 2.7 High-electron-mobility transistor (HEMT).

The multiple quantum well (MQW) light modulator is another device realized by precise wavefunction control using MBE. Figure 2.8a shows schematic illustrations of band diagrams and electron orbitals in a bulk and a quantum well structure. When an electron is excited by a photon, an electron-hole pair, namely, an exciton, is generated. Due to the attractive force between the electron and the hole, the energy level of the exciton is located below the conduction band edge by E_B, which is the binding energy of the exciton. In the quantum well, the excited electron is confined to a narrow space, and then, deformation in the electron orbital is caused, making the average distance between the electron and the hole smaller comparing to the case of the bulk. Consequently, in the quantum well structure, the binding energy increases, and the dissociation probability of the exciton decreases to increase the life time τ of the exciton [6].

Uncertainty principle tells us that, with an increase in τ, the uncertainty of energy decreases, namely, the width of the exciton absorption band becomes narrow. Thus, as schematically depicted in Figure 2.8b, sharpening of the exciton absorption band is realized in the quantum well structure. The sharpening is favorable for light modulation as described later. A blue shift of the absorption band in the quantum well structure is attributed to an increase in energy gap due to the quantum confinement of the electron/hole, as shown in Figure 2.8a.

When an electric field is applied to the quantum well structure, a slope appears in the energy bands, as illustrated in Figure 2.9a. Then, the energy level for the electron moves downward while that for the hole upward, resulting in a red shift of the absorption band, as illustrated in Figure 2.9b. The absorption band shift enables light modulation. For the same amount of the absorption band wavelength shift, the sharper the absorption band becomes, the larger the modulation contrast becomes.

In addition, by applying an electric field to the quantum well structure, the electron wavefunction and the hole wavefunction are moved to opposite directions to reduce the wavefunction overlap between the ground state and the excited state. This causes a decrease in transition probability for the electron excitation, inducing an absorption peak height decrease.

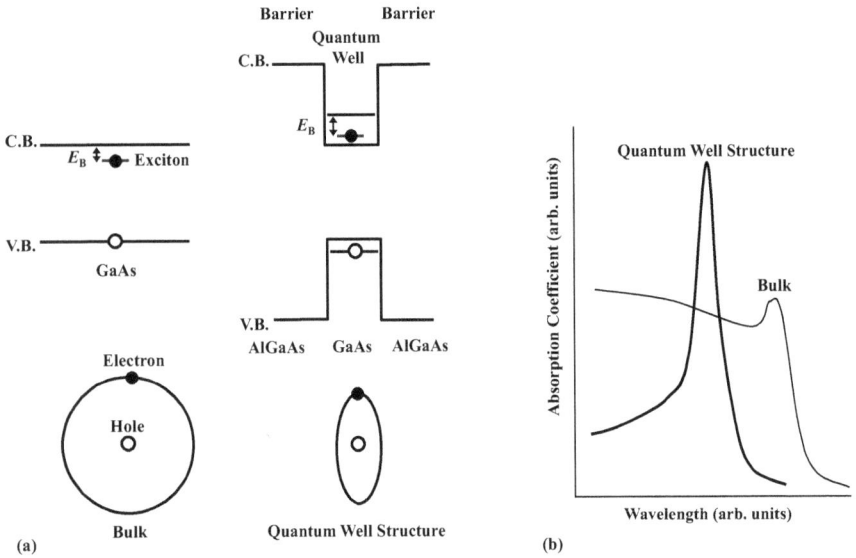

FIGURE 2.8 (a) Schematic illustrations of (a) band diagrams and electron orbitals and (b) absorption spectra in a bulk and a quantum well structure.

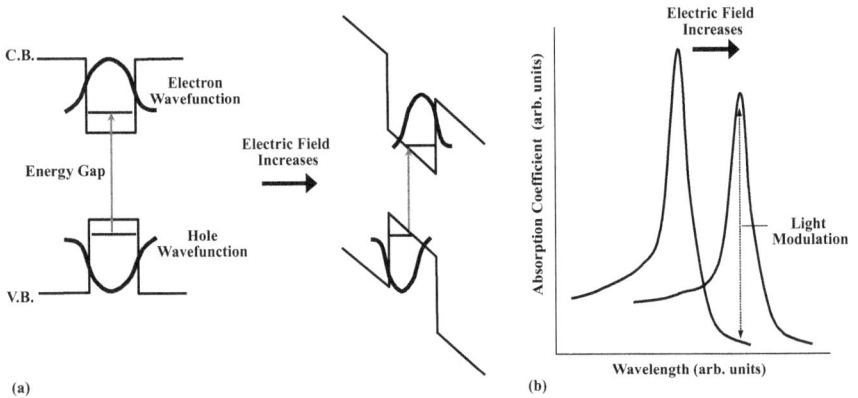

FIGURE 2.9 Operation principle of the multiple quantum well (MQW) light modulator. Effects of applied electric fields on (a) band diagrams and (b) absorption spectra of the quantum well structure.

2.4 AMORPHOUS SUPERLATTICES FABRICATED BY PLASMA CHEMICAL VAPOR DEPOSITION

2.4.1 TRANSFER DOPING IN A-SIN$_x$:H/A-SI:H AMORPHOUS SUPERLATTICES

One of the interesting phenomena observed in amorphous superlattices is the transfer doping [7–9]. Figure 2.10 depicts a model for an a-SiN$_x$:H/a-Si:H superlattice.

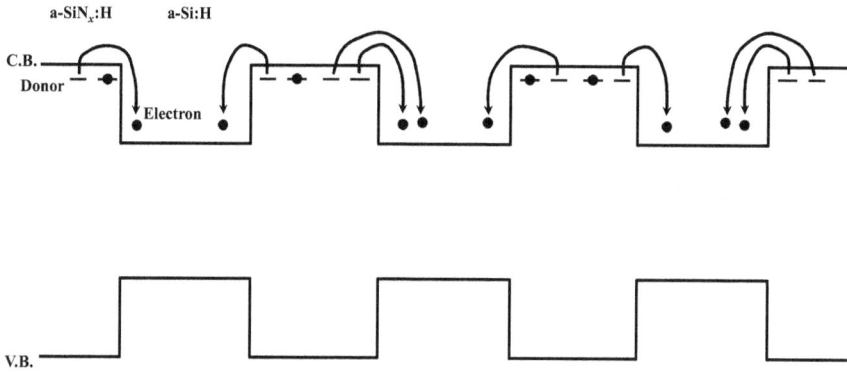

FIGURE 2.10 Model for the transfer doping in a-SiN$_x$:H/a-Si:H superlattices.

Electrons from donors in a-SiN$_x$:H layers are transferred to adjacent a-Si:H layers to increase in-plane electrical conductivity. This situation is similar to that of HEMT shown in Figure 2.7. The author's research group controlled the transfer doping by varying the layer composition and thickness.

Amorphous superlattices consisting of 12 a-SiN$_x$:H layers (thickness: L_N) and 13 a-Si:H layers (thickness: L_S) were fabricated by plasma CVD by changing the NH$_3$/SiH$_4$ gas composition periodically. The nitrogen content x of the a-SiN$_x$:H layer was 0.75, 0.85, 1.0, and 1.15. The in-plane conductivity was measured using slit-type electrodes with 1-mm gap under 100-V bias. The samples were illuminated by a yellow light-emitting diode (LED) (583 nm, 1 W/m²) to measure photocurrents [10,11].

Dependence of optical parameters of a-SiN$_x$:H on x is shown in Figure 2.11. The optical gap E_g was obtained from the relationship $h\nu\alpha = B(h\nu - E_g)^2$, where α, ν, and h are, respectively, absorption coefficient, frequency of light, and the Planck constant, and B is a material-related constant. In a range of $x > 0.85$, E_g increases dramatically with increasing x. B decreases with increasing x for $0 < x < 0.85$, and then, increases for $x > 0.85$. These results indicate that a principal structural change occurs around $x = 0.85$, namely, from a tetrahedral structure to an Si$_3$N$_4$-like structure [12].

As Figure 2.12a shows, with increasing x, dark- and photo-conductivity in a-SiN$_x$: H/a-Si:H superlattices increase by several orders of magnitude and reach the maximum at $x = 0.85$. For $x > 0.85$, they decrease rapidly with x. These results indicate that the transfer doping is most remarkable at $x = 0.85$. The dark- and photo-conductivity for $x = 0.85$ increase with increasing L_N as Figure 2.12b shows. This indicates that, as depicted in Figure 2.10, donors of the whole part of a-SiN$_x$:H layers contribute to the transfer doping [10,11].

The reason why the transfer doping becomes most remarkable at $x = 0.85$ can be explained as follows. In a range of $0 < x < 0.85$, a-SiN$_x$:H has a tetrahedral structure, and hence, the concentration of donor states originating from nitrogen atoms

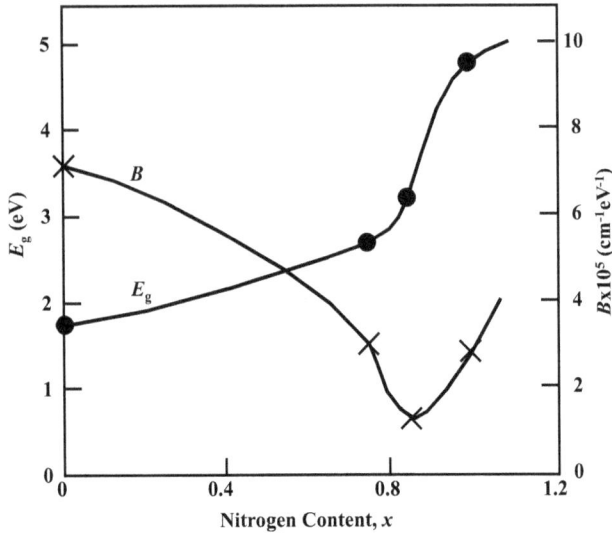

FIGURE 2.11   Dependence of optical parameters of a-SiN$_x$:H on nitrogen content x of the a-SiN$_x$:H layer. (Reprinted with permission from Yoshimura et al. [10].)

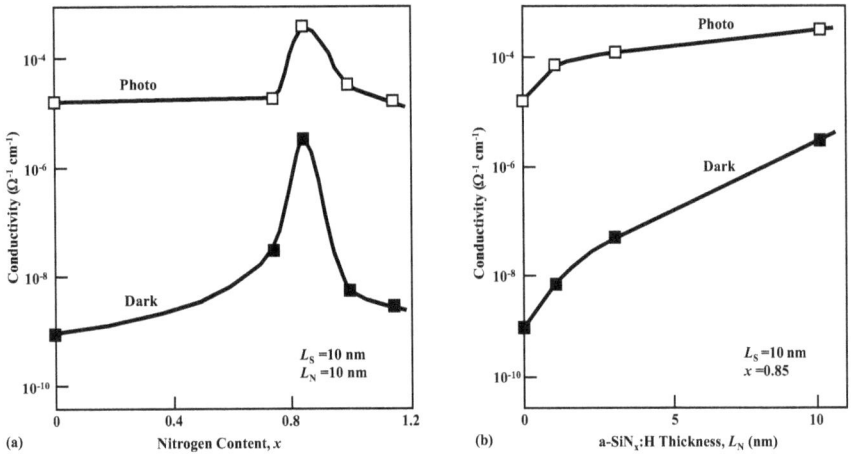

FIGURE 2.12   Dark- and photo-conductivity in a-SiN$_x$:H/a-Si:H superlattices as a function of (a) x and (b) a-SiN$_x$:H thickness L_N. (Reprinted with permission from Yoshimura et al. [10].)

increases with x. For $x > 0.85$, a-SiN$_x$:H changes toward an Si$_3$N$_4$-like structure, in which nitrogen atoms do not act as donors.

As shown in Figure 2.13a and b, the photocurrent decay becomes slow with increasing L_N, and the half-decay time $\tau_{1/2}$ tends to be saturated around $L_N \sim 10$ nm. These results imply that the interaction depth for the electron-trapping phenomena is ~10 nm in the a-SiN$_x$:H layer.

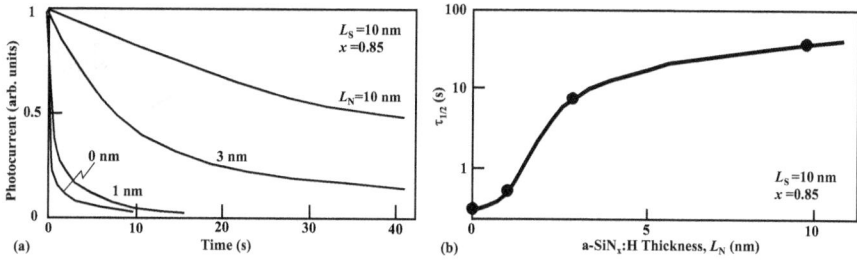

FIGURE 2.13   (a) Decay curves of photocurrents for various L_N and (b) dependence of the half-decay time $\tau_{1/2}$ on L_N. (Reprinted with permission from Yoshimura et al. [10].)

2.4.2   INFLUENCE OF A-SiN$_x$:H GATE INSULATOR COMPOSITION ON THIN-FILM TRANSISTOR PERFORMANCE

Hiranaka et al. investigated the influence of the a-SiN$_x$:H gate insulator composition on the performance of a-Si thin-film transistors (TFTs) by photoluminescence (PL) measurement.

A sample for PL measurement was prepared by depositing a 300-nm-thick a-SiN$_x$:H layer and a 100-nm-thick a-Si:H layer successively on a ground quartz substrate by plasma CVD. As depicted in Figure 2.14a, PL measurement was performed using excitation light of He-Cd laser (325 nm) or a pulsed N$_2$ laser (337 nm). PL was detected by a cooled Ge pin photodiode (response time: ~200 ns). A considerable fraction of the excitation light passes through the a-SiN$_x$:H layer and is absorbed in the a-Si:H layer near the interface with a penetration depth of about 10 nm. Thus, PL spectra of a-Si:H near the a-SiN$_x$:H/a-Si:H interface are obtained. A TFT of $L/W = 20/300$ μm was fabricated on a glass substrate, as shown in Figure 2.14b [13].

Figure 2.15a shows time-resolved PL spectra of a-Si:H near the a-SiN$_x$:H/a-Si:H interface for $x = 1.24$ at 77K. After pulsed excitation, two broad bands are observed. The band on the high-energy side is denoted by the H-band and that on the low-energy side by the L-band. The H-band decays more rapidly than the L-band.

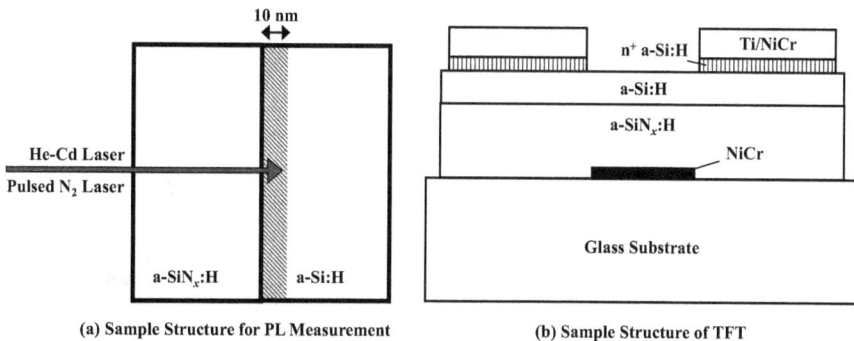

(a) Sample Structure for PL Measurement   (b) Sample Structure of TFT

FIGURE 2.14   Sample structures (a) for PL measurement and (b) for a-Si TFTs.

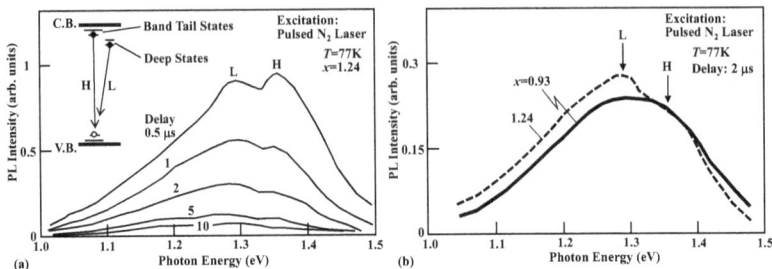

FIGURE 2.15 (a) Time-resolved PL spectra of a-Si:H near the a-SiN$_x$:H/a-Si:H interface for $x = 1.24$ at 77 K. Pulsed N$_2$ laser is used for excitation light. The insert is a model of the deep states in the a-Si:H layer. (b) PL spectra of a-Si:H near the a-SiN$_x$:H/a-Si:H interface at 2 μs after pulsed excitation for $x = 0.93$ and 1.24. (Reprinted with permission from Yoshimura et al. [11].)

The H-band with a peak near 1.35 eV was ascribed to transitions from band tail states [14]. The L-band is supposed to relate to some deep states. In Figure 2.15b, PL spectra at 2 μs after pulsed excitation are compared for $x = 1.24$ and 0.93. The L-band is more dominant in $x = 1.24$ than in $x = 0.93$. Cw PL measurement performed by He-Cd laser excitation revealed that the L-band dominates the PL spectra with increasing x, which shifts the peak position of the PL spectra, denoted by E_{PL}, toward the low-energy side. Thus, it is concluded that the deep states in the a-Si:H layer increase with increasing x of the underlying a-SiN$_x$:H layer [13].

It can be seen from Figure 2.16 that the field effect mobility of TFTs tends to decrease as x of the a-SiN$_x$:H gate insulator increases, which well correlates with the low-energy

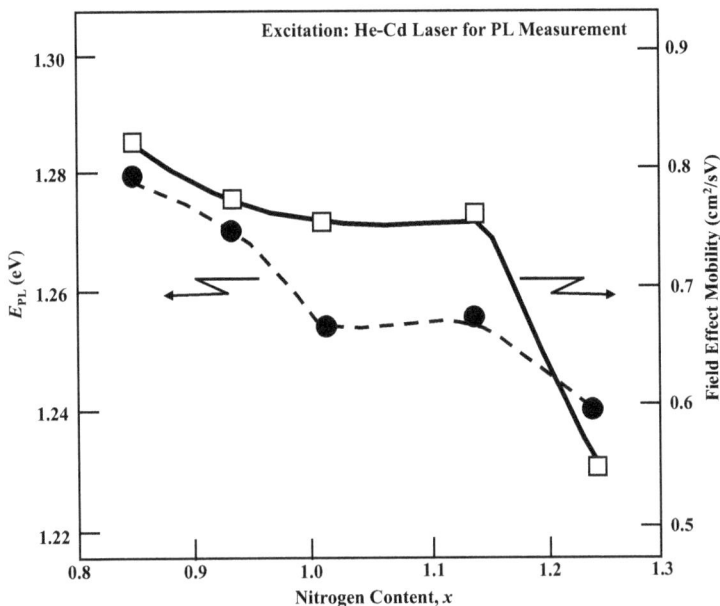

FIGURE 2.16 Dependence of E_{PL} and field effect mobility in a-Si TFT on the nitrogen content x of the a-SiN$_x$:H gate insulator. (Reprinted with permission from Yoshimura et al. [11].)

shift of E_{PL}. This correlation can be explained as follows. The low-energy shift of E_{PL} indicates an increase in the deep states in the a-Si:H layer. With an increase in the deep states, the probability of electron trapping by the deep states increases, resulting in the decrease in the field effect mobility. The origin of the deep state formation might be the lattice strain induced by lattice mismatch [15]. Since the nearest neighbor interatomic distance of a-SiN$_x$:H decreases from 0.24 to 0.17 nm as x increases from 0 to 4/3 [16], the difference in the nearest interatomic distance between a-SiN$_x$:H and a-Si:H becomes large as x increases. In addition, the tetrahedral structure of a-SiN$_x$:H is ordered to a Si$_3$N$_4$-like structure as x increased. This would increase the lattice mismatch between the a-SiN$_x$:H layer and the a-Si:H layer [13].

2.5 WAVEFUNCTION CONTROL IN ELECTROCHROMIC THIN FILMS BY SPUTTERING

The author's research group investigated the influence of film compositions and structures on the small-polaron absorption in WO$_x$ electrochromic thin films deposited by sputtering [17,18]. The small-polaron absorption was found to be enhanced by controlling the polaron wavefunction. The insight into the small-polaron absorption contributes to the discussion on the quantum confinement of electrons in polymer multiple quantum dots (MQDs) described in Chapter 6.

Figure 2.17 shows a typical structure of an electrochromic device (ECD) [19,20]. In this example, an electrochromic layer of an n-type WO$_3$ thin film and an electrolyte layer of a Ta$_2$O$_5$ thin film are stacked. By applying negative voltage to the indium tin oxide (ITO) electrode, electrons and protons are injected into the WO$_3$ thin film to color it. The injected charge is stored in the film, providing a memory function. When the ITO and counter electrodes are connected, the electrons and protons are ejected to bleach the film.

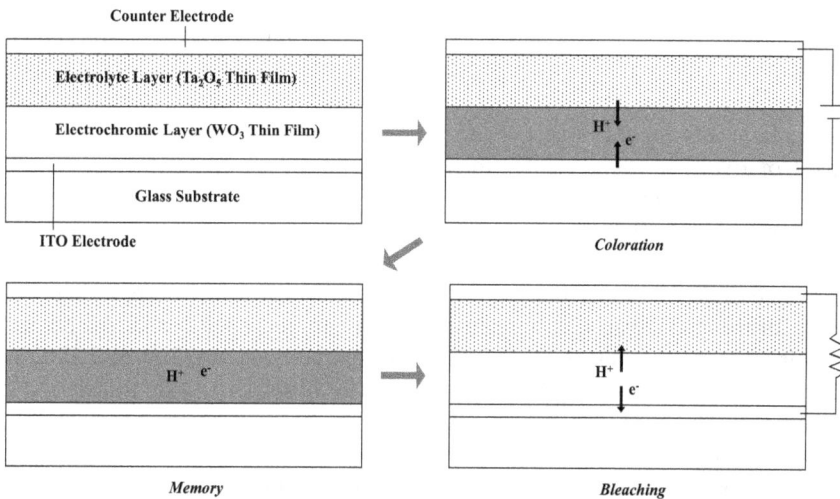

FIGURE 2.17 Typical structure of an electrochromic device (ECD) and coloring/bleaching mechanism.

Electron Trapping

$$W_A^{6+} + W_B^{6+} + e^- \longrightarrow W_A^{5+} + W_B^{6+}$$

Electron Transfer by Photon Absorption

$$W_A^{5+} + W_B^{6+} + h\nu \longrightarrow W_A^{6+} + W_B^{5+}$$

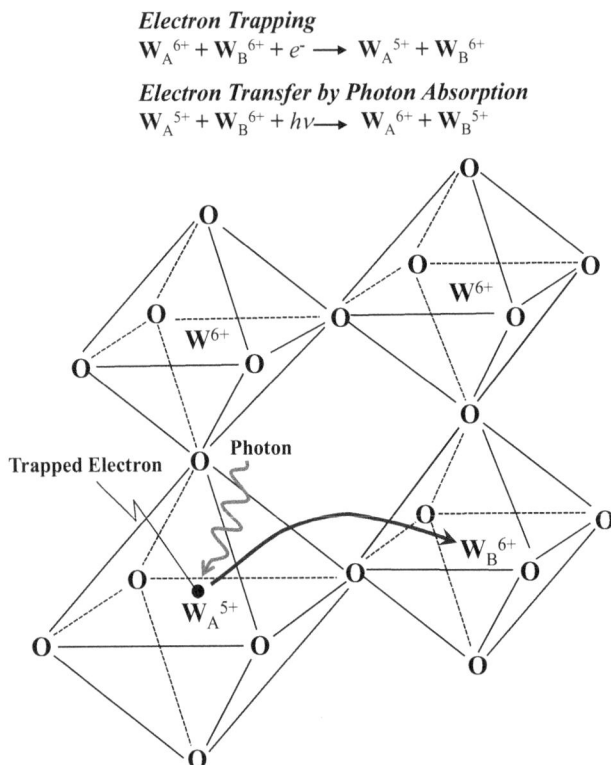

FIGURE 2.18   Model for the small-polaron absorption in WO_3. (Reprinted with permission from Yoshimura [18].)

By inserting a p-type semiconductor like an IrO_x thin film between the counter electrode and the electrolyte layer to construct a structure such as $WO_3//Ta_2O_5/IrO_x$, ECD operation is stabilized due to the complementary electrochemical reaction [20]. Since a large amount of charge, corresponding to an electron/proton count of several tens % of a W atom count, can be stored in WO_3 thin films, the ECD is expected to be a battery. By employing a multilayer-stacked structure of ECDs, high-capacity battery might be available.

A model for the small-polaron absorption in WO_3 is shown in Figure 2.18 [21–24]. In WO_3, each W site is surrounded by six O sites. W atoms are ionized to be W^{6+}. When an injected electron is trapped at the W_A site, the electron distorts the surrounding lattices, forming a small polaron. When the electron is transferred to the neighboring W_B site by absorbing light, the lattice distortion is also transferred to the W_B site. This means that the small polaron hops from the W_A site to the W_B site. The light absorption accompanied with the polaron hopping is called the small polaron absorption, which contributes to the WO_3 coloration.

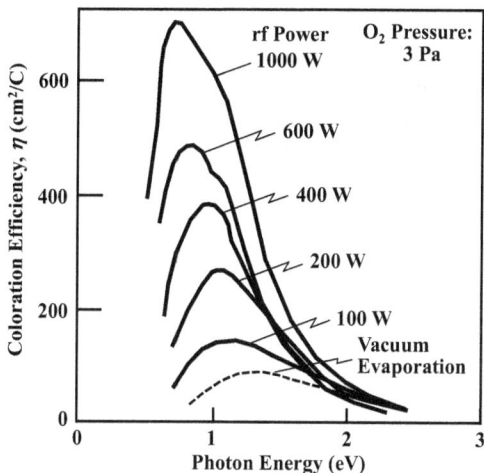

FIGURE 2.19   Coloration efficiency spectra of WO_x thin films deposited by vacuum evaporation and by rf reactive magnetron sputtering in oxygen gas with a pressure of 3 Pa at various rf powers. (Reprinted with permission from Yoshimura [18].)

WO_x thin films were deposited by rf reactive magnetron sputtering in oxygen gas with a pressure of 3 Pa at various rf powers as well as by conventional vacuum evaporation. As can be seen in Figure 2.19, the coloration efficiency for the small-polaron absorption is enhanced as rf power increases. The peak position of the spectra, E_0, shifts to the low-energy side with an increase in efficiency [17,18].

Lattice constants obtained by X-ray diffraction measurement for WO_x thin films are summarized in Figure 2.20 together with the reference powder data of monoclinic WO_3, $W_{20}O_{58}$, and $W_{18}O_{49}$. The vacuum-evaporated film annealed at 400°C in air has a series of lattice constants similar to those for WO_3 powder. The film

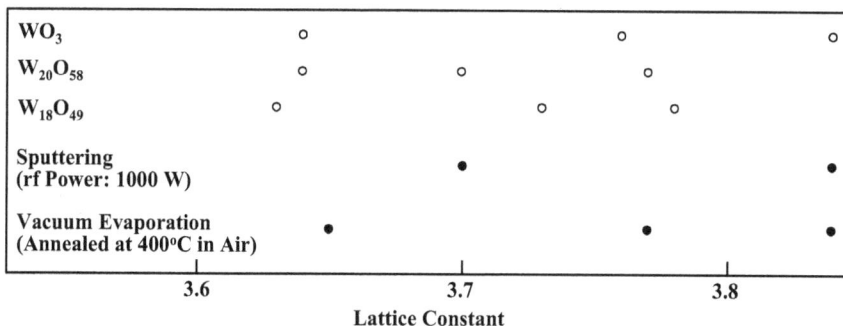

FIGURE 2.20   Lattice constants obtained by X-ray diffraction measurement for WO_x thin films deposited by vacuum evaporation and by rf reactive magnetron sputtering. Lattice constants from the powder data are also plotted for monoclinic WO_3, $W_{20}O_{58}$, and $W_{18}O_{49}$.

deposited by sputtering at 1000 W has a lattice constant of 3.7 that appears in $W_{20}O_{58}$ powder, suggesting that the film contains the $W_{20}O_{58}$-like structure due to a lack of oxygen. X-ray photoelectron spectroscopy (XPS) measurement revealed that the oxygen content in the WO_x thin films decreases with increasing rf power, which is consistent with the development of the $W_{20}O_{58}$ structure at 1000 W. These results indicate that the small-polaron absorption enhancement in the WO_x thin films arises from a lack of oxygen.

Faughnan et al. determined oscillator strength of the small-polaron absorption in a vacuum-evaporated WO_3 thin film to be $f_0 = 0.1$ [21]. Then, oscillator strength can be estimated for various WO_x thin films using the expression $f = f_0(I/I_0)$. Here, I and I_0 are, respectively, the integrated intensities of the coloration efficiency spectra for WO_x thin films and that for a vacuum-evaporated WO_3 thin film. In Table 2.2, the oscillator strength and the peak position of the coloration efficiency spectra are listed. Oscillator strength increases with increasing the rf power.

The mechanism of the oscillator strength enhancement is explained from the view point of the degree of polaron extension [17,18]. Faugnan et al. reported that f, E_0 in eV, and the dipole strength D in nm are related as follows [21]:

$$f = 9E_0D^2 \tag{2.1}$$

Considering that each W^{5+} site is surrounded by six neighboring W^{6+} sites, D is given by

$$D = \sqrt{6}\xi r \tag{2.2}$$

TABLE 2.2
Peak Position E_0, Oscillator Strength f, and Delocalization Parameter $x \rightarrow \xi$ for Coloration Efficiency Spectra of Various WO_x Thin Films

	E_0 (eV)	f	x
(Sputtering) rf Power (W)			
1000	0.7	0.51	0.22
600	0.85	0.42	0.18
400	1.0	0.32	0.14
200	1.1	0.26	0.12
100	1.2	0.16	0.09
(Vacuum evaporation) as-deposited	1.3	0.1 [21]	0.07

Source: Reprinted with permission from Yoshimura [18].

where r is the nearest W–W distance in nm, and ξ is the delocalization parameter of the polaron wavefunction [21]. ξ has the following physical meaning: when an electron is trapped at a W site, the wavefunction of the trapped electron is not completely localized at the site, but a fraction of the wavefunction, ξ, penetrates to the neighboring W sites. Substituting Equation (2.2) into Equation (2.1), f is expressed in terms of E_0 and ξ as follows:

$$f = 54 E_0 \xi^2 r^2 \qquad (2.3)$$

ξ is determined by substituting E_0, f, and r into Equation (2.3). The nearest W–W distance in the crystalline WO_3, 0.54 nm, is used for r as a zeroth approximation. The obtained values of ξ are listed in Table 2.2.

It is found from Table 2.2 that f increases from 0.1 to 0.51 and E_0 shifts to the low-energy side as ξ increases from 0.07 to 0.22. The relationship between f and ξ can be explained by the small-polaron hopping model presented in Figure 2.21 [17,18]. Oscillator strength of the small-polaron absorption increases with an increase in the transition probability for the electron to hop from the W_A site to the W_B site. If the trapped electron is highly localized, that is, ξ is small, transition probability is small because the overlap between the initial state wavefunction (trapped state at the W_A site) and the final state wavefunction (trapped state at the W_B site) is small. This situation corresponds to Case I. With an increase in penetration of the wavefunction to the neighboring W_B site, that is, with an increase in ξ, as in Case II, the wavefunction overlap increases to increase the transition probability, namely, oscillator strength.

The relationship between E_0 and ξ can be explained by configurational coordinate diagrams presented in Figure 2.21 [17,18]. Q_A and Q_B, respectively, represent the equilibrium position of the surrounding lattice distortion for an electron trapped at the W_A site and trapped at the W_B site. In the case that the initial state is the trapped state at the W_A site, the potential energy is given by the curve with a minimum point at Q_A. When the electron is excited by absorbing a photon with energy E_0, it transfers to the final state (trapped state at the W_B site), whose potential energy is given by the curve with a minimum point at Q_B. After the transition of the electron, relaxation of the surrounding lattices occurs along the potential energy curve to reach the minimum point at Q_B. So, $\Delta Q = Q_B - Q_A$ represents the change of lattice distortion accompanied by the electron hopping from the W_A site to the W_B site. When the trapped electron is highly localized (Case I), the lattice distortion is large and is localized around the W site, causing a large lattice distortion change ΔQ accompanied by the electron hopping. With delocalization of the trapped electron (Case II), the lattice distortion becomes small and spread over the neighboring sites, causing small ΔQ. Therefore, ΔQ decreases with an increase in the delocalization of the trapped electron, namely, ξ. This indicates that the potential energy curve for the W_B site shifts from the solid curve toward the dotted curve and the peak position of the absorption spectrum shifts from E_0 to E_0', namely, E_0 shifts to the low-energy side, with increasing ξ.

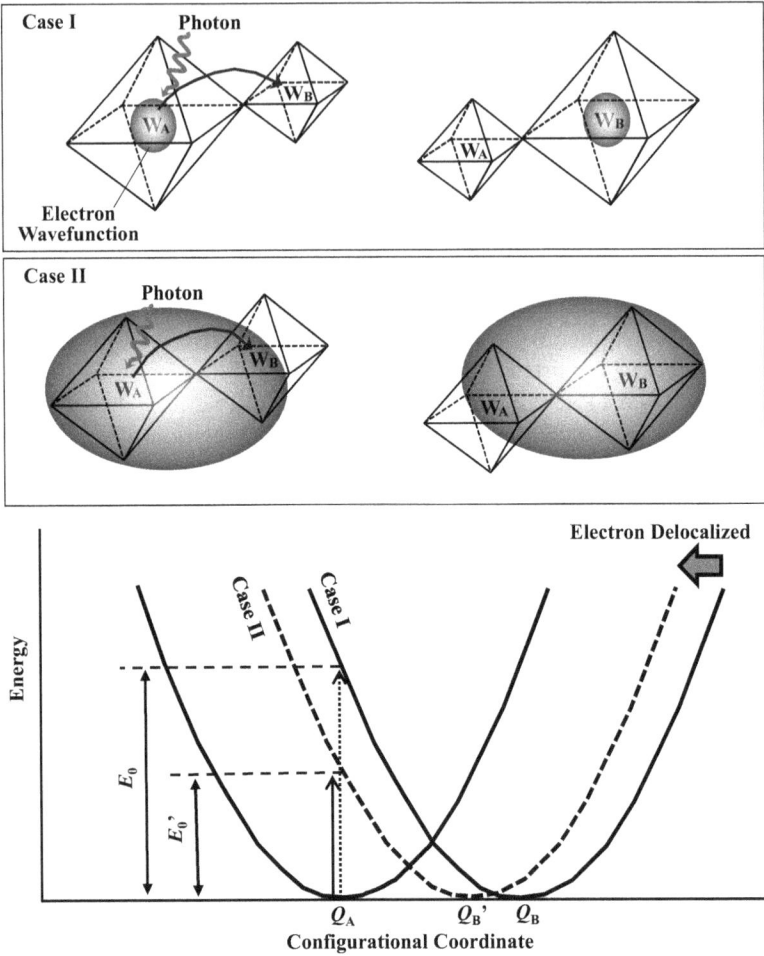

FIGURE 2.21 Schematic illustrations for the small-polaron hopping with small wavefunction delocalization (Case I) and large wavefunction delocalization (Case II), and configurational coordinate diagrams for the small-polaron absorption.

REFERENCES

1. G. Binnig and H. Rohrer, Scanning tunneling microscope, U.S. Patent 4,343,993 (1982).
2. D. M. Eigler and E. K. Schweizer, "Positioning single atoms with a scanning tunneling microscope," *Nature* **344**, 524–526 (1990).
3. M. F. Crommie, C. P. Lutz, and D. M. Eigler, "Confinement of electrons to quantum corrals on a metal surface," *Science* **262**, 218–220 (1993).
4. H. C. Manoharan, C. P. Lutz, and D. M. Eigler, "Quantum mirages formed by coherent projection of electronic structures," *Nature* **403**, 512–515 (2000).
5. T. Mimura, S. Hiyamizu, T. Fujii, and K. Nanbu, "A new field-effect transistor with selectively doped GaAs/n-$Al_xGa_{1-x}As$ heterojunctions," *Jpn. J. Appl. Phys.* **19**, L225–L227 (1980).

6. D. S. Chemla, D. A. B. Miller, and P. W. Smith, "Nonlinear optical properties of GaAs/GaAlAs multiple quantum well material: Phenomena and applications," *Opt Eng.* **24**, 556–564 (1985).

7. C.B. Roxlo, B. Abeles, and T. Tiedje, "Amorphous semiconductor superlattices," *Phys, Rev. Lett.* **51**, 2003–2006 (1983).

8. T. Tiedje and B. Abeles, "Charge transfer doping in amorphous semiconductor superlattice," *Appl. Phys. Lett.* **45**, 179–181 (1984).

9. J. Kakalios, H. Fritzsche, N. Ibaraki, and S.R. Ovshinsky, "Properties of amorphous semiconducting multilayer films," *J. Non-Cryst. Solids* **66**, 339–344 (1984).

10. T. Yoshimura, K. Hiranaka, T. Yamaguchi, and S. Yanagisawa, "Influence of a-SiN$_x$:H composition on transfer-doping and electron trapping effects in a-SiN$_x$:H/a-Si:H superlattices," *Philos. Mag. B* **55**, 409–416 (1987).

11. T. Yoshimura, K. Hiranaka, T. Yamaguchi, S. Yanagisawa, and K. Asama, "Characterization of a-SiN$_x$:H/a-Si:H interface by photoluminescence spectra," *Mater. Res. Soc. Symp. Proc.* **70**, 373–378 (1986).

12. T. V. Herak, R. D. McLeod, K. C. Kao, H. C. Card, H. Watanabe, K. Katoh, M. Yasui, and Y. Shibata, "Undoped amorphous SiN$_x$: H alloy semiconductors: Dependence of electronic properties on composition," *J. Non-Cryst. Solids* **69**, 39–48 (1984).

13. K. Hiranaka, T. Yoshimura, and T. Yamaguchi, "Influence of an a-SiN$_x$:H gate insulator on an amorphous silicon thin-film transistor," *J. Appl. Phys.* **62**, 2129–2135 (1987).

14. C. Tsang and R.A. Street, "Recombination in plasma-deposited amorphous Si:H luminescence decay," *Phys. Rev. B* **19**, 3027–3040 (1979).

15. C. B. Roxlo, B. Abeles, and T. Tiedje, "Evidence for lattice-mismatch-induced defects in amorphous semiconductor hetero-junctions," *Phys. Rev. Lett.* **52**, 1994–1997 (1984).

16. K. Tanaka, *Fundamentals of Amorphous Semiconductors*, Ohmusha, Tokyo (1982) [in Japanese].

17. T. Yoshimura, M. Watanabe, Y. Koike, K. Kiyota, and M. Tanaka, "Enhancement in oscillator strength of color centers in electrochromic thin films deposited from WO_2," *J. Appl. Phys.* **53**, 7314–7320 (1982).

18. T. Yoshimura, "Oscillator strength of small-polaron absorption in WO_x ($x \leq 3$) electrochromic thin films," *J. Appl. Phys.* **57**, 911–919 (1985).

19. S. K. Deb, "Optical and photoelectric properties and colour centres in thin films of tungsten oxide," *Pholos. Mag.* **27**, 801–822 (1973).

20. Y. Takahashi, T. Niwa, "All-solid-state electrochromic device," *Eng. Mater.* **31**, 48–51 (1983) [in Japanese].

21. B. W. Faughnan, R. S. Crandall, and D. Heyman, "Electrochromism in WO_3 amorphous films," *RCA Rev.* **36**, 177–197 (1975).

22. B. W. Faughnan and R. S. Crandall, "Optical properties of mixed-oxide WO_3/MoO_3 electrochromic films," *Appl. Phys. Lett.* **31**, 834–835 (1977).

23. O. F. Schirme and E. Salje, "Conducting bi-polarons in low-temperature crystalline WO_{3-x}," *J. Phys. C* **13**, 1067–1072 (1980).

24. E. Salje and G. Hoppmann, "Small-polaron absorption in $W_x Mo_{1-x} O_3$," *Philos. Mag. B* **43**, 105–114 (1981).

3 Molecular Layer Deposition (MLD)

In the present section, after a brief explanation of atomic layer deposition (ALD) and the beginning of molecular layer deposition (MLD) inspired by ALD, the basic concept of MLD is described. MLD grows polymer wires and molecular wires with monomolecular steps to form organic thin films utilizing the self-limiting effect. By switching source molecules with intended sequences, tailored organic thin films with designated molecular arrangements are fabricated. MLD is performed in vapor phase (Knudsen-cell (K-cell)-type MLD and carrier-gas-type MLD) or liquid phase (liquid-phase MLD (LP-MLD)). As a future challenge, MLD in gas-flow circuits or fluidic circuits is proposed. To understand the "MLD window," where the self-limiting growth is normally achieved, the growth mechanism is discussed by considering qualitative molecular dynamics based on configurational coordinate diagrams. The proof of concepts of the monomolecular-step growth by MLD is demonstrated for polyamic acid (polyimide) wires and polyazomethine conjugated wires. LP-MLD of stacked structures of p-type and n-type dye molecules on ZnO surfaces is also demonstrated. In addition, high-rate MLD, which is performed by employing purge gas curtains to make separated domains for individual source molecular gases, is described.

3.1 ATOMIC LAYER DEPOSITION AND MOLECULAR LAYER DEPOSITION

3.1.1 ATOMIC LAYER DEPOSITION

3.1.1.1 Concept of Atomic Layer Deposition

As explained in Section 2.3, molecular beam epitaxy (MBE) is an important thin-film growth method in the photonic/electronic industry. However, MBE has the following problems: insufficient conformality, thickness controllability, and mass productivity. Atomic layer deposition (ALD) overcomes these to be an ideal growth method for inorganic thin films.

In ALD, by introducing two kinds of precursors onto a substrate surface alternately, ultra-thin/conformal films are grown with monoatomic-layer steps [1–3]. Figure 3.1 shows an ALD process for ZnS thin-film growth. (1) On a substrate surface, $ZnCl_2$ is introduced to form a $ZnCl_2$ monolayer. Once the surface is covered with $ZnCl_2$, additional $ZnCl_2$ re-evaporates and the film growth automatically stops because $ZnCl_2$ and $ZnCl_2$ cannot be connected, giving the self-limiting effect. (2)–(3) After removing $ZnCl_2$, H_2S is introduced. H_2S reacts with $ZnCl_2$ on the substrate surface to form an S monolayer on a Zn monolayer. Once the surface is

DOI: 10.1201/9781003094012-3

$$ZnCl_2 + H_2S \rightarrow ZnS + 2HCl$$

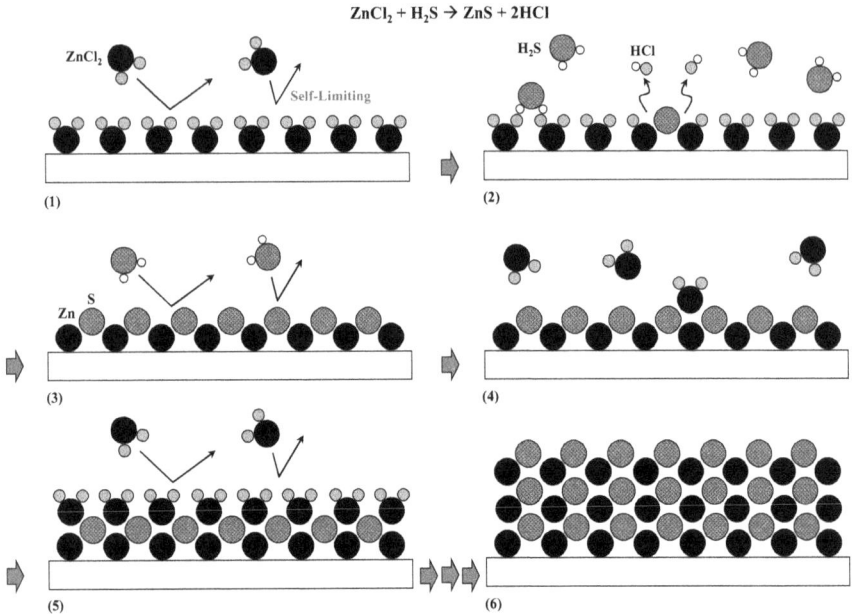

FIGURE 3.1   ALD process for ZnS thin-film growth.

covered with S, the film growth stops because no chemical reaction occurs between H_2S and S. (4)–(5) After removing H_2S, $ZnCl_2$ is again introduced to form a $ZnCl_2$ monolayer. (6) By repeating the same process, layer-by-layer growth of a ZnS thin film proceeds.

ALD was developed by V. B. Aleskovsky and S. I. Koltsov in the Soviet Union, and T. Suntola in Finland. Suntola initially applied ALD to inorganic thin-film electroluminescent (EL) devices [2,3]. To date, ALD has received much attention for a wide range of applications due to the feature that it can deposit ultra-thin/conformal films on surfaces with arbitrary shapes such as trenches, steps, holes, cylinders, fibers, and particles in addition to flat surfaces. Typical materials deposited by ALD are ZnS, ZnO, SiO_2, HfO_2, ZrO_2, Al_2O_3, TiO_2, and various metals.

In Figure 3.2a, a metal-oxide-semiconductor field-effect transistor (MOSFET) with an ultra-thin/conformal Al_2O_3 gate insulator deposited by ALD is shown [4]. Comparing with conventional deposition methods, ALD reduced leakage currents drastically. Figure 3.2b depicts hole filling by ALD. Figure 3.2c shows ultra-thin/conformal multilayers deposited on trench surfaces by ALD in a high-density capacitor with large effective areas [5]. The same structure can be applied to electrochemical batteries with ultra-thin solid electrolytes [6]. Many other applications of ALD have been reported, including solar cells, fuel cells, catalysts, gratings, thin-film transistors, and so on, as well as batteries. Surface treatment of cotton fibers and tissue paper was also reported [7].

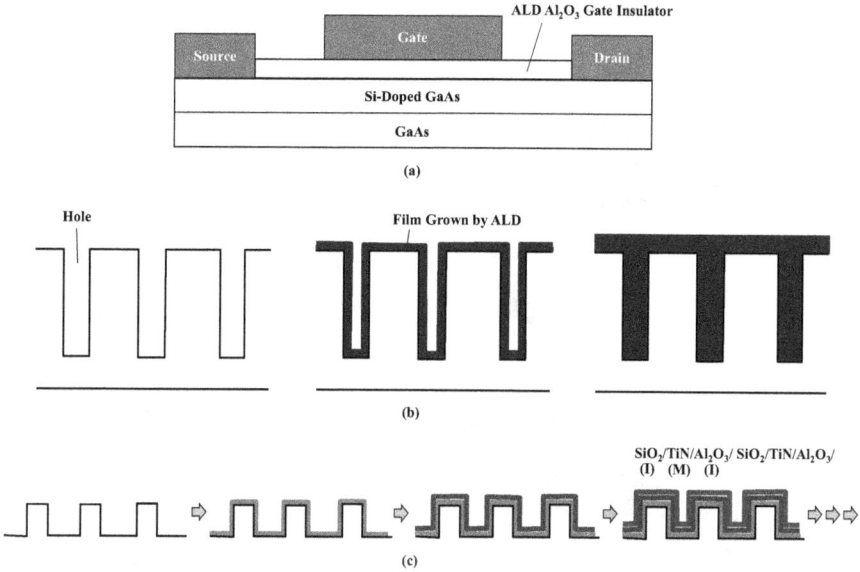

FIGURE 3.2 (a) Ultra-thin/conformal gate insulator deposited by ALD in a MOS FET, (b) hole filling by ALD, and (c) ultra-thin/conformal multilayers deposited on trench surfaces by ALD in a high-density capacitor.

3.1.1.2 Optoelectronic Materials and Devices Expected to be Realized by ALD

In Fujitsu Laboratories, the author was engaged in research on tailored inorganic optoelectronic (OE) materials and devices from the late 1970s to the mid-1980s under a policy that "improve material performance by controlling electron wavefunctions via artificial assembling of constituent atoms." The author thought that ALD was an ideal method to do this when information about ALD was obtained [2]. OE materials and devices expected to be realized by arranging various kinds of atoms utilizing ALD are summarized in Figure 3.3. Figure 3.3 (a) and (b), respectively, depict a quantum well and a superlattice that can be constructed without thickness monitoring, which is required in MBE. Figure 3.3 (c) shows an all-electronic electrochromic device (ECD), where monoatomic n-type layers, monoatomic p-type layers, and monoatomic electron-blocking layers are stacked. By applying electric fields, electrons are transferred from the p-type layers (e.g., IrO_x) to the n-type layers (e.g., WO_x) to induce the small-polaron absorption explained in Section 2.5. Figure 3.3 (d) shows an artificial nonlinear optical material. Optical nonlinearity like the EO effect is determined by electron wavefunction shapes, as described in Chapter 7. ALD can grow thin-film materials, in which atoms are assembled so as to optimize the wavefunction shapes. When electric fields are applied to the materials, the wavefunctions are distorted to induce refractive index change. Unfortunately, since no ALD equipment was available in Fujitsu Laboratories those days, no experimental work on these tailored inorganic materials/devices was done.

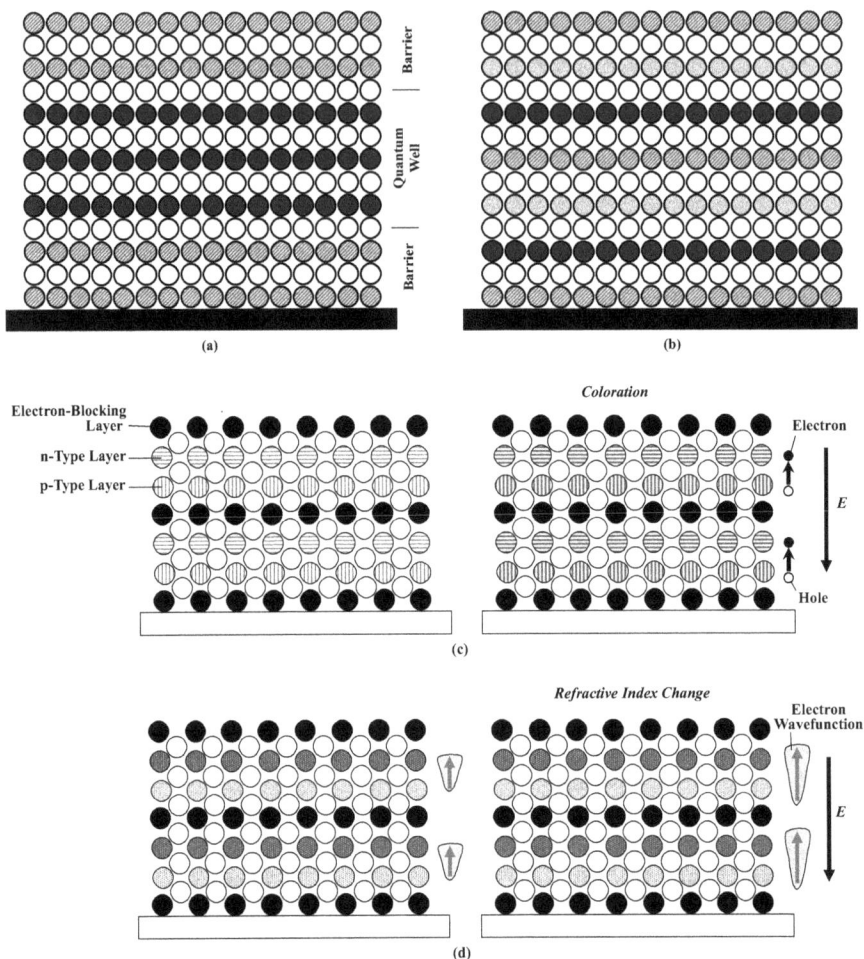

FIGURE 3.3 OE materials and devices expected to be realized by arranging various kinds of atoms utilizing ALD. (a) Quantum well, (b) superlattice, (c) all-electronic ECD, and (d) artificial nonlinear optical material.

3.1.2 BEGINNING OF MLD

In the late 1980s, the author studied organic nonlinear optical materials for light modulators and optical switches in optical interconnects within computers. In order to enhance the optical nonlinearity, electron wavefunction shapes should be optimized by assembling constituent molecules within the organic materials in designated arrangements. To do this, the development of MLD was initiated [8–10].

MLD was inspired by ALD [1–3], which is performed using the self-limiting effect to form inorganic thin films. Meanwhile, from the mid-1980s, vacuum deposition polymerization (VDP) had emerged [11–17]. VDP is a method to deposit polymer films by evaporating two kinds of reactive molecules in a vacuum chamber and

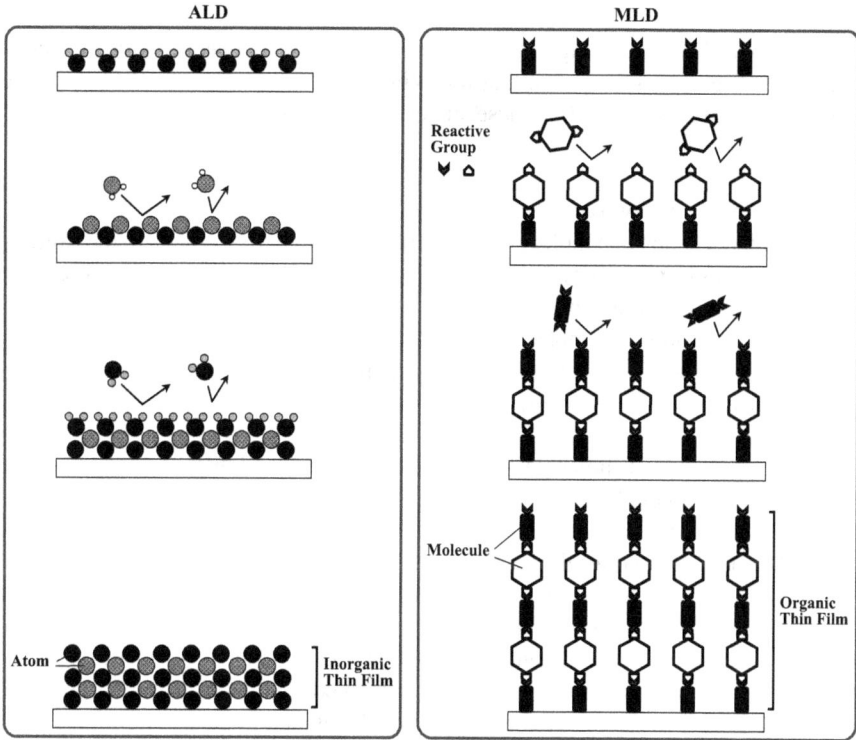

FIGURE 3.4 Comparison between ALD and MLD.

is regarded as an organic version of MBE. MLD utilizes the self-limiting effect in ALD and chemical reactions in VDP to achieve molecule-by-molecule growth of tailored organic thin films. MLD is regarded as the organic version of ALD.

The author called the process "molecular layer epitaxy (MLE)" in the first patent [8] because the ALD process was called "atomic layer epitaxy (ALE)" in the early years. However, the term "epitaxy" is not suitable for polymer thin films. So, the author adopted "MLD" instead of "MLE" to cover a wider scope.

In Figure 3.4, ALD and MLD are compared. In ALD, precursors are sequentially provided onto a substrate surface to form an inorganic thin film with monoatomic-layer growth steps. In MLD, source molecules having reactive groups are sequentially provided onto a substrate surface. Through reaction between the reactive groups, the source molecules are connected to each other to form an organic thin film with monomolecular growth steps.

3.2 CONCEPT OF MLD

3.2.1 Basic Concept of MLD

MLD is performed utilizing selective chemical reaction or repulsive/attractive electrostatic force between source molecules.

3.2.1.1 MLD Utilizing Chemical Reaction

The concept of MLD utilizing selective chemical reaction between source molecules is shown in Figure 3.5 for a case that four kinds of source molecules, Molecules A, B, C, and D, are used [8–10]. These molecules are prepared based the following guidelines:

- A molecule has two or more reactive groups.
- Reactive groups in a molecule cannot react with reactive groups in the same kind of molecules. For example, reactive groups in Molecule A cannot react with reactive groups in Molecule A.
- Reactive groups in a molecule can react with reactive groups in other kinds of molecules. For example, reactive groups in Molecule A can react with reactive groups in Molecule B.

Thus, the same kind of molecules cannot be connected, and different kinds of molecules can be connected through the reaction between the groups. In the case shown in Figure 3.5, strong chemical reaction occurs in combinations of Molecules A-B, B-C, C-D, and D-A, and weak chemical reaction in combinations of Molecules A-A, B-B, C-C, D-D, A-C, and B-D. This enables us to carry out sequential self-limiting reactions, like ALD, by switching molecules as follows.

By providing Molecule A onto a surface, a monomolecular layer of Molecule A, denoted by a monomolecular layer A, is formed. Once the surface is covered with Molecule A, the deposition of the molecules is automatically terminated due to the small reactivity between the same kind of molecules, giving rise to thickness saturation. When molecules are switched from A to B, Molecule B is connected to Molecule A to form a monomolecular layer B on the monomolecular layer A.

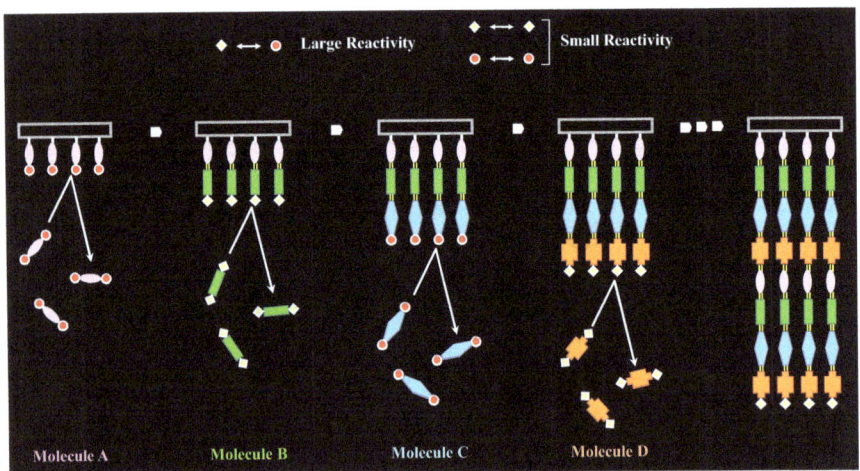

FIGURE 3.5 Concept of MLD utilizing selective chemical reaction between source molecules.

During the formation of the monomolecular layer B, a rapid thickness increase followed by saturation is induced, exhibiting monomolecular-step growth. When molecules are switched from B to C, Molecule C is connected to Molecule B to form a monomolecular layer C. Similarly, by switching molecules from C to D, a monomolecular layer D is formed. By repeating the switching with designated sequences, artificial polymer wires with a molecular sequence like ABCD.... are obtained.

3.2.1.2 MLD Utilizing Electrostatic Force

The concept of MLD utilizing repulsive/attractive electrostatic force between source molecules is shown in Figure 3.6 [18–21]. Attraction works in combinations of Molecules A-B, B-C, C-D, and D-A, and repulsion in combinations of Molecules A-A, B-B, C-C, D-D, A-C, and B-D.

When Molecule A is provided onto a surface, a monomolecular layer A is formed. Once the surface is covered with Molecule A, the deposition of the molecules is automatically terminated due to the repulsion between the same kind of molecules, giving rise to thickness saturation. By switching molecules from A to B, a monomolecular layer B is formed on the monomolecular layer A, exhibiting monomolecular-step growth. By repeating the switching with designated sequences, artificial molecular wires with a molecular sequence like ABCD.... are obtained.

3.2.2 Variations of MLD

3.2.2.1 MLD Involving Steps for Side Molecule Attachment and Cross-Linking

Figure 3.7a shows a variation of MLD, in which steps for side molecule attachment are involved. This process was developed by Zhou and Bent for fabrication of advanced

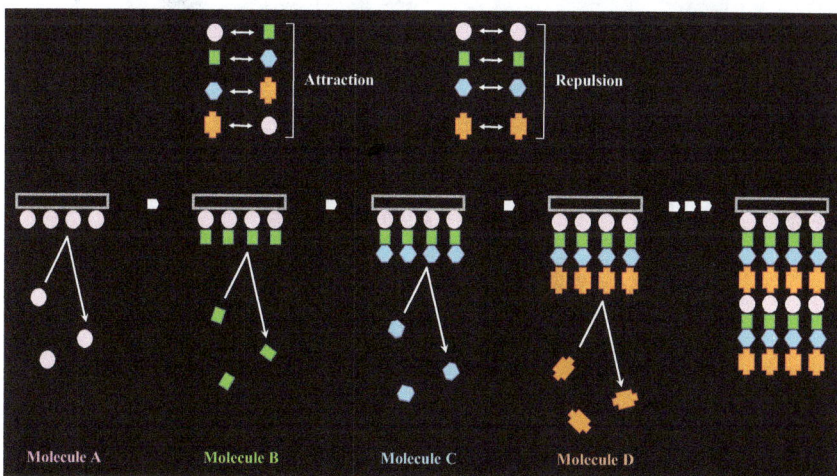

FIGURE 3.6 Concept of MLD utilizing repulsive/attractive electrostatic force between source molecules.

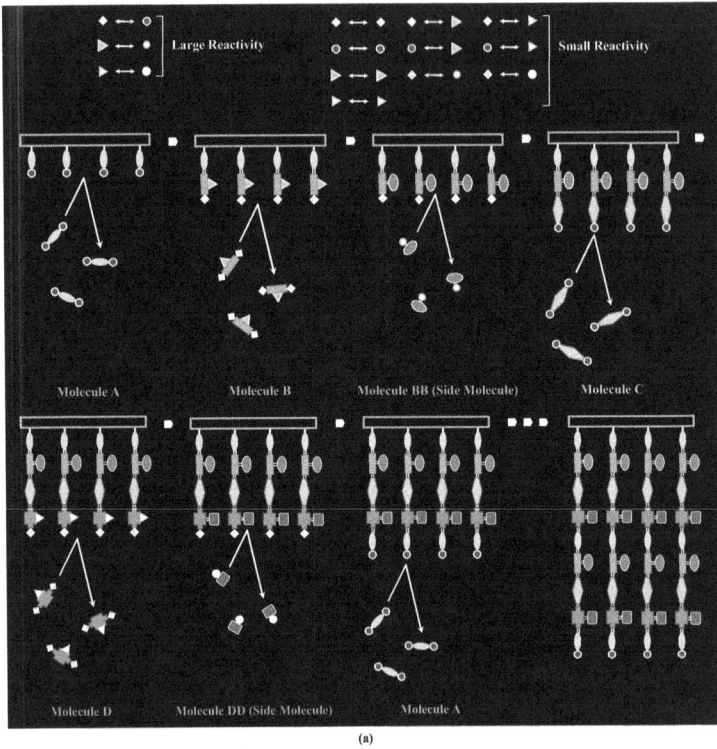

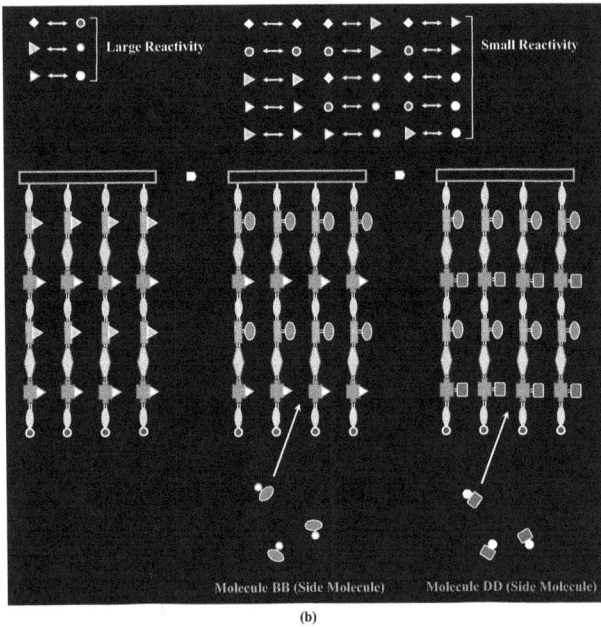

FIGURE 3.7   Concept of MLD involving steps for (a) side molecule attachment (I), (b) side molecule attachment (II), and (c) cross-linking.

(*Continued*)

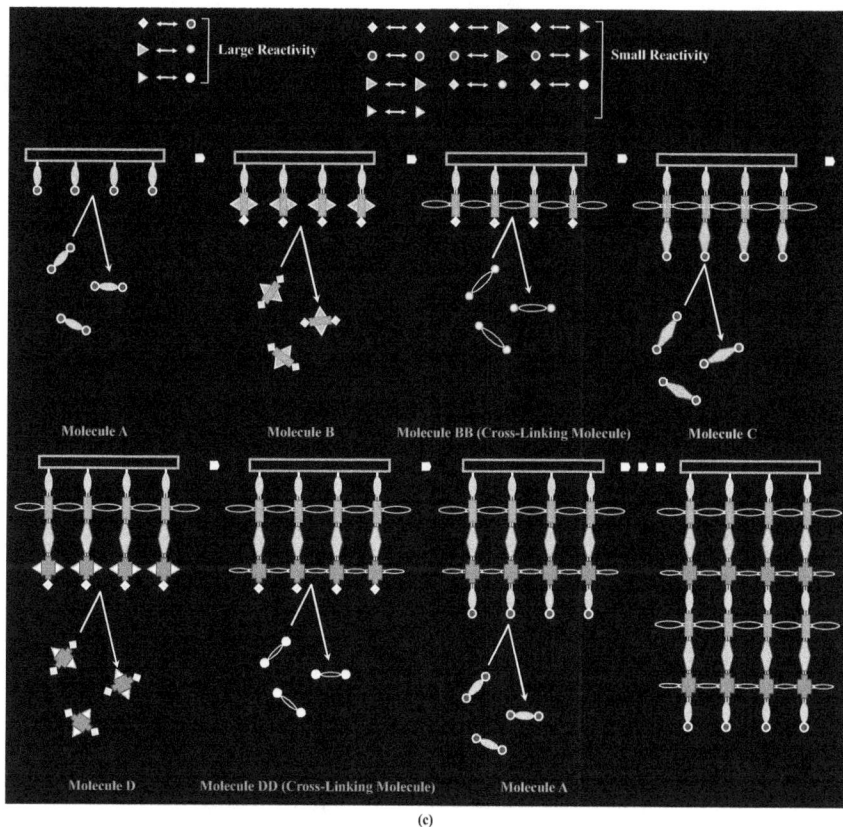

FIGURE 3.7 (*Continued*) Concept of MLD involving steps for (a) side molecule attachment (I), (b) side molecule attachment (II), and (c) cross-linking.

resist thin films, as described in Section 14.5.1 [22]. After a monomolecular layer A is formed by providing Molecule A, Molecule B having a reactive group for side molecule attachment is provided to form a monomolecular layer B. Next, Molecule BB (side molecule) having a reactive group, which exhibits large reactivity with the reactive group for side molecule attachment in Molecule B, is provided to connect Molecule BB to Molecule B as a side molecule. Then, Molecule C is provided to form a monomolecular layer C. The similar process proceeds for Molecules D, DD (side molecule), A, and so on to grow polymer wires with side molecules. This MLD process is effective to produce functional organic thin films. For example, in the case that the side molecules have donor or acceptor characteristics, donors and acceptors can be distributed in designated arrangements in the polymer wires, enabling enhancement of optical nonlinearity of the thin films, as described in Chapter 7. Figure 3.7b shows another variation of MLD. In this case, after whole of the main polymer wires are grown, the side molecules are provided in the order of Molecules BB and DD. In Figure 3.7c, MLD, in which steps for cross-linking are involved, is depicted. By using cross-linking molecules for Molecules BB and DD, cross-links between polymer wires are formed.

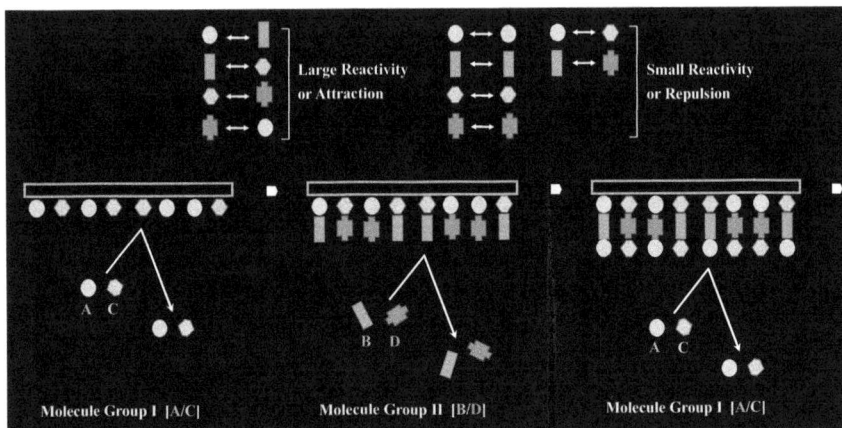

FIGURE 3.8 Concept of MLD utilizing molecule groups, where several kinds of molecules are provided at each MLD step.

3.2.2.2 MLD Utilizing Molecule Groups

Figure 3.8 depicts MLD utilizing molecule groups. Several kinds of molecules are provided in each MLD step. Molecule Group I containing Molecules A and C and Molecule Group II containing Molecules B and D are prepared, for example. By providing molecules onto a surface in the order of Molecule Groups I, II, and I, molecular wires of ABA, ABC, ADA, ADC, CBA, CBC, CDA, and CDC are formed in parallel. This process is also applicable to the growth of polymer wires. Such a feature that molecular wires or polymer wires with different molecular arrangements are grown simultaneously is useful in some cases, as described in Section 12.2.4.

3.2.2.3 Both-Side and Branching Wire Growth

As depicted in Figure 3.9a, when Molecule A is adsorbed on a surface horizontally, Molecule B is connected to both sides of Molecule A. Once both sides of Molecule A are covered with Molecule B, the deposition of the molecules is automatically terminated due to the self-limiting effect, forming a BAB structure. By switching molecules from B to C, Molecule C is connected to both sides of the BAB structure to form a CBABC structure, and by switching molecules from C to D, a DCBABCD structure is formed. By repeating the switching with designated sequences, artificial polymer wires with a molecular sequence like ….DCBABCD…. are obtained with monomolecular growth steps.

If a source molecule having three reactive groups is used, as shown in Figure 3.9b, three-way branching wires can be constructed. Similarly, wires with more than four-way branching can be constructed by using a source molecule with more than four reactive groups. Bending of wires is also possible by using a source molecule with two reactive groups substituted in appropriate configurations.

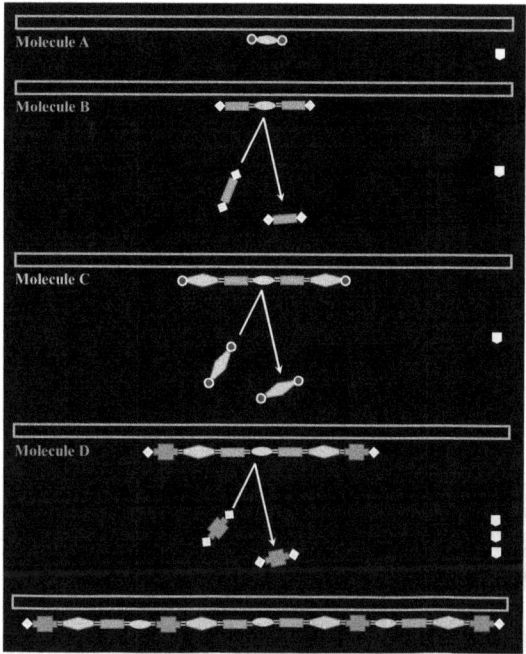

(a)

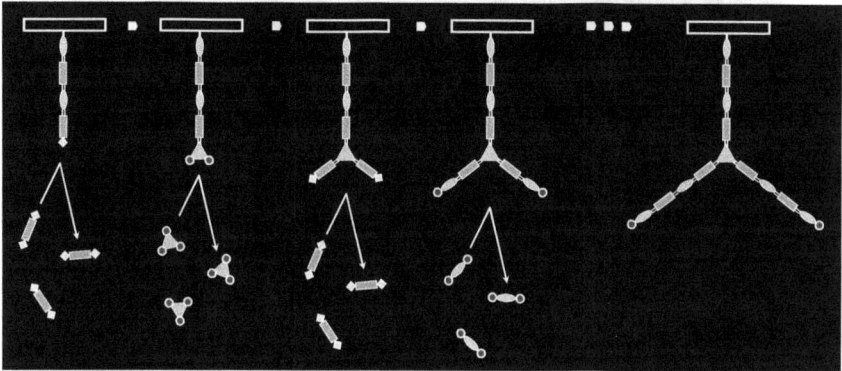

(b)

FIGURE 3.9 Concept of MLD with (a) both-side wire growth and (b) branching wire growth.

3.3   MLD EQUIPMENT

In MLD, source molecules are supplied onto a substrate surface in vapor phase or liquid phase. The former is vapor-phase MLD like Knudsen-cell (K-cell)-type MLD and carrier-gas-type MLD, and the latter is LP-MLD.

3.3.1   Vapor-Phase MLD

3.3.1.1   K-Cell-Type MLD

Figure 3.10 shows a schematic illustration and a photograph of K-cell-type MLD equipment [8–10]. Source molecules are provided onto a substrate surface in a vacuum chamber using low-temperature K-cells, in which source molecules are loaded. Source-molecule switching is performed by shutters. When the vapor pressure of the source molecules is high, they leak from gaps between the K-cells and the shutters. In such cases, valves are used for the source-molecule switching.

3.3.1.2   Carrier-Gas-Type MLD

Figure 3.11 shows a schematic illustration and a photograph of carrier-gas-type MLD equipment [23,24], where the carrier gas is employed to introduce source molecules in temperature-controlled molecular cells onto a substrate surface. Source-molecule

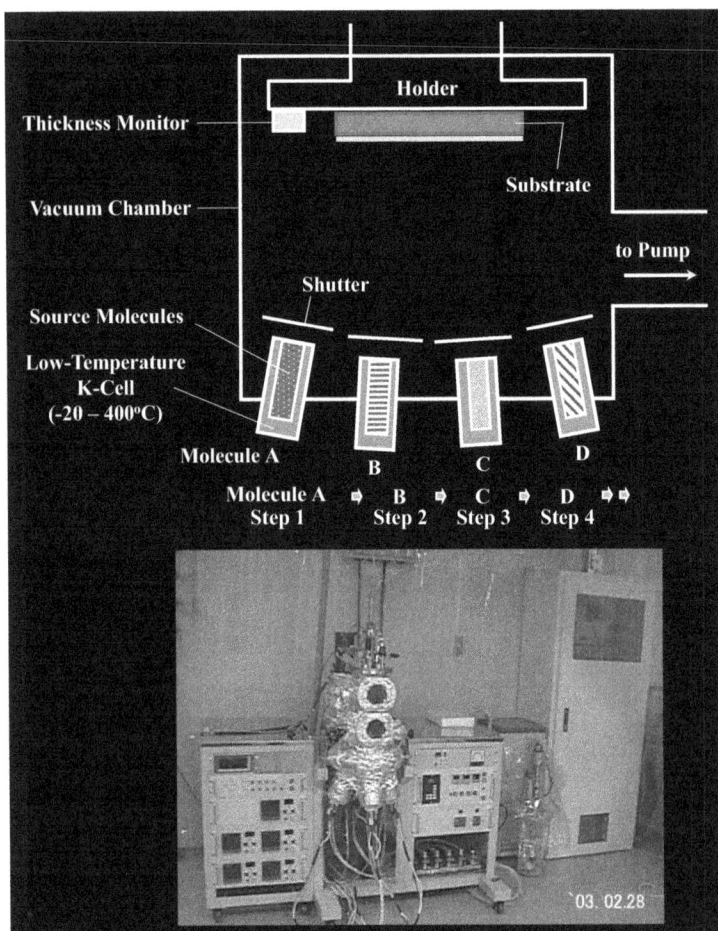

FIGURE 3.10   K-cell-type MLD equipment.

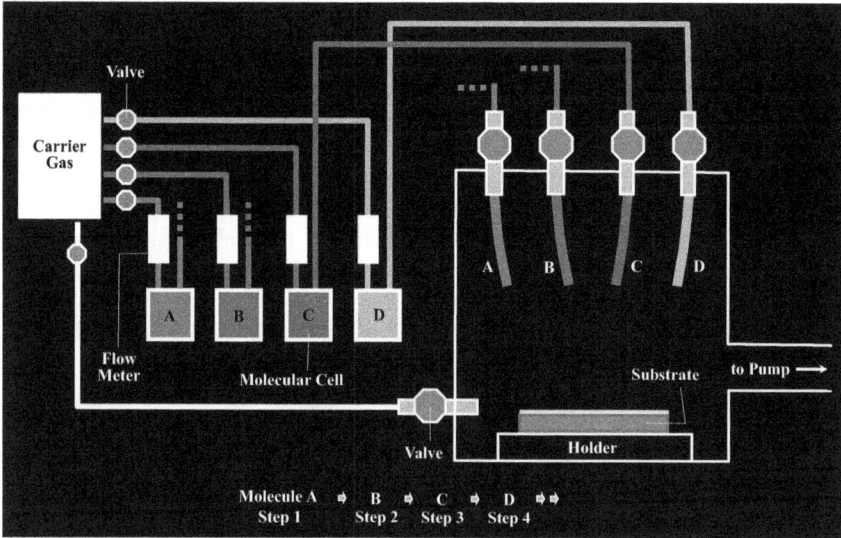

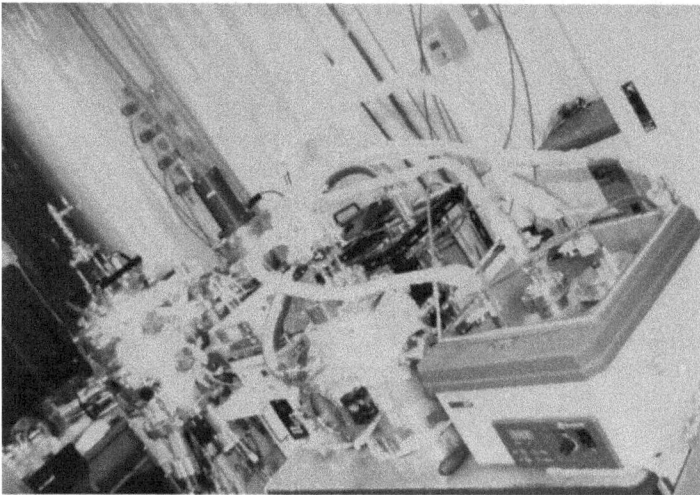

FIGURE 3.11 Carrier-gas-type MLD equipment.

switching is performed by valves. Excess source molecules are removed with the carrier gas to a pump.

Carrier-gas-type MLD has the following advantages:

1. By-products generated by chemical reactions are swept away from the surface by a carrier gas blow.
2. The carrier gas prevents contaminations in the surrounding region from reaching the substrate surface.
3. Stable supplying rates of source molecules can be maintained.

Items 1 and 2 keep the surface clean, which is essential in MLD. If by-products remain near the surface and source molecules are adsorbed on them, polymer/molecular wires will grow randomly from the by-product sites. Furthermore, by-products may terminate wire growth when their presence is near the wire edges. Contaminations such as remaining source molecules used in the previous steps also prevent MLD from performing normal molecule-by-molecule growth.

The carrier-gas-type MLD does not need high vacuum because the carrier gas flow isolates the substrate surface from outside to maintain desirable reactions between reactive groups of molecules.

3.3.2 Liquid-Phase MLD

Figure 3.12 shows a schematic illustration of LP-MLD [18–21]. Source molecules are solved in solvents. Solution A containing Molecule A is provided onto a substrate surface in a flow cell to form a monomolecular layer A. A cleaning solvent is injected into the cell to rinse the substrate and the cell. Then, Solution B containing Molecule B, a cleaning solvent, Solution C containing Molecule C, a cleaning solvent, and so on, are successively injected into the cell to perform MLD.

LP-MLD is particularly effective for MLD with source molecules that are hard to be evaporated. It is also useful when catalysis assists the MLD process. LP-MLD can be performed in a human body for medical applications. For example, cancer cells and the body, respectively, correspond to substrates and a chamber. By introducing different kinds of source molecules into the body with a designated sequence, tailored materials will be synthesized on or in the cancer cells, as proposed in Chapter 13.

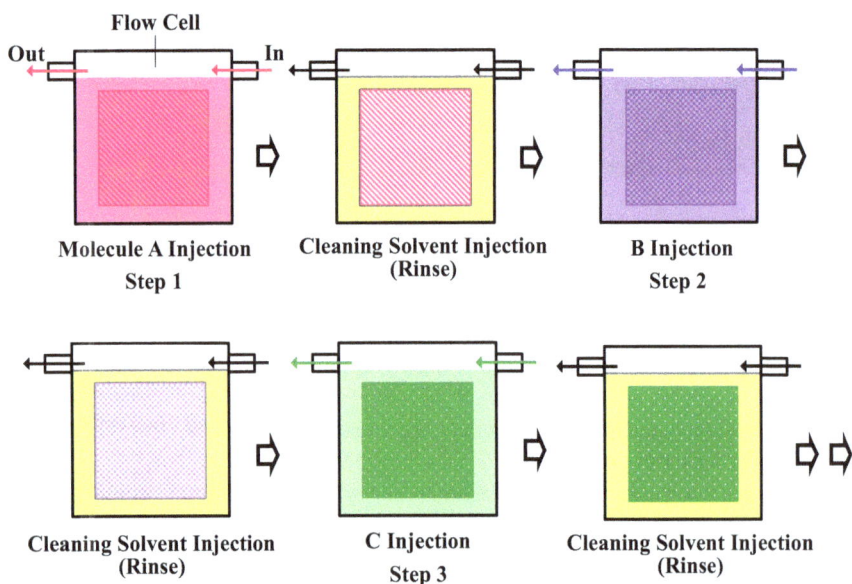

FIGURE 3.12 Liquid-phase MLD.

3.3.3 MLD Utilizing Gas-Flow Circuits and Fluidic Circuits

Figure 3.13a shows gas-flow-circuit-type MLD [19]. On a substrate, a cover is placed to build a gas-flow circuit. Source molecules are carried by carrier gas into the

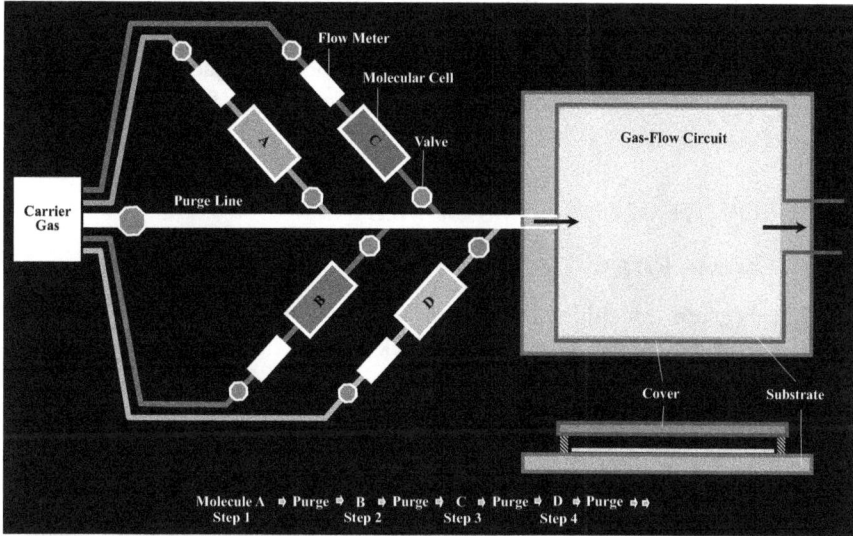

(a)

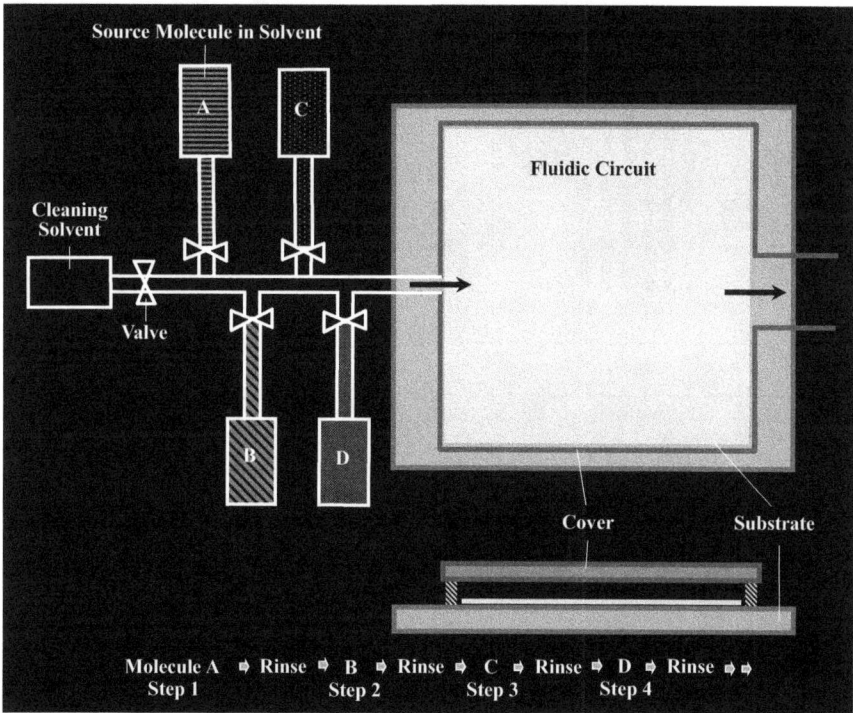

(b)

FIGURE 3.13   (a) Gas-flow-circuit-type MLD and (b) fluidic-circuit-type LP-MLD.

gas-flow circuit. The source molecules are switched by valves. To ensure the removal of remaining source molecules, purge steps are inserted in the source-molecule switching processes. Gas-flow-circuit-type MLD is expected to reduce the gas volume and the switching duration.

Figure 3.13b shows fluidic-circuit-type LP-MLD [19]. On a substrate, a cover is placed to build a fluidic circuit on the substrate surface. Source molecules are carried by a solvent into the fluidic circuit to be supplied onto the surface. The source molecules are switched by valves. Cleaning solvent injection steps for rinses are inserted in the switching processes.

3.4 GROWTH MECHANISMS OF MLD

3.4.1 GROWTH RATES

In MLD, because thin films are formed with monomolecular growth steps, the growth rate, which is defined as the film thickness increase per one MLD cycle, is determined by lengths of constituent molecules. In Figure 3.14a, schematic polymer film growth structures are depicted for the case that four kinds of source molecules are used. θ represents the tilted angle of polymer wires.

For $\theta = 0°$, as drawn by a solid line in Figure 3.14b, when Molecule A is supplied in Step 1, the film thickness increases rapidly just after the source molecules reach the surface, and is saturated. The thickness increase is L_A, which is the length of

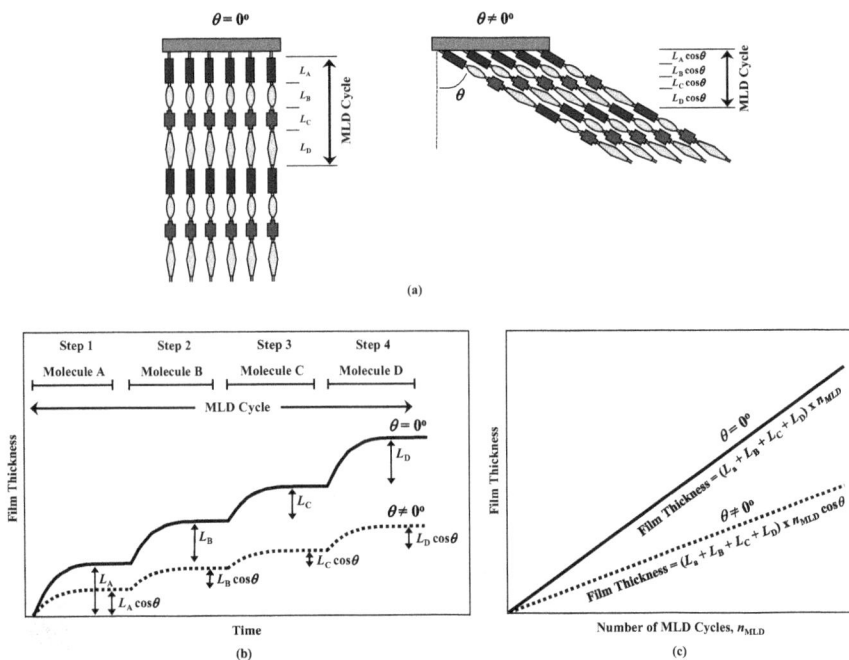

FIGURE 3.14   (a) Film growth structures, (b) stepwise film growth, and (c) film thickness vs. the number of MLD cycle, n_{MLD}, characteristics in MLD with four kinds of source molecules.

Molecule A. The film thickness is maintained after the source-molecule supply is stopped. By switching the source molecules sequentially in the order of Molecules A (Step 1), B (Step 2), C (Step 3), and D (Step 4), the film thickness correspondingly increases by L_A, L_B, L_C, and L_D, achieving stepwise film growth. Here, L_B, L_C, and L_D, are, respectively, the lengths of Molecules B, C, and D. In the example depicted in Figure 3.14, one MLD cycle involves Steps 1, 2, 3, and 4, and consequently, the growth rate equals $L_A + L_B + L_C + L_D$. The film thickness is given by [growth rate] × [number of MLD cycles n_{MLD}] as indicated by a solid line in Figure 3.14c.

For the case that polymer wires grow along a direction with a tilted angle of θ, the growth rate is reduced as drawn by dotted lines in Figure 3.14b and c because [length of molecule] × cos θ contributes to the film thickness.

3.4.2 MLD WINDOWS

The MLD window means a substrate temperature range, where MLD is carried out normally. In Figure 3.15, qualitative molecular dynamics in MLD are schematically shown for three substrate temperatures. In general, by increasing the temperature, residence time of incident source molecules decreases while reactivity between the incident source molecules and molecules at the surface increases. In the optimum temperature within the MLD window, incident molecules are physically adsorbed on the surface followed by migration/rotation until they are chemically adsorbed by forming bonds to the molecules at the surface. This is similar to the dynamics for MBE shown in Figure 2.6. Some of the incident molecules re-evaporate before forming chemical bonds. In such a condition, the MLD process proceeds normally to grow a monomolecular layer. When the temperature is too high, the residence time is reduced, and almost all of the incident molecules re-evaporate before they are fixed at the surface by chemical bonds. This destroys the MLD operation. When the temperature is too low, there might be two situations, which break the normal MLD process. In Situation 1, incident molecules cannot form chemical bonds due to the reduced reactivity, causing re-evaporation. In Situation 2, migration/rotation cannot occur due to the reduced kinetic energy, causing condensation of incident molecules. Therefore, it is concluded that optimization of the substrate temperature is important for MLD.

Based on the molecular dynamics for MLD described above, qualitative relationship between the MLD window and the shape of configurational coordinate diagrams is discussed. In Figure 3.16, growth rates are schematically drawn as a function of the substrate temperature T_s, and corresponding configurational coordinate diagrams are presented. The first potential well with a depth of E_W gives rise to physical adsorption, allowing incident molecules to migrate and rotate on the surface. The second potential well gives rise to chemical adsorption with chemical bonds, and it is deeper than the first well. There is a potential barrier with a height of E_{aB} between the first and the second wells. (H), (M), and (L) indicate that T_s is high, medium, and low, respectively. The MLD window width might depend on E_W and E_{aB}.

Figure 3.16a presents qualitative dependence of the MLD window width on E_W. E_W decreases in the order of Cases I, II, and III. In Case I, incident molecules are trapped with the physical adsorption in the first potential well. At high T_s, the residence time

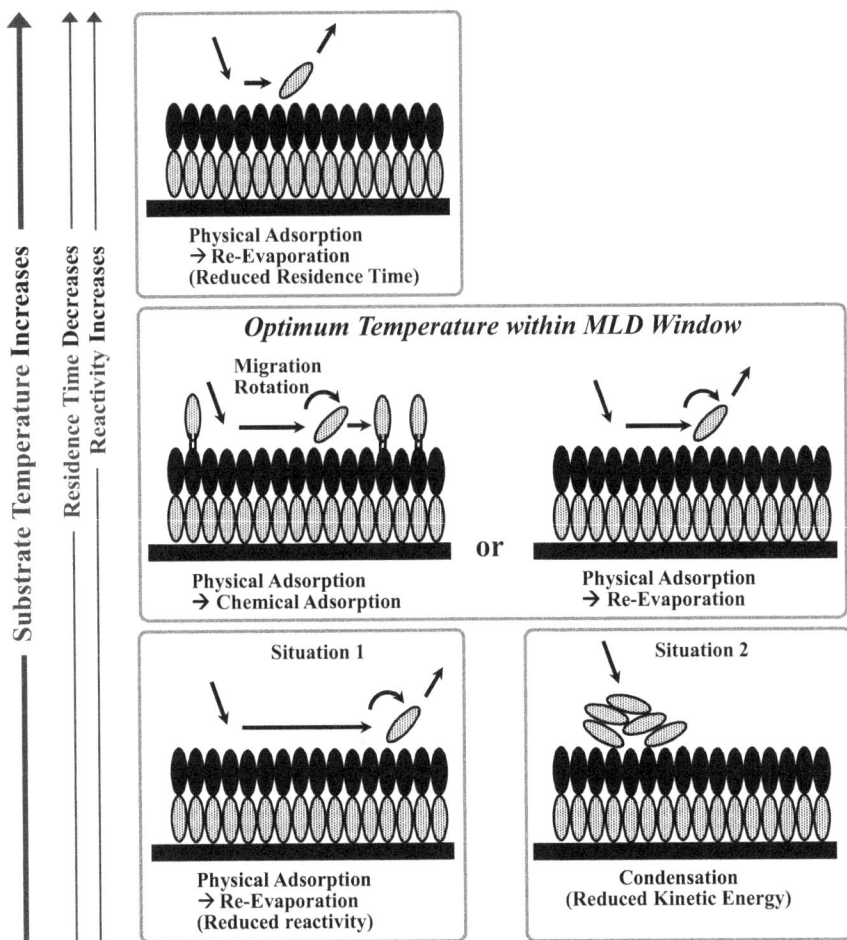

FIGURE 3.15 Qualitative molecular dynamics in MLD for three substrate temperatures.

is short and re-evaporation of the incident molecules becomes dominant to decrease the growth rate. At medium T_s, because the residence time increases, some trapped incident molecules climb over the barrier with a height of E_{aB} from the first well to the second well to be bonded to the surface molecules, forming a monomolecular layer. This temperature range is within the MLD window. At low T_s, because trapped incident molecules cannot climb over the E_{aB}-height barrier, condensation of them occurs. In Case II, E_W is smaller than in Case I. This causes severe re-evaporation of the incident molecules at high T_s, decreasing the growth rate. At medium T_s, due to a decrease in the residence time, in other words, due to an increase in the re-evaporation probability, the higher temperature side of the MLD window is eroded. At low T_s, although the trapped incident molecules in the first well cannot climb over the E_{aB}-height barrier, they can re-evaporate because of the reduced E_W, preventing

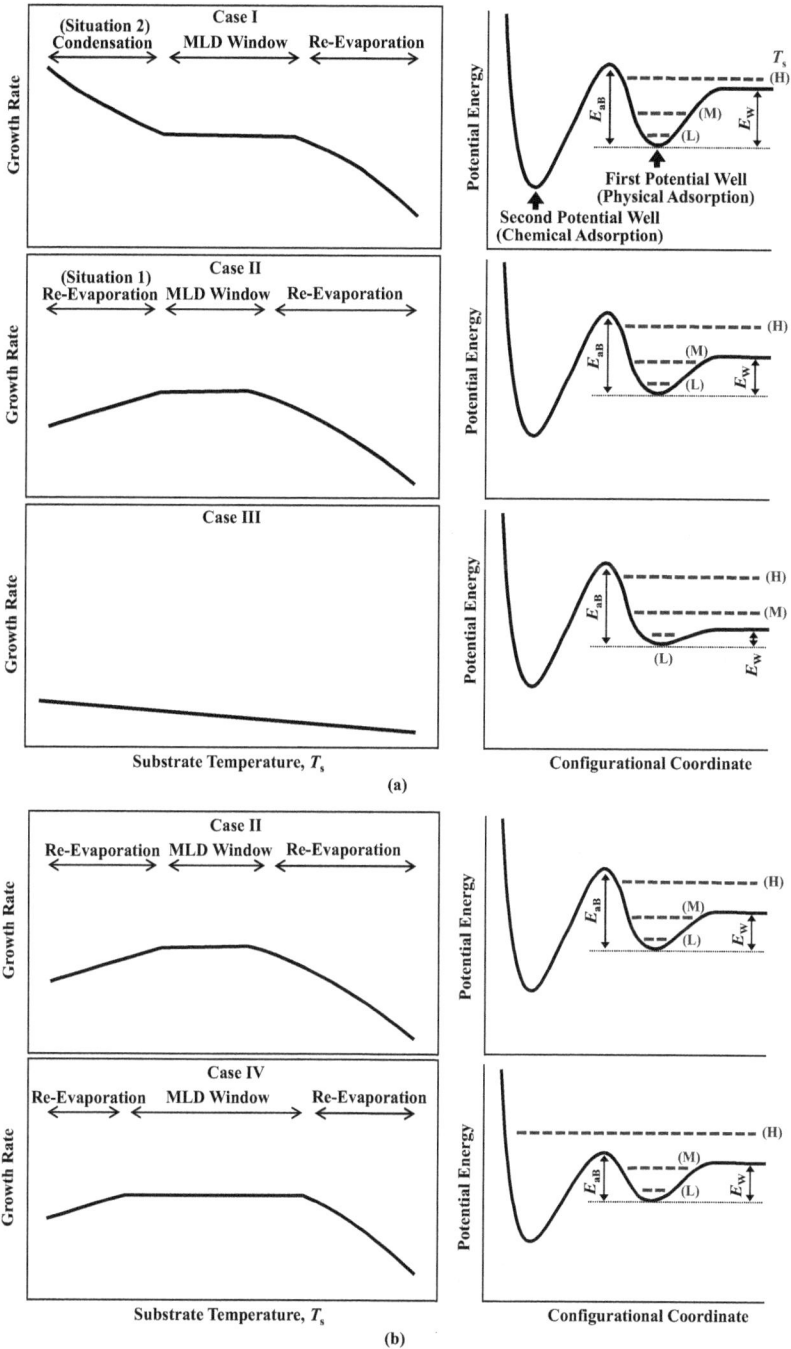

FIGURE 3.16 Qualitative dependence of the MLD window width on (a) the first potential well depth, E_W, and (b) the barrier height, E_{aB}, from the first well to the second well.

the condensation. In Case III, because E_W is further reduced, the residence time drastically decreases. This causes growth rate reduction in all of the temperature ranges, and the MLD window disappears. Thus, the MLD window is expected to be widened by adjusting the first potential well depth.

Figure 3.16b presents qualitative dependence of the MLD window width on E_{aB}. In Case IV, E_{aB} is smaller than in Case II. Then, the probability that trapped incident molecules climb over the E_{aB}-height barrier from the first well to the second well becomes large, enhancing the chemical adsorption of the incident molecules. Thus, the MLD window is expected to be widened by reducing E_{aB}.

3.5 PROOF OF CONCEPTS OF MONOMOLECULAR-STEP GROWTH IN MLD

3.5.1 POLYMER WIRES GROWN BY MLD UTILIZING CHEMICAL REACTION

Typical reactive groups used in MLD utilizing chemical reaction are summarized in Figure 3.17. The reactivity between a reactive group in Reactive Group I and that in Reactive Group II is large while the reactivity between reactive groups in the same Reactive Group is small. By selecting the combination of the reactive groups, a variety of polymers with covalent bonds can be grown, including polyamic acid, polyimide, polyamide, polyurea, polyazomethine (poly-AM), and polyoxadiazole (poly-OXD). In the present section, the monomolecular-step growth performed by the K-cell-type MLD is described for polyamic acid, polyimide, and poly-AM.

3.5.1.1 Polyamic Acid Made from PMDA/DNB

For source molecules, Molecule A of pyromellitic dianhydride (PMDA) and Molecule B of 2,4-diaminonitrobenzene (DNB) were used. As shown in Figure 3.18, these are

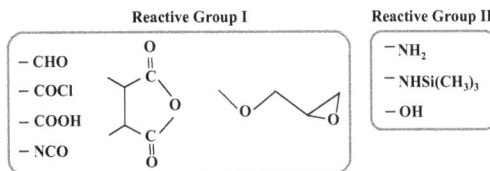

FIGURE 3.17 Typical reactive groups used in MLD utilizing chemical reaction.

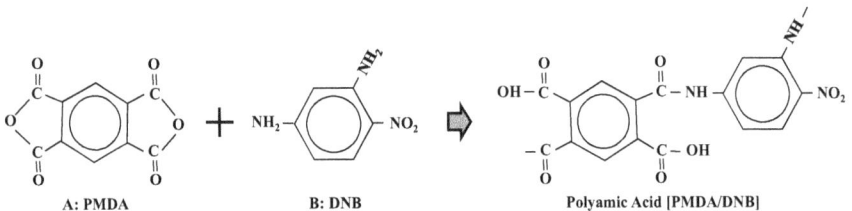

FIGURE 3.18 Reaction between PMDA and DNB for polyamic acid [PMDA/DNB] wire growth.

connected by reaction between carbonyl-oxy-carbonyl groups in PMDA and amino groups in DNB to form a wire of polyamic acid [PMDA/DNB] [10].

PMDA and DNB were evaporated from temperature-controlled K-cells in the K-cell-type MLD equipment shown in Figure 3.10. The supplied molecules were selected by shutters. Background pressure was 5×10^{-6} Pa and source molecular gas pressure was $1–5 \times 10^{-3}$ Pa. The thickness of deposited films was measured by a quartz oscillator thickness monitor, assuming that the film density is 1 g/cm^3. The monitor head contacted the heated holder and was secured by an aluminum belt. The substrate temperature, T_s, was measured using a thermocouple on the holder. It is suspected that the actual temperature of the thickness monitor head was lower than the displayed values.

Figure 3.19a shows thickness change during DNB deposition on PMDA. Firstly, PMDA molecules were introduced onto the surface by opening the shutter for PMDA

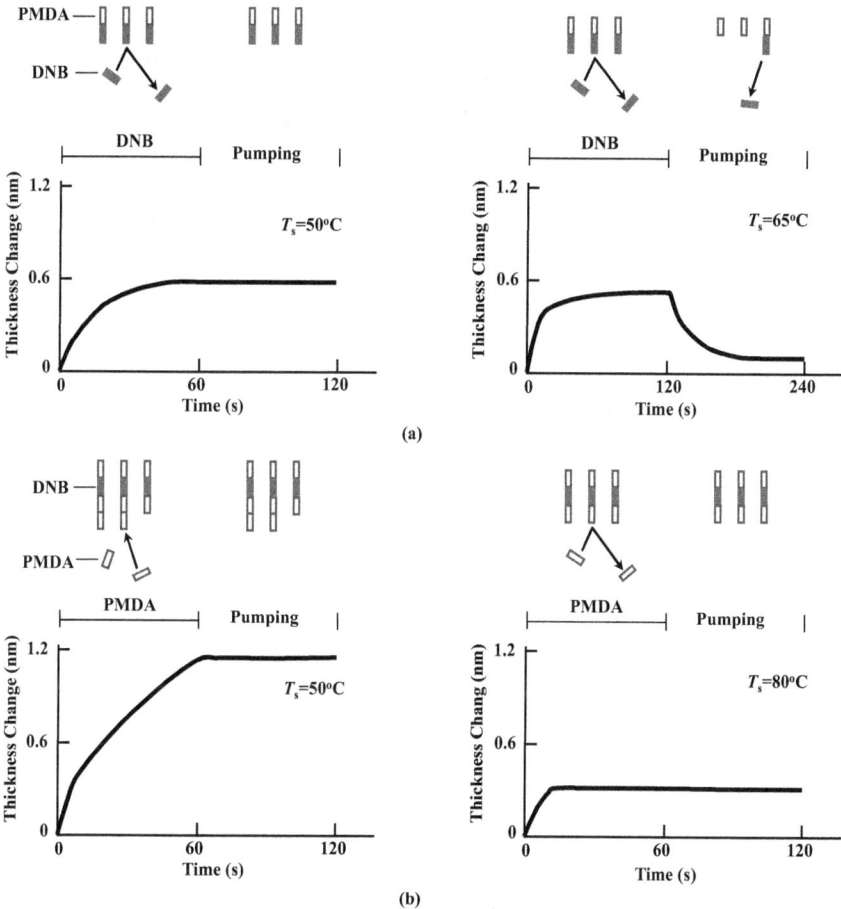

FIGURE 3.19 Thickness change with time in MLD of polyamic acid [PMDA/DNB] for (a) DNB deposition on PMDA and (b) PMDA deposition on DNB. (Reprinted with permission from Yoshimura et al. [10].)

to form a PMDA surface, and the shutter was closed to remove remaining PMDA molecules from the vacuum chamber for a removing duration, t_{rm}. Then, DNB molecules were supplied onto the PMDA surface by opening the shutter for DNB.

When DNB is supplied to the PMDA surface at $T_s = 50°C$, the film thickness rapidly increases and is saturated in 40 s after the shutter for DNB is opened. The thickness is maintained after the shutter is closed to stop the DNB supply. The change in thickness is 0.6 nm, which is close to the molecular size of DNB. The film thickness change was not affected by t_{rm} from 5 to 60 s, which excludes the possibility that the thickness change is attributed to the remaining PMDA molecules. These results indicate that a monomolecular layer of DNB is formed on the PMDA surface as follows. The binding energy between PMDA and DNB (i.e., carbonyl-oxy-carbonyl group and amino group) is larger than that between DNB and DNB (i.e., amino group and amino group). So, the DNB molecules can be connected onto the PMDA molecules. Once the surface is covered with the DNB molecules, additional DNB molecules cannot be connected onto the surface to terminate the film growth automatically, achieving the monomolecular-step growth. This corresponds to the situation of *optimum temperature within MLD window* in Figure 3.15.

When DNB is supplied to the PMDA surface at $T_s = 65°C$, a similar film thickness increase and saturation as at $T_s = 50°C$ is observed. However, the thickness decreases after the DNB supply is stopped, getting close to the initial value. This indicates that DNB molecules re-evaporate from the surface before strong chemical bonds between DNB molecules and PMDA molecules are formed. This corresponds to the situation at high temperature in Figure 3.15. When we consider that average residence time for DNB to remain at the surface equals the time constant of the thickness decay, the residence time is about 30 s at $T_s = 65°C$.

Figure 3.19b shows thickness change during PMDA deposition on DNB. Firstly, DNB molecules were introduced onto the surface by opening the shutter for DNB to form a DNB surface, and the shutter was closed to remove the remaining DNB molecules. Then, PMDA molecules were supplied onto the DNB surface by opening the shutter for PMDA.

When PMDA is supplied to the DNB surface at $T_s = 50°C$, an initial rapid increase in film thickness is observed. Thereafter, the thickness continues to increase and is not saturated. This indicates that PMDA molecules condense on the surface, which corresponds to the situation at low temperature (Situation 2) in Figure 3.15. When the substrate temperature is raised to 80°C, the film thickness saturation is observed. This indicates that PMDA condensation is avoided at 80°C and a monomolecular layer of PMDA is formed on the DNB surface to achieve the monomolecular-step growth.

As described, the optimum temperature for the monomolecular-step growth in "DNB on PMDA" differs from that in "PMDA on DNB." This might be attributed to the difference in vapor pressure between PMDA and DNB. In such a case, to fabricate polymer wires by MLD, the substrate temperature must be elevated synchronously with molecular switching.

3.5.1.2 Polyimide Made from PMDA/DDE

As illustrated in Figure 3.20, by using PMDA and 4,4'-diaminodiphenyl ether (DDE) for Molecules A and B, respectively, polyamic acid [PMDA/DDE] is produced. By heating the polyamic acid in vacuum, polyimide is obtained. Figure 3.21 shows an

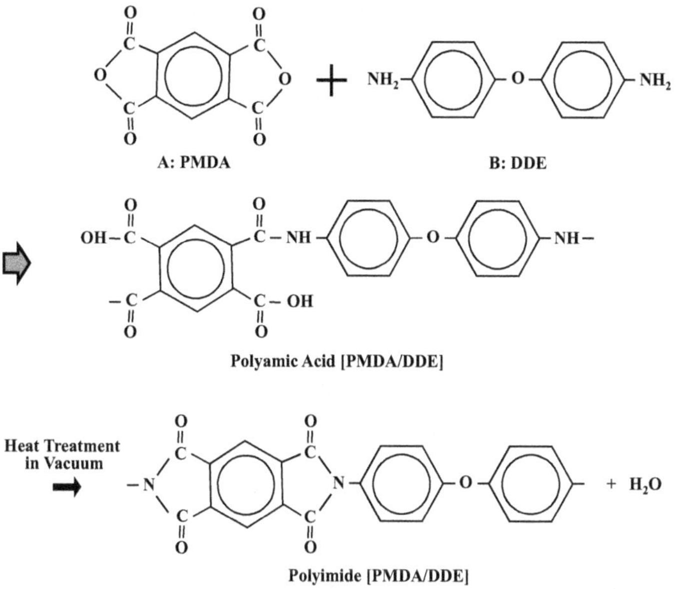

FIGURE 3.20 Reaction between PMDA and DDE for polyamic acid [PMDA/DDE] wire growth, and conversion reaction from polyamic acid to polyimide.

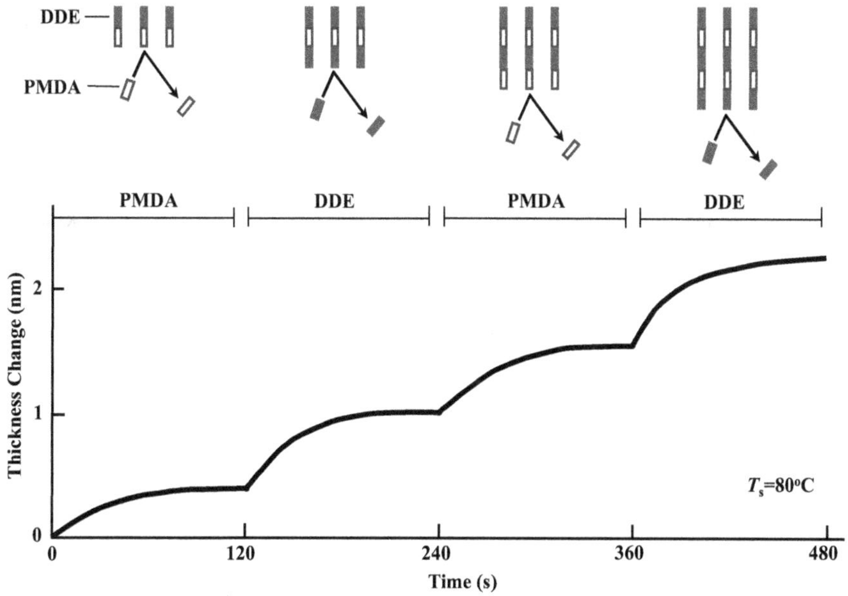

FIGURE 3.21 MLD of polyamic acid [PMDA/DDE] with source molecules of PMDA and DDE. Monomolecular-step growth is observed at $T_s = 80°C$ for both "DDE on PMDA" and "PMDA on DDE." (Reprinted with permission from Yoshimura et al. [10].)

example of MLD with PMDA and DDE. Monomolecular-step growth is observed at $T_s = 80°C$ for both "DDE on PMDA" and "PMDA on DDE," meaning that MLD is carried out without elevating the substrate temperature. This might be attributed to the fact that vapor pressures of PMDA and DDE are not largely different. A 10-nm-thick polymer film of polyamic acid was deposited by MLD with 15 steps. By annealing the film in vacuum, the polyamic acid was converted into polyimide [10].

3.5.1.3 Conjugated Polymers

Iijima et al. reported that terephthalaldehyde (TPA) and *p*-phenylenediamine (PPDA) react through –CHO in TPA and –NH$_2$ in PPDA to form –C=N– producing a π-conjugated system [14]. By employing this chemical reaction, MLD can be applied to monomolecular-step growth of conjugated polymers. As illustrated in Figure 3.22, by using TPA for molecule A and PPDA for molecule B, a conjugated polymer of poly-AM [TPA/PPDA] is grown. When oxalic acid (OA) and oxalic dihydrazide (ODH) are, respectively, used for Molecules A and B, a conjugated polymer of poly-OXD is grown with the assistance of heating in vacuum [25].

Results of MLD for poly-AM using TPA and PPDA is shown in Figure 3.23 [26]. When PPDA is provided on a TPA surface with a molecular gas pressure of $1-8 \times 10^{-2}$ Pa at $T_s = 25°C$, a stepwise growth of ~0.7 nm that is close to the molecular size is observed. After removing PPDA from the vacuum chamber, by providing TPA on PPDA at the top surface with a molecular gas pressure of 10^{-1} Pa or more, stepwise growth is observed again. These results indicate that monomolecular-step growth of a conjugated polymer is achieved by MLD. The film thickness is found to be proportional to the number of MLD cycles. The monomolecular-step growth characteristics of MLD enable us to obtain the desired properties of conjugated polymers with controlled π-electron wavefunction shapes.

3.5.2 MOLECULAR STACKED STRUCTURES GROWN BY MLD UTILIZING ELECTROSTATIC FORCE

MLD can be applied to the growth of molecular wires, in which different kinds of molecules are aligned with designated sequences. In the present section, formation of

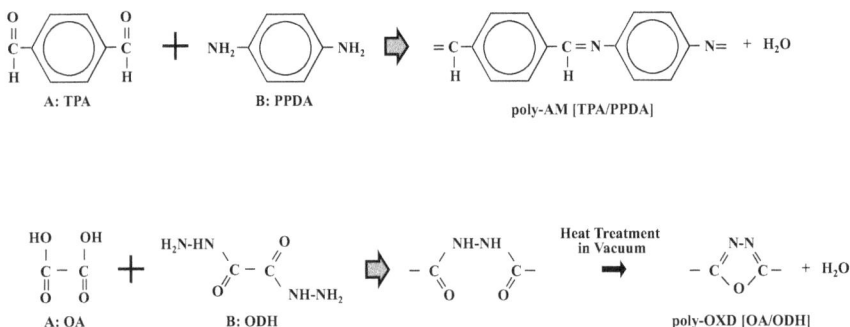

FIGURE 3.22 Source molecules and reaction for wire growth of conjugated polymers, poly-AM and poly-OXD.

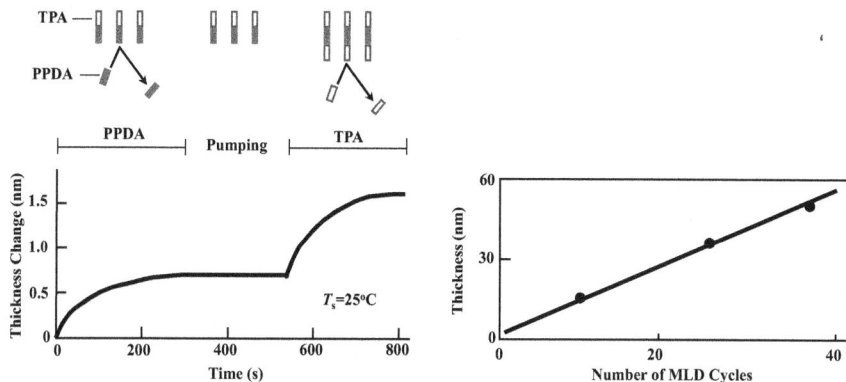

FIGURE 3.23 MLD of poly-AM [TPA/PPDA] with source molecules of TPA and PPDA: Thickness change with time and thickness vs. MLD cycles characteristics. (Reprinted with permission from Yoshimura et al. [26].)

stacked structures of p-type and n-type dye molecules on ZnO surfaces by LP-MLD utilizing electrostatic force is presented. In addition, monomolecular-step growth of charge-transfer complexes in the vapor phase is briefly mentioned.

3.5.2.1 Stacked Structures of p-Type and n-Type Dye Molecules on ZnO Surfaces

Meier reported that p-type dyes tend to accept electrons and have negative charge while n-type dyes tend to donate electrons and have positive charge [27]. This suggests that p-type and n-type dyes are candidates of source molecules in MLD utilizing the electrostatic force.

Kohei Kiyota et al. succeeded in self-assembling p-type dye molecules and n-type dye molecules on n-type ZnO surfaces to form p/n stacked structures [28]. To analyze the molecular assembling structures, the surface potential was measured by an electrostatic voltmeter [29]. The samples were prepared as follows. Firstly, a 20-μm-thick ZnO powder layer consisting of ZnO particles was spread on a glass substrate with an SnO$_2$ electrode. For the first dye adsorption, 0.01-mL alcohol solution of the first dye molecules with concentration of 1.6×10^{-3} mol/L was dropped onto the ZnO powder layer. For the second dye adsorption, alcohol solution of the second dye molecules was similarly dropped onto the surface, where the first dye molecules had already been adsorbed. Dye molecules are summarized in Figure 3.24. Fluorescein (FL), rose bengal (RB), eosine (EO), and bromophenol blue (BPB) are p-type, and crystal violet (CV), malachite green (MG), and brilliant green (BG) are n-type.

Figure 3.25a shows the surface potential of a plain ZnO powder layer and ZnO powder layers adsorbing the first dye molecules. In the figure, as well as in Figures 3.25b, 3.26, 3.28, 3.29, 3.31a, and 3.32a, for simplicity, only adsorbed dye molecules at the top surface of the ZnO powder layer are drawn, although actually the dye molecules are also adsorbed on ZnO particle surfaces in inner parts of the layer. The surface potential of the plain ZnO powder layer is about −200 mV. When p-type dye molecules are adsorbed on the ZnO powder layer, the surface potential becomes

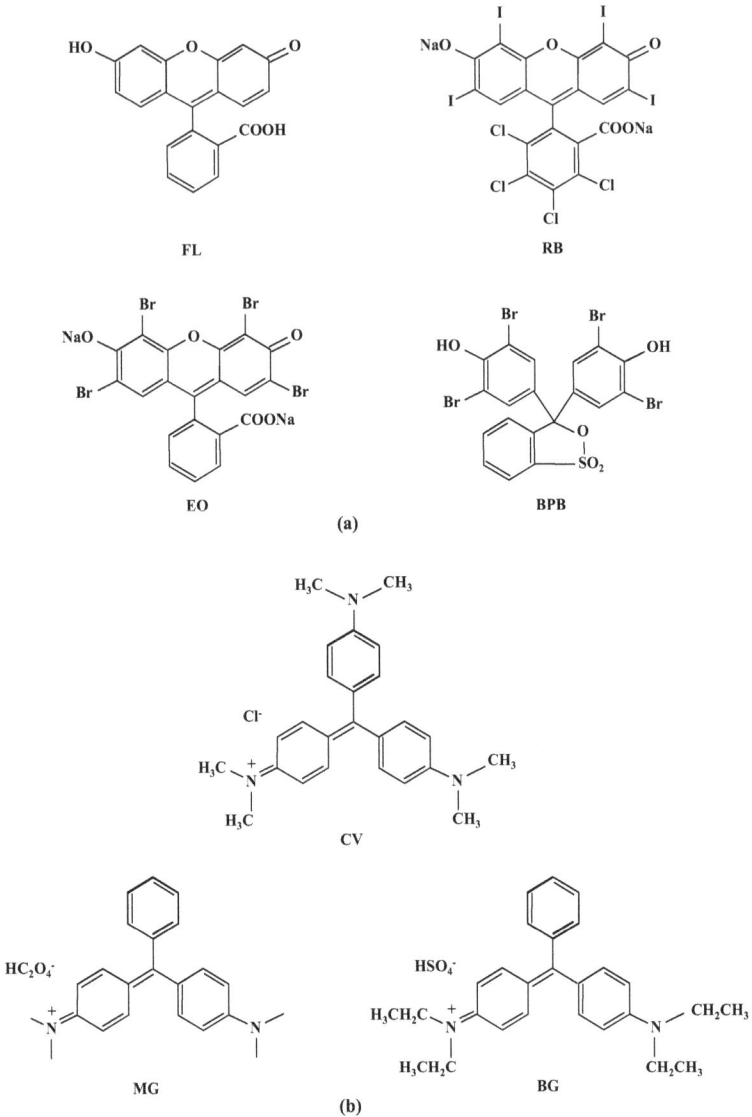

FIGURE 3.24 (a) P-type dye molecules: fluorescein (FL), rose bengal (RB), eosine (EO), and bromophenol blue (BPB). (b) N-type dye molecules: crystal violet (CV), malachite green (MG), and brilliant green (BG).

more negative than that of the plain ZnO powder layer. Conversely, when n-type dye molecules are adsorbed, the surface potential becomes less negative.

These results can be explained as follows. In plain ZnO, zinc atoms in interstitial positions donate electrons to oxygen adsorbed on the ZnO surface to generate electric dipoles, causing negative surface potential. When p-type dye molecules are adsorbed on the ZnO surface, electrons in ZnO are transferred to the dye molecules.

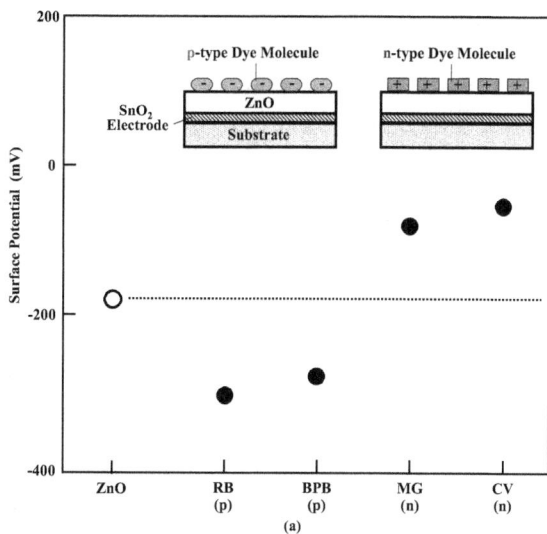

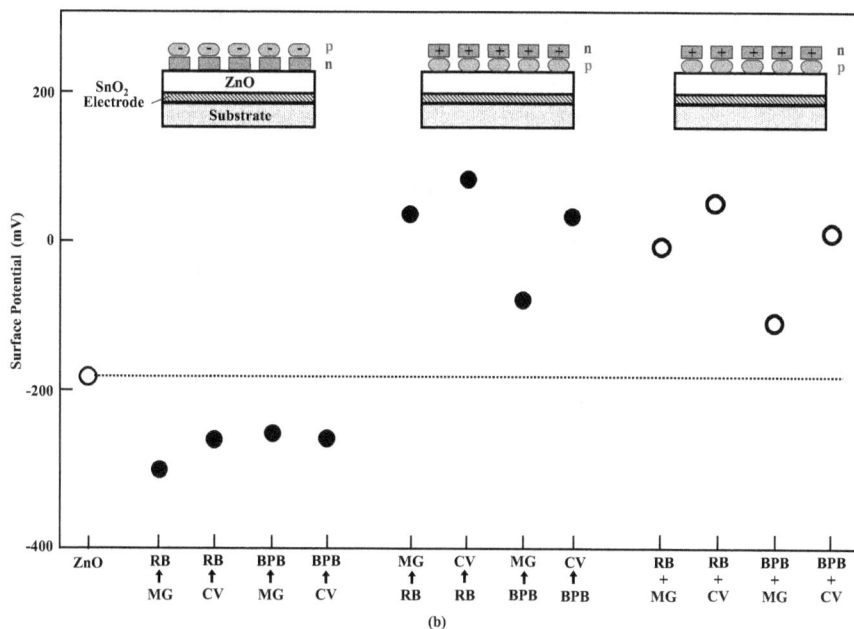

FIGURE 3.25 (a) Surface potential of a plain ZnO powder layer and ZnO powder layers adsorbing the first dye molecules. (b) Surface potential of ZnO powder layers adsorbing the first and the second dye molecules successively, and surface potential of ZnO powder layers coadsorbing p-type and n-type dye molecules simultaneously. (Reprinted with permission from Yoshimura et al. [29]. Copyright (1979) The Physical Society of Japan and The Japan Society of Applied Physics.)

This generates additional electric dipoles with the same direction as that of the dipoles arising from the oxygen to make the surface potential more negative. When n-type dye molecules are adsorbed on the ZnO surface, electrons in the dye molecules are transferred to ZnO. This generates electric dipoles with the opposite direction to make the surface potential less negative. Kokado et al. found that conductivity of a ZnO layer is reduced by p-type dye adsorption and is raised by n-type dye adsorption. The results of the surface potential measurement are consistent with their findings.

Figure 3.25b shows surface potential of ZnO powder layers adsorbing the first and the second dye molecules successively. Here, MG → RB, for example, indicates that the first dye is MG and the second dye is RB. Comparison of results shown in Figure 3.25a and b reveals that variation of the surface potential is mainly determined by the second dye. This implies that the surface potential is determined by differences in Fermi levels between ZnO and the second dye molecules on the top.

In Figure 3.25b, the surface potential of ZnO powder layers coadsorbing p-type and n-type dye molecules simultaneously is also shown. Here, RB + MG, for example, indicates coadsorption of RB and MG. For the coadsorption, 0.01-mL alcohol solution containing both p-type and n-type dye molecules at an equal concentration of 1.6×10^{-3} mol/L is dropped onto the ZnO powder layer. It is found that the surface potential of "p-type dye molecule + n-type dye molecule" is very similar to that of "n-type dye molecule → p-type dye molecule." This suggests that, as depicted in Figure 3.26, dye molecules are self-assembled in the order of the p-type dye and the n-type dye on the n-type ZnO to form a stacked structure of "n-type ZnO/p-type dye molecule/n-type dye molecule" when p-type and n-type dye molecules are provided simultaneously. The self-assembling arises from electron transfers from ZnO to p-type dye molecules as well as from n-type dye molecules to p-type dye molecules, resulting in strong connections formed by the electrostatic force between ZnO and p-type dye molecules and between p-type dye molecules and n-type dye molecules. The self-assembling effect supports the viability of MLD utilizing electrostatic force that works between source molecules of p-type and n-type dyes.

It should be noted that the concept of the self-assembled structure consisting of p-type and n-type molecules gives the base for the self-assembled monolayer (SAM) developed in the 1980s. As Figure 3.27 shows, in the case of SAM, two reactive groups having different reactivity are combined into a single molecule to achieve the self-assembling.

3.5.2.2 Dye Adsorption Strength

In order to perform LP-MLD on a substrate, selective attractive/repulsive force induced between source molecules as well as between source molecules and the

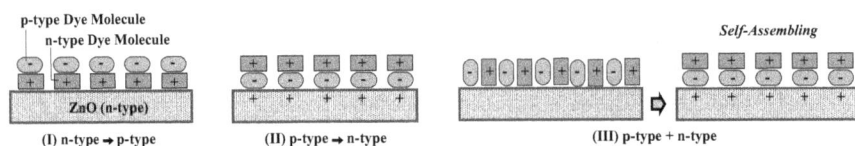

FIGURE 3.26   Assembled structures of p-type and n-type dye molecules on ZnO surfaces.

$$H_2N \left(CH_2 \right)_n S - H$$

Amino-alkanethiol

$\downarrow$

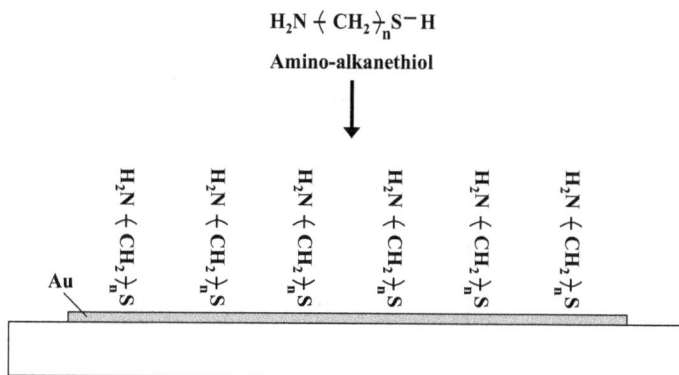

FIGURE 3.27 Self-assembled monolayer (SAM).

substrate is required. Tateno et al. confirmed that the p-type dye and n-type dye molecules, and a ZnO substrate satisfy the condition as follows [21].

When an isopropyl alcohol (IPA) solution containing p-type RB with concentration of 1.6×10^{-3} mol/L is dropped onto a ZnO powder layer with a thickness of ~5 µm on a 5 cm × 5 cm glass substrate, the IPA solution is spread along the layer. Because p-type RB is strongly connected to n-type ZnO, as depicted in Figure 3.28a, all the RB molecules are adsorbed on ZnO before IPA is evaporated. Consequently, a small-radius dyed area appears within a trace of a circumference caused by the IPA evaporation, whose radius is larger than that of the dyed area. The circumference is indicated by an arrow in the photograph. When p-type EO is used instead of RB, the same observation is obtained, as shown in Figure 3.28b. When IPA containing n-type CV is dropped onto a ZnO powder layer, as shown in Figure 3.28c, CV is spread along the layer together with IPA until IPA is evaporated because CV is weakly connected to n-type ZnO. The dyed-area radius becomes large and coincides with the radius of the circumference caused by the IPA evaporation. At the edge of the dyed area, a deeply-colored circumference due to the CV aggregation appears.

When IPA containing n-type CV is dropped onto a ZnO powder layer that has already adsorbed p-type RB, as shown in Figure 3.28d, the radius of the blue dyed area of CV becomes small because CV is strongly connected to RB. A trace of a circumference caused by the IPA evaporation can be seen as a pink line, whose radius is larger than that of the dyed area of CV. When IPA containing p-type EO is put on a ZnO powder layer with p-type RB adsorption, as shown in Figure 3.28e, because EO is weakly connected to RB, the radius of the dyed area of EO becomes large and a deeply-colored circumference due to the EO aggregation caused by the IPA evaporation appears at the edge of the dyed area.

Thus, it is confirmed that the dye adsorption strength is strong for a combination of p-n, and weak for combinations of p-p and n-n. Requirements for the source molecules in LP-MLD are satisfied if dyes are provided with a sequence of p-n-p-n--- on n-type ZnO.

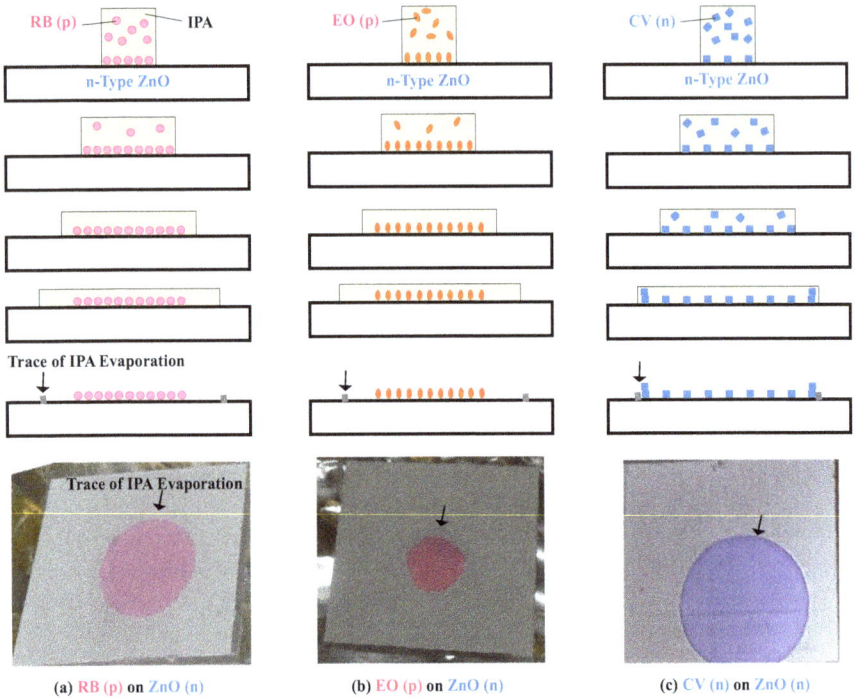

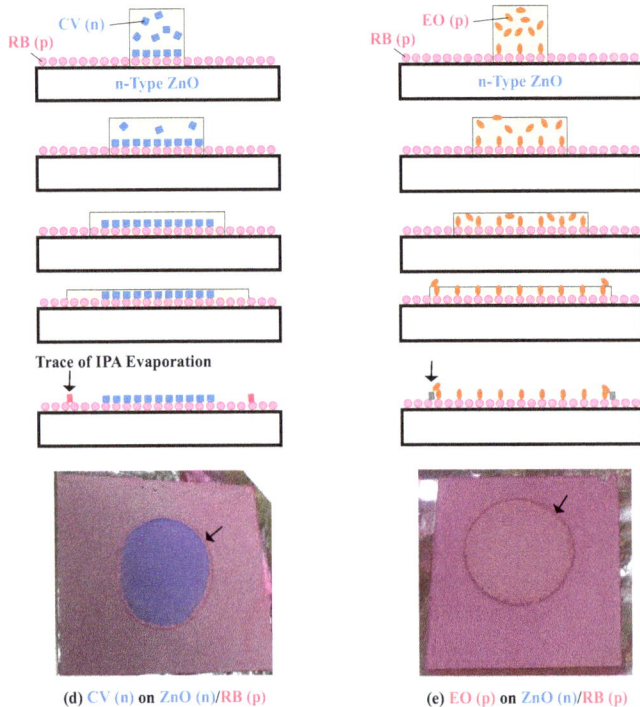

FIGURE 3.28 Dye adsorption strength estimated by dyed areas of ZnO powder layers. The substrate size is 5 cm × 5 cm. (Reprinted with permission from Yoshimura et al. [21].)

3.5.2.3 LP-MLD with p-Type and n-Type Dye Source Molecules

Figure 3.29 shows an LP-MLD process investigated by Yoshino et al. to grow three-dye-stacked structures using RB, CV, and EO as source molecules, and ZnO powder layers as substrates. In Step 1, p-type RB solved in IPA is introduced to connect RB to n-type ZnO. Once the surface is covered with RB, the RB layer growth is automatically terminated by the self-limiting effect to form a monomolecular layer structure of [ZnO/RB]. After residual RB is removed by rinsing with IPA, in Step 2, n-type CV solved in IPA is introduced to connect CV to RB. Once the surface is covered with CV, the CV layer growth is automatically terminated to form a two-dye-stacked structure of [ZnO/RB/CV]. In Step 3, p-type EO solved in IPA is introduced similarly to connect EO to CV, completing a three-dye-stacked structure of [ZnO/RB/CV/EO] [21,30].

Here, it should be noted that ZnO powder layers consisting of ZnO particles are used in the experiments. Although only adsorbed dye molecules at the top surface of the layer are drawn in Figure 3.29, actually the dye-containing IPA flows in channels within the ZnO powder layer to provide source molecules to surfaces of ZnO particles located at the inner parts. Due to the capability of MLD to grow ultra-thin/conformal organic films on particle surfaces, the multi-dye-stacked structure is formed with monomolecular growth steps on the surfaces of ZnO particles in the whole of the powder layer, as schematically drawn in Figure 3.30.

Figure 3.31a shows the surface potential of ZnO powder layers in each step of the LP-MLD process depicted in Figure 3.29. The ZnO powder layer is formed on a glass substrate with an indium tin oxide (ITO) electrode. The surface potential of a plain ZnO powder layer before LP-MLD was found to be about −200 mV, which

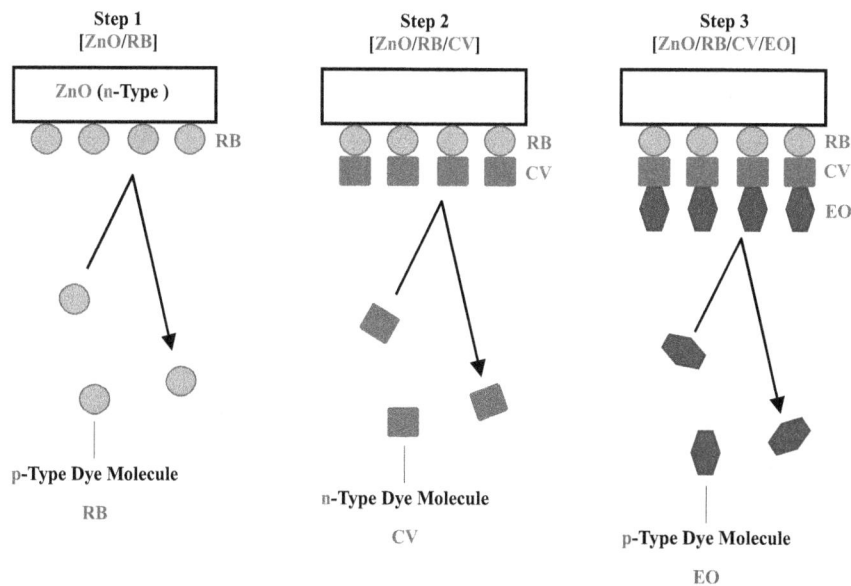

FIGURE 3.29 LP-MLD process to grow three-dye-stacked structures on ZnO powder layers with source molecules of RB, CV, and EO. (Reprinted with permission from Yoshimura et al. [21].)

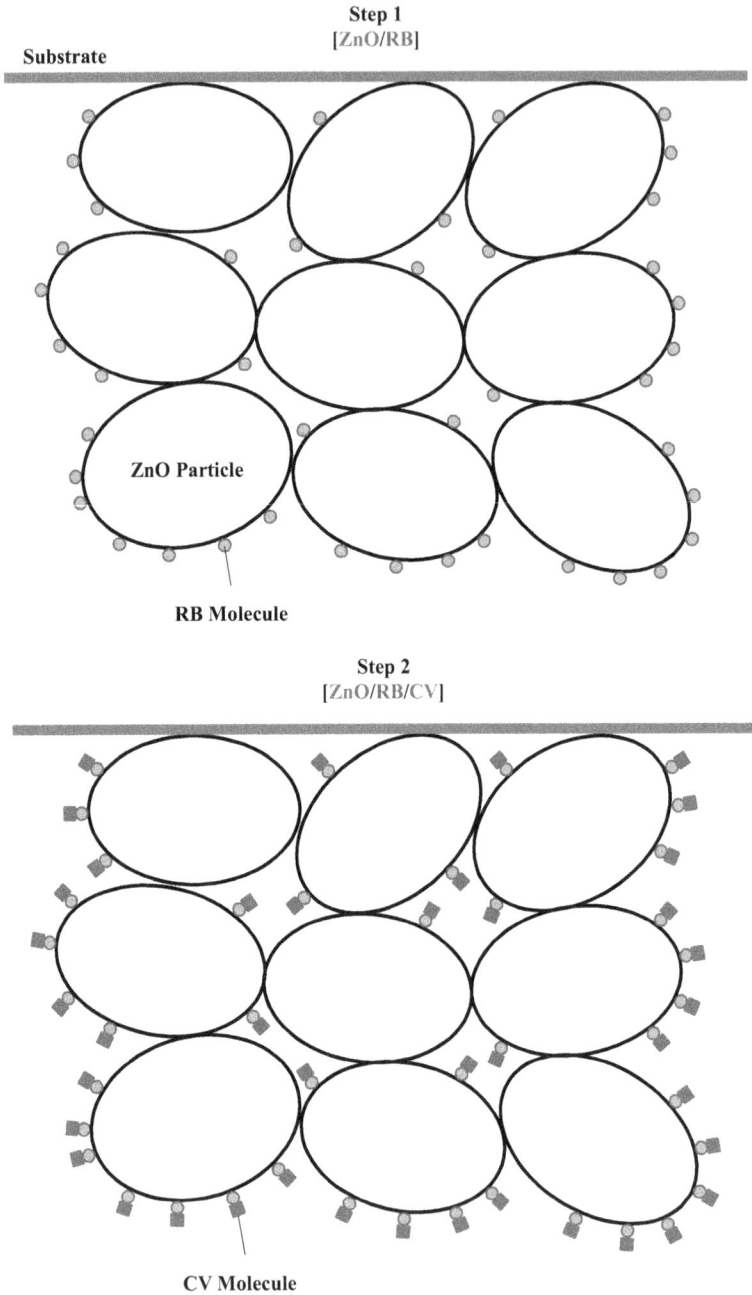

FIGURE 3.30 Conformal monomolecular-step growth of multi-dye-stacked structures on ZnO particle surfaces in a ZnO powder layer by LP-MLD. (Reprinted with permission from Yoshimura et al. [21].)

(Continued)

Step 3
[ZnO/RB/CV/EO]

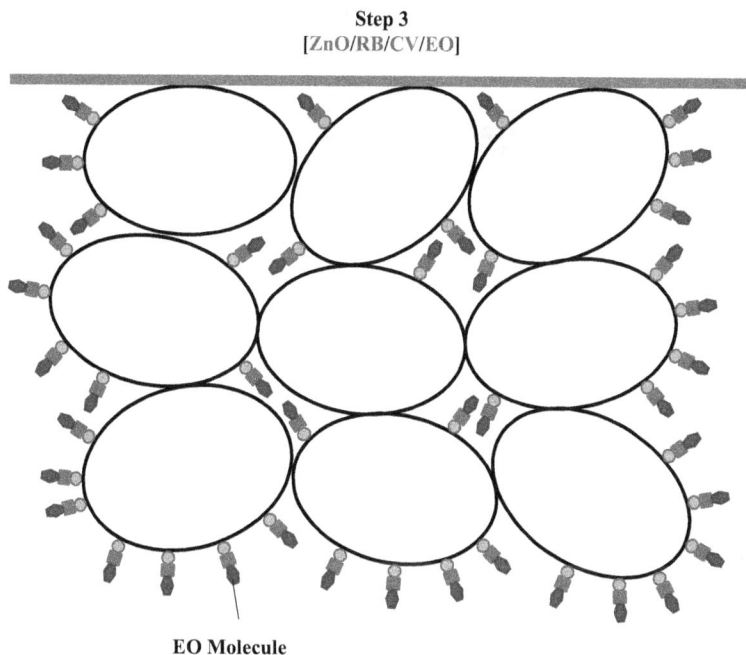

EO Molecule

FIGURE 3.30 (*Continued*) Conformal monomolecular-step growth of multi-dye-stacked structures on ZnO particle surfaces in a ZnO powder layer by LP-MLD. (Reprinted with permission from Yoshimura et al. [21].)

is attributed to negative electric dipole moments generated by adsorbed oxygen, as mentioned in the explanation for Figure 3.25. In Step 1, p-type RB is supplied on the ZnO powder layer. Then, the surface potential becomes more negative. This indicates that RB with negative charge is adsorbed on the ZnO surface to form the [ZnO/RB] structure. In Step 2, n-type CV is supplied on the [ZnO/RB] structure. Then, the surface potential moves to the positive side, indicating that CV with positive charge is adsorbed on the RB layer to form the [ZnO/RB/CV] structure. In Step 3, p-type EO is supplied on the [ZnO/RB/CV] structure. Then, the surface potential moves to the negative side. The surface potential was found to be close to that of [ZnO/EO] structure. This indicates that EO with positive charge is adsorbed on the CV layer to form the [ZnO/RB/CV/EO] structure.

Figure 3.31b shows photoluminescence (PL) spectra in each LP-MLD step. The excitation light wavelength is 532 nm. The PL spectrum of the [ZnO/RB/CV] structure extends to a longer wavelength region comparing to that of the [ZnO/RB] structure. This is attributed to CV, whose absorption peak appears at a longer wavelength region comparing to the peak of RB as found in Figure 12.6. The PL spectrum of CV is superposed on that of RB. In the [ZnO/RB/CV/EO] structure, the PL spectrum of ZnO/EO is added on that of the [ZnO/RB/CV] structure. Thus, it is confirmed that the three-dye-stacked structure, which is regarded as a short molecular wire, is formed on ZnO by LP-MLD using p-type and n-type dye source molecules.

(a)

(b)

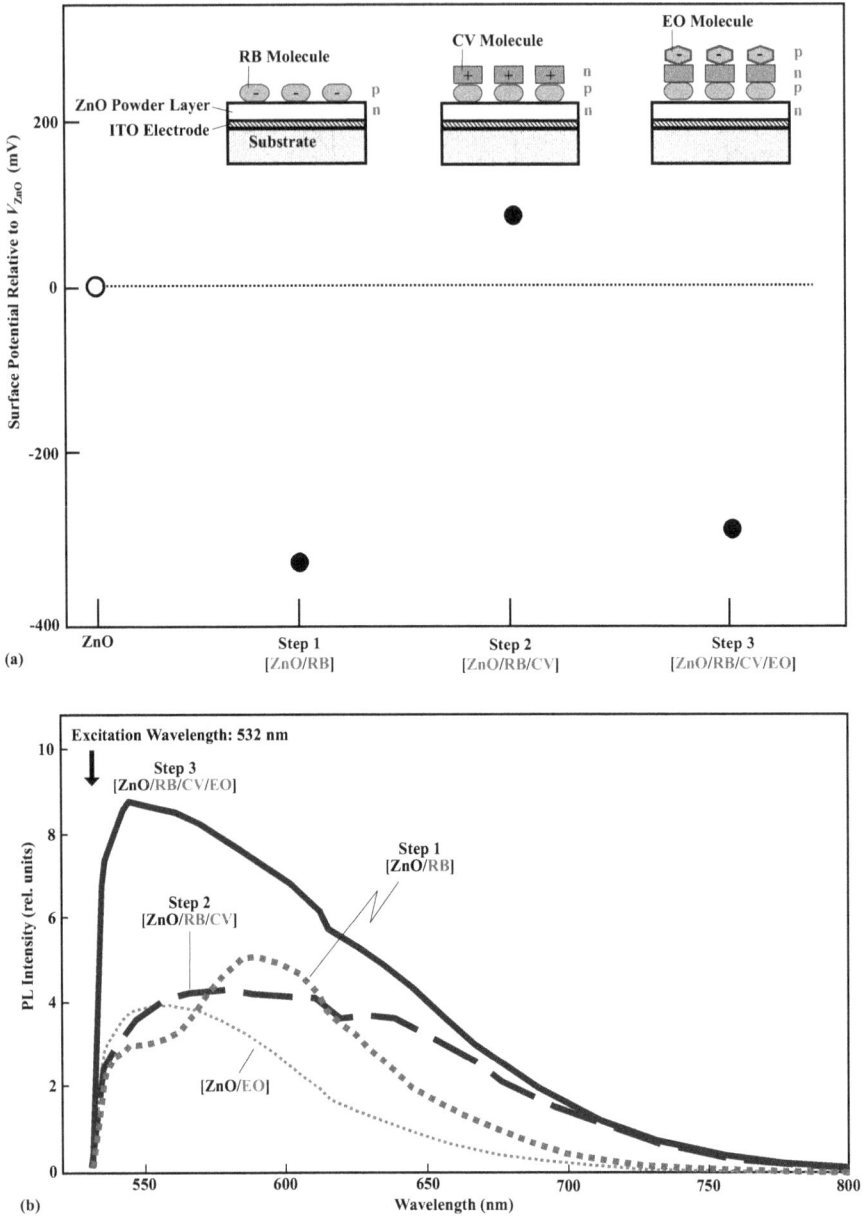

FIGURE 3.31   (a) Surface potential and (b) PL spectra of ZnO powder layers in each step of the LP-MLD process depicted in Figure 3.29. (Reprinted with permission from Yoshimura et al. [21].)

3.5.2.4 LP-MLD Utilizing Molecule Groups

Figure 3.32a shows an LP-MLD process investigated by Liu et al. using Molecule Group I containing p-type dye molecules (RB, EO) and Molecule Group II containing

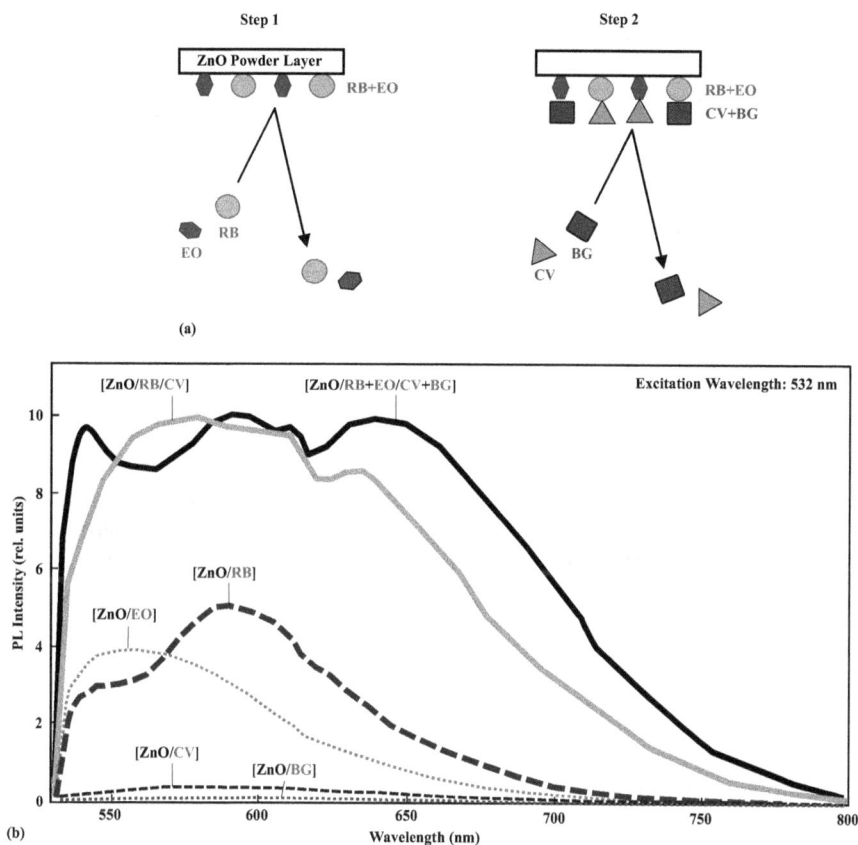

FIGURE 3.32 (a) LP-MLD process using Molecule Group I containing RB and EO and Molecule Group II containing CV and BG, and (b) PL spectra of ZnO powder layers adsorbing dye molecules.

n-type dye molecules (CV, BG). In Step 1, p-type RB and EO are deposited on a ZnO powder layer to form a monomolecular layer RB+EO. Once the surface is covered with RB and EO, the dye molecule deposition is automatically terminated. After removing residual RB and EO, n-type CV and BG are deposited in Step 2 to connect them to RB or EO, resulting in a stacked structure containing four kinds of dye molecules. Figure 3.32b shows PL spectra of ZnO powder layers adsorbing dye molecules. The two-dye-stacked structure of [ZnO/RB/CV] exhibits a broad spectrum compared with the single dye structure of [ZnO/RB]. The four-dye-containing structure of [ZnO/RB+EO/CV+BG] exhibits a broader spectrum than [ZnO/RB/CV] as expected [30,31].

3.5.2.5 Monomolecular-Step Growth of Charge-Transfer Complexes

It is known that tetrathiafulvalene (TTF) and tetracyanoquino-dimethane (TCNQ) combine through charge transfer between them to form molecular crystals of charge-transfer complexes. By using the charge-transfer property, it is expected that MLD

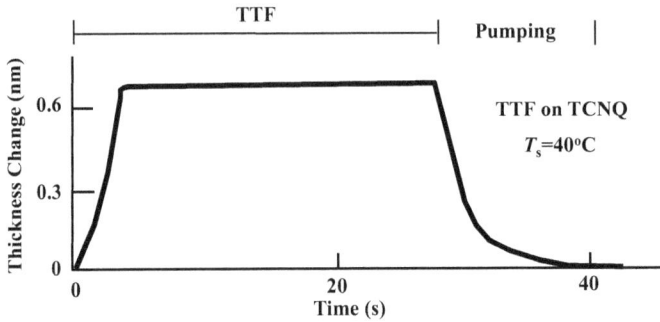

FIGURE 3.33 Monomolecular-step growth of TTF on TCNQ.

utilizing electrostatic force can be performed in the vapor phase. It is found from Figure 3.33 that a monomolecular-step growth of TTF on TCNQ is achieved by the K-cell-type MLD. Unfortunately, the film thickness decreases due to re-evaporation of TTF when the shutter for TTF is closed, suggesting that the connecting force between TTF and TCNQ is not sufficient at room temperature. Since the residence time increases with decreasing substrate temperature, it might be possible to apply MLD to charge-transfer complexes by optimizing the temperature.

3.6 HIGH-RATE MLD

In MLD, gas injection and ejection are necessary for molecular gas switching, resulting in an increase in process time. To solve the problem, Kim et al. developed the domain-isolated MLD [32,33].

3.6.1 Influence of Molecular Gas Flow on Polymer Film Growth

It was found that carrier gas blocks molecules to come into the region, where the carrier gas flows. Figure 3.34 shows the influence of molecular gas flow on polymer film growth in the carrier-gas-type organic chemical vapor deposition (organic CVD) [34]. Two types of molecular gas-flow configurations, I and II, were examined using TPA and PPDA as source molecules. In configuration I, where the substrate is tilted by 45° and the TPA gas flow overlaps with the PPDA gas flow, poly-AM is formed over a wide region on the substrate. Conversely, in configuration II, where the TPA gas flow parallels the PPDA gas flow, poly-AM grows only near the boundary between the TPA region and the PPDA region. The polymer does not grow just under the nozzles, where only one kind of source molecules, that is, TPA or PPDA, exists. This indicates that the carrier gas blocks penetration of molecules from the other regions.

3.6.2 Domain-Isolated MLD

The molecular blocking capability of the carrier gas enables us to develop high-rate MLD called the domain-isolated MLD. In Figure 3.35a, schematic illustrations for

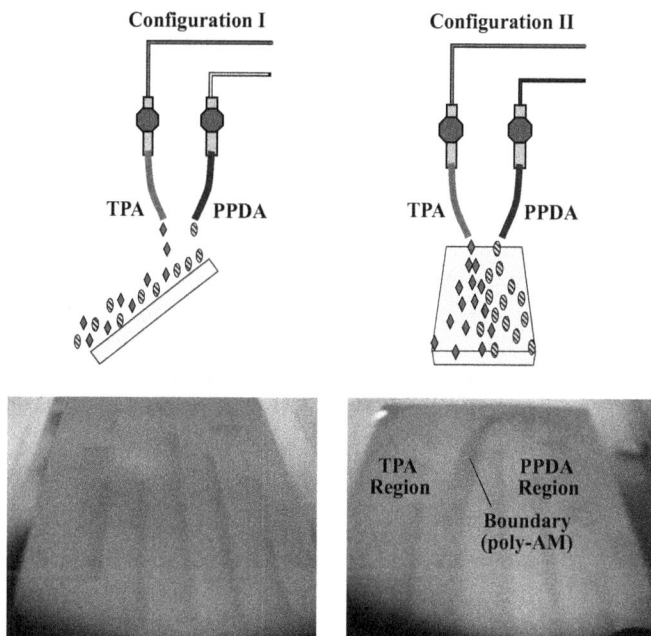

FIGURE 3.34 Influence of molecular gas flow on polymer film growth in carrier-gas-type organic CVD. (Reprinted with permission from Matsumoto et al. [34].)

the rotation-type domain-isolated MLD are presented. For two kinds of source molecules, a nozzle for Molecule A (denoted by A) and a nozzle for Molecule B (denoted by B) are set near a substrate holder. Between the two nozzles, a nozzle for purge gas (denoted by P) is located. Substrates are put on the holder. By blowing gases containing Molecules A and B, and purge gas onto the holder from these nozzles, Molecule A domain, Molecule B domain, and a purge gas curtain (PGC) are generated. PGC isolates the two domains. By rotating the holder, each substrate passes through the two domains alternately to achieve monomolecular-step growth of polymer films. For four kinds of source molecules, nozzles for Molecules A, B, C, and D are set around the center of the substrate holder. Between the four nozzles, nozzles for purge gas are placed. By blowing gases onto the holder from these nozzles, domains for Molecules A, B, C, and D, and PGCs are generated. By rotating the holder, each substrate passes through the four domains sequentially to achieve MLD. When the gas blowing is switched by opening and closing the nozzle valves, arbitrary molecular sequences are available.

Figure 3.35b shows train-type domain-isolated MLD with four kinds of source molecules. Nozzles for Molecules A, B, C, and D are set in parallel, and PGCs are inserted between them. By moving substrates on a conveyor, each substrate passes through the domains sequentially to achieve MLD. By switching the gas blowing, arbitrary molecular sequences are available. By bringing the substrates back to the starting point to repeat MLD, the number of required nozzles can be reduced.

In order to examine PGC's blocking ability against molecular mixing, TPA/N_2 gas and PPDA/N_2 gas were simultaneously introduced onto a glass substrate in the rotation-type domain-isolated MLD without rotation. By increasing the flow rate of N_2 for the PGC from 2.0 to 8.0 NL/min, light absorption due to poly-AM films deposited by TPA/PPDA mixing was drastically suppressed, indicating that the PGC has enough blocking ability to isolate the molecular domains.

Using PGC with 8.0 NL/min, domain-isolated MLD of poly-AM was carried out with rotation counts of 9 and 20 (step counts of 18 and 40). The rotation speed of the spin stage is 0.25 rpm. As can be seen in Figure 3.36a, the peak height ratio of poly-AM absorption spectra is height (18 steps): height (40 steps)=0.45:1. This ratio is close to the step count ratio of 18:40=0.51:1. These results suggest that the thickness of poly-AM films is approximately proportional to the step count with the assumption that the peak height ratio of poly-AM absorption spectra is proportional to the

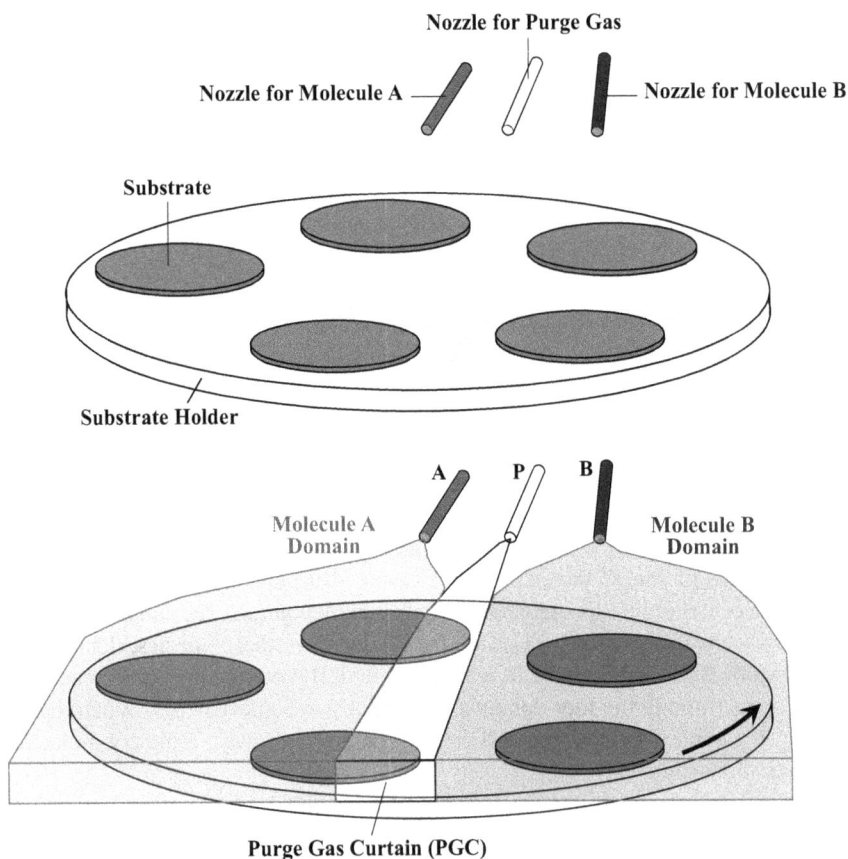

FIGURE 3.35 Concepts of domain-isolated MLD. (a) Rotation type and (b) train type.

(Continued)

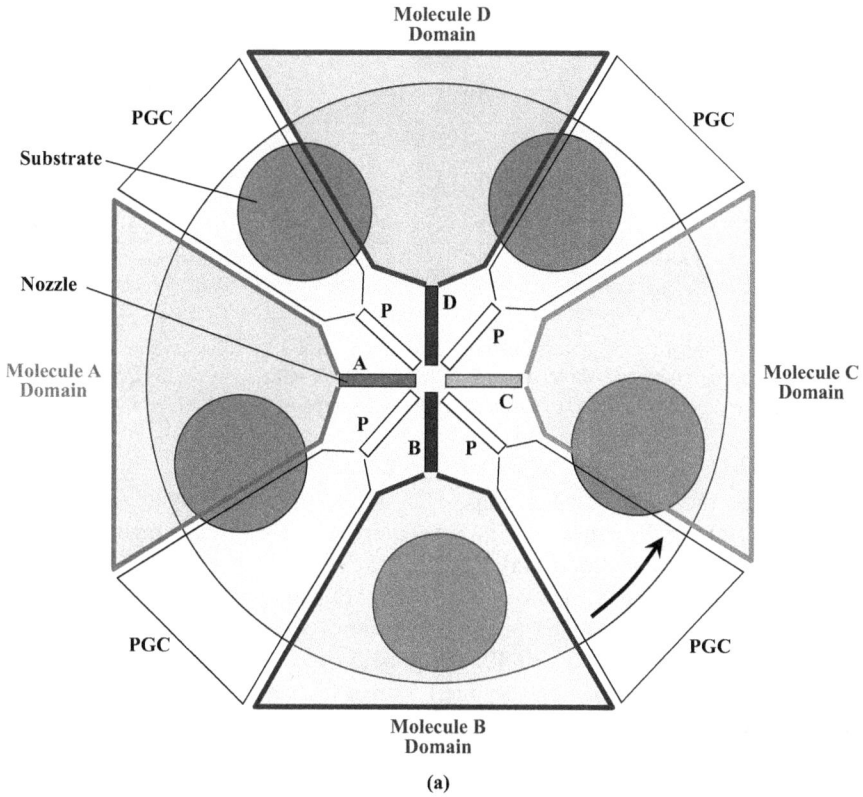

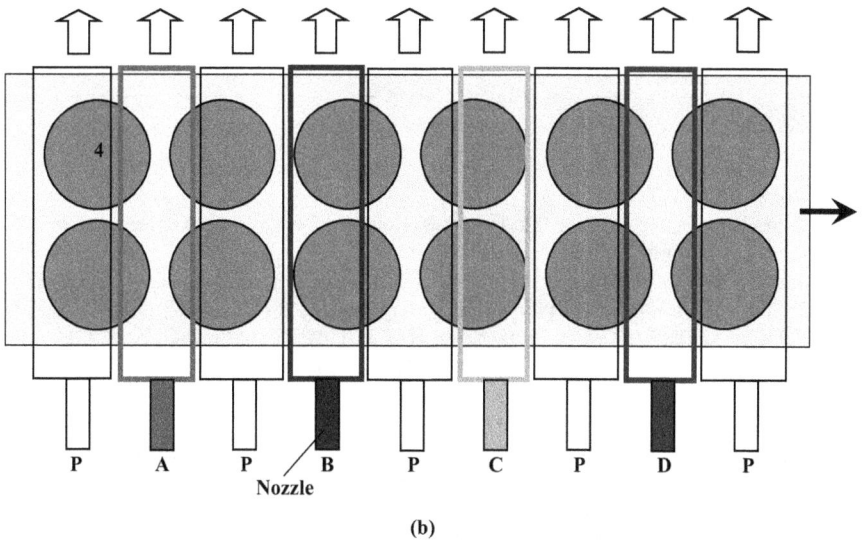

FIGURE 3.35 (*Continued*) Concepts of domain-isolated MLD. (a) Rotation type and (b) train type.

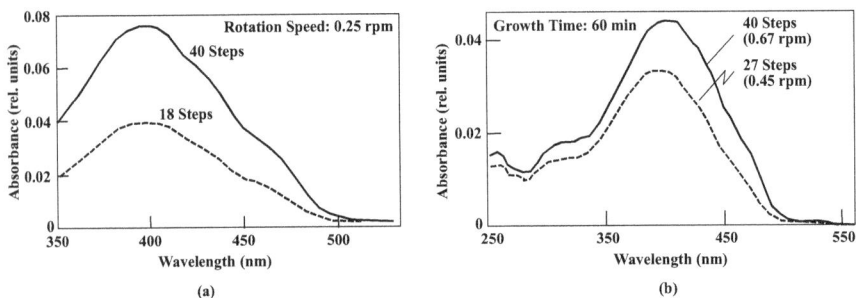

FIGURE 3.36 Absorption spectra of poly-AM films grown by rotation-type domain-isolated MLD (a) using source molecules of TPA and PPDA with steps of 18 and 40, and (b) using source molecules of TPA, PPDA, and ODH with steps of 27 and 40.

film thickness ratio. The time to complete the 40-step growth is about 80 min while it is about 200 min in the conventional MLD equipment. The process time is saved by a factor of 2.5 by the domain-isolated MLD.

The domain-isolated MLD using three kinds of source molecules, TPA, PPDA, and ODH, was also carried out. Figure 3.36b shows absorption spectra of the films grown with step counts of 27 and 40. The growth time was fixed at 60 min for both cases, implying that the rotation speed is 1.5 times higher for a step count of 40 than for a step count of 27. It is found that the absorption peak height of the film, which is proportional to the film thickness, increases by 1.3 times when the step count increases from 27 to 40. The film thickness ratio of 1.3 is fairly close to the step count ratio of $40/27 = 1.5$, suggesting the feasibility of the domain-isolated MLD for three kinds of source molecules.

REFERENCES

1. T. Suntola and J. Antson, Method for producing compound thin films, U.S. Patent 4,058,430 (1977).
2. T. Suntola, J. Antson, A. Pakkala, and S. Lindfors, "Atomic layer epitaxy for producing EL-thin films," *SID International Symposium in San Diego, California, 29 April–1 May 1980, Digest of Technical Papers*, SID, 108–109 (1980).
3. T. Suntola, "Atomic layer epitaxy," *Mater. Sci. Rep.* **4**, 261–312 (1989).
4. P. D. Ye, G. D. Wilk, J. Kwo, B. Yang, H.-J. L. Gossmann, M. Frei, S. N. G. Chu, J. P. Mannaerts, M. Sergent, M. Hong, K. K. Ng, and J. Bude, "GaAs MOSFET with oxide gate dielectric grown by atomic layer deposition," *IEEE Electron Device Lett.* **24**, 209–211 (2003).
5. F. Roozeboom, W. Dekker, K. B. Jinesh, J. H. Klootwijk, M. A. Verheijen, H.-D. Kim, and D. Blin, "Ultrahigh-density trench decoupling capacitors comprising multiple MIM layer stacks grown by atomic layer deposition," *AVS, 8th International Conference on Atomic Layer Deposition*, Bruges, Belgium, MonA1-1 (2008).
6. J. H. Shim, J. S. Park, J. An, T. M. Gur, and F. B. Prinz, "Atomic layer Deposition of Proton Conducting Y:BaZrO$_3$," *AVS, 9th International Conference on Atomic Layer Deposition*, Monterey, California, 76 (2009).

7. G. N. Parsons, Q. Peng, J. C. Spagnola, G. Scarel, G. K. Hyde, B. Gong, C. Devine, K. Lee, J. Jur, K. Roberts, and J. S. Na, "Modification of Fibers and Nonwoven Fiber Mats using Atomic Layer Deposition," *AVS, 9th International Conference on Atomic Layer Deposition*, Monterey, California, 29 (2009).

8. T. Yoshimura, Organic nonlinear optical material, U.S. Patent 5,194,548 (1993).

9. T. Yoshimura, E. Yano, S. Tatsuura, and W. Sotoyama, Organic functional optical thin film, fabrication and use thereof, U.S. Patent 5,444,811 (1995).

10. T. Yoshimura, S. Tatsuura, and W. Sotoyama, "Polymer films formed with monolayer growth steps by molecular layer deposition," *Appl. Phys. Lett.* **59**, 482–484 (1991).

11. Y. Takahashi, M. Iijima, K. Inagawa, and A. Itoh, "Synthesis of aromatic polyimide in vacuum," *Shinku* **28**, 440–442 (1985) [in Japanese].

12. J. R. Salem, F. O. Sequeda, J. Duran, and W. Y. Lee, "Solventless polyimide films by vapor deposition," *J. Vac. Sci. Technol. A* **4**, 369–374 (1986).

13. Y. Takahashi, M. Iijima, K. Inagawa, and A. Itoh, "Synthesis of aromatic polyimide film by vacuum deposition polymerization," *J. Vac. Sci. Technol. A* **5**, 2253–2256 (1987).

14. M. Iijima, Y. Takahashi, and E. Fukuda, "Vacuum deposition polymerization," Nikkei New Materials, 93–100 (1989) [in Japanese].

15. Y. Ito, M. Hikita, T. Kimura, and T. Mizutani, "Effect of degree of imidization in polyimide thin films prepared by vapor deposition polymerization on the electrical conduction," *Jpn. J. Appl. Phys.* **29**, 1128–1131 (1990).

16. A. Kubono, N. Okui, K. Tanaka, and S. Umemoto, "Highly oriented polyamide thin films prepared by vapor deposition polymerization," *Thin Solid Films* **199**, 385–393 (1991).

17. M. S. Weaver and D. D. C. Bradley, "Organic electroluminescence devices fabricated with chemical vapour deposited polyazomethine films," *Synth. Met.* **83**, 61–66 (1996).

18. T. Yoshimura, "Liquid phase deposition," Japanese Patent, Tokukai Hei 3–60487 (1991) [in Japanese].

19. T. Yoshimura, "Growth methods of polymer wires and thin films," Japanese Patent Tokukai 2008–216947 (2008) [in Japanese].

20. T. Yoshimura, H. Watanabe, and C. Yoshino, "Liquid-phase molecular layer deposition (LP-MLD): Potential applications to multi-dye sensitization and cancer therapy," *J. Electrochem. Soc.* **158**, 51–55 (2011).

21. T. Yoshimura, S. Bai, H. Tateno, C. Yoshino, "In situ photocurrent spectra measurements during growth of three-dye-stacked structures by the liquid-phase molecular layer deposition," *J. Appl. Phys.* **122**, 015309 (2017).

22. H. Zhou and S. F. Bent, "Novel photoresist thin films with in-situ photoacid generator by molecular layer deposition," *Proc. SPIE* **8682**, 86820U (2013).

23. T. Yoshimura, S. Ito, T. Nakayama, and K. Matsumoto, "Orientation-controlled molecule-by-molecule polymer wire growth by the carrier-gas-type organic chemical vapor deposition and the molecular layer deposition," *Appl. Phys. Lett.* **91**, 033103 (2007).

24. T. Yoshimura, R. Ebihara, A. Oshima, "Polymer wires with quantum dots grown by molecular layer deposition of three source molecules for sensitized photovoltaics," *J. Vac. Sci. Technol. A* **29**, 051510 (2011).

25. H. Murata, S. Ukishima, H. Hirano, and T. Yamanaka, "A Novel fabrication technique and new conjugated polymers for multilayer polymer light-emitting diodes," *Polym. Adv. Technol.* **8**, 459–464 (1996).

26. T. Yoshimura, S. Tatsuura, W. Sotoyama, A. Matsuura, and T. Hayano, "Quantum wire and dot formation by chemical vapor deposition and molecular layer deposition of one-dimensional conjugated polymer," *Appl. Phys. Lett.* **60**, 268–270 (1992).

27. H. J. Meier, "Sensitization of electrical effects in solids," *Phys. Chem.* **69**, 719–729 (1965).

28. K. Kiyota, T. Yoshimura, and M. Tanaka, "Electrophotographic behavior of ZnO sensitized by two dyes," *Photogr. Sci. Eng.* **25**, 76–79 (1981).

29. T. Yoshimura, K. Kiyota, H. Ueda, and M. Tanaka, "Contact potential difference of ZnO layer adsorbing p-type dye and n-type dye," *Jpn. J. Appl. Phys.* **18**, 2315–2316 (1979).

30. T. Yoshimura, "Thin-film molecular nanophotonics," in *Photonics, Vol. 2: Nanophotonic Structures and Materials* (Ed. D. L. Andrews), 261–310, Wiley, Hoboken, NJ (2015).

31. T. Liu and T. Yoshimura, "Sensitization of ZnO in stacked structures containing multiple dyes grown using liquid phase molecular layer deposition," *ECS Tran.* **69**, 217–222 (2015).

32. D. Kim and T. Yoshimura, "The Domain-Isolated Molecular Layer Deposition (DI-MLD) for fast polymer wire growth," *AVS, 9th International Conference on Atomic Layer Deposition,* Monterey, California, 243 (2009).

33. T. Yoshimura and D. Kim "A molecular-controlling deposition method and its applications to solar cells, electro-luminescent devices and optical switches," Japanese Patent, 2009–102408 (1991) [in Japanese].

34. K. Matsumoto and T. Yoshimura, "Electro-optic waveguides with conjugated polymer films fabricated by the carrier-gas-type organic CVD for chip-scale optical interconnects," *Proc. SPIE* **6899**, 68990E (2008).

4 Selective MLD

As described in Chapter 3, molecular layer deposition (MLD) achieves monomolecular-step growth of tailored organic thin films. By implementing selective growth functions into MLD, the application area is spread. In the case of organic thin films, "selective" has two meanings: "area selective" and "orientation selective." In the present chapter, various types of selective growth are demonstrated by vacuum deposition polymerization (VDP), organic chemical vapor deposition (CVD), and MLD. Since the mechanisms of selective growth are common to these three methods, the selective growth demonstrated by VDP and organic CVD are also applicable to MLD. First, area-selective growth of polyazomethine (poly-AM) thin films on surfaces treated to form designed hydrophilic/hydrophobic patterns is demonstrated. Area-selective MLD of poly-AM-based multiple quantum dots (MQDs) is also shown. Next, horizontally-aligned selective growth of poly-AM and polyimide thin films on patterned obliquely-evaporated dielectric films is presented. The driving force for the polymer wire alignment is the interaction between the source molecules and the atomic-scale anisotropic structures of the dielectric films. Electric-field-assisted selective growth of polyamic acid thin films is also described. Finally, vertically-aligned selective growth from seed cores is demonstrated by MLD, and the monomolecular-step growth of poly-AM thin films from the seed cores is confirmed.

4.1 CLASSIFICATION OF SELECTIVE GROWTH

Figure 4.1 presents three types of selective growth of organic thin films. In the area-selective growth, organic wires, namely, polymer wires and molecular wires, are grown in designated areas. In the horizontally-aligned selective growth, organic wires are grown horizontally with designated directions in designated areas. In the vertically-aligned selective growth, organic wires are grown vertically in designated areas. By using seed cores, the area-/orientation-controlled selective growth, which involves both horizontally-aligned and vertically-aligned growth, is expected to be realized as described in Sections 5.2 and 5.3.

4.2 AREA-SELECTIVE GROWTH ON HYDROPHILIC/ HYDROPHOBIC SURFACES

4.2.1 CONCEPT

Concept of the area-selective growth of polymer thin films using source molecules of Molecules A and B is shown in Figure 4.2. Surface treatment is applied to a substrate with designed patterns. Molecules are connected by chemical reaction to grow polymer wires. In the present example, adsorption strength of molecules on the surface is stronger in the surface-treated region than in the untreated region. This implies that

DOI: 10.1201/9781003094012-4

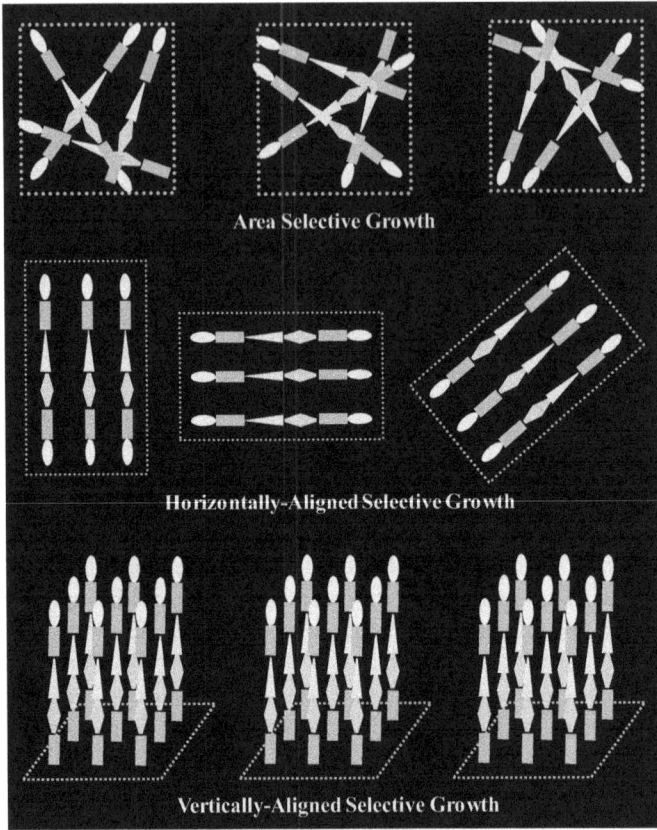

FIGURE 4.1 Types of selective growth of organic thin films.

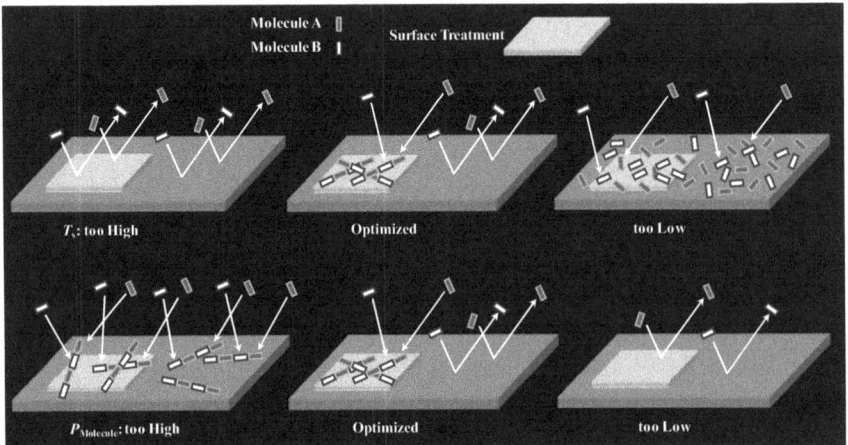

FIGURE 4.2 Concept of the area-selective growth of polymer thin films.

residence time $\tau_{\text{Residence}}$ for molecules to migrate and rotate at the surface is longer in the surface-treated region than in the untreated region.

For the area-selective growth, optimization of substrate temperature T_s and molecular gas pressure P_{Molecule} is required. When T_s is too high, $\tau_{\text{Residence}}$ becomes short, resulting in re-evaporation of molecules in the whole surface regions before they combine with other molecules. When T_s is too low, the molecules aggregate all over the surface. At optimized T_s, molecules re-evaporate in the untreated region. In the treated region, because $\tau_{\text{Residence}}$ is longer than in the untreated region, collision between molecules is enhanced. Then, the molecules can be combined to make new molecules with larger molecular weight, preventing the re-evaporation, and consequently, polymer wires selectively grow in the treated region.

For P_{Molecule}, when it is too high, Molecules A and B are combined before their re-evaporation in the whole surface regions because the collision between the molecules occur frequently, resulting in polymer wire growth. When P_{Molecule} is too low, molecules re-evaporate before the collision happens, resulting in no polymer wire growth. At optimized P_{Molecule}, molecules re-evaporate in the untreated region while polymer wires grow in the treated region, where $\tau_{\text{Residence}}$ is long enough to induce the collision between molecules.

A process example of the area-selective growth is depicted in Figure 4.3. First, a hydrophobic treatment is applied to a substrate surface. Then, a hydrophilic treatment is applied on the surface, forming a hydrophilic/hydrophobic pattern. Finally, Molecules A and B are introduced onto the surface to grow a polymer thin film using chemical reaction between the molecules. When the hydrophilic surface is more favorable for molecules to be adsorbed than the hydrophobic surface, reaction between the molecules occurs only on the hydrophilic region, resulting in area-selective growth of a polymer thin film.

4.2.2 AREA-SELECTIVE GROWTH OF POLY-AM ON SIO/ HEXAMETHYL-DISILAZANE (HMDS) SURFACES

By patterned hydrophilic/hydrophobic surface treatment of a glass substrate, experimental demonstration of the area-selective growth of poly-AM was performed utilizing VDP. For hydrophobic treatment, the substrate surface was coated with HMDS for 1 h, and then, it was put into dichloromethane to remove excess HMDS, forming a hydrophobic surface. Next, a 100-nm-thick SiO film was deposited with a designed pattern on the surface by vacuum evaporation. Since SiO is hydrophilic, a hydrophilic/hydrophobic pattern was formed. Finally, terephthalaldehyde (TPA) and

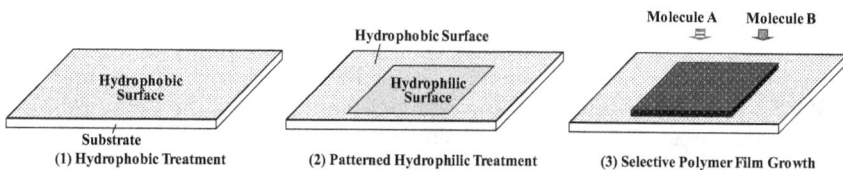

FIGURE 4.3 Process example of the area-selective growth.

p-phenylenediamine (PPDA) were supplied to the substrate in a vacuum chamber for 150 min with molecular gas pressure around 0.1 Pa at $T_s = 25°C$. Because the hydrophilic SiO surface is more favorable for TPA and PPDA molecules to be adsorbed than the hydrophobic HMDS surface, reaction between the molecules occurs only on SiO to grow a poly-AM thin film [1].

A yellow rectangle (a dark rectangle in the monochromatic photograph) appearing in Figure 4.4a corresponds to the SiO region, where an 11-nm-thick poly-AM [TPA/PPDA] thin film is grown. In the surrounding HMDS region, the substrate exhibits no color, indicating little sign of poly-AM thin-film growth. Thus, the area-selective growth of a poly-AM thin film in the SiO region is achieved. Figure 4.4b is a magnified photograph around the boundary. As Figure 4.4c shows, an absorption band of poly-AM with a peak around 410 nm appears in the SiO region, while there is no absorption in the HMDS region, confirming the area-selective growth.

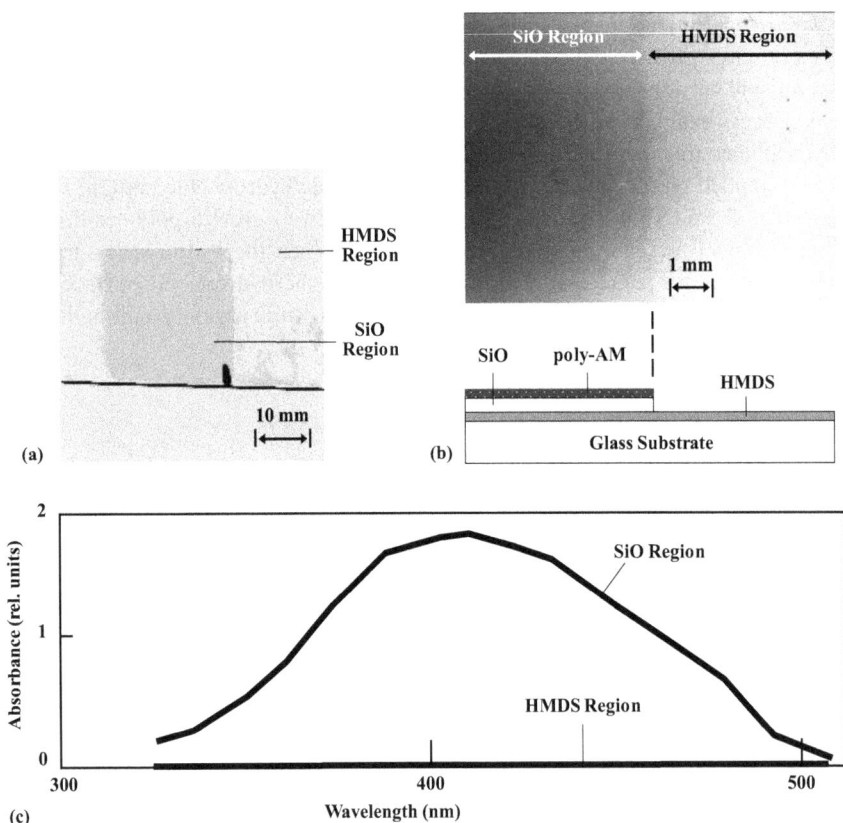

FIGURE 4.4   (a) Observation of area-selective growth of a poly-AM [TPA/PPDA] thin film on a substrate surface with a hydrophilic/hydrophobic pattern formed by SiO and HMDS, (b) magnified photograph and schematic diagram near the boundary between the hydrophilic and the hydrophobic regions, and (c) absorption spectra in the SiO region and the HMDS region. (Reprinted with permission from Yoshimura et al. [1].)

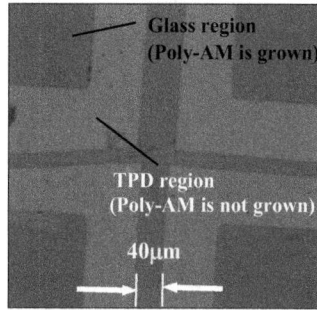

FIGURE 4.5 Observation of area-selective growth of a poly-AM thin film on a substrate surface with a hydrophilic/hydrophobic pattern formed by a glass substrate and patterned TPD films. (Reprinted with permission from Matsumoto and Yoshimura [2].)

4.2.3 AREA-SELECTIVE GROWTH OF POLY-AM ON SiO₂/ TRIPHENYLDIAMINE (TPD) SURFACES

A patterned hydrophobic TPD film was deposited on a glass substrate by vacuum evaporation through a metal mask, and a poly-AM thin film was grown by carrier-gas-type organic CVD with TPA and PPDA on the substrate. As a result, as shown in Figure 4.5, a poly-AM thin film selectively grows on the hydrophilic glass surface [2].

4.2.4 AREA-SELECTIVE MLD OF POLYMER MQDS ON TiO₂/ZnO SURFACES

Ishii et al. succeeded in the area-selective MLD of an OT polymer MQD film. Details of the OT polymer MQD are described in Section 6.4. It is found from Figure 4.6a that the OT polymer MQD film grows on a TiO₂ powder layer while no film grows on a ZnO powder layer, indicating that the area-selective MLD is achieved by using TiO₂ and ZnO. The inhibition of the film growth on ZnO might be attributed to weak hydrophilic characteristics of ZnO. The OT polymer MQD film grown on a TiO₂ powder layer and that grown on a glass substrate exhibits similar PL spectra, as depicted in Figure 4.6b [3].

4.3 HORIZONTALLY-ALIGNED SELECTIVE GROWTH ON ANISOTROPIC ATOMIC-SCALE STRUCTURES

In polymer functional devices, it is important to control polymer wire alignment. For example, if polymer wires are aligned selectively within a stripe with a width of micron or submicron scale, a channel optical waveguide can be constructed without etching because the selective alignment induces a refractive index difference between the stripe region and the surrounding region. The polymer wire alignment is favorable to enhance optical nonlinearity. Carrier mobility in polymer thin-film transistors (TFTs) might also be increased by the alignment.

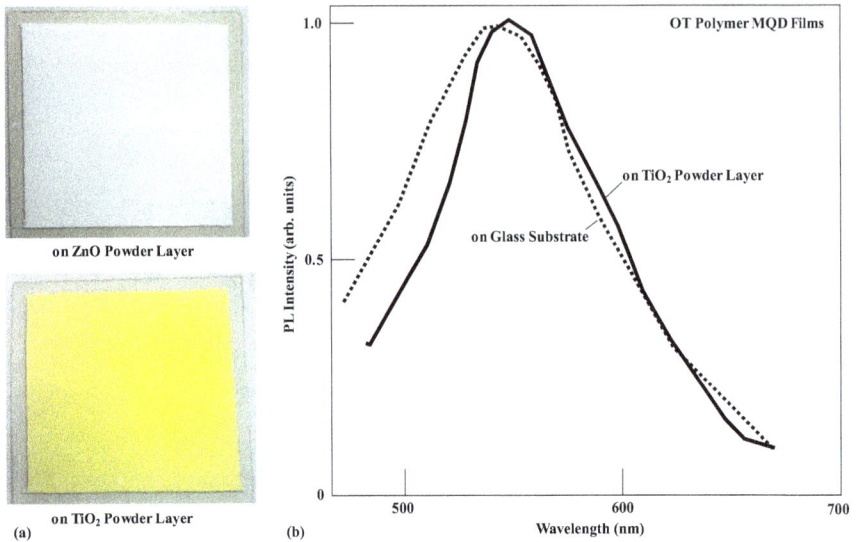

FIGURE 4.6 (a) Area-selective MLD of OT polymer MQD film. (b) PL spectra of OT polymer MQD films on a TiO_2 powder layer and on a glass substrate.

Kanetake et al. reported that an aligned polydiacetylene film is grown by vacuum evaporation on rubbing polydiacetylene films [4]. Patel et al. demonstrated selective alignment of spin-coated polydiacetylene films on a rubbing polymer underlayer [5]. The author's research group attempted the horizontally-aligned selective growth of poly-AM and polyimide thin films on surfaces with anisotropic structures by an all-dry process as described below.

4.3.1 ALL-DRY PROCESS FOR POLYMER WIRE ALIGNMENT

The all-dry process for polymer wire alignment is illustrated in Figure 4.7. First, a dielectric film is deposited by oblique vacuum evaporation with a designed pattern on a substrate, which is tilted along the y-axis by θ. Here, θ is a tilting angle of the surface-normal direction from the evaporant beam direction. The obliquely-evaporated dielectric film has ultrafine anisotropic atomic-scale structures along the y-axis. Next, Molecules A and B are introduced onto the surface to grow polymer wires. The wires grown on the obliquely-evaporated dielectric film are oriented in a direction parallel to the y-axis while the wires grown on the substrate surface directly are oriented in random directions. Thus, the horizontally-aligned selective growth is achieved [6].

The driving force for the wire alignment might be the interaction between the source molecules and the atomic-scale anisotropic structures of the dielectric film. The molecules tend to be placed on the dielectric film in a configuration at potential energy minimum. It should be noted that any surfaces inducing anisotropic potential energy distributions can be used for the horizontally-aligned growth.

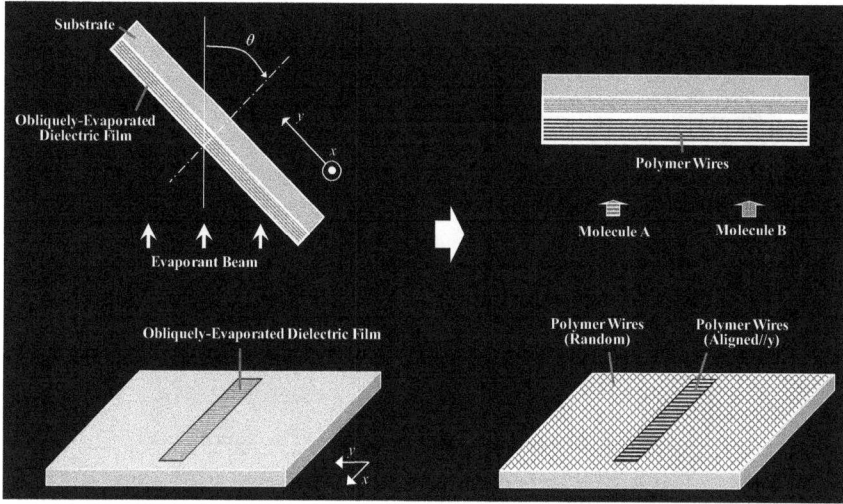

FIGURE 4.7   All-dry process for the horizontally-aligned selective growth on anisotropic atomic-scale structures.

4.3.2   POLY-AM ON OBLIQUELY-EVAPORATED SiO₂ UNDERLAYERS

The horizontally-aligned selective growth of poly-AM thin films was performed on a patterned obliquely-evaporated SiO_2 film. The procedure is illustrated in Figure 4.8a [6].

1. A photoresist pattern with a window of a 10-µm-wide stripe is formed on a quartz substrate.
2. An 80-nm-thick SiO_2 film is deposited on it by oblique vacuum evaporation with $\theta = 45°$ or $0°$. The substrate tilting direction, namely, the y-axis direction, is perpendicular to the stripe.
3. The photoresist is removed by the liftoff, forming a 10-µm-wide stripe of an obliquely-evaporated SiO_2 underlayer.
4. A 160-nm-thick poly-AM thin film is deposited over the entire substrate by VDP with TPA and PPDA.

An expected cross-section structure of the sample is schematically drawn in Figure 4.8b. The direction of the poly-AM wires on the obliquely-evaporated SiO_2 underlayer is parallel to the y-axis while that of the wires on the quartz substrate surface is random.

Microscopic images of poly-AM thin films were observed using polarized light, as shown in Figure 4.9. Here, light polarization parallel to the y-axis and perpendicular to the y-axis are, respectively, represented by $E_{Opt}//y$ and $E_{Opt} \perp y$. In the case of $\theta = 45°$, the stripe region is darker than the surrounding region for $E_{Opt}//y$ while it is brighter for $E_{Opt} \perp y$. The result indicates that the polymer wires on the

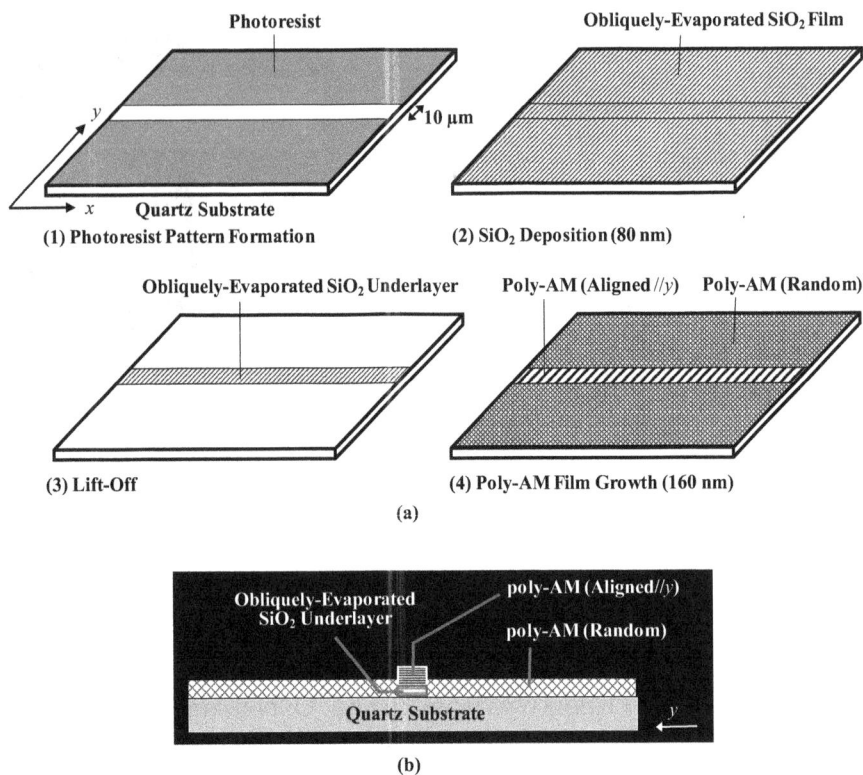

(1) Photoresist Pattern Formation

(2) SiO₂ Deposition (80 nm)

(3) Lift-Off

(4) Poly-AM Film Growth (160 nm)

(a)

(b)

FIGURE 4.8 (a) Procedure of the horizontally-aligned selective growth and (b) schematic illustration of an expected cross-section structure of the sample. (Reprinted with permission from Yoshimura et al. [6]. Copyright (1992) The Physical Society of Japan and The Japan Society of Applied Physics.)

obliquely-evaporated SiO_2 underlayer are selectively aligned along the y-axis, as predicted in Figure 4.8b. In the case of $\theta = 0°$, on the other hand, no light polarization dependence of the contrast appears, indicating that the wire direction is random over the entire surface. These results indicate that the horizontally-aligned selective growth is realized by the obliquely-evaporated SiO_2 underlayers. The selective alignment was also observed in thick poly-AM films with a thickness of 500 nm or more. By increasing θ from $0°$ to $60°$, the degree of the alignment was increased.

As Figure 4.10 shows, the poly-AM thin film exhibits large dichroism for $\theta = 45°$, namely, absorption for $E_{Opt}//y$ is about 10 times larger than that for $E_{Opt} \perp y$. For $\theta = 0°$, on the other hand, it exhibits no dichroism. The conjugated polymer like poly-AM has large absorption for light polarization along the polymer wire direction. Therefore, the results shown in Figure 4.10 reconfirm that the wires are aligned along the y-axis in the poly-AM thin film on the obliquely-evaporated SiO_2 underlayer.

In the experiment presented in Figure 4.11, after an obliquely-evaporated SiO_2 film is deposited on a substrate by tilting the substrate along the x-axis, a stripe of

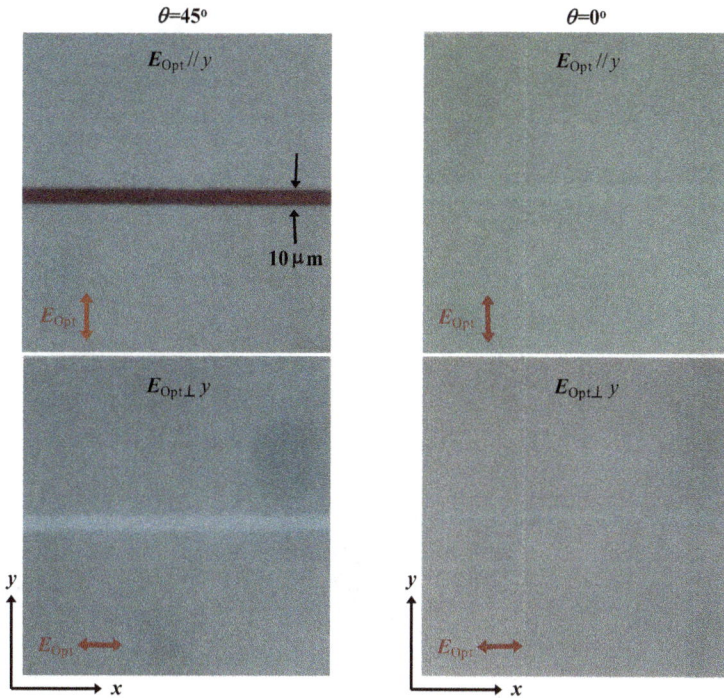

FIGURE 4.9 Microscopic images of poly-AM thin films fabricated by the horizontally-aligned selective growth for $E_{Opt}//y$ and $E_{Opt}\perp y$. (Reprinted with permission from Yoshimura et al. [6]. Copyright (1992) The Physical Society of Japan and The Japan Society of Applied Physics.)

100-nm-thick obliquely- evaporated SiO_2 film is deposited by tilting the substrate along the y-axis. And then, a poly-AM thin film is grown on the substrate. It is found that the contrast between the stripe region and the surrounding region is completely inverted by rotating polarization from $E_{Opt}//y$ to $E_{Opt}//x$. This indicates that the polymer wires in the stripe region are aligned along the y-axis while those in the surrounding region are aligned along the x-axis. This result leads us to believe that the polymer wire direction is determined only by the SiO_2 layer coming in contact with the polymer wires, suggesting that the wire direction can freely be controlled with designated patterns on a substrate by stacking SiO_2 underlayers obliquely-evaporated with different tilting directions.

In Figure 4.9, in the case of $\theta = 45°$, it is expected that the refractive index is larger for $E_{Opt}//y$ than for $E_{Opt}\perp y$ in the stripe region, where poly-AM wires are aligned along the y-axis. To examine the birefringence induced by the aligned poly-AM wires, microscopic images were observed under the cross-Nicol condition. As Figure 4.12 shows, a bright 10-μm-wide stripe is observed, indicating that birefringence occurs selectively in the stripe region. In the case of $\theta = 0°$, no birefringence is observed.

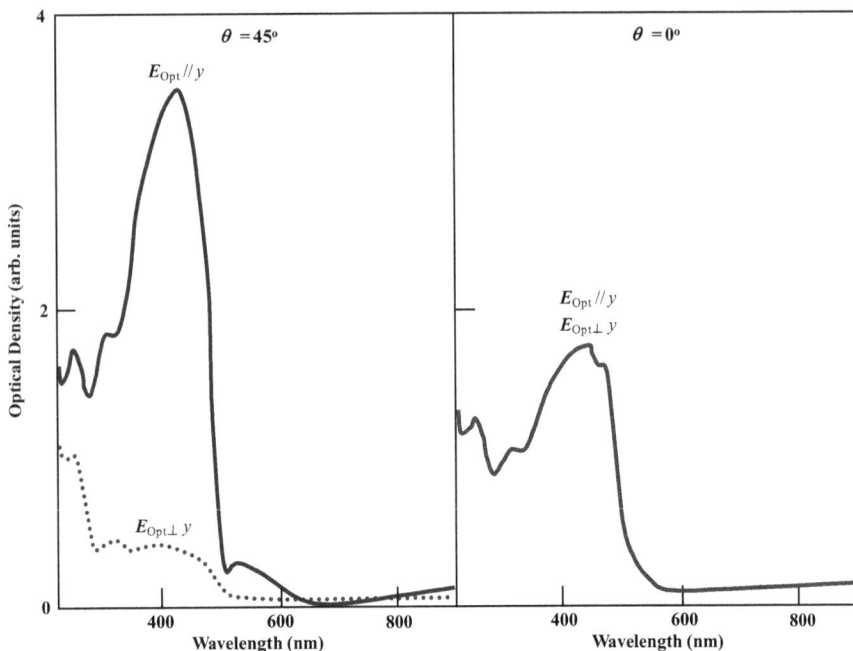

FIGURE 4.10 Absorption spectra of poly-AM thin films fabricated by the horizontally-aligned selective growth for $E_{Opt}//y$ and $E_{Opt}\perp y$. (Reprinted with permission from Yoshimura et al. [6]. Copyright (1992) The Physical Society of Japan and The Japan Society of Applied Physics.)

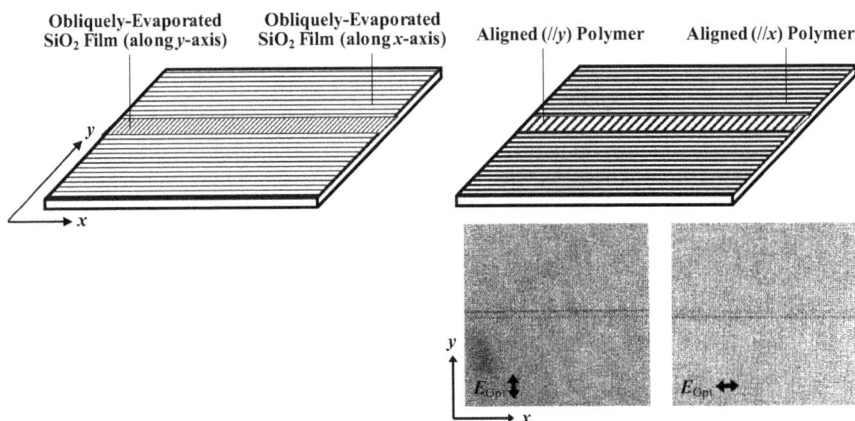

FIGURE 4.11 Microscopic images of poly-AM thin films fabricated by the horizontally-aligned selective growth for $E_{Opt}//y$ and $E_{Opt}\perp y$. The thin films are grown on stacked obliquely-evaporated SiO_2 films. (Reprinted with permission from Yoshimura et al. [6]. Copyright (1992) The Physical Society of Japan and The Japan Society of Applied Physics.)

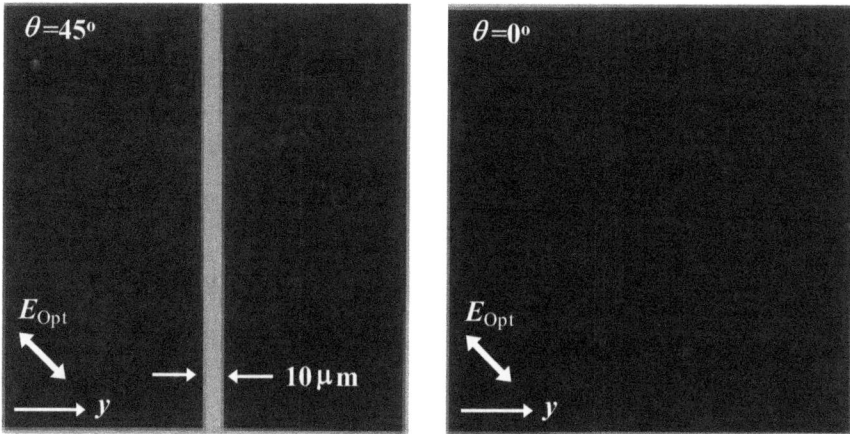

FIGURE 4.12 Microscopic images observed under the cross-Nicol condition for poly-AM thin films fabricated by the horizontally-aligned selective growth.

4.3.3 POLY-AM ON OBLIQUELY-EVAPORATED SiO UNDERLAYERS

Figure 4.13 shows microscopic images under the cross-Nicol condition for a poly-AM thin film. The thin film was grown by carrier-gas-type organic CVD on a glass substrate with stripe patterns of 200-nm-thick SiO films, which were obliquely-evaporated with $\theta = 45°$ tilting along the y-axis [2]. Source molecules were TPA and PPDA, and the carrier gas was nitrogen. Growth time was 20 min, and gas pressure was around 100 Pa. For $E_{Opt}//y$, the microscopic image is dark. When the substrate is rotated by 45°, the stripe regions on the obliquely-evaporated SiO underlayers become bright due to birefringence. This result indicates that poly-AM wires are selectively aligned by the obliquely-evaporated SiO films.

Observation of an obliquely-evaporated SiO film for the underlayer and poly-AM thin films was carried out by atomic force microscope (AFM) [7]. As can be seen from Figure 4.14a, in an obliquely-evaporated SiO film, a grain morphology along the y-axis is observed, revealing that the film has an anisotropic structure. In a poly-AM thin film on the obliquely-evaporated SiO film, as shown in Figure 4.15b, line-shaped patterns along the y-axis are observed. This suggests that polymer wires are aligned along the y-axis. In a poly-AM thin film on the glass substrate, as shown in Figure 4.15c, structures with a definite orientation are not observed.

4.3.4 POLYIMIDE ON OBLIQUELY-EVAPORATED SiO₂ UNDERLAYERS

As Figure 4.15 shows, a polyimide thin film grown by VDP with pyromellitic dianhydride (PMDA) and 4,4′-diaminodiphenyl ether (DDE) on a glass substrate with a 10-μm-wide stripe of SiO₂ film obliquely evaporated with $\theta = 45°$ exhibits birefringence. This indicates that the horizontally-aligned selective growth is effective in polyimide thin films.

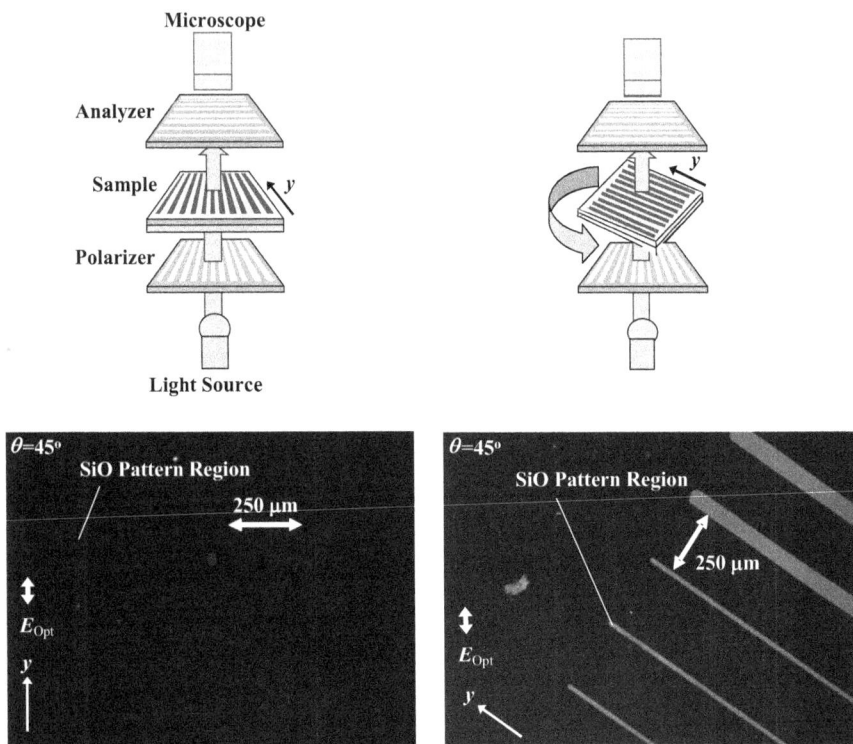

FIGURE 4.13   Microscopic images observed under the cross-Nicol condition for a poly-AM thin film grown by carrier-gas-type organic CVD on a patterned obliquely-evaporated SiO film with $\theta = 45°$. (Reprinted with permission from Matsumoto and Yoshimura [2].)

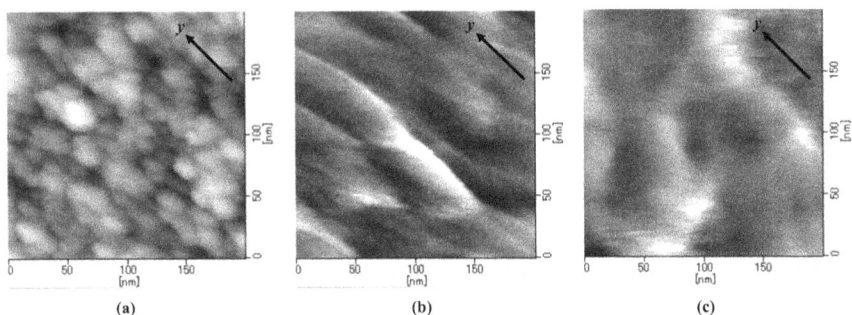

FIGURE 4.14   AFM images of (a) an obliquely-evaporated SiO film, (b) a poly-AM thin film grown on the obliquely-evaporated SiO film, and (c) a poly-AM thin film grown on a glass substrate. (Reprinted with permission from Suzuki et al. [7].)

As described, the horizontally-aligned selective growth is applicable to various kinds of polymers, and various dielectric films can be used for the underlayers.

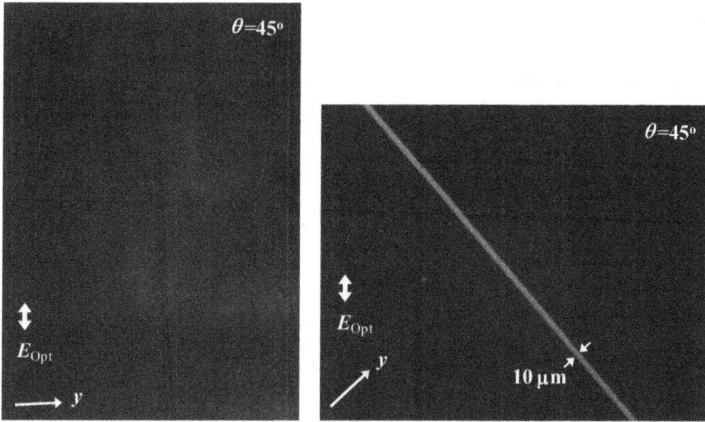

FIGURE 4.15 Microscopic images under the cross-Nicol condition for a polyimide thin film fabricated by the horizontally-aligned selective growth.

4.4 ELECTRIC-FIELD-ASSISTED SELECTIVE GROWTH

Electric-field-assisted selective growth, in which constituent molecules with electric dipoles are aligned by electric fields to a favorite direction, is expected to produce poled polymers for electro-optic (EO) materials and waveguides at low temperature without the conventional poling process and solvent/contamination.

4.4.1 CONCEPT

Concept of the electric-field-assisted selective growth with Molecules A and B is depicted in Figure 4.16. Here, at least one of the two kinds of molecules has an electric dipole. In the present case, Molecule B has an electric dipole. Molecules are

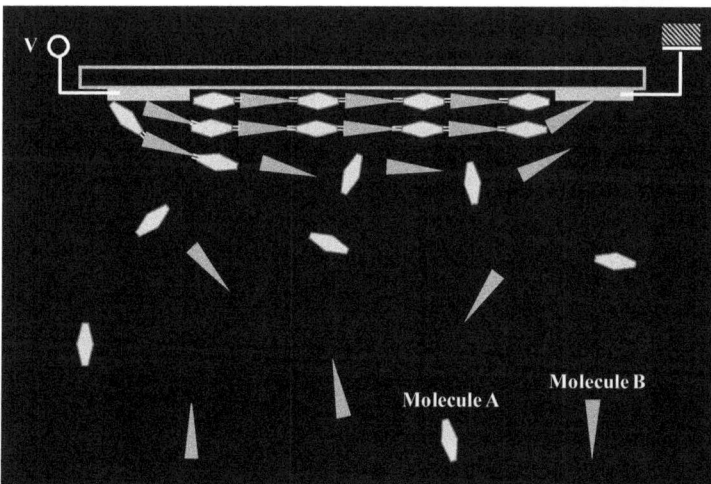

FIGURE 4.16 Concept of the electric-field-assisted selective growth.

provided onto a substrate surface, on which slit-type electrodes are formed. When Molecule B comes near the surface, it tends to be aligned along the direction of the electric field generated between the electrodes. Because the molecular alignment is completed before the molecules are tightly inserted into polymer wires, the resistance against molecular rotation is small and molecules are aligned without raising temperature above the glass transition temperature. Thus, it is expected that polymer wires, in which Molecule B tends to have a unique orientation, grow in a direction parallel to the electric field [8].

4.4.2 POLYAMIC ACID

The electric-field-assisted selective growth of polyamic acid thin films was demonstrated by VDP using PMDA and DNB evaporated from temperature-controlled K-cells. The background pressure was 5×10^{-6} Pa. An electric field of 0.78 MV/cm was applied to a 10-μm gap of slit-type electrodes on a glass substrate during film deposition [8].

As preliminary experiments, film deposition was attempted in the following three ways: supplying only PMDA, only DNB, and PMDA+DNB, which implies that PMDA and DNB are co-evaporated. As schematically illustrated in Figure 4.17, at $T_s = 25°C$, these films appear frosted, indicating that aggregates of molecules exist in the films. At $T_s = 65$ and $110°C$, the deposition rates are very low for only PMDA and only DNB owing to re-evaporation. For PMDA+DNB, PMDA and DNB are connected on the substrate to grow polymer wires, and residual molecules re-evaporate, forming a stoichiometric clear polyamic acid thin film.

Figure 4.18a shows transmission spectra of a polyamic acid [PMDA/DNB] thin film deposited at $T_s = 65°C$ and of methanol solutions of PMDA and DNB. In DNB, an absorption peak appears around 400 nm in wavelength. In the polyamic acid thin film, an absorption band also appears around 400 nm, confirming that DNB molecules are inserted into the wires. Figure 4.18b shows absorption spectra of the polyamic acid thin film in the gap region between electrodes. Film thickness

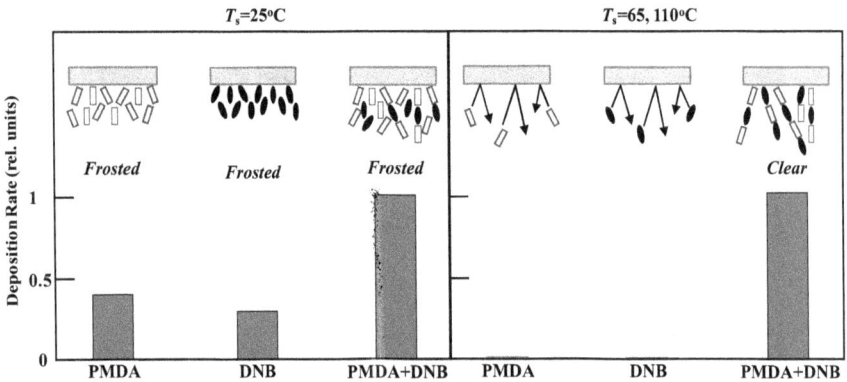

FIGURE 4.17 Deposition rates and film appearance for cases of supplying only PMDA, only DNB, and PMDA+DNB.

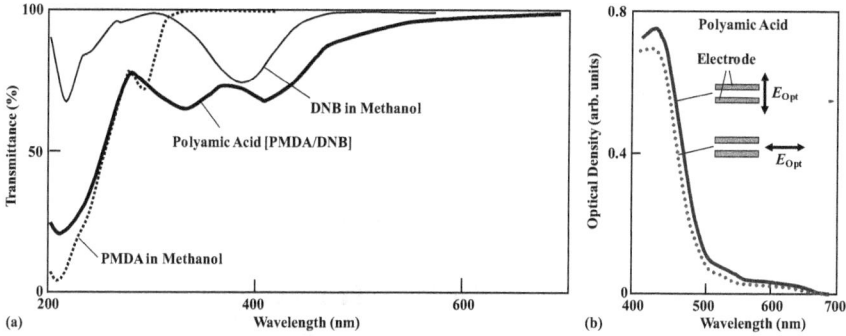

FIGURE 4.18 (a) Transmission spectra of PMDA, DNB, and polyamic acid [PMDA/DNB] deposited at $T_s = 65°C$ and (b) absorption spectra of the polyamic acid thin film for polarized light in the gap region between electrodes. (Reprinted with permission from Yoshimura et al. [8].)

is 160 nm. The absorption for light polarization in the direction of the electric field applied during film deposition is larger than that for light polarization perpendicular to the electric field direction, suggesting the electric-field-assisted alignment.

4.5 VERTICALLY-ALIGNED SELECTIVE GROWTH FROM SEED CORES

In order to fabricate advanced organic photonic/electronic devices and nanosystems, vertically-aligned polymer wire growth in designated areas is necessary because most of the devices have vertical configurations, in which sandwich-type electrodes are employed. Furthermore, it will be extended to the area-/orientation-controlled selective growth, which enables polymer wire growth to arbitrary directions, realizing three-dimensional polymer wire networks. In the present section, the vertically-aligned selective growth of poly-AM thin films from seed cores by MLD is described [9,10].

4.5.1 VERTICALLY-ALIGNED SELECTIVE MLD FROM SEED CORES WITH SELF-ASSEMBLED MONOLAYERS (SAMs)

4.5.1.1 Surface-Active Materials for Seed Cores

Kubono et al. reported that a polyamide thin film with wire orientation perpendicular to the substrate grows by VDP when it is deposited on a 1,10-diaminodecane layer with the perpendicular orientation [11]. Such epitaxial effects work in MLD. In Figure 4.19, two examples of surface-active materials are presented [12]. When a substrate surface is treated with a silane-based surface-active material having terminal amino protons, polymer wires will grow vertically from the substrate. Self-assembled monolayers (SAMs) such as amino-alkanethiol are also candidates of surface-active materials. These can be applied to seed cores.

NH$_2$ NH$_2$ NH$_2$
$(CH_2)_n(CH_2)_n(CH_2)_n$
S   S   S

NH$_2$ NH$_2$ NH$_2$

Silane-Based Surface-Active Material   **Self-Assembled Monolayer (SAM)**

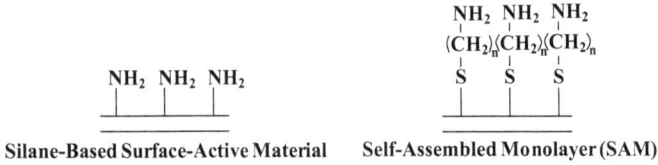

FIGURE 4.19   Examples of surface-active materials for seed cores.

4.5.1.2 Vertically-Aligned Selective MLD of Poly-AM from SAMs

Ito et al. demonstrated the vertically-aligned selective MLD of a poly-AM thin film from a seed core by carrier-gas-type MLD. The process is schematically drawn in Figure 4.20. In Step 0, a patterned Au film for a seed core is deposited on a glass substrate. A SAM of amino-alkanethiol is formed on the Au film by putting the substrate in ethanol solution of 11-amino-1-undecanethiol, hydrochloride, and rinsing/drying it to fabricate a seed core. In Step 1, TPA molecules supplied from a molecular cell kept at 50°C are introduced onto the substrate surface with nitrogen gas for 5 min to connect TPA molecules to the seed core molecules of amino-alkanethiol. In Step 2, after removing excess TPA molecular gas by pumping for 5 min, PPDA molecules from a molecular cell kept at 50°C are introduced onto the surface for 5 min to connect PPDA molecules to TPA molecules. By repeating the molecular gas switching, poly-AM wires are grown from the seed core surface vertically to form a poly-AM thin film. In the present experiments, the number of MLD cycles was 10, namely, the step count m was 20. The process was carried out at room temperature. For the

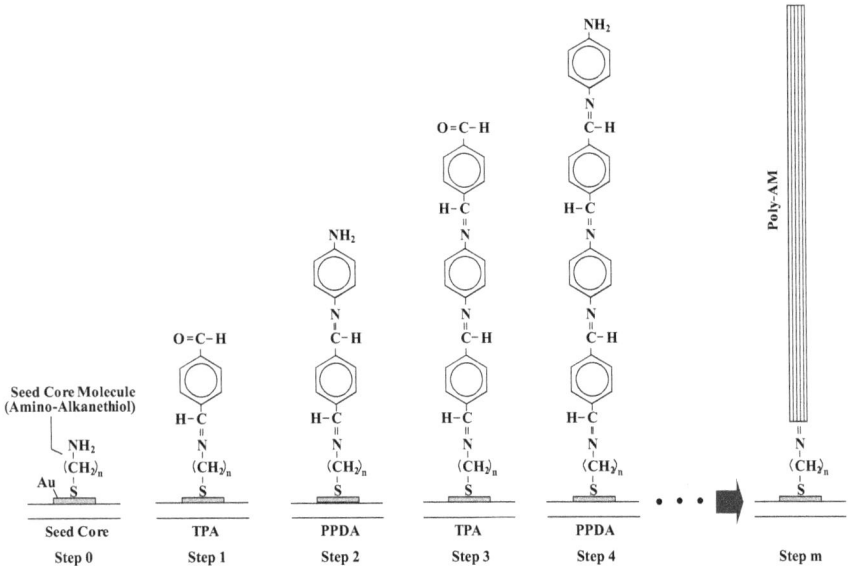

FIGURE 4.20   Process of the vertically-aligned selective MLD from a seed core. (Reprinted with permission from Yoshimura et al. [9].)

purpose of comparison, a poly-AM thin film was prepared on an Au film without the SAM by the same procedure described above [9,10].

Figure 4.21 shows Fourier transform infrared reflection absorption spectroscopy (FTIR-RAS) spectra of TPA-adsorbed surfaces corresponding to Step 1. The absorption peaks attributed to TPA in wavenumber regions around 820 and 780 cm^{-1} are much larger in the sample prepared using a surface with SAM (denoted by "with SAM") than in the sample prepared using a surface without SAM (denoted by "without SAM"). By raising the sample temperature to 45°C, the peak height in "without SAM" decreases while that in "with SAM" does not decrease. These results indicate that more TPA molecules exist on the surface with stronger bonds between TPA and SAM in "with SAM" than in "without SAM."

Figure 4.22 shows FTIR-RAS spectra of poly-AM thin films in Step 20. The absorbance for "with SAM" is larger than that for "without SAM" in almost whole of the wavenumber regions, indicating that the poly-AM thin film is thicker in "with SAM" than in "without SAM." Absorption peaks in wavenumber regions around 1200–1300 and 850 cm^{-1} are, respectively, attributed to the vibration of in-plane direction (denoted by "in-plane") and the vibration of surface-normal direction (denoted by "surface-normal") in benzene rings. The absorbance ratio of "in-plane" to "surface-normal" ($A_{\text{In-Plane}}/A_{\text{Surface-Normal}}$) is larger by a factor of ~2 in "with SAM"

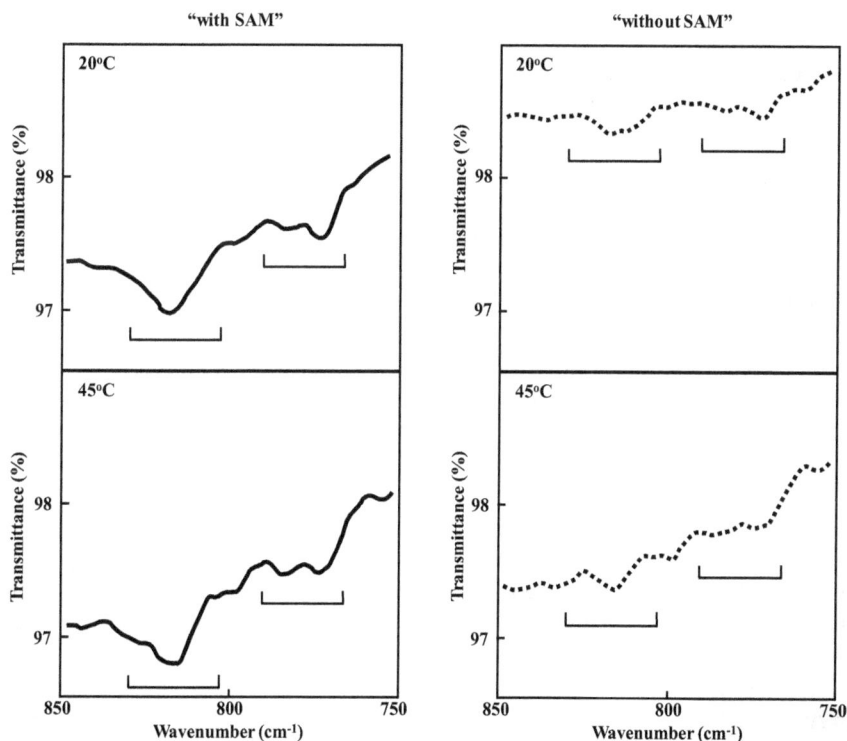

FIGURE 4.21 FTIR-RAS spectra of TPA-adsorbed surfaces corresponding to Step 1. (Reprinted with permission from Yoshimura et al. [9].)

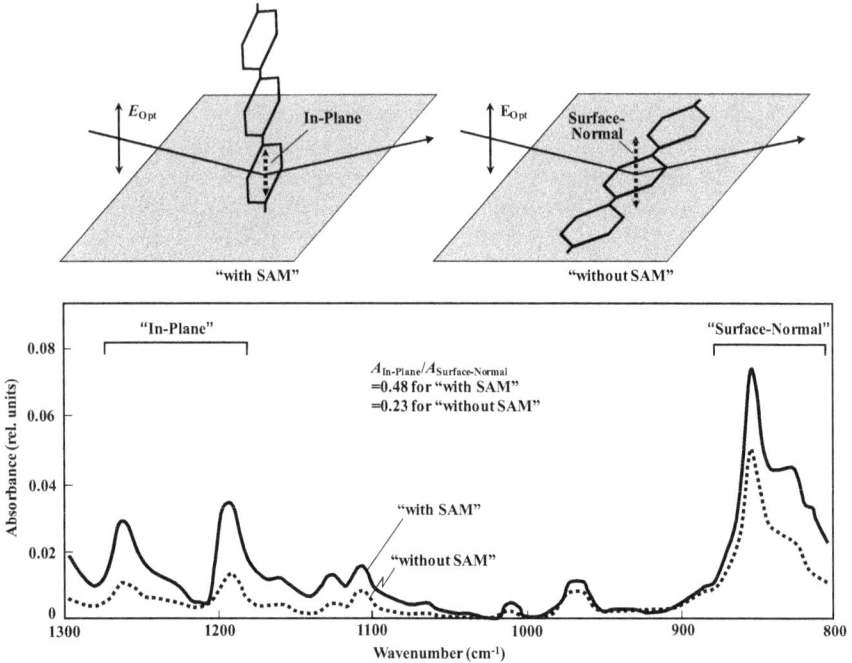

FIGURE 4.22 FTIR-RAS spectra of poly-AM thin films grown by carrier-gas-type MLD for Step 20. (Reprinted with permission from Yoshimura et al. [9].)

than in "without SAM." This suggests that polymer wires tend to grow along a vertical direction in "with SAM" as the model drawn in Figure 4.20, while the wires tend to grow along a horizontal direction in "without SAM."

4.5.1.3   Carrier-Gas-Type Organic CVD of Poly-AM from SAMs

Nakayama et al. grew poly-AM thin films by carrier-gas-type organic CVD on a surface with SAM and a surface without SAM. The molecular cell temperature and the carrier gas flow rate were 50°C and 4 NL/min, respectively. As shown in Figure 4.23, for "without SAM," film thickness increases at the initial stage, followed by saturation. For "with SAM," on the other hand, film thickness increases linearly with growth time [9].

These results can be explained in terms of the reactive site density on the surface. Reactive groups of –CHO and –NH_2 correspond to the reactive sites. Since these groups are hydrophilic, TPA and PPDA molecules are preferably adsorbed to –CHO and –NH_2 groups at the edges of poly-AM wires while they are not adsorbed to the body parts of poly-AM wires that are hydrophobic. As illustrated in Figure 4.24, in the case of "without SAM," TPA and PPDA molecules are adsorbed horizontally on an Au-coated substrate, resulting in horizontal poly-AM wire growth. The density of reactive sites existing at the poly-AM wire edges decreases as poly-AM wires grow, and consequently, the growth rate of the poly-AM film decreases, resulting in the film thickness saturation. Here, body parts of the wires contain CH=N double bonds,

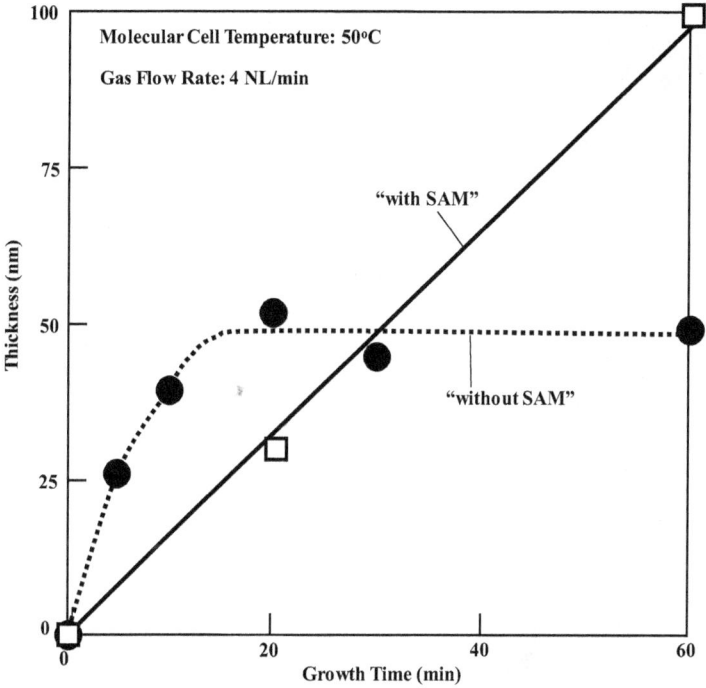

FIGURE 4.23 Dependence of poly-AM film thickness on growth time in carrier-gas-type organic CVD for "with SAM" and "without SAM." (Reprinted with permission from Yoshimura et al. [9].)

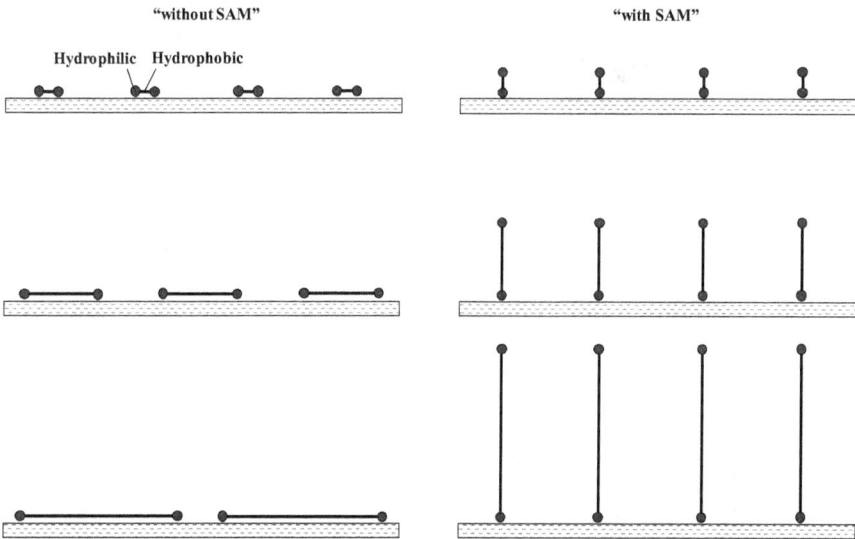

FIGURE 4.24 Models for poly-AM growth by the carrier-gas-type organic CVD for "with SAM" and "without SAM." (Reprinted with permission from Yoshimura et al. [9].)

which exhibit H-bond acceptor characteristics. However, their polarity is smaller than that of –CHO and –NH$_2$ groups. So, the model shown in Figure 4.24 is still viable as the 0th approximation.

As Figure 4.25 shows, the contact angle of a water droplet is 44.6° on SiO$_2$ surface. The angle increases after poly-AM is grown on the surface. The contact angle increases with growth time followed by saturation, indicating surface change from hydrophilic states to hydrophobic states. This corresponds to the reduction of the hydrophilic reactive site density at the surface.

In the case of "with SAM," on the other hand, because TPA and PPDA molecules are connected to the substrate vertically as depicted in Figure 4.24, poly-AM wires grow vertically, keeping the reactive site density constant. Therefore, poly-AM film thickness is not saturated.

These results reveal that nearly vertical growth of poly-AM wires is achieved on substrates with SAM by carrier-gas-type MLD. This might enable three-dimensional self-organized wire growth by using seed cores, which are ultra-small, e.g., nm-scale Au blocks with SAM on the top, and/or sidewall surfaces, as described in Section 5.3.

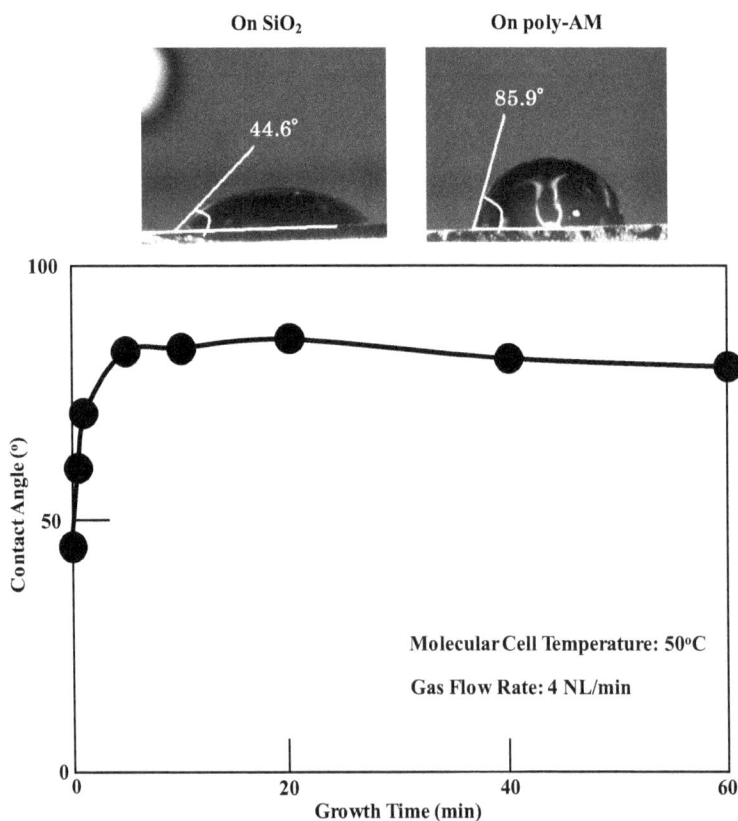

FIGURE 4.25   Growth time dependence of the contact angle of water droplets. (Reprinted with permission from Yoshimura et al. [9].)

By distributing them with designed configurations on a substrate, polymer wires are expected to grow vertically from the top surfaces and horizontally from the sidewall surfaces to form polymer wire networks by providing molecular gases sequentially to the substrate.

4.5.2 MONOMOLECULAR-STEP POLYMER WIRE GROWTH FROM SAMs

Kudo et al. investigated precise processes of monomolecular-step growth of poly-AM wires from seed core molecules of amino-alkanethiol in SAMs on Au films using carrier-gas-type MLD [10].

The MLD process is the same as that drawn in Figure 4.20. A simulation based on the molecular orbital method (Fujitsu WinMOPACK developed by Azuma Matsuura and Tomoaki Hayano) reveals that a poly-AM wire grows from the seed core molecule with slight helix waving as Figure 4.26 shows.

Carrier-gas-type MLD was performed in the following conditions: substrate temperature of 25°C; molecular cell temperatures of 25°C and 50°C for TPA and PPDA, respectively; blowing duration of molecular gases onto the surface of 1 or 5 min; pumping duration for removing residual molecular gases of 5 min; and a carrier gas flow rate of 4 NL/min.

In Figure 4.27a, FTIR-RAS spectra of poly-AM thin films grown by carrier-gas-type MLD with a molecular gas blowing duration of 5 min are shown for Step 6. The absorbance ratio of $A_{\text{In-Plane}}/A_{\text{Surface-Normal}}$ is larger in "with SAM" than in "without SAM," indicating that poly-AM wires tend to grow in upward directions in "with

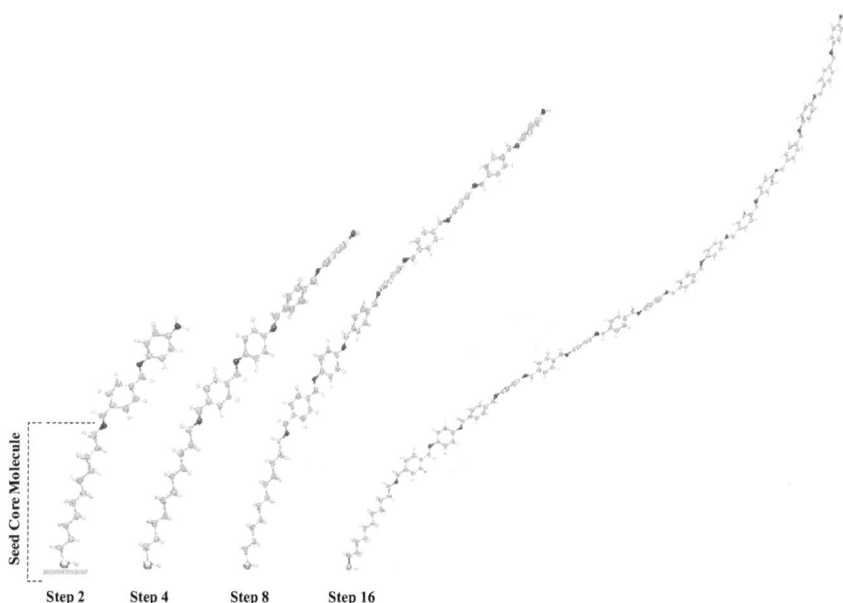

FIGURE 4.26 Poly-AM wire growth from a seed core molecule simulated by the molecular orbital method.

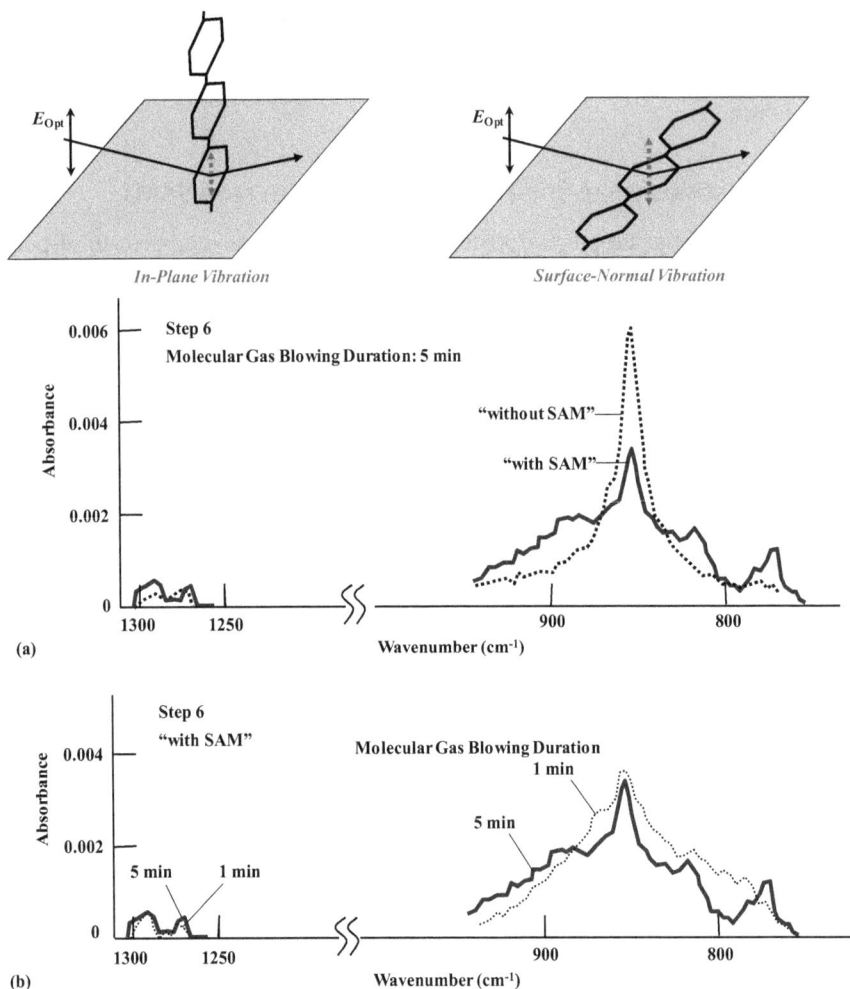

FIGURE 4.27 FTIR-RAS spectra of poly-AM thin films grown by carrier-gas-type MLD (a) with a molecular gas blowing duration of 5 min for "with SAM" and "without SAM" and (b) with a molecular gas blowing duration of 1 and 5 min for "with SAM." (Reprinted with permission from Yoshimura and Kudo [10]. Copyright (2009) The Japan Society of Applied Physics.)

SAM" while the wires tend to grow in horizontal directions in "without SAM." It can be seen from Figure 4.27b that absorption spectra in "with SAM" for a blowing duration of 1 min are almost the same as those for a blowing duration of 5 min, suggesting that 1-min blowing is enough to complete the self-limiting monomolecular-step growth.

Figure 4.28a shows step count dependence of peak absorbance in the region around 850 cm^{-1} for poly-AM wires in "with SAM." Until Step 12, the absorbance linearly increases with step count. This indicates that perfect monomolecular-step growth is achieved in a step count range from 1 to 12, which approximately corresponds to wire lengths of ~0.5 to ~6 nm. In Step 18, however, the absorbance deviates from the linear relationship, implying that the MLD process causes growth imperfection in long wires. One possible reason is the double reaction at the top surface during growth, as suggested by George [13], and the details are described in Section 14.3.1. Another possible reason is "tangling" of poly-AM wires, as indicated by Liskola [14]. The tangling, which is caused by the helix-shaped growth characteristics of poly-AM wires shown in Figure 4.26, might make the wire growth direction disordered.

Figure 4.28b shows step count dependence of absorbance ratio $A_{In\text{-}Plane}/A_{Surface\text{-}Normal}$ for poly-AM wires in "with SAM" and "without SAM." Until Step 12, the absorbance ratio is larger in "with SAM" than in "without SAM" by a factor of 2~3. The result indicates that, in a step count range from 1 to 12, poly-AM wires tend to grow in upward directions in "with SAM" while in "without SAM" the wires tend to grow in horizontal directions. In Step 18, the absorbance ratio in "with SAM" decreases, getting close to the absorbance ratio in "without SAM." This result suggests that the wires in "with SAM" are tilted. Possible reasons for the tilt might be abovedescribed double reaction and the tangling.

In order to achieve perfect monomolecular-step upward growth from seed core molecules, precise optimization of the substrate temperature might be necessary. Iijima and Takahashi reported that DDE molecules are adsorbed on a surface vertically at a substrate temperature of 40°C, while they are adsorbed horizontally at 10°C [15]. If TPA and PPDA molecules have the same tendency, namely, they tend to be adsorbed on the surface vertically at high substrate temperature, it is expected that the double reaction can be suppressed by raising the substrate temperature because the probability that the source molecules react with two reactive sites at the surface simultaneously is reduced when the source molecules stand vertically. The gas flow rate, which is one of the important parameters in MLD, should also be optimized.

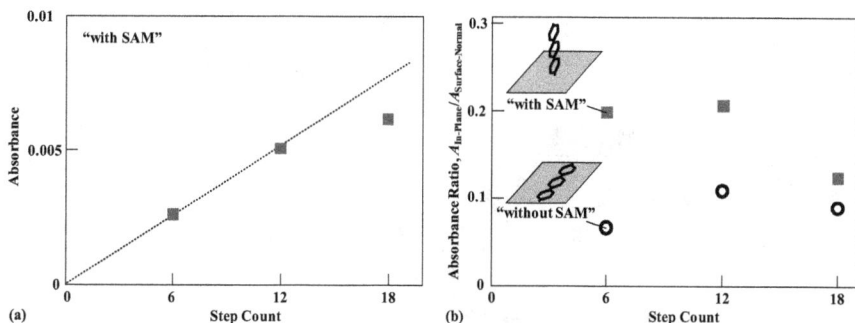

FIGURE 4.28 (a) Step count dependence of peak absorbance in the region around 850 cm^{-1} for poly-AM wires in "with SAM." (b) Step count dependence of absorbance ratio $A_{In\text{-}Plane}/A_{Surface\text{-}Normal}$ for poly-AM wires in "with SAM" and "without SAM." (Reprinted with permission from Yoshimura and Kudo [10]. Copyright (2009) The Japan Society of Applied Physics.)

4.6 IMPROVED AREA-SELECTIVE RESOLUTION

Schematical drawing for the selective growth by MLD is depicted in Figure 4.29. The surface treatment film itself is hydrophilic/hydrophobic, or its surface is covered with the surface-active materials. Polymer wires grow vertically from the top surface of the surface treatment film. At the same time, because MLD enables conformal film growth, polymer wires grow along in-plane directions from the sidewall surface of the film. This might degrade the selective resolution.

To overcome the problem, there are two ways presented in Figure 4.30a and b. The first approach uses blocking cover layers. By embedding the surface treatment film in blocking cover layers, source molecules for MLD are prevented from coming onto the sidewall surfaces to suppress the in-plane polymer wire growth. The second

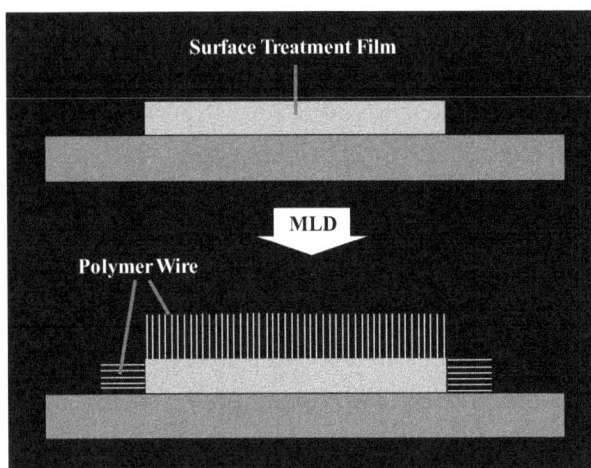

FIGURE 4.29   Schematical drawing for the selective growth by MLD.

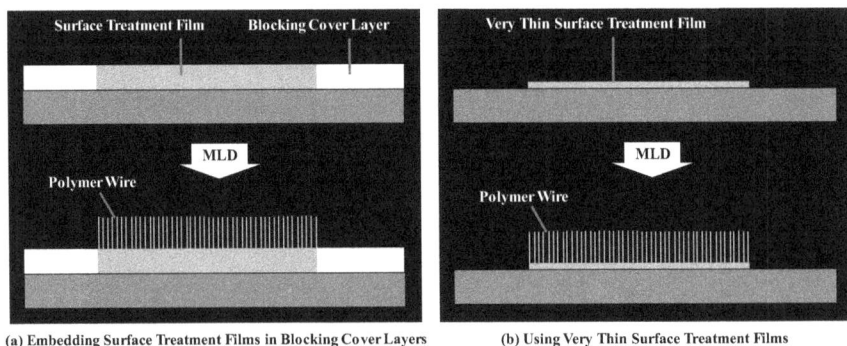

(a) Embedding Surface Treatment Films in Blocking Cover Layers   (b) Using Very Thin Surface Treatment Films

FIGURE 4.30   Two approaches to suppress the resolution degradation in the selective MLD. One is (a) embedding surface treatment films in blocking cover layers, and the other is (b) using very thin surface treatment films.

approach is to decrease the thickness of the surface treatment film. This might also suppress the in-plane growth.

REFERENCES

1. T. Yoshimura, N. Terasawa, H. Kazama, Y. Naito, Y. Suzuki, and K. Asama, "Selective growth of conjugated polymer thin films by the vapor deposition polymerization," *Thin Solid Films* **497**, 182–184 (2006).
2. K. Matsumoto and T. Yoshimura, "Electro-optic waveguides with conjugated polymer films fabricated by the carrier-gas-type organic CVD for chip-scale optical interconnects," *Proc. SPIE* **6899**, 98990E (2008).
3. T. Yoshimura, "Molecular layer deposition (MLD): Monomolecular-step growth of polymers with designated sequences," *Macromol. Symp.* **361**, 141–148 (2016).
4. T. Kanetake, K. Ishikawa, T. Koda, Y. Tokura and K. Takeda, "Highly oriented polydiacetylene films by vacuum deposition," *Appl. Phys. Lett.* **51**, 1957–1959 (1987).
5. J. S. Patel, S. D. Lee, G. L. Baker, and J. A. Shelburne, "Epitaxial growth of aligned polydiacetylene films on anisotropic orienting polymers," *Appl. Phys. Lett.* **56**, 131–133 (1990).
6. T. Yoshimura, K. Motoyoshi, S. Tatsuura, W. Sotoyama, A. Matsuura, and T. Hayano, "Selectively aligned polymer film growth on obliquely evaporated SiO$_2$ pattern by chemical vapor deposition," *Jpn. J. Appl. Phys.* **31**, L980–L982 (1992).
7. Y. Suzuki, H. Kazama, N. Terasawa, Y. Naito, T. Yoshimura, Y. Arai, and K. Asama, "Selective growth of conjugated polymer thin film with nano scale controlling by chemical vapor depositions (CVD) toward "nanonics"," *Proc. SPIE* **5938**, 59310G (2005).
8. T. Yoshimura, S. Tatsuura, and W. Sotoyama, "Chemical vapor deposition of polyamic acid thin films stacking non-linear optical molecules aligned with a strong electric field," *Thin Solid Films* **207**, 9–11 (1992).
9. T. Yoshimura, S. Ito, T. Nakayama, and K. Matsumoto, "Orientation-controlled molecule-by-molecule polymer wire growth by the carrier-gas-type organic chemical vapor deposition and the molecular layer deposition," *Appl. Phys. Lett.* **91**, 033103 (2007).
10. T. Yoshimura and Y. Kudo, "Monomolecular-step polymer wire growth from seed core molecules by the carrier-gas type molecular layer deposition (MLD)," *Appl. Phys. Express* **2**, 015502 (2009).
11. A. Kubono, N. Okui, K. Tanaka, S. Umemoto, and T. Sakai, "Highly oriented polyamide thin films prepared by vapor deposition polymerization," *Thin Solid Films* **199**, 385 (1991).
12. T. Yoshimura, E. Yano, S. Tatsuura, and W. Sotoyama, "Organic functional optical thin film, fabrication and use thereof," U.S. Patent 5,444,811 (1995).
13. D. Seghete, B. Yoon, A. S. Cavanagh, and S. M. George, "New approaches to molecular layer deposition using ring-opening and heterobifunctional reactants," *AVS, 8th International Conference on Atomic Layer Deposition*, WedM1b-3, Bruges, Belgium (2008).
14. Private comment in AVS, 8th International Conference on Atomic Layer Deposition, Bruges, Belgium (2008).
15. M. Iijima and Y. Takahashi, "Techniques for vapor deposition polymerization -preparation of organic functional thin films," *Oyo Buturi* **66**, 1084–1088 (1997) [in Japanese].

5 Expected Impacts of MLD on Photonics/ Electronics Field

Molecular layer deposition (MLD) has featured capabilities of *tailored, ultrathin/ conformal*, and *selective* organic material growth. In the present chapter, expected impacts given by MLD on the photonics/electronics field are discussed. In electro-optic (EO) devices, by growing aligned conjugated polymer wires between electrodes with artificially-controlled molecular sequences, the nonlinear optical effect will be enhanced. In light-emitting diodes (LEDs), photodiodes (PDs), and photovoltaic devices, by growing conjugated polymer wires, in which p-type and n-type regions are involved, the carrier mobility will be improved. In thin-film transistors (TFTs), by growing conjugated polymer wires between source and drain electrodes, the carrier mobility will increase. In electrochromic devices (ECDs), by arranging p-type, n-type, and electron-blocking molecules in polymer wires, the coloration/ bleaching function will appear by electric-field-induced polarization. Selective MLD utilizing seed cores will realize molecular nanosystems and molecular nanoduplication (MND), which is a mass production process for organic devices. By combining MND with photolithographic packaging with selectively occupied repeated transfer (PL-Pack with SORT), low-cost integration of organic devices will be achieved.

5.1 FEATURED CAPABILITIES OF MLD

MLD exhibits three featured capabilities that are useful in the photonics/electronics field, as summarized in Figure 5.1: (1) *tailored* organic material growth, (2) *ultra-thin/conformal* organic material growth, and (3) *selective* organic material growth. The self-limiting effect in MLD contributes to capabilities (1) to produce organic materials with artificially-controlled molecular sequences, and (2) to form ultrathin/ conformal organic films on three-dimensional surfaces with arbitrary structures like

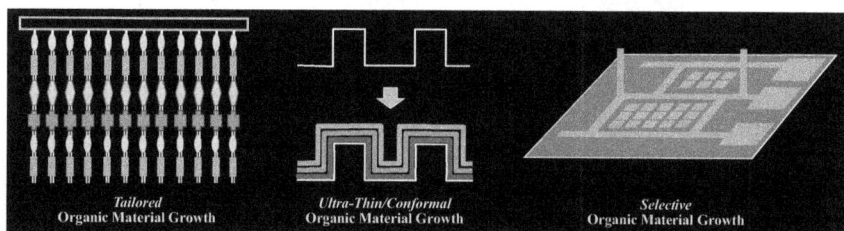

FIGURE 5.1   Three featured capabilities of MLD.

DOI: 10.1201/9781003094012-5

trenches, holes, particles, and rough/porous/tortuous objects as well as flat surfaces. An example of the ultrathin/conformal film deposition is depicted in Figure 3.30 for a layer made from particles. Capability (3) is brought by selective MLD described in Chapter 4.

5.2   PHOTONIC/ELECTRONIC DEVICES REALIZED BY MLD

5.2.1   Advanced Photonic/Electronic Devices

Due to the three featured capabilities described in Section 5.1, MLD is expected to realize advanced photonic/electronic devices [1–3]. Various devices consisting of aligned conjugated polymer wires grown by MLD are schematically shown in Figure 5.2. For comparison, conventional devices consisting of random polymer wires fabricated by spin coating are illustrated on the left.

For nonlinear optical devices like electro-optic (EO) devices utilizing the Pockels effect, in the conventional structure, nonlinear optical molecules cannot be perfectly aligned, reducing the optical nonlinearity. In the advanced structure fabricated by MLD, perfectly aligned conjugated polymer wires, in which molecular sequences are controlled to optimize the wavefunction shapes, are formed between electrodes. This structure will enhance the Pockels effect. For photorefractive devices, in which carrier traps are added in the nonlinear optical materials with photoconductive characteristics, the performance will be enhanced by using aligned conjugated polymer wires grown by MLD similarly to the nonlinear optical devices.

For LEDs, PDs, and photovoltaic devices, in the conventional structure, polymer wires have random structures to reduce the carrier mobility. In the advanced structure, conjugated polymer wires, which involve p-type and n-type regions in the individual wires, are aligned. This will improve the carrier mobility in the devices. For TFTs, the conjugated polymer wires are aligned from the source to drain electrodes in the advanced structure. This configuration will improve the carrier mobility.

For ECDs, in the conventional structure, long-distance transportation of electrons and ions is required for coloration and bleaching. In the advanced structure, p-type and n-type electrochromic (EC) molecules and electron-blocking molecules are arranged in a polymer wire repeatedly. This structure will generate electrons and holes by electric-field-induced polarization to enable coloration without the long electrons/ion transportation.

5.2.2   MLD Processes for Advanced Photonic/Electronic Devices

5.2.2.1   Devices with Conjugated Polymer Wires
          Containing Donors/Acceptors

For most of the advanced organic photonic/electronic devices presented in Section 5.2.1, conjugated polymer backbones are employed. Polyazomethine (poly-AM) is a candidate of the conjugated backbones because MLD of the poly-AM wires has already been demonstrated experimentally as mentioned in Chapters 3, 4, and 6. Polyoxadiazole (poly-OXD) is also expected to be the conjugated backbones [1].

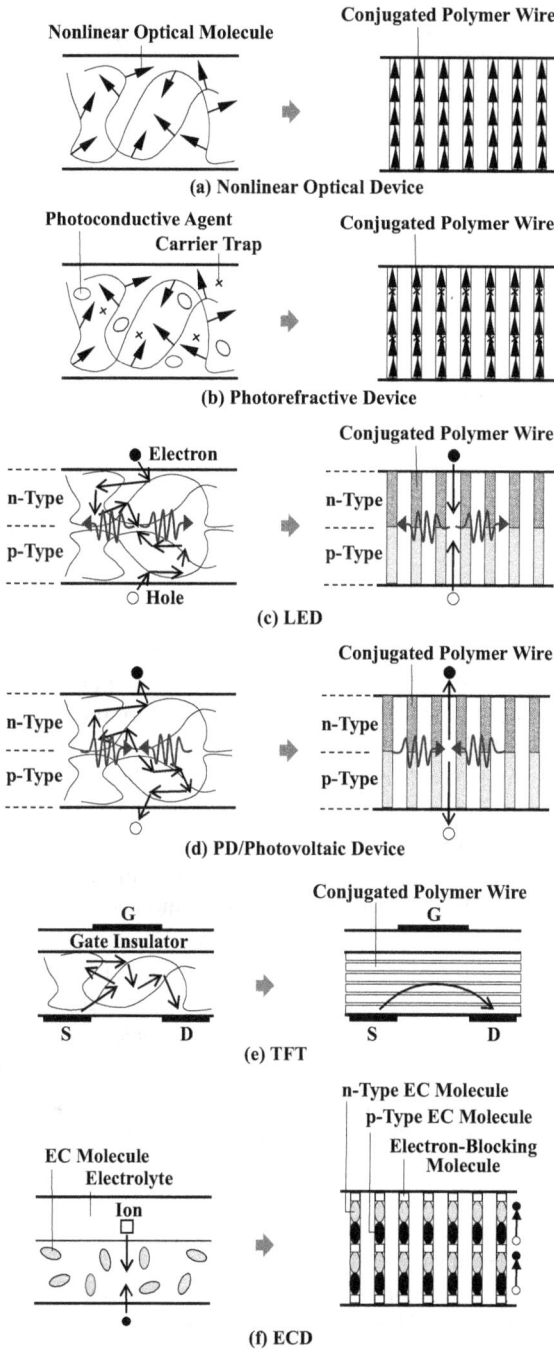

FIGURE 5.2 Advanced photonic/electronic devices consisting of aligned conjugated polymer wires grown by MLD. For comparison, conventional devices consisting of random polymer wires fabricated by spin coating are illustrated.

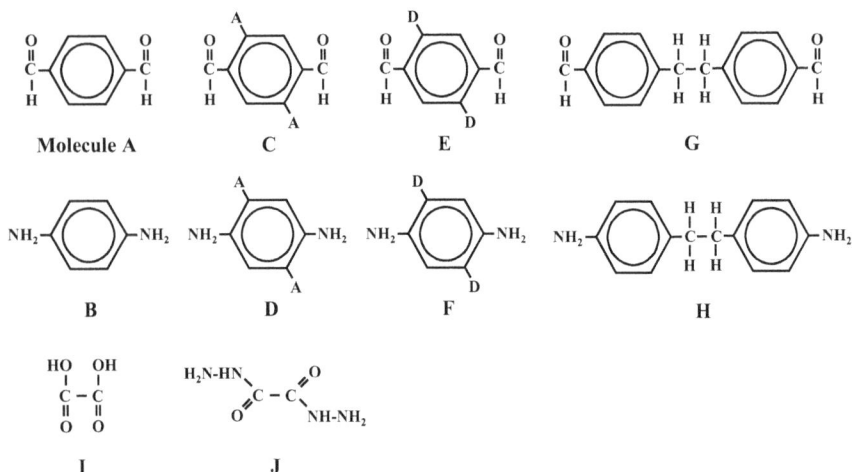

FIGURE 5.3 Examples of source molecules for conjugated polymer wire growth by MLD.

Figure 5.3 depicts examples of source molecules for conjugated polymer wire growth by MLD. A and D in structural formula, respectively, represent acceptor and donor groups. Molecule A is terephthalaldehyde (TPA) and Molecule B is *p*-phenylenediamine (PPDA). Molecules A-F are used for the growth of conjugated backbones of poly-AM. Molecules G and H are for terminating the conjugated systems by three single bonds in the molecule. Molecule J is also used for the conjugated system termination. Molecules I and J grow conjugated backbones of poly-OXD. As shown in Figure 5.4, by connecting molecules using MLD in a sequence of Molecules G, D, A, F, and G, a poly-AM wire containing a quantum dot (QD) with donors and acceptors is grown. The donors and acceptors in the conjugated backbone induce nonlinear optical effects like the Pockels effect.

Figure 5.5 shows a schematic representation of the MLD process illustrated in Figure 5.4 for nonlinear optical material growth. A and D in schematically drawn molecules, respectively, represent acceptor and donor groups, and H represents neutral units, which have neither donor nor acceptor characteristics. Figure 5.6a depicts the schematic representation of an organic nonlinear optical device, in which the conjugated portion is divided into two by a barrier consisting of a couple of single bonds. Each conjugated portion, namely QD, contains donor and acceptor groups to generate the Pockels effect. By inserting carrier-trapping molecules (CTMs) having electron traps (ETs) and hole traps (HTs) in the polymer wire, as depicted in Figure 5.6b, organic photorefractive devices will be available.

Figure 5.6c depicts an example of pin junction construction. The control of n-type and p-type characteristics of the conjugated polymer wires might be performed by donor and acceptor groups. Electrons and holes flow along the conjugated polymer wires. The strength of n-type and p-type characteristics might be adjusted by the amount of the donor and acceptor groups. These devices function as LEDs, PDs, and

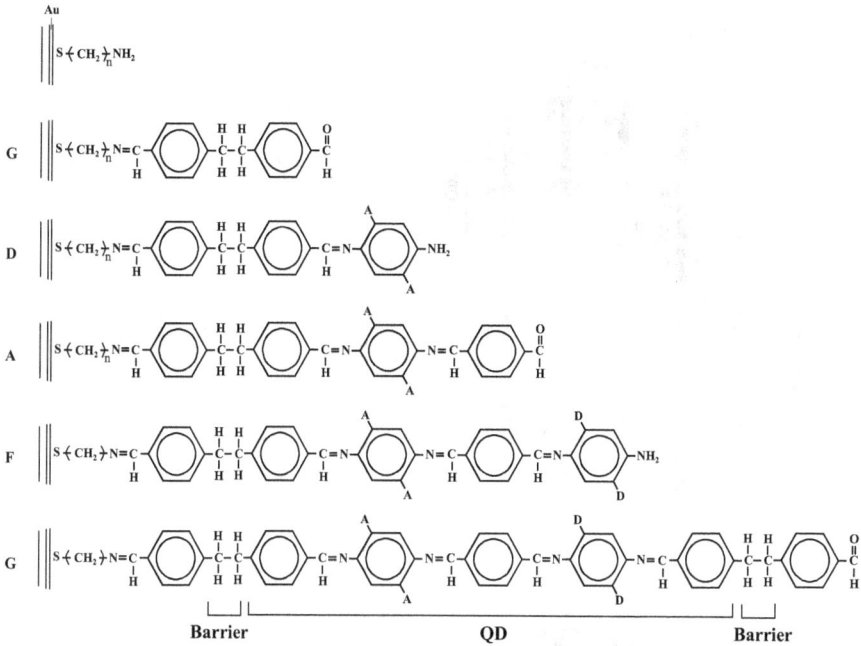

FIGURE 5.4 MLD process for growth of a poly-AM wire containing a QD with donors and acceptors, which induces nonlinear optical effects like the Pockels effect.

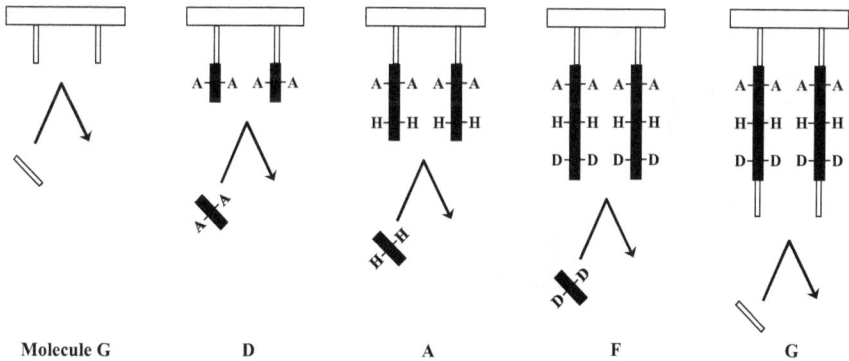

FIGURE 5.5 Schematic representation of the MLD process illustrated in Figure 5.4 for nonlinear optical material growth.

photovoltaic devices. By placing luminescent molecules (LMs) between p-type and n-type portions, as depicted in Figure 5.6d, adjustment of emitted light wavelength might be possible.

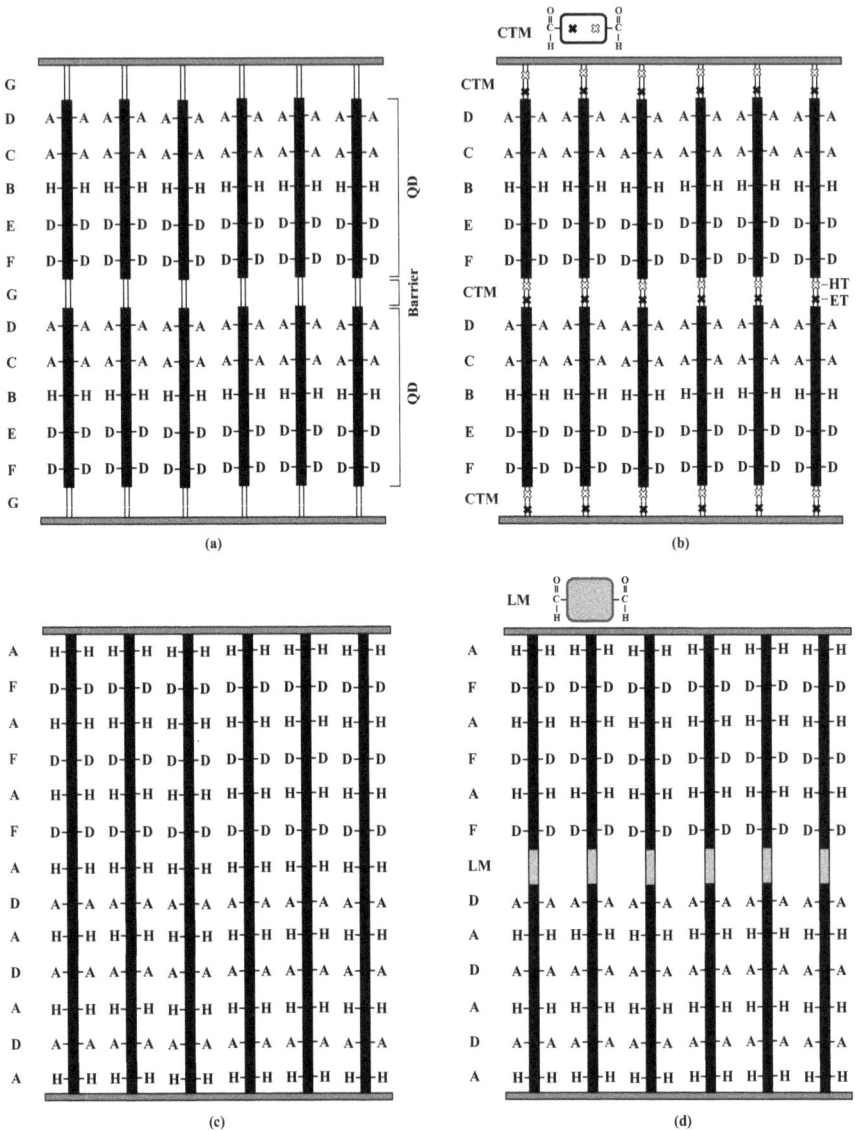

FIGURE 5.6 Schematic representation of (a) an organic nonlinear optical device, in which the conjugated portion is divided into two by a barrier, (b) an organic photorefractive device (c), pin junctions in conjugated polymer wires, and (d) pin junctions in conjugated polymer wires containing luminescent molecules.

5.2.2.2 Area-/Orientation-Controlled Selective MLD for Devices with Conjugated Polymer Wires

To enhance the performance of the photonic/electronic devices, control of polymer wire orientation is essential because the photonic/electronic property parallel to the wire direction greatly differs from that perpendicular to the wire direction.

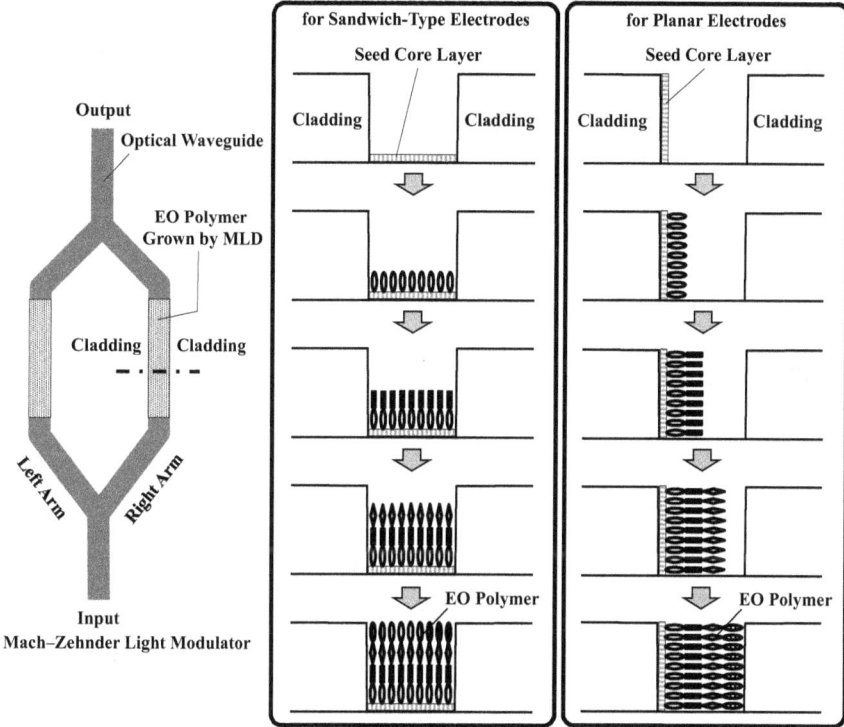

FIGURE 5.7 Mach–Zehnder light modulator, in which aligned EO conjugated polymer wires are grown by MLD from a seed core layer.

Area-/orientation-controlled selective MLD based on the vertically aligned selective growth from seed cores described in Section 4.5 gives the solution [4].

Figure 5.7 shows an example for a Mach–Zehnder light modulator. An input optical waveguide branches into the left arm and the right arm, and these arms are combined to be connected to an output optical waveguide. The arm portions involve optical waveguides made of EO conjugated polymers that exhibit the Pockels effect. By applying electric fields to the arm portions, the refractive index is changed to induce a phase difference between the light wave passing through the left arm and that passing through the right arm, causing light beam intensity modulation. When sandwich-type electrodes are used, the EO conjugated polymers aligned along the vertical direction exhibit a large refractive index change. In this case, the polymer wires are grown by MLD from a seed core layer deposited at the bottom of the waveguide core region. When planar electrodes are used, the EO conjugated polymers aligned along the planar direction exhibit a large refractive index change. In this case, polymer wires are grown from a seed core layer deposited on the side wall of the waveguide core region.

Figure 5.8 shows an example of a ring resonator light modulator, whose detail is described in Section 9.7. EO polymer wires are grown in the vertical direction by MLD from a seed core layer deposited with the ring waveguide shape at the bottom to form an EO polymer ring waveguide. By applying electric fields

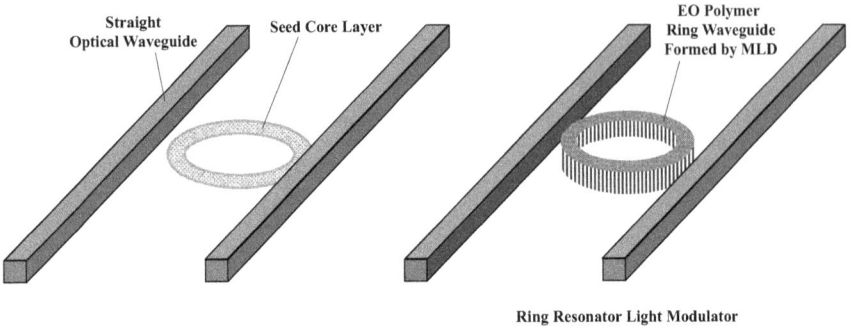

FIGURE 5.8 Ring resonator light modulator, in which aligned EO conjugated polymer wires are grown by MLD from a seed core layer.

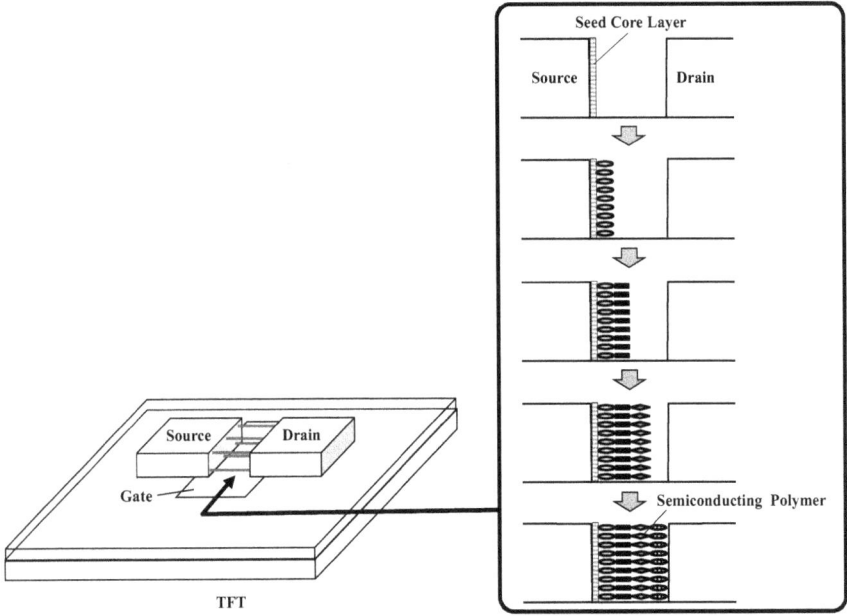

FIGURE 5.9 TFT, in which aligned semiconducting conjugated polymer wires are grown by MLD from a seed core layer.

to the ring portion, the refractive index is changed to adjust the resonant condition, switching light beams back and forth between the two straight optical waveguides.

In the case of a TFT, as depicted in Figure 5.9, it is desirable to grow semiconducting polymer wires between source and drain electrodes from a seed core layer deposited on the side wall of the source/drain electrodes.

5.3 PHOTONIC/ELECTRONIC MOLECULAR NANOSYSTEMS FABRICATED BY MLD

In future, conjugated polymer wires are expected to construct molecular nanosystems like molecular circuits schematically illustrated in Figure 5.10. The molecular circuits consist of molecular lines/electrodes, molecular PDs/photovoltaic devices, and molecular LEDs, for example. They are made of conjugated polymer wires with pin junctions that are sandwiched by a pair of conjugated polymer wires for electrodes [2,3,5].

Figure 5.11 presents tentative examples of conjugated polymer wires with pin junctions. For the donor/acceptor substitution-type pin junction, donors and acceptors are expected to generate n-type and p-type regions. For the backbone-type pin junction, an n-type backbone and a p-type backbone are connected. In the tentative example, poly-AM and poly-OXD are connected to make a junction. Although p-/n-type characterization of poly-AM and poly-OXD are uncertain, if poly-AM and poly-OXD are formed within a single polymer wire following a molecular sequence like -OA-ODH-OA-ODH-TPA-PPDA-TPA- using the reactions shown in Figure 3.22, some kind of pin-like junction of poly-AM/poly-OXD might be obtained. Here, OA and ODH are, respectively, oxalic acid and oxalic dihydrazide. For LED operation, electrons and holes are combined in the intrinsic, i, region under a forward bias condition to emit light. For PD operation,

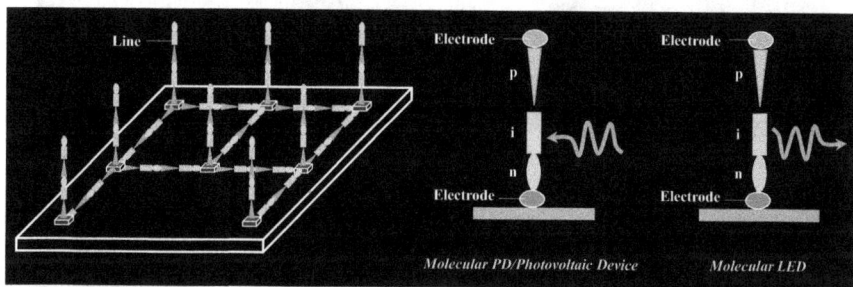

FIGURE 5.10 Concept of molecular circuits consisting of molecular lines/electrodes, PDs/photovoltaic devices, and LEDs.

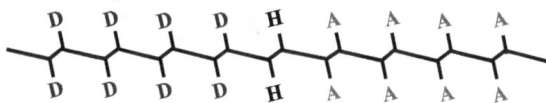

(a) Donor/Acceptor-Substitution Type pin Junction

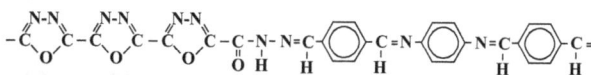

(b) Backbone Type pin Junction

FIGURE 5.11 Tentative examples of conjugated polymer wires with pin junctions.

electrons and holes are generated by absorbing light in the i region under a backward bias condition to induce currents and voltage. Photovoltaic operation is also available in the same structure as that of the PD.

To construct molecular nanosystems, self-organized growth of polymer/molecular wire networks is preferable. It could be performed by using the vertically aligned selective MLD from seed cores described in Section 4.5. The concept is shown in Figure 5.12 for polymer wire networks. On the top surfaces and/or sidewall surfaces of seed cores, surface-active material layers like SAMs are deposited. For vertical

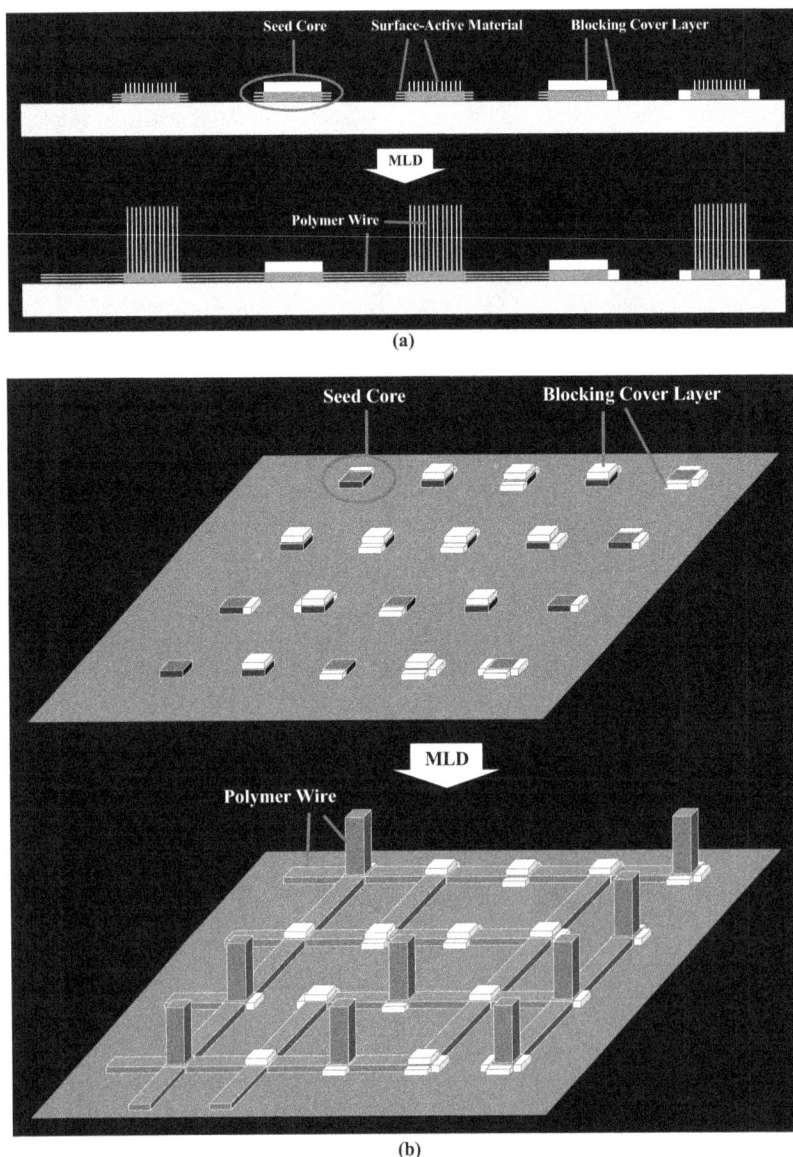

FIGURE 5.12 Concept of self-organized growth of polymer wire networks by the vertically aligned selective MLD. (a) Side view and (b) bird's eye view.

growth, surface-active materials are put on the top of the seed cores while for horizontal growth they are put on the sidewalls. Blocking cover layers prevent the surface-active material layers from being deposited in the regions, where polymer wires should not grow. By distributing the seed cores with designated patterns, polymer wires are expected to grow in designated configurations by MLD, constructing polymer wire networks.

5.4 MOLECULAR NANODUPLICATION (MND)

Molecular MND [2,3,5,6] is a mass production process of organic devices fabricated by MLD. The concept is depicted in Figure 5.13. First, to form seed cores, nano-patterned/micro-patterned layers of seed core molecules such as SAMs are deposited on a growth substrate by lithography. Next, removable molecules, which form disconnectable bonds to the seed core molecules and Molecule A for MLD, are connected to the seed core molecules, and polymer thin films are grown on the removable molecules by MLD. Then, after attaching a pick-up substrate onto the surface, the disconnectable bonds are cut by chemical agents to separate the thin films from the seed core molecule layers. Thus, nano-patterned/micro-patterned polymer thin films are transferred onto the pick-up substrate. Since the growth substrate with patterned layers of seed core molecules is reusable, further lithography processes are not required, enabling the nanoscale/microscale pattern duplication. In Figure 5.14, a conceptual process flow for the repeated MND is summarized.

Figure 5.15 shows an example of MND. On a growth substrate, Au thin films are formed in designed patterns for seed cores by photolithography, electron beam (EB) lithography, or scanning tunneling microscopy (STM). Seed core molecules of amino-alkanethiol are connected to the Au thin films to form seed cores with SAMs. Next, after removable molecules of (COOH)R(COOH) are connected to the seed core molecules, PPDA and TPA are sequentially provided to grow poly-AM thin films by MLD. Then, a pick-up substrate is attached onto the surface and the removable molecules are removed by hydrolysis to transfer the poly-AM thin films to the pick-up substrate.

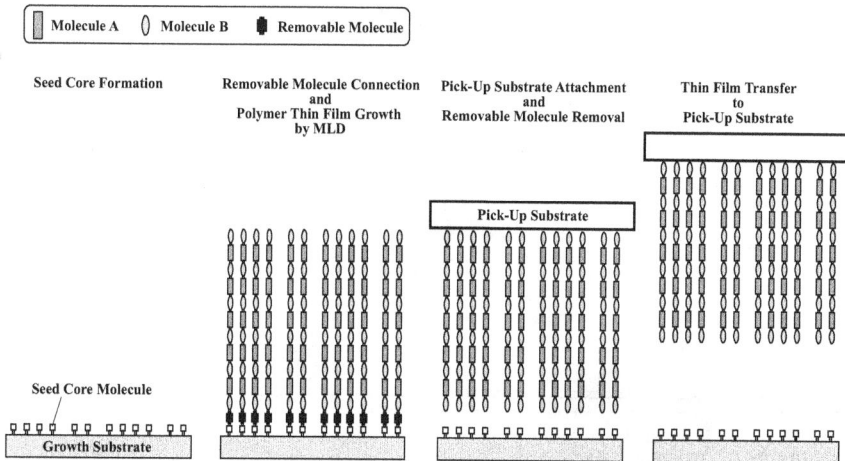

FIGURE 5.13 Concept of molecular nanoduplication (MND) [5].

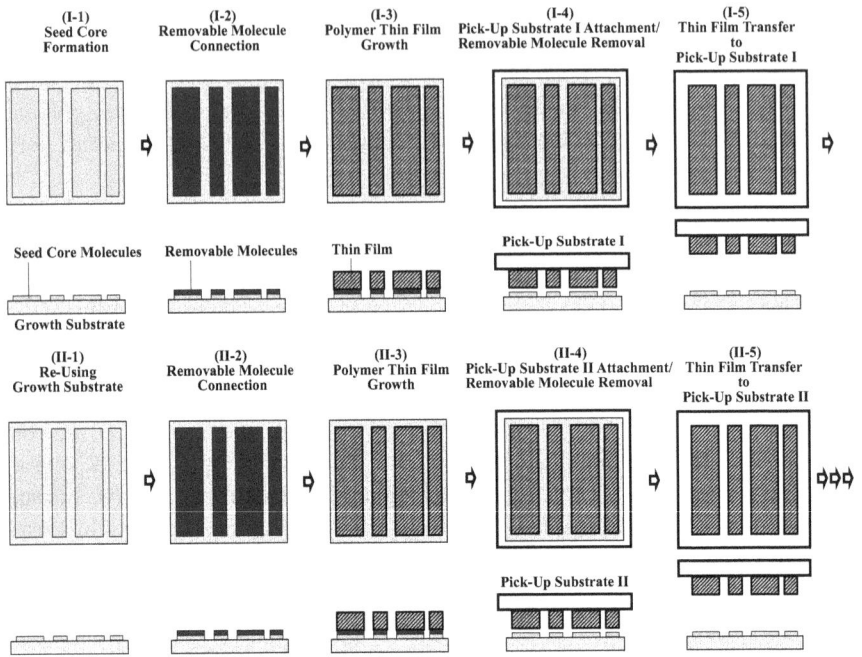

FIGURE 5.14   Conceptual process flow for repeated MND.

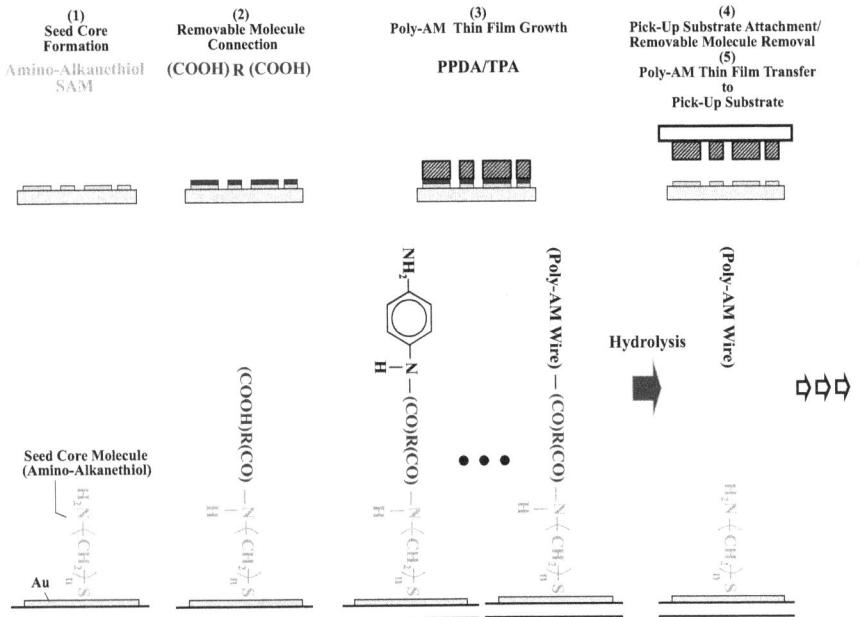

FIGURE 5.15   Example of MND.

The growth substrate with seed cores is reused repeatedly. For the seed core molecule and the removable molecule, other combinations could be employed, for example, $-(CH_2)_n-OH + (COOH)R(COOH) \rightarrow -(CH_2)_n- (COO)R(COOH)$.

By combining MND with photolithographic packaging with selectively occupied repeated transfer (PL-Pack with SORT) [7–11], a low-cost integration process for nanoscale/microscale devices will be realized. PL-Pack with SORT is a resource-saving heterogeneous integration process with multi-step device transfers using the all-photolithographic method, which is superior to the conventional flip-chip assembly packaging. Figures 5.16 and 5.17 show examples for fabrication of nanoscale integrated optical circuits consisting of high-index contrast (HIC) waveguides and photonic crystals. In Figure 5.16, HIC waveguides are formed on a growth substrate with patterned SAMs by MLD. The waveguides are picked up onto a pick-up substrate by MND. The waveguides are transferred to a final substrate by SORT via a supporting substrate. In Figure 5.17, HIC waveguides and photonic crystals are integrated into an optical waveguide film by PL-Pack with SORT.

When the nanoscale optical circuits are used for chip-scale optical interconnects, the circuits are distributed with a pitch of ~100 µm corresponding to the pad pitch of the large-scale integrated circuit (LSI) chips. In this case, since the sizes of the nanoscale optical circuits are in the order of 10 µm, the optical circuits exist with very low density in the interconnect systems. PL-Pack with SORT, in which the nanoscale optical circuits fabricated in dense two-dimensional arrays are distributed

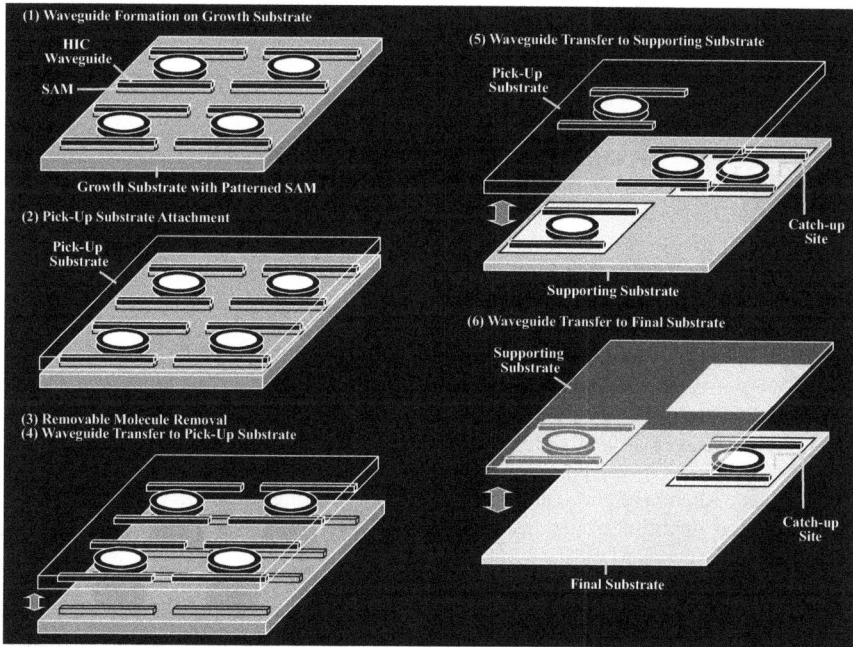

FIGURE 5.16   PL-Pack with SORT combined with MND for fabrication of nanoscale integrated optical circuits [5].

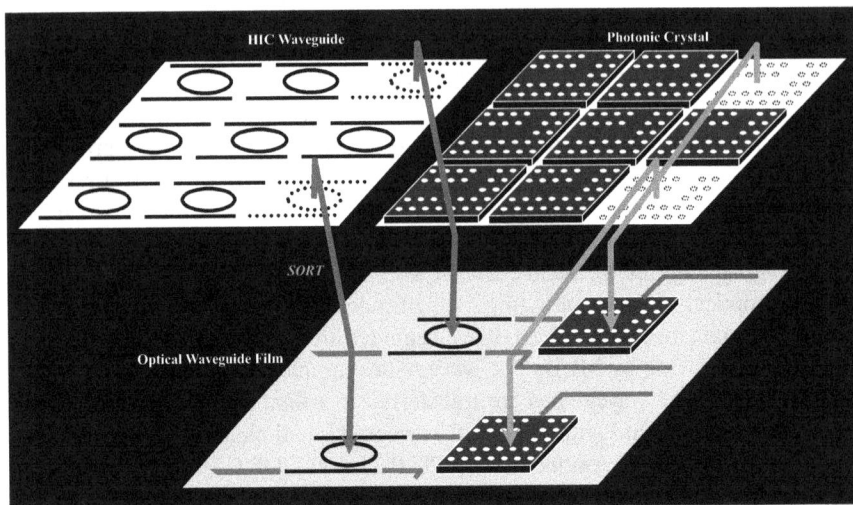

FIGURE 5.17   Heterogeneous integration of nanoscale optical circuits using PL-Pack with SORT combined with MND [5].

to necessary sites, enables us to save materials and process costs for the nanoscale optical circuit integration [11].

REFERENCES

1. T. Yoshimura, E. Yano, S. Tatsuura, and W. Sotoyama, "Organic functional optical thin film, fabrication and use thereof," U.S. Patent 5,444,811 (1995).
2. T. Yoshimura, "Self-organized growth of polymer wire networks with designed molecular sequences for wavefunction-controlled nano systems," in *Trends in Thin Solid Films Research* (Ed. A. R. Jost), Nova Science Publishers, New York (2007).
3. T. Yoshimura, *Molecular Nano Systems: Applications to Optoelectronic Computers and Solar Energy Conversion*, Corona Publishing Co., Ltd., Tokyo (2007) [in Japanese].
4. T. Yoshimura, W. Sotoyama, S. Tatsuura, T. Ishizuka, K. Tsukamoto, and K. Motoyoshi, "Integrated optical circuits, optical circuit waveguide devices," Japanese Patent, Tokukai Hei 8–262246 (1996) [in Japanese].
5. T. Yoshimura, Y. Suzuki, N. Shimoda, T. Kofudo, K. Okada, Y. Arai, and K. Asama, "Three-dimensional chip-scale optical interconnects and switches with self-organized wiring based on device-embedded waveguide films and molecular nanotechnologies," *Proc. SPIE* **6126**, 612609 (2006).
6. T. Yoshimura, H. Tomita, T. Inoguchi, T. Yamamoto, S. Moriya, Y. Suzuki, and K. Asama, "Micro/nano devices/systems, self-organized lightwave networks, and molecular nano duplication method," Japanese Patent, Tokukai 2005–258376 (2005) [in Japanese].
7. T. Yoshimura, J. Roman, W-C. Wang, M. Inao, and M. T. McCormack, "Device transfer method," U.S. Patent 6,669,801 (2003).
8. T. Yoshimura, M. Ojima, Y. Arai, and K. Asama, "Three-dimensional self-organized micro optoelectronic systems for board-level reconfigurable optical interconnects: Performance modeling and simulation," *IEEE J. Select. Top. Quantum Electron.* **9**, 492–511 (2003).

9. T. Yoshimura and Y. Arai, "3D optoelectronic micro system," U.S. Patent 20060261432A1 (2006).

10. T. Yoshimura, K. Kumai, T. Mikawa, and O. Ibaragi, "Photolithographic packaging with selectively occupied repeated transfer (PL-Pack with SORT) for scalable optical link multi-chip-module (S-FOLM)," *IEEE Trans. Electron. Packag. Manufact.* **25**, 19–25 (2002).

11. T. Yoshimura, *Self-Organized 3D Integrated Optical Interconnects: With All-Photolithographic Heterogeneous Integration*, Jenny Stanford Publishing, Singapore (2021).

6 Polymer Multiple Quantum Dots (MQDs) Fabricated by MLD

Organic multiple quantum dots (MQDs), namely, polymer MQDs and molecular MQDs, are typical tailored organic thin-film materials fabricated by molecular layer deposition (MLD). They contribute to the photonics/electronics field and the energy conversion field. As described in Chapter 7, the polymer MQDs are expected to be high-performance electro-optic (EO) materials for light modulators, optical switches, and tunable wavelength filters, which are key devices in optical interconnects and optical switching systems. In the present chapter, after the concept of the organic MQD is explained, details of polymer MQDs with polyazomethine (poly-AM) backbones fabricated by MLD are described. It is confirmed that the degree of electron confinement in the quantum dots (QDs) depends on the QD length, which agrees with results obtained by calculation based on the free-electron model and calculation based on the molecular orbital (MO) method. The quantum confinement effect is discussed using the configurational coordinate diagrams. It is also demonstrated that polymer MQDs containing three kinds of QDs with different lengths in a single wire can be fabricated by MLD with three kinds of source molecules.

6.1 QUANTUM DOTS

Dimensionality control is an important issue to improve optical and electronic properties of materials. Figure 6.1 presents schematic illustrations of three-dimensional (3-D), two-dimensional (2-D), one-dimensional (1-D), and zero-dimensional (0-D) systems, which, respectively, correspond to bulks, quantum wells, quantum wires, and QDs. The dimensionality affects the energy dependence of density of states [1], which gives rise to variation of material properties.

Figure 6.2 shows shell shapes for given energy in 3-D, 2-D, and 1-D systems. In the 3-D system, the shell is a spherical surface with radius of k and thickness of dk. Here, k represents the wavenumber. In the 2-D system, the shell is a circle with radius of k and thickness of dk, and in the 1-D system, the shell is two short lines with length of dk.

From these shell shapes, density of states can be calculated based on the free-electron model, where wavefunctions and energy of an electron are expressed as:

$$\psi(\mathbf{r}) = \left(\frac{1}{L^3}\right)^{1/2} e^{i\mathbf{k}\cdot\mathbf{r}}, \qquad (6.1)$$

DOI: 10.1201/9781003094012-6

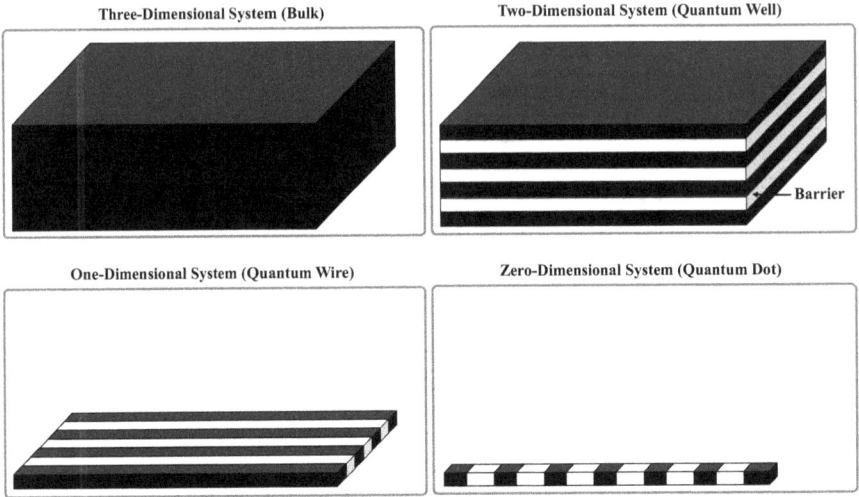

FIGURE 6.1   Schematic illustrations of three-dimensional (3-D), two-dimensional (2-D), one-dimensional (1-D), and zero-dimensional (0-D) systems, which, respectively, correspond to bulks, quantum wells, quantum wires, and quantum dots (QDs).

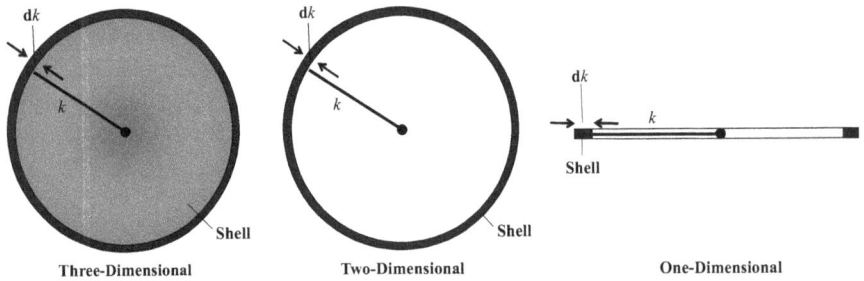

FIGURE 6.2   Shell shapes for given energy in 3-D, 2-D, and 1-D systems.

$$E_{\mathbf{k}} = \frac{\hbar^2}{2m} k^2. \tag{6.2}$$

Here, $\mathbf{r}$ is the position vector of an electron, $\mathbf{k}$ is the wavenumber vector, m is the mass of an electron, L^3 is the volume, where the electron exists, and $\hbar$ is Planck constant divided by 2π. When L is sufficiently large and k is regarded as continuous quantity, Equations (6.3) and (6.4) are derived from Equation (6.2) by replacing $E_{\mathbf{k}}$ with E.

$$k = \left(\frac{2m}{\hbar^2} E \right)^{1/2} \tag{6.3}$$

$$k dk = \frac{m}{\hbar^2} dE \tag{6.4}$$

For the 3-D system, the number of states, N_{3D}, contained in the shell is given by the following expression, considering that up and down spin states exist in a volume of $(2\pi/L)^3$ in the k-space.

$$N_{3D} = \left[2/(2\pi/L)^3\right]\int_{\text{shell}} dk = \left[2/(2\pi/L)^3\right]4\pi k^2 dk \qquad (6.5)$$

By substituting Equations (6.3) and (6.4) into Equation (6.5), Equations (6.6) and (6.7) are obtained.

$$N_{3D} = D(E)dE, \qquad (6.6)$$

$$D(E) = \left(L^3/2\pi^2\right)\left(2m/\hbar^2\right)^{3/2}\sqrt{E} \qquad (6.7)$$

Here, $D(E)$ represents density of states at energy E. For the 2-D system, the number of states, N_{2D}, contained in the shell is given by the following expression, considering that two states exist in an area of $(2\pi/L)^2$.

$$N_{2D} = \left[2/(2\pi/L)^2\right]\int_{\text{shell}} dk = \left[2/(2\pi/L)^2\right]2\pi k dk \qquad (6.8)$$

By substituting Equations (6.3) and (6.4) into Equation (6.8), density of states is obtained as follows:

$$D(E) = \left(L^2/\pi\right)\left(m/\hbar^2\right) \qquad (6.9)$$

For the 1-D system, the number of states, N_{1D}, contained in the shell is given by the following expression, considering that two states exist in a length of $2\pi/L$.

$$N_{1D} = \left[2/(2\pi/L)\right]\int_{\text{shell}} dk = \left[2/(2\pi/L)\right]2dk \qquad (6.10)$$

By substituting Equations (6.3) and (6.4) into Equation (6.10), density of states is obtained as follows:

$$D(E) = (L/\pi)\sqrt{2m/\hbar^2}/\sqrt{E} \qquad (6.11)$$

Figure 6.3 shows schematic diagrams of density of states for the 3-D, 2-D, and 1-D systems. In the 3-D system, density of states monotonically increases with energy. In the 2-D system, density of states does not depend on energy. In the 1-D system, density of states monotonically decreases with energy.

In Figure 6.4, dependence of electron confinement and density-of-state profiles on dimensionality is summarized. In the 3-D system, an electron freely moves in three directions, which causes the density of states to be small near the band edge and to increase with being apart from the band edge. In the 2-D system, translational motion freedom of an electron is reduced from three to two directions, which suppresses the increase in density of states with being apart from the edge. In the 1-D system,

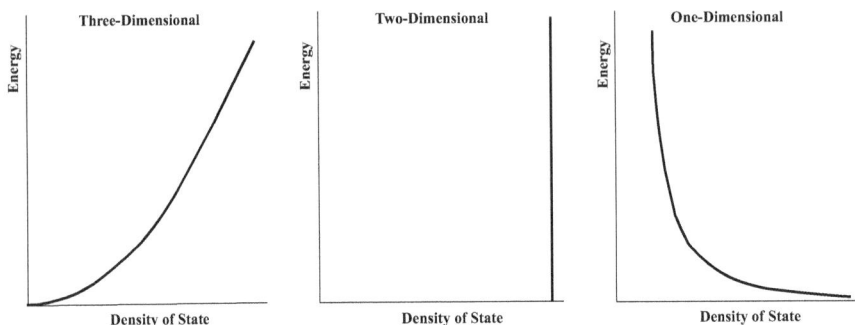

FIGURE 6.3   Schematic diagrams of density of states for 3-D, 2-D, and 1-D systems.

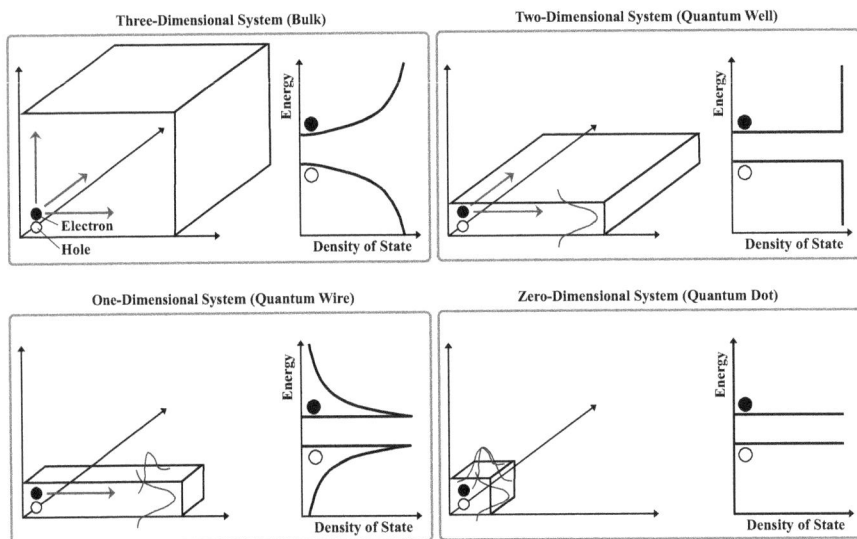

FIGURE 6.4   Dependence of electron confinement and density-of-state profiles on dimensionality.

electron translational motion is allowed only along one direction, which causes the density of states to be the largest at the band edge and to decrease with being apart from the edge. As a result, the absorption band arising from the electron transition becomes sharp in the order of 3-D, 2-D, and 1-D. In the 0-D system, which provides a situation similar to that of atoms or small molecules, electron translational motion is not allowed, which causes a δ-function-like density-of-states profile, resulting in an extremely-sharp absorption band.

As described in Chapter 1, QDs correspond to dimensionality between 0 and 0.5. The 0.5-dimensional (0.5-D) system corresponds to a long QD, and the 0-D system to a short QD. Dimensionality control between 0 and 0.5, that is, control of the QD size, can be achieved by inserting barriers for π-electrons in conjugated wires.

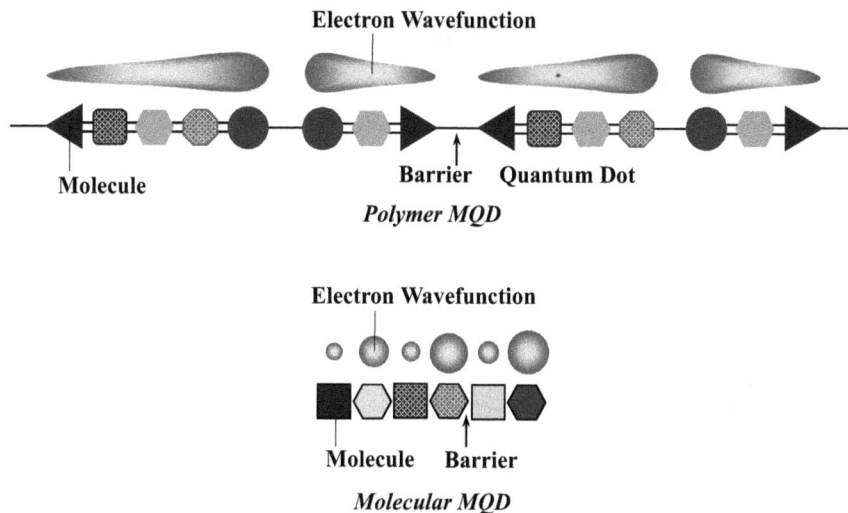

FIGURE 6.5 Control of electron wavefunction shapes of organic MQDs by placing constituent molecules in designated arrangements.

6.2 POLYMER MQDs AND MOLECULAR MQDs

In order to optimize optical and electronic properties, electron wavefunction shapes in organic MQDs should be controlled by placing constituent molecules in designated arrangements, as schematically depicted in Figure 6.5 [2]. For example, noncentrosymmetric wavefunctions can be realized in QDs by placing molecules noncentrosymmetrically. The barriers can be formed by inserting disconnecting molecular units or by modulating energy gaps. A similar wavefunction control might also be achieved in a molecular wire consisting of molecules connected to each other by the electrostatic force. In this case, the individual molecules could be regarded as QDs and the interfaces between the molecules as barriers.

6.3 POLYMER MQDs FABRICATED BY VDP

Prior to polymer MQD fabrication by MLD, polymer MQD fabrication by vacuum deposition polymerization (VDP) was attempted using two kinds of source molecules shown in Figure 6.6 [3]. Molecule A is terephthalaldehyde (TPA) and Molecule B is p-phenylenediamine (PPDA), 4,4'-diaminodiphenyl sulfide (DDS), 4,4'-diaminodiphenyl ether (DDE), or 1,4-diaminodiphenylmethane (DDM). Molecules A and B are connected by double bonds through -CHO and $-NH_2$ groups to form wires of poly-AM.

Source molecules were introduced into a vacuum chamber in the gas phase from K-cells. Molecular gas pressure was $1-8 \times 10^{-1}$ Pa, which was controlled by the K-cell temperature ranging from 120°C to 150°C. The thickness of deposited films was measured, assuming that the film density is 1 g/cm^3, using a quartz oscillator thickness monitor contacting the substrate holder. The substrate temperature T_s was measured using a thermocouple on the holder.

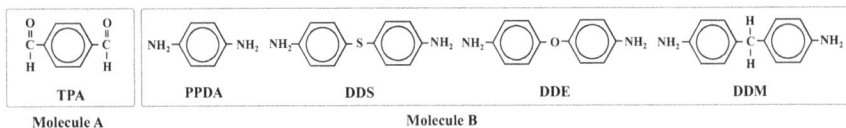

FIGURE 6.6 Source molecules for fabricating polymer MQDs with poly-AM backbones by VDP.

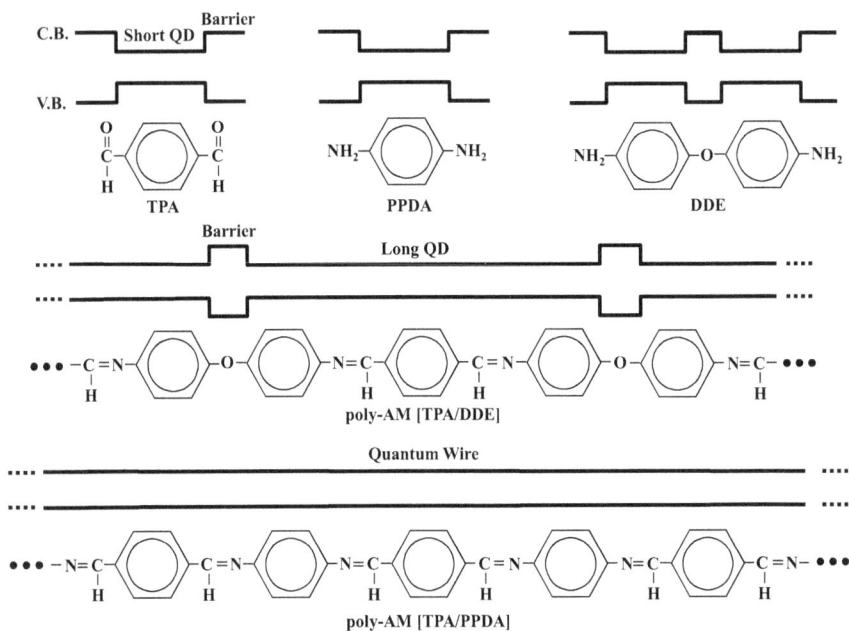

FIGURE 6.7 Structures of molecules and polymer wires, and their corresponding electronic potential energy curves. (Reprinted with permission from Yoshimura et al. [3].)

Figure 6.7 shows structures of molecules and polymer wires with their corresponding electronic potential energy curves. Benzene rings in TPA, PPDA, and DDE are regarded as short QDs of ~0.5-nm long. Poly-AM [TPA/DDE] is a polymer MQD fabricated by combining TPA and DDE. The π-conjugated part is regarded as a long QD with barriers of –O–. The QD contains three benzene rings and the length is ~2 nm. In poly-AM [TPA/PPDA] formed by combining TPA and PPDA, the π-conjugation is spread all along the wire to provide a quantum wire.

Figure 6.8 shows absorption spectra of poly-AM [TPA/PPDA] thin films. The deposition rate was adjusted by molecular gas pressure. The deposition rate increased with increasing the gas pressure. The spectra were measured at room temperature with incident light angles of 0° and 60°. In each spectrum, an absorption band appears in a wavelength region between 400 and 500 nm, while TPA and PPDA exhibit little absorption in the visible region (see Figure 6.9). This indicates that long conjugated systems, that is, quantum wires, are produced in poly-AM [TPA/PPDA]. The sharp

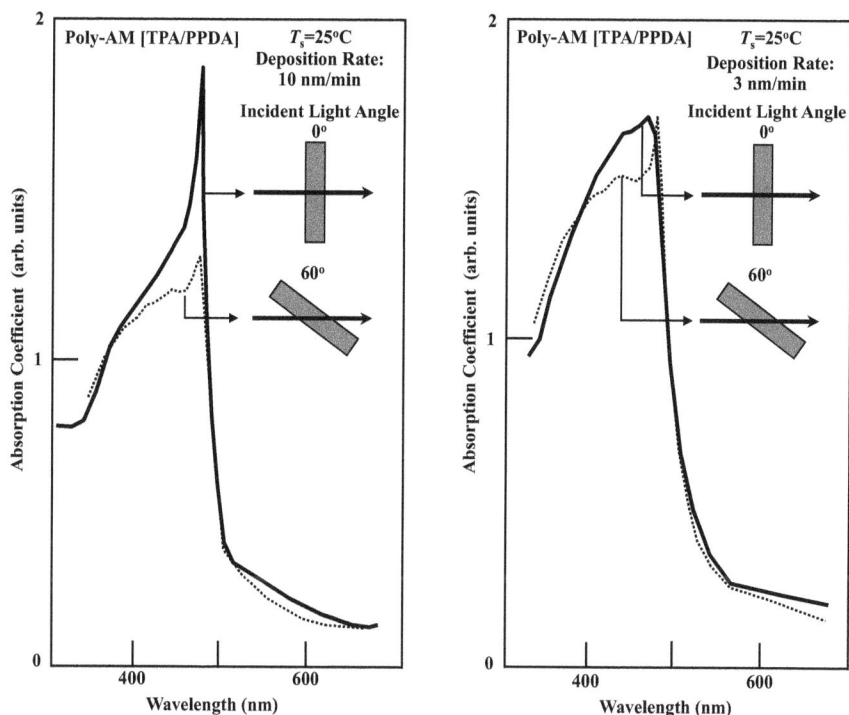

FIGURE 6.8   Absorption spectra of poly-AM [TPA/PPDA] thin films measured with incident light angles of 0° and 60°. (Reprinted with permission from Yoshimura et al. [3].)

peaks observed near 500 nm are attributed to excitons. The half width of the exciton absorption lines for the long-wavelength side is about 50 meV. Assuming that the exciton in poly-AM is similar in size to that in polydiacetylene, the conjugated length is estimated to be more than the exciton length, namely, 5 nm [4].

In the film deposited at 10 nm/min, the exciton peak is stronger for 0° than for 60°. In the film deposited at 3 nm/min, the exciton peak is more dominant for 60° than for 0°. In general, exciton absorption is strong for light polarization in the wire direction. These results suggest that polymer wires stretch to the in-plane direction for 10 nm/min while wires tend to stretch out of the in-plane direction for 3 nm/min. Scanning electron microscope (SEM) observation with magnification of ×25000 revealed that a belt-like texture appears on the surface for 10 nm/min while no such structure appears for 3 nm/min. The difference in the surface morphology might be attributable to the wire orientation.

Note that exciton absorption becomes strong with increasing the deposition rate. No exciton peak was observed for a slow deposition rate of 0.05 nm/min. It was also found that the exciton absorption is enhanced by maintaining the substrate temperature to −10°C during deposition. These results imply that, to obtain long conjugated systems, the deposition condition must be carefully optimized. It is speculated that thermal disturbance would induce disorder in the conjugated wires.

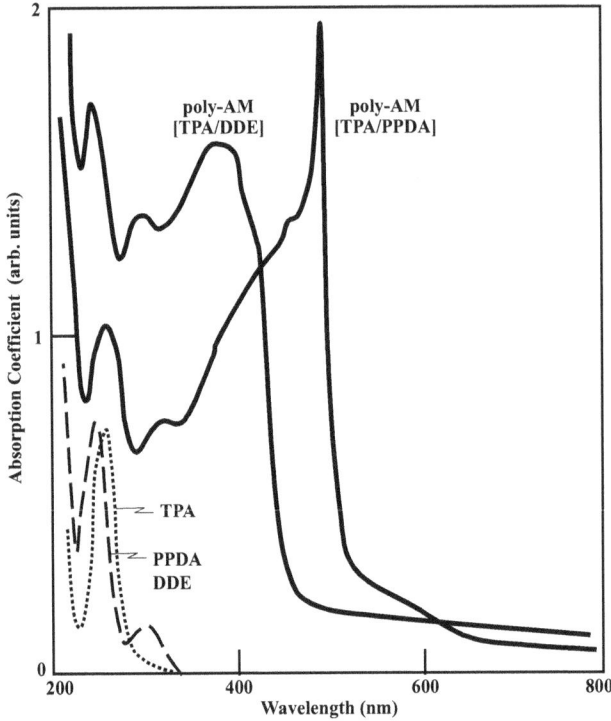

FIGURE 6.9   Absorption spectra of TPA, PPDA, DDE, poly-AM [TPA/DDE], and poly-AM [TPA/PPDA]. (Reprinted with permission from Yoshimura et al. [3].)

Figure 6.9 shows absorption spectra of TPA, PPDA, DDE, poly-AM [TPA/DDE], and poly-AM [TPA/PPDA]. Here, the spectra of TPA, PPDA, and DDE were measured in methanol solutions. In poly-AM [TPA/PPDA], as mentioned above, a sharp peak of excitons is observed near 500 nm in wavelength. In poly-AM [TPA/DDE], an absorption band appears in the shorter-wavelength (higher-energy) side of that in poly-AM [TPA/PPDA] due to electron confinement in QDs of ~2-nm long. In TPA, PPDA, and DDE, the absorption bands exhibit further higher-energy shift, indicating strong electron confinement in short QDs.

By using DDS and DDM as Molecule B, poly-AM [TPA/DDS] and poly-AM [TPA/DDM] were fabricated. In the former, –S– makes a barrier, and in the latter, –CH$_2$– makes a barrier as –O– does in poly-AM [TPA/DDE]. Figure 6.10 shows the influence of barrier bonds on the absorption peak energy. In the order of –S–, –O–, and –CH$_2$–, the absorption peak shifts to the high-energy side, indicating that the barrier height increases in this order to produce strong electron confinement, as schematically depicted in Figure 6.11.

Energy of electrons confined in a QD with infinite potential barrier height is expressed as follows in the free-electron model.

$$E_n = \hbar^2\pi^2 n^2 / \left(2m^*m_0L^2\right). \tag{6.12}$$

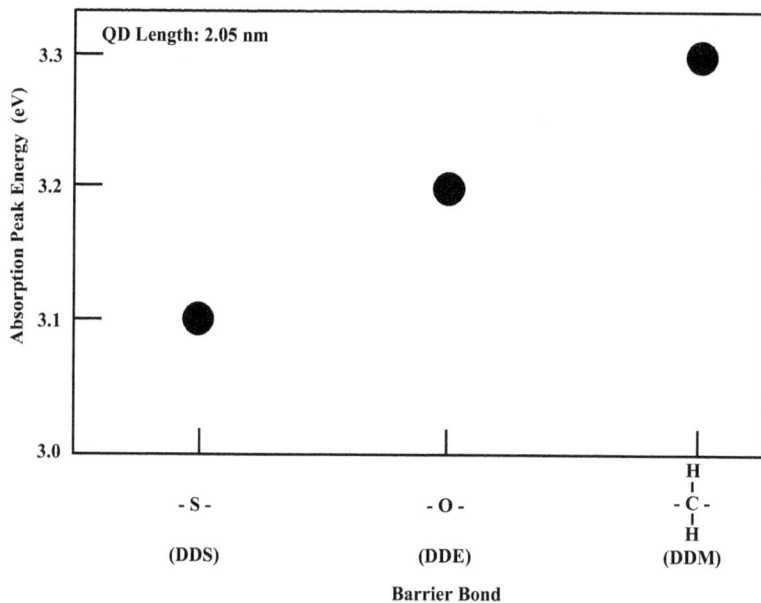

FIGURE 6.10 Influence of barrier bonds on the absorption peak energy. (Reprinted with permission from Yoshimura et al. [3].)

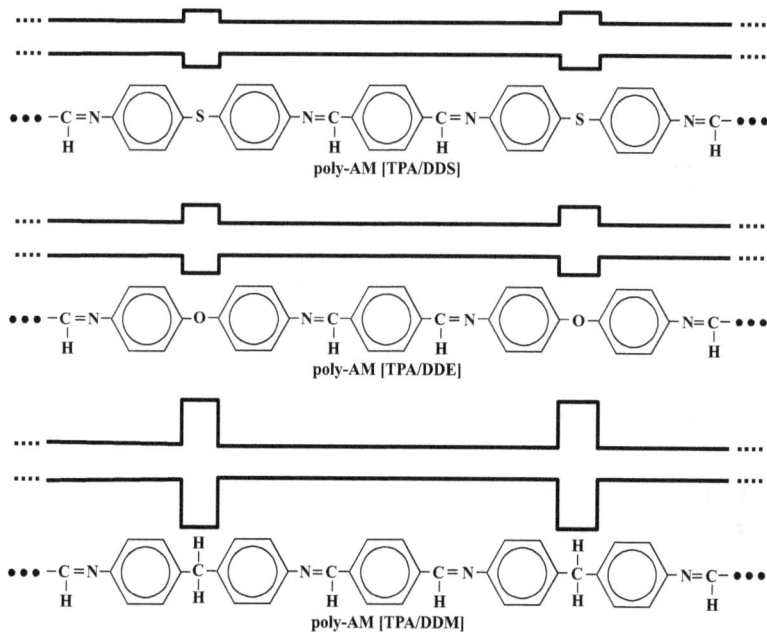

FIGURE 6.11 Structures of polymer wires and their corresponding electronic potential energy curves for barrier bonds of –S–, –O–, and –(CH$_2$)–.

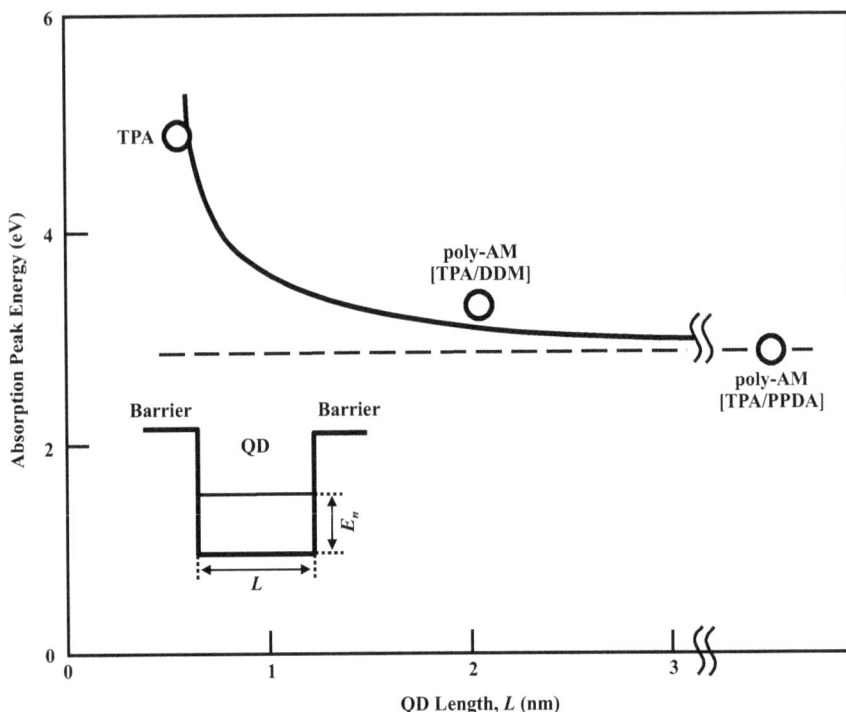

FIGURE 6.12 Dependence of absorption peak energy on QD lengths. (Reprinted with permission from Yoshimura et al. [3].)

Here, $n = 1, 2, \ldots$, m_0 is the electron mass, m^* is the effective mass in units of m_0, and L is the QD length. Assuming that the barrier height is infinite and the effective mass is 1 as the 0th approximation, and the absorption peak energy for $L = \infty$ equals that of poly-AM [TPA/PPDA], the absorption peak energy for the lowest allowed transition is calculated as a function of L as drawn with a solid curve in Figure 6.12. The calculated results and experimental results plotted by circles agree fairly well, confirming the electron confinement in the QDs.

These results indicate the possibility of controlling dimensionality in polymer MQDs to enhance the optical nonlinearity in conjugated polymer wires and to optimize electron confinement effects for photovoltaic devices.

6.4 POLYMER MQDs FABRICATED BY MLD WITH THREE KINDS OF SOURCE MOLECULES

As described in Section 6.3, polymer MQDs with simple periodical structures can be fabricated by VDP. However, to construct functional polymer MQDs, assembling of more than three kinds of source molecules is required. In such cases, MLD is essential. In the present section, polymer MQDs with poly-AM backbones, which are fabricated by MLD utilizing three kinds of source molecules, is described [2,5–8].

6.4.1 STRUCTURES OF POLYMER MQDS

Figure 6.13 shows three source molecules, TPA, PPDA, and oxalic dihydrazide (ODH) for MLD, and reactions between them. TPA and PPDA are connected with a double bond, allowing the wavefunction of π-electrons to be spread over both the molecules. The connection of TPA and ODH gives a series of single bonds, which cut the wavefunction. These bond characteristics enable us to construct QDs with various lengths in polymer wires.

Figure 6.14 shows an example of MLD process for fabricating a polymer MQD. Molecules are connected in a sequence of ---ODH-TPA-PPDA-TPA-ODH---.

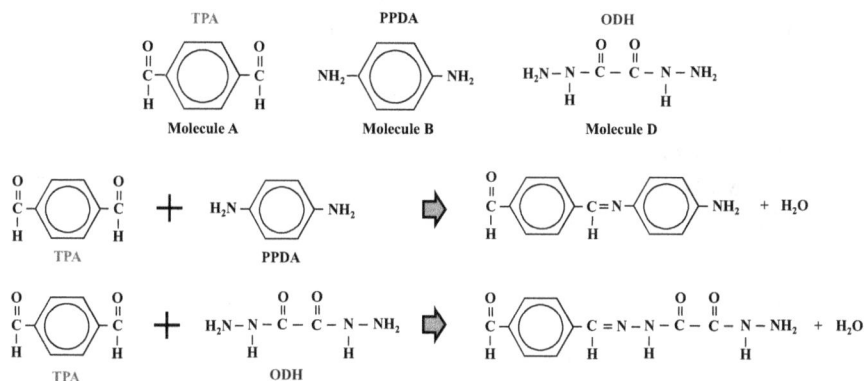

FIGURE 6.13 Three source molecules, TPA, PPDA, and ODH, for MLD, and reactions between them.

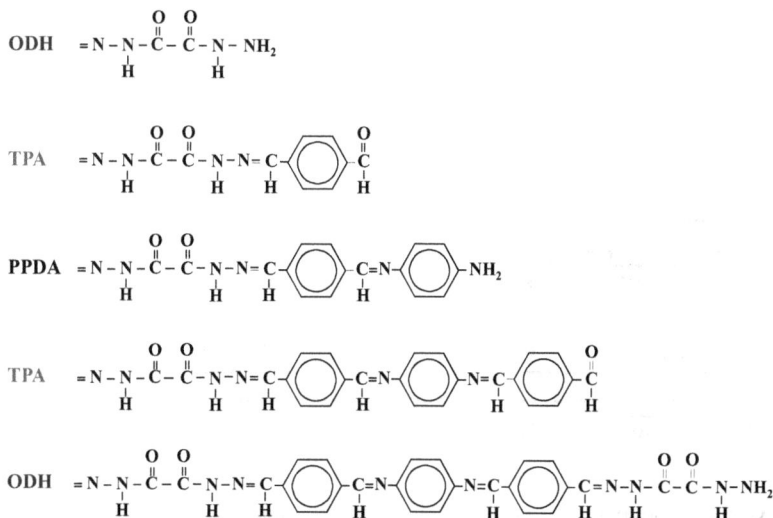

FIGURE 6.14 An example of MLD process for fabricating a polymer MQD. (Reprinted with permission from Yoshimura et al. [7].)

The region between two ODHs is a QD since the π-electron wavefunction is spread through the region.

In Figure 6.15, structures of a poly-AM quantum wire, and polymer MQDs of OTPTPT, OTPT, OT, and 3Q are presented. In the poly-AM quantum wire, TPA and PPDA are alternately connected. Wavefunctions of π-electrons are delocalized along the polymer wire, giving a quantum wire. In OTPTPT, molecules are connected in a sequence of ---ODH-TPA-PPDA-TPA-PPDA-TPA-ODH---. The region between two ODHs is a long QD, in which π-electron wavefunctions are delocalized over five benzene rings. The QD length is estimated from the theoretical structure to be ~3.1 nm. In OTPT, molecules are connected in a sequence of ---ODH-TPA-PPDA-TPA-ODH--- to construct QDs. Each QD contains three benzene rings, and the length is ~2 nm. This structure corresponds to that shown in Figure 6.14. In OT, ODH and TPA are alternately connected to construct short QDs. Each QD consists of a single benzene ring, and the length is ~0.9 nm. 3QD represents a polymer MQD containing three kinds of QDs, namely, OT-like QD [OT], OTPT-like QD [OTPT], and OTPTPT-like QD [OTPTPT], which are inserted in a single polymer wire. 3QD is grown by MLD with a molecular switching sequence of -ODH-TPA-ODH-TPA-PPDA-TPA-ODH-TPA-PPDA-TPA-PPDA-TPA-ODH---, for example.

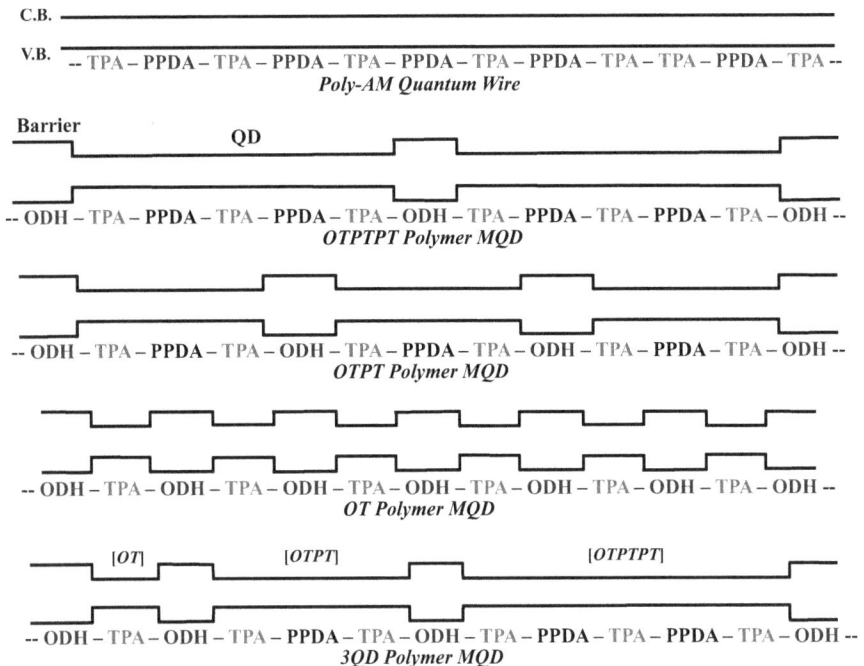

FIGURE 6.15 Structures of poly-AM quantum wire, and polymer MQDs of OTPTPT, OTPT, OT, and 3Q, and their corresponding electronic potential energy curves.

6.4.2 CARRIER-GAS-TYPE MLD FOR POLYMER MQD FABRICATION

A schematic diagram of the carrier-gas-type MLD equipment for polymer MQD fabrication is shown in Figure 6.16 [5–7]. N_2 carrier gas is used to introduce source molecules from temperature-controlled molecular cells into the MLD chamber, giving the source molecular gas flow across the substrate. Source-molecule switching is carried out by valves. The source molecules are combined through chemical reactions on the substrate surface to perform MLD. Excess molecules are removed with the carrier gas using a rotary pump.

Poly-AM quantum wires, OTPTP, OTPT, and OT were fabricated on glass substrates at room temperature. The molecular cell temperature was 50°C for PPDA and ODH, and 25°C for TPA. The substrate surface was exposed to source molecular gas introduced with N_2 carrier gas at a flow rate of 4 NL/min for 5 min at each step. To switch the source molecules, after removing the previous source molecular gas for 5 min, the valve for the next source molecular gas was opened. Switching step count was around 20.

For fabrication of 3QD, the N_2 gas blocking is employed. If contaminants such as remaining source molecules and by-products reach the surface during the source-molecule switching period, they interfere with the normal MLD process. Purges by inert gas like N_2 can protect the surface from the contamination. An example of a process involving the N_2 gas blocking is depicted in Figure 6.17 for a case of MLD with TPA and PPDA. To ensure the N_2 gas blocking effect, the flow time for the source molecular gas and the N_2 purge gas partially overlaps.

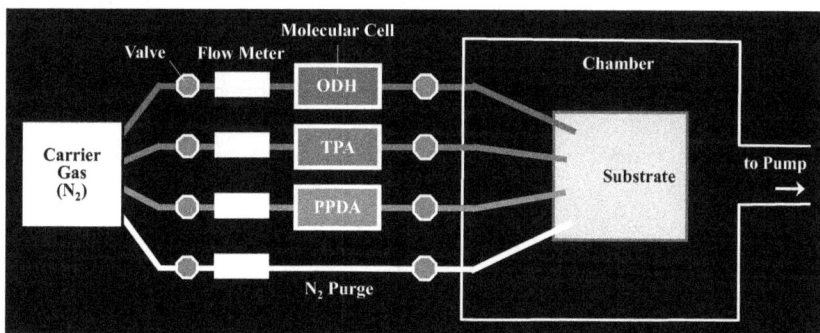

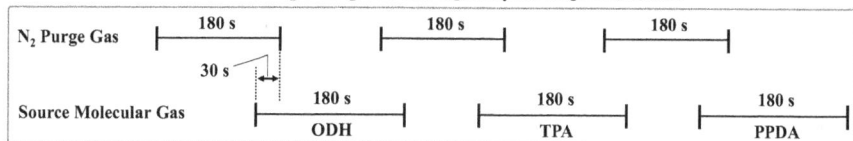

FIGURE 6.16 Carrier-gas-type MLD equipment for polymer MQD fabrication, and an actual gas flow time chart for the 3QD fabrication.

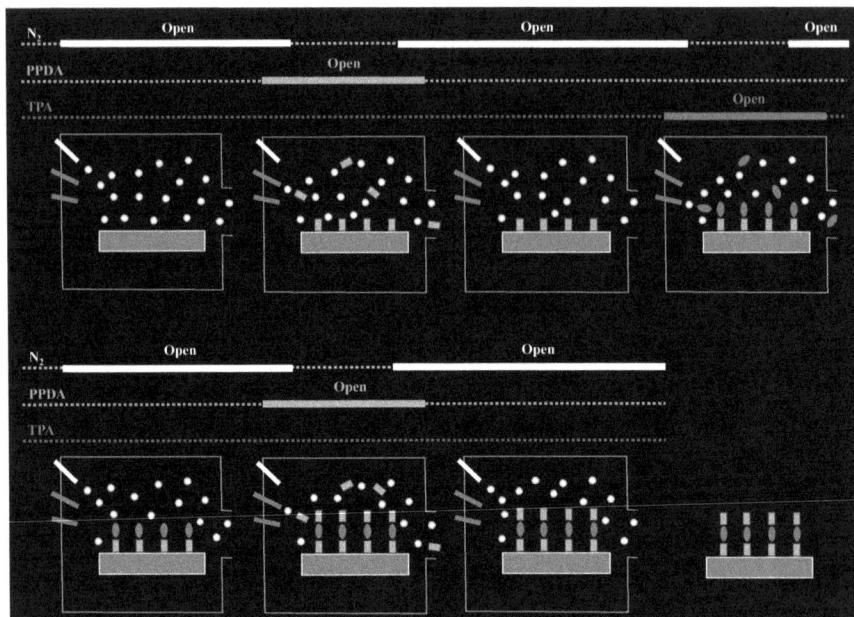

FIGURE 6.17   Example of a process involving the N_2 gas blocking for MLD with TPA and PPDA.

An actual gas flow time chart for the 3QD polymer MQD fabrication is depicted at the bellow part of Figure 6.16. The substrate surface is exposed to source molecular gas for 180 s at each step. For switching source molecules, the valve for the previous source molecular gas is closed, and pumping is performed for 120 s with the N_2 purge, and then, the valve for the next source molecular gas is opened. The overlapping time of the source molecular gas flow and the N_2 purge gas flow is 30 s. The switching step count is 12–24.

6.4.3   QUANTUM CONFINEMENT OF ELECTRONS IN POLYMER MQDs

Ohshima et al. fabricated poly-AM quantum wires, OTPTPT, OTPT, and OT by MLD, and measured their absorption spectra to obtain results shown in Figure 6.18 [5,6]. The absorption peak of poly-AM quantum wire is located around 430 nm. In the order of OTPTPT, OTPT, and OT, namely, with decreasing the QD length, the absorption peak shifts to the short-wavelength side. The absorption peak shift can be attributed to the changing degree of quantum confinement of π-electrons in the QDs.

To confirm the origin of the absorption peak shift, absorption peak energy for poly-AM quantum wire, OTPTPT, OTPT, and OT were calculated by the MO method using WinMOPACK, which is the MO calculation software developed by Azuma Matsuura and Tomoaki Hayano of Fujitsu Laboratories, and is currently involved in SCIGRESS. Molecular structures after the structural optimization are presented in Figure 6.19. Wire-like shapes are obtained for all the structures.

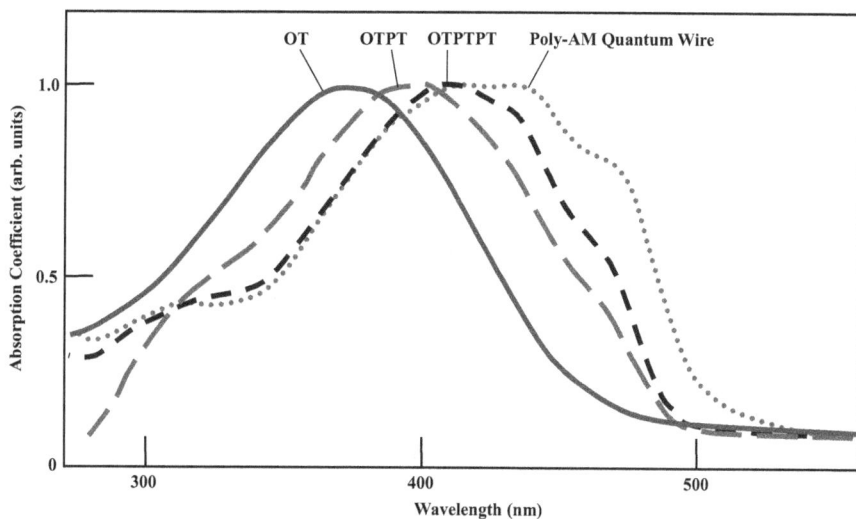

FIGURE 6.18 Absorption spectra of poly-AM quantum wire, OTPTP, OTPT, and OT. (Reprinted with permission from Yoshimura et al. [7].)

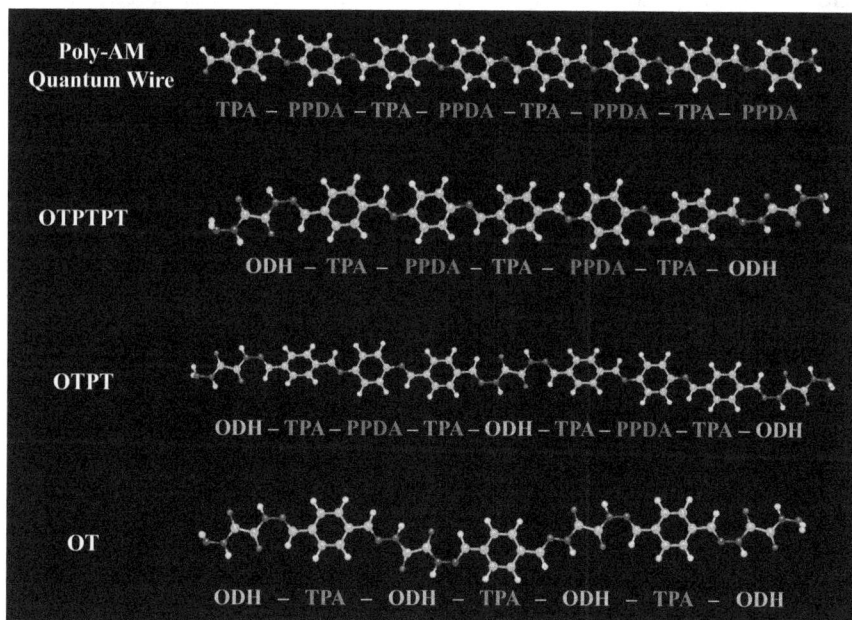

FIGURE 6.19 Molecular structures of poly-AM quantum wire, OTPTPT, OTPT, and OT after the structural optimization performed by the MO method using WinMOPACK.

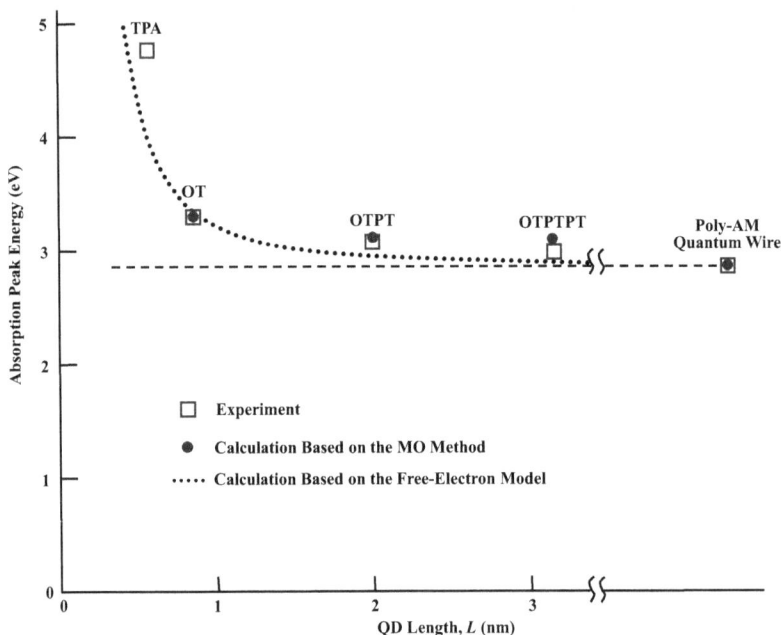

FIGURE 6.20 Dependence of experimentally-measured and calculated absorption peak energy on the QD length. (Reprinted with permission from Yoshimura et al. [7].)

In Figure 6.20, absorption peak energy obtained by the experiment and by the calculation based on the MO method is, respectively, plotted with square dots and circle dots as a function of the QD length. Since there was a 0.3-eV offset between the results of the experiment and the calculation, the calculated results are plotted by taking the experimentally-measured peak energy of the poly-AM quantum wire as the standard. It is found that, with decreasing the QD length, the absorption peak energy shifts toward the high-energy side both in the experiment and in the calculation.

Absorption peak energy derived from the free-electron model using Equation (6.12) is also presented in Figure 6.20 with a dotted line. The calculation was carried out for the lowest allowed transition, assuming that the absorption peak energy for $L = \infty$ is that of the poly-AM quantum wire. Experimental values and calculated values are in good agreement. This indicates that the π-electrons in the QDs can be treated with the free-electron model, and the high-energy shift of the absorption peak arising from a decrease in the QD length is attributed to the quantum confinement of π-electrons in the MQDs.

From the abovementioned argument, it is concluded that the molecular arrangements in polymer wires with poly-AM backbones have successfully been controlled by artificially assembling three kinds of molecules using MLD to fabricate polymer MQDs having QDs of designated lengths.

6.4.4 POLYMER MQDs CONTAINING DIFFERENT-LENGTH QUANTUM DOTS IN A SINGLE WIRE

Ebihara et al. fabricated 3QD containing [OT], [OTPTP], and [OTPTPT] in a single poly-AM wire (see Figure 6.15) by MLD with source molecules of TPA, PPDA, and ODH, and carried out absorption spectrum measurement [7]. The measured spectrum of 3QD is drawn by a solid line in Figure 6.21. A broad absorption band extending from ~480 to ~300 nm is apparent, which is attributed to the superposition of component absorption bands of [OT], [OTPT], and [OTPTPT].

The absorption spectrum profile of 3QD is expected to be reproduced by summing up the spectra of OT, OTPT, and OTPTPT. As can be seen from Figure 6.15, the ratio of the number of QDs contained in a unit polymer wire length is approximately $N_{OT}:N_{OTPT}:N_{OTPTPT} = 6:3:2$, where N_{OT}, N_{OTPT}, and N_{OTPTPT} are the numbers of QDs per unit length in OT, OTPT, and OTPTPT, respectively. In 3QD, the number ratio of [OT], [OTPT], and [OTPTPT] is 1:1:1. Therefore, the absorption spectrum of 3QD can be predicted by adding the spectra of OT, OTPT, and OTPTPT with weightings of 1/6, 1/3, and 1/2, respectively. The spectra with the weightings are drawn by dashed lines for OT, OTPT, and OTPTPT, and the predicted spectrum of 3QD is presented by a dotted line in Figure 6.21. The predicted spectrum is fairly coincident with the measured one. This confirms that the widening of the absorption band in 3QD is due to the superposition of absorption bands of [OT], [OTPT], and [OTPTPT], and that the MLD process with the three kinds of source molecules is successfully achieved.

Using SCIGRESS, the molecular structure of 3QD fabricated on amino-alkanethiol and electron density distribution in the structure were calculated. The results are shown in Figure 6.22a and b, where brighter color indicates higher electron density. The density is high in the QD regions of [OT], [OTPT], and [OTPTPT].

For poly-AM-based polymer MQDs, it is difficult to absorb light of wavelengths longer than ~500 nm because the band gap of poly-AM is located around 500 nm. For some applications of polymer MQDs, polymers with narrower band gaps should instead be used as wire backbones.

Dense polymer wire growth is preferable when the polymer MQD is used for photonic/electronic devices. In such cases, the inter-wire interaction would sometimes interfere with the normal operation of the MQD. In future, the inter-wire control might become important. Loscutoff et al. demonstrated an approach to fix polyurea wires utilizing hydrogen bonds [9]. Zhou et al. fabricated cross-linked polyurea wires [10]. The MLD process involving steps for cross-linking, which is shown in Figure 3.7c, might be another way to control the inter-wire properties.

6.4.5 EXPLANATION OF QUANTUM CONFINEMENT EFFECTS BASED ON CONFIGURATIONAL COORDINATE DIAGRAMS

Ishii et al. measured photoluminescence (PL) spectra of poly-AM quantum wire, OTPTPT, OTPT, and OT grown by MLD using 365-nm excitation light at room temperature [8]. The results are presented in Figure 6.23. Background PL from the glass

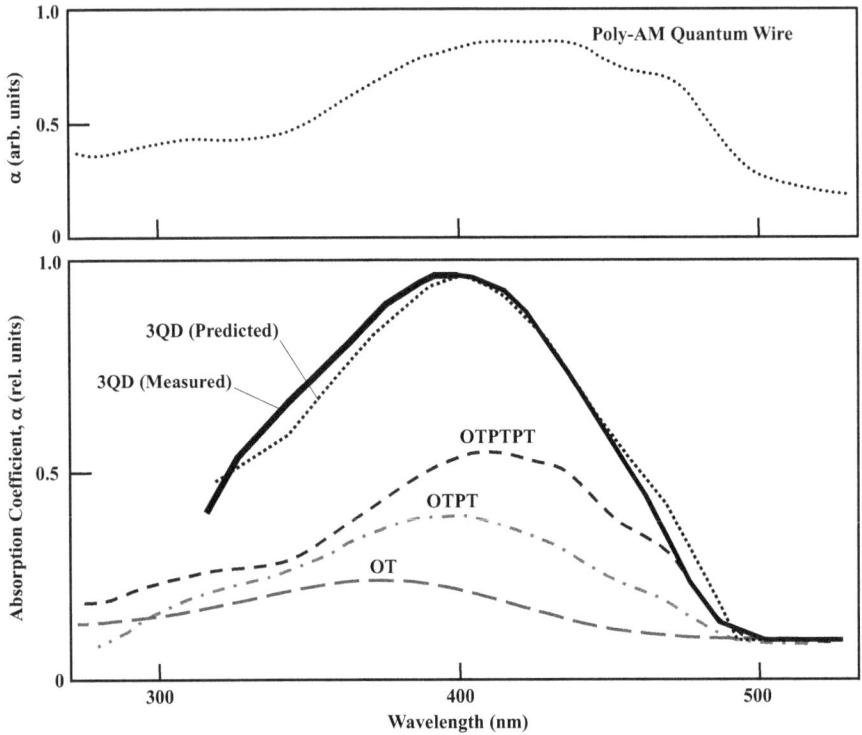

FIGURE 6.21 Measured and predicted absorption spectra of 3QD containing [OT], [OTPTP], and [OTPTPT] in a single poly-AM wire. (Reprinted with permission from Yoshimura et al. [7].)

substrates was subtracted from all the data. The PL peak shifts toward the longer-wavelength side in the order of poly-AM quantum wire, OTPTPT, OTPT, and OT, namely, with decreasing the QD length.

In Figure 6.24, the PL spectra shown in Figure 6.23 are drawn with photon energy for the horizontal axis. PL peak positions for the spectra are indicated by arrows. Since the PL peak position E_{PL} is ambiguous because of broad spectrum width, it is defined as follows:

$$E_{PL} = (E_{90L} + E_{90H})/2, \tag{6.13}$$

where E_{90L} and E_{90H} are, respectively, low-energy-side and high-energy-side photon energy, at which the PL intensity is 90% of the peak intensity. The PL peak of the poly-AM quantum wire is around 2.5 eV, and the PL peak shifts to the lower-energy side in the order of OTPTPT, OTPT, and OT, namely, with decreasing the QD length.

Since there are only two kinds of molecules alternately arranged in the poly-AM quantum wire and OT, they can be grown by either MLD or organic chemical vapor deposition (CVD). In Figure 6.25, PL spectra of the poly-AM quantum wire and OT grown by MLD are compared with those of poly-AM quantum wire and OT grown

(a)

(b)

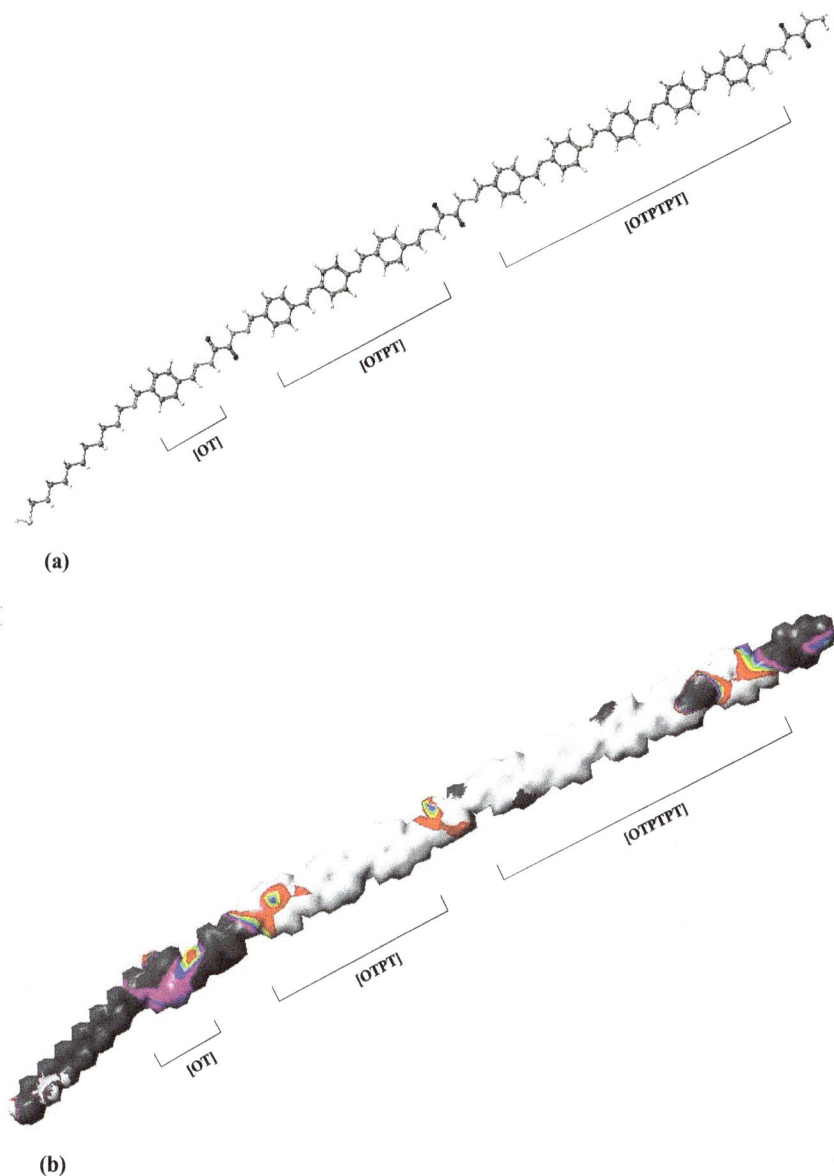

FIGURE 6.22 (a) Molecular structure of 3QD fabricated on amino-alkanethiol and (b) electron density distribution in the structure. (Reprinted with permission from Yoshimura et al. [7].)

by organic CVD. Profiles of the PL spectra for MLD are found to be similar to those for organic CVD.

In Figure 6.26, PL peak positions are plotted as a function of the QD length together with absorption peak positions presented in Figure 6.20. By decreasing the

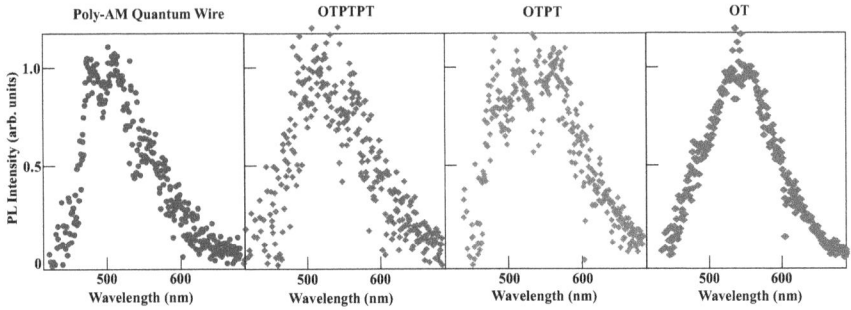

FIGURE 6.23   PL spectra of poly-AM quantum wire, OTPTPT, OTPT, and OT grown by MLD.

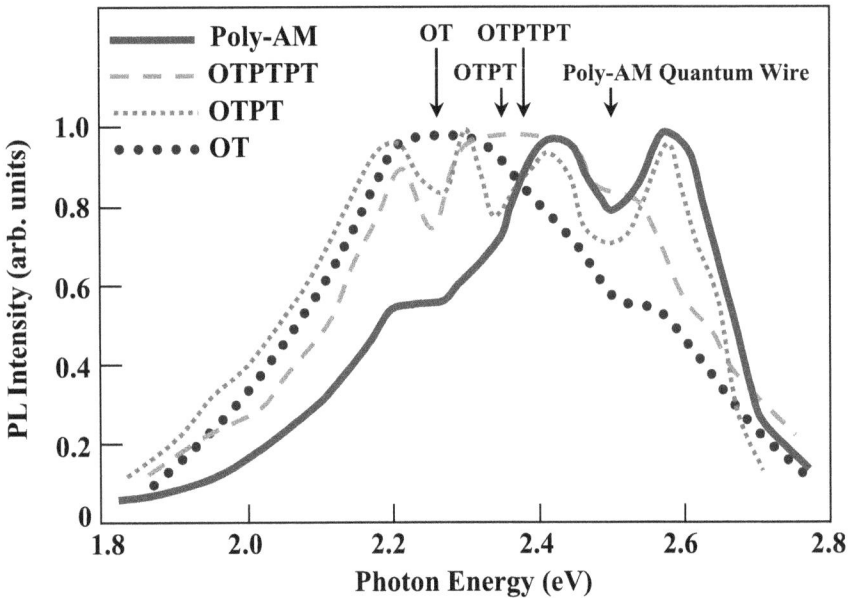

FIGURE 6.24   PL spectra of poly-AM quantum wire, OTPTPT, OTPT, and OT grown by MLD. The horizontal axis is photon energy. (Reprinted with permission from Yoshimura and Ishii [8].)

QD length, the absorption peak exhibits a higher-energy shift due to the quantum confinement of π-electrons, whereas the PL peak exhibits a lower-energy shift due to the Stokes shift.

The low-energy shift of the PL peak can be explained using a configurational coordinate model analogous to the model for the small-polaron absorption in WO_x shown in Figure 2.21. As mentioned in Section 2.5, the small-polaron absorption model leads us to the following conclusion: "as an electron is highly localized, the lattice distortion change accompanied with the electron transition becomes large,

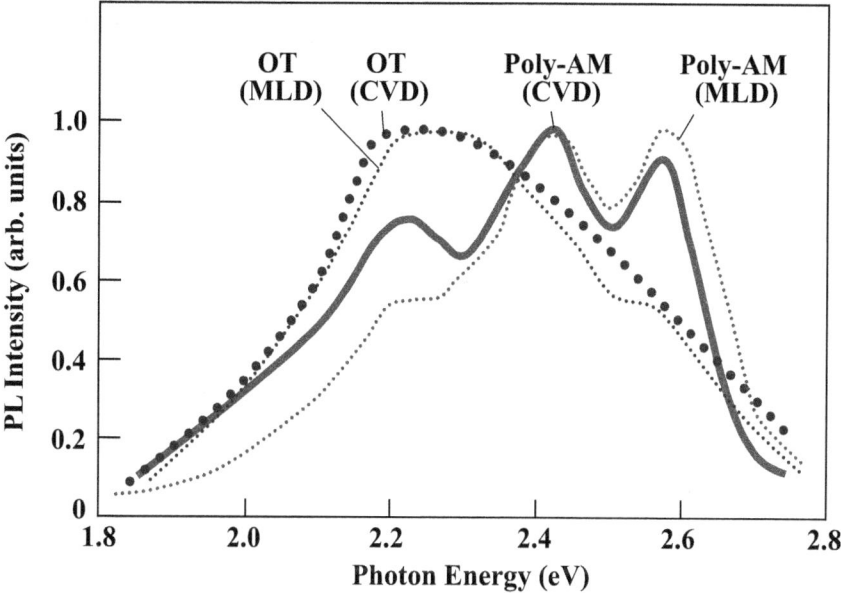

FIGURE 6.25 PL spectra of poly-AM quantum wire and OT grown by MLD, and those of poly-AM quantum wire and OT grown by organic CVD. (Reprinted with permission from Yoshimura and Ishii [8].)

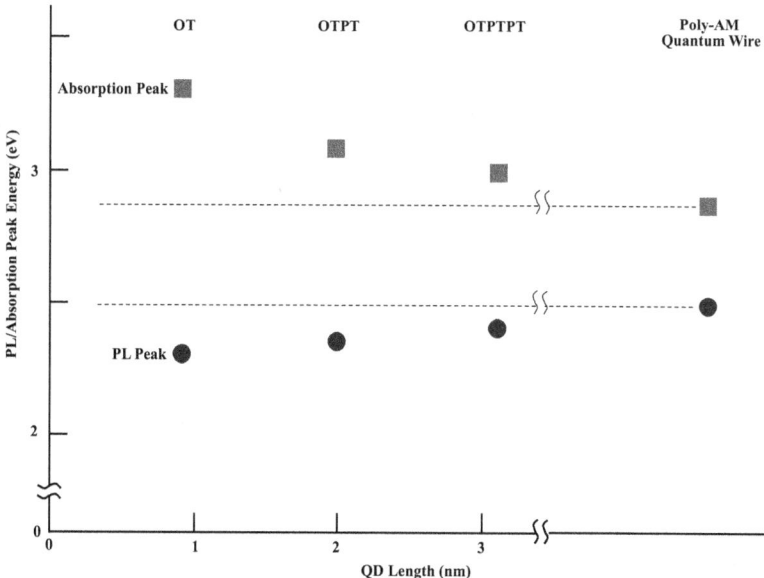

FIGURE 6.26 PL peak positions plotted as a function of the QD length together with absorption peak positions presented in Figure 6.20. (Reprinted with permission from Yoshimura and Ishii [8].)

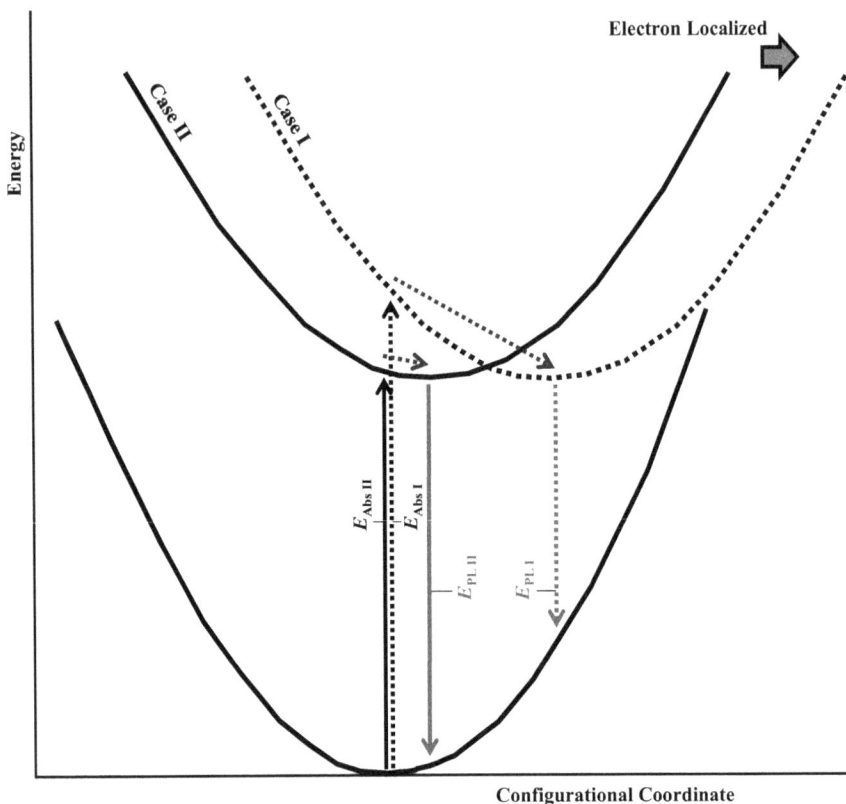

FIGURE 6.27 Configurational coordinate diagram for absorption and PL in polymer MQDs. (Reprinted with permission from Yoshimura and Ishii [8].)

namely, the equilibrium position change of the configurational coordinate curve becomes large." Based on this argument, a configurational coordinate diagram for absorption and PL in polymer MQDs can be postulated as in Figure 6.27. When an electron is delocalized (Case II), the horizontal shift of the configuration coordinate curve for the excited state is small, resulting in a small Stokes shift. When an electron is highly localized (Case I), the horizontal shift of the configuration coordinate curve becomes large, resulting in a large Stokes shift. Consequently, photon energy of PL in Case I, $E_{PL\,I}$, becomes smaller than photon energy of PL in Case II, $E_{PL\,II}$. Thus, by decreasing the QD length, the degree of the electron localization increases, resulting in the low-energy shift of the PL peak, as shown in Figure 6.26.

REFERENCES

1. T. Yoshimura, *Molecular Nano Systems: Applications to Optoelectronic Computers and Solar Energy Conversion*, Corona Publishing Co., Ltd., Tokyo (2007) [in Japanese].
2. T. Yoshimura, "Thin-film molecular nanophotonics," in *Photonics, Vol. 2: Nanophotonic Structures and Materials* (Ed. D. L. Andrews), Wiley, Hoboken, NJ (2015).

3. T. Yoshimura, S. Tatsuura, W. Sotoyama, A. Matsuura, and T. Hayano, "Quantum wire and dot formation by chemical vapor deposition and molecular layer deposition of one-dimensional conjugated polymer," *Appl. Phys. Lett.* **60**, 268–270 (1992).

4. B. I. Greene, J. Orenstein, R. R. Millard, and L. R. Williams, "Nonlinear optical response of excitons confined to one dimension," *Phys. Rev. Lett.* **58**, 2750–2753 (1987).

5. A. Oshima and T. Yoshimura, "Controlling sequences of three molecules and quantum dot lengths in conjugated polymer wires by molecular layer deposition," *AVS, 9th International Conference on Atomic Layer Deposition*, Monterey, California, 149 (2009).

6. T. Yoshimura, A. Oshima, D. Kim, and Y. Morita, "Quantum dot formation in polymer wires by three-molecule molecular layer deposition (MLD) and applications to electro-optic/photovoltaic devices," *ECS Transactions 25 (4) "Atomic Layer Deposition Applications 5," from the 216th ECS Meeting*, Vienna, Austria, 15–25 (2009).

7. T. Yoshimura, R. Ebihara, and A. Oshima, "Polymer wires with quantum dots grown by molecular layer deposition of three source molecules for sensitized photovoltaics," *J. Vac. Sci. Technol. A* **29**, 051510 (2011).

8. T. Yoshimura and S. Ishii, "Effect of quantum dot length on the degree of electron localization in polymer wires grown by molecular layer deposition," *J. Vac. Sci. Technol. A* **31**, 031501 (2013).

9. P. W. Loscutoff, H. Zhou, and S. F. Bent, "Molecular Layer Deposition of Multicomponent Organic Films for Nanoelectronics," *AVS, 9th International Conference on Atomic Layer Deposition*, Monterey, California, 148 (2009).

10. H. Zhou, M. F. Toney, and S. F. Bent, "Cross-linked ultrathin polyurea films via molecular layer deposition," *Macromolecules* **46**, 5638–5643 (2013).

7 Theoretically-Predicted Electro-Optic (EO) Effect in Polymer MQDs

Polymer multiple quantum dots (MQDs) fabricated by molecular layer deposition (MLD) are expected to be applied to electro-optic (EO) materials for light modulators, optical switches, tunable wavelength filters, and so on. In the present chapter, considering the fact that the EO effect, which is one of the nonlinear optical effects, is determined by electron wavefunction shapes of the ground state and the excited states, qualitative guidelines to enhance the EO effect are derived. Based on the guidelines, the EO effect in polymer MQDs is predicted theoretically for models with polydiacetylene and polyazomethine (poly-AM) backbones using the molecular orbital (MO) method. The calculation reveals that the EO effect is enhanced by controlling the donor/acceptor group distribution in a polymer wire to optimize the wavefunction shapes, and at the same time, by controlling the polymer wire length to optimize the wavefunction dimensionality. The calculation also reveals that the energy gap of polymer wires depends on the donor/acceptor group distribution. This implies that polymer MQDs can be produced by varying the donor/acceptor group arrangements in a polymer wire to modulate the energy gap.

7.1 NONLINEAR OPTICAL PHENOMENA

Nonlinear optical phenomena can be explained using the spring analogy shown in Figure 7.1a [1]. Displacement x of a particle attached to a spring is proportional to external force F for small F. With increasing F, the spring stretches rapidly with a nonlinear relationship between x and F. Nonlinear optical phenomena is similarly described by replacing x with polarization P and F with electric field E. In the model shown in Figure 7.1a, when E equals 0, the center of an electron cloud coincides with the position of a nucleus. With increasing E, the center of the cloud is displaced inducing P as follows:

$$P = \varepsilon_0 \chi^{(1)} E + \varepsilon_0 \chi^{(2)} E^2 + \varepsilon_0 \chi^{(3)} E^3 + \cdots \tag{7.1}$$

Here, ε_0 is dielectric constant in vacuum, $\chi^{(1)}$ is linear optical susceptibility, and $\chi^{(2)}$ and $\chi^{(3)}$ are, respectively, second-order and third-order nonlinear optical susceptibilities. When E is small, P is proportional to E. With increasing E, the contribution of the E^2 and E^3 terms become dominant. The E^2 and E^3 terms induce second-order and third-order nonlinear optical effects, respectively.

DOI: 10.1201/9781003094012-7

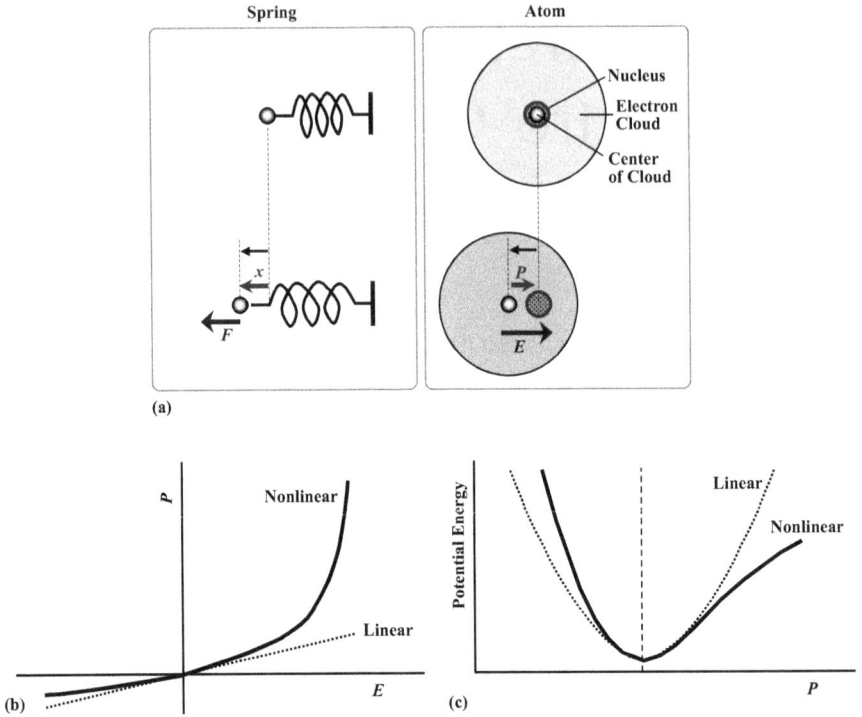

FIGURE 7.1 Explanation for nonlinear optical phenomena. (a) An analogy between a model for a particle attached to a spring and a model for an electron cloud bound to a nucleus. (b) Relationship between P and E. (c) Relationship between electronic potential energy and P.

In the case that an electric field of light with an angular frequency of ω, $E_{\mathrm{opt}} = E_0 e^{-j\omega t}$, and an externally-applied electric field E_{DC} exist in a space, the total electric field E is expressed as follows:

$$E = E_{\mathrm{opt}} + E_{\mathrm{DC}} \tag{7.2}$$

The term of E^2 gives E_{opt}^2, $E_{\mathrm{opt}}E_{\mathrm{DC}}$, E_{DC}^2 that are origins of the second-order nonlinear optical effects, and the term of E^3 gives E_{opt}^3, $E_{\mathrm{opt}}^2 E_{\mathrm{DC}}$, $E_{\mathrm{opt}}E_{\mathrm{DC}}^2$, E_{DC}^3 that are origins of the third-order nonlinear optical effects. When we focus on the term of E_{opt}^2, Equation (7.1) is written as

$$P = \varepsilon_0 \chi^{(1)} E_0 e^{-j\omega t} + \varepsilon_0 \chi^{(2)} E_0^2 e^{-j2\omega t}. \tag{7.3}$$

Polarization oscillating at 2ω generates light with an angular frequency of 2ω. This is the second harmonic generation (SHG). When we focus on the term of E_{opt}^3, Equation (7.1) is written as

$$P = \varepsilon_0 \chi^{(1)} E_0 e^{-j\omega t} + \varepsilon_0 \chi^{(3)} E_0{}^3 e^{-j3\omega t}. \tag{7.4}$$

Polarization oscillating at 3ω generates light with an angular frequency of 3ω. This is the third harmonic generation (THG).

When we focus on the term of $E_{opt} E_{DC}$, Equation (7.1) is written as

$$P = \varepsilon_0 \chi^{(1)} E_{opt} + \varepsilon_0 \chi^{(2)} E_{opt} E_{DC} = \left(\varepsilon_0 \chi^{(1)} + \varepsilon_0 \chi^{(2)} E_{DC} \right) E_{opt}. \tag{7.5}$$

Since refractive index n is a function of $\varepsilon_0 \chi^{(1)} + \varepsilon_0 \chi^{(2)} E_{DC}$, n is expressed as follows:

$$n = f\left(\varepsilon_0 \chi^{(1)} + \varepsilon_0 \chi^{(2)} E_{DC} \right) \tag{7.6}$$

In usual cases, there is a relationship of $\varepsilon_0 \chi^{(1)} \gg \varepsilon_0 \chi^{(2)} E_{DC}$. n is rewritten using the Taylor expansion as

$$n = f\left(\varepsilon_0 \chi^{(1)} \right) + f'\left(\varepsilon_0 \chi^{(1)} \right) \varepsilon_0 \chi^{(2)} E_{DC}. \tag{7.7}$$

By putting $n_0 = f\left(\varepsilon_0 \chi^{(1)} \right)$ and const $r = f'\left(\varepsilon_0 \chi^{(1)} \right) \varepsilon_0 \chi^{(2)}$

$$n = n_0 + \text{const } r\, E_{DC} \tag{7.8}$$

is obtained. This formula implies that the refractive index changes linearly with externally-applied electric field. This is the Pockels effect. r is called the Pockels coefficient. When we focus on the term of $E_{opt} E_{DC}{}^2$, Equation (7.1) is written as

$$P = \left(\varepsilon_0 \chi^{(1)} + \varepsilon_0 \chi^{(3)} E_{DC}{}^2 \right) E_{opt}. \tag{7.9}$$

By a procedure similar to that for the Pockels effect,

$$n = n_0 + \text{const } R\, E_{DC}{}^2 \tag{7.10}$$

is derived. This implies that the refractive index changes in proportion with $E_{DC}{}^2$, which corresponds to the Kerr effect. R is the Kerr coefficient. The Pockels effect and the Kerr effect are classified into the EO effect, and are used for various kinds of EO devices such as light modulators, optical switches, and tunable filters.

The relationship between P and E can be translated into the electronic potential energy vs. P curve as Figure 7.1b and c shows. When P and E are linearly related, the potential energy curve is parabolic. This corresponds to a situation wherein the potential energy of a particle attached to a spring is proportional to x^2 in the spring model. When nonlinear terms predominate, the potential energy curve deviates from the parabola. The asymmetric curve shown in Figure 7.1c corresponds to a situation wherein a particle attached to a spring easily moves to the positive direction while it hardly moves to the negative direction.

Nonlinear optical effects are understood visually by illustrations drawn in Figure 7.2. For SHG, because it originates from the term of E^2 in Equation (7.1), the potential energy vs. P curve becomes asymmetric as shown in Figure 7.2a.

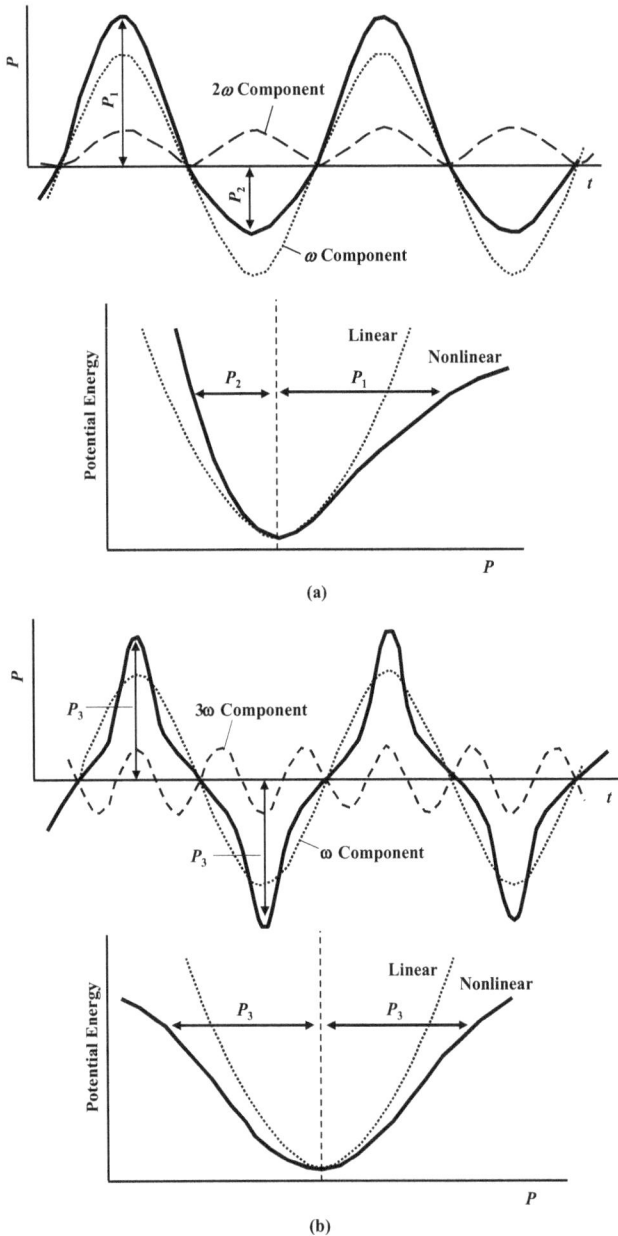

FIGURE 7.2   Visual explanation for (a) SHG, (b) THG, and (c) the Pockels effect.

(*Continued*)

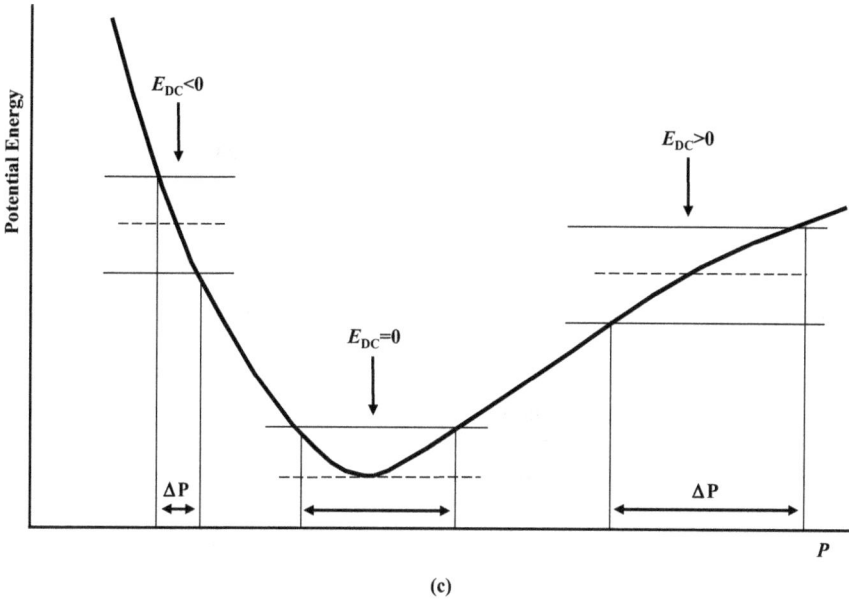

(c)

FIGURE 7.2 (*Continued*)   Visual explanation for (a) SHG, (b) THG, and (c) the Pockels effect.

When light with ω is introduced, an asymmetric response of P is induced, namely, P increases more largely toward the positive direction than the negative direction. Consequently, a 2ω component of P is induced to generate light with 2ω. For THG, which originates from the term of E^3, the potential energy curve deviates from the parabola symmetrically as shown in Figure 7.2b. When light with ω is introduced, a 3ω component of P is induced to generate light with 3ω.

For the Pockels effect, which originates from the term of E^2, the potential energy vs. P curve deviates from the parabola asymmetrically. In Figure 7.2c, a potential energy curve is drawn for a case that const r in Equation (7.8) is positive. When light is introduced, the total electric field is modulated around E_{DC} by the electric field of the light to result in P disturbance, ΔP, which implies the polarization change induced by light. Because the slope of the curve is smaller for $E_{DC} > 0$ than for $E_{DC} < 0$, larger ΔP is induced for $E_{DC} > 0$ than for $E_{DC} < 0$. Large ΔP gives rise to a large refractive index. Thus, with increasing E_{DC}, the refractive index increases. For a case that const r is negative, the slope of the curve is smaller for $E_{DC} < 0$ than for $E_{DC} > 0$, and the refractive index increases with decreasing E_{DC}.

Second-order optical nonlinearity for a molecule is given by molecular second-order nonlinear optical susceptibility β, and third-order optical nonlinearity by molecular third-order nonlinear optical susceptibility γ. Macroscopic second-order optical nonlinearity for aggregates of many molecules is given by $\chi^{(2)}$ originating from β, and macroscopic third-order optical nonlinearity by $\chi^{(3)}$ originating from γ. EO coefficient r, which is proportional to $\chi^{(2)}$, is used as the measure for the Pockels effect, and EO coefficient R, which is proportional to $\chi^{(3)}$, is used as the measure for the Kerr effect.

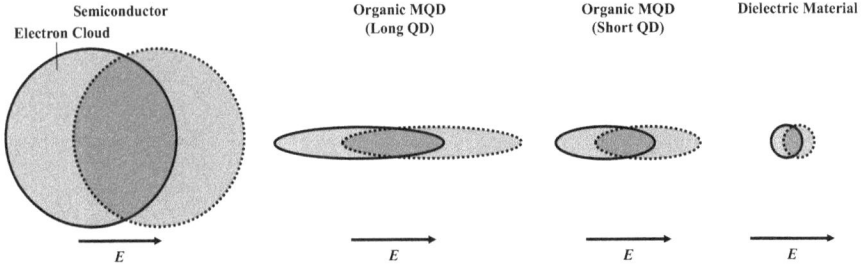

FIGURE 7.3 Schematic illustrations for electron-cloud polarization induced by an external electric field.

In general, the potential energy curve deviation from the parabola tends to become large in highly polarizable systems to enhance the optical nonlinearity. Figure 7.3 shows schematic illustrations for the electron-cloud polarization induced by an external electric field [1,2]. In dielectric materials, electron clouds are localized near atoms. This situation is regarded as a strong spring, and would not induce large electronic polarization. Semiconductors have widely-spread electron clouds, which give a chance to induce large polarization. Polymer MQDs have electron clouds delocalized one-dimensionally along polymer wires, which is favorable to induce large electronic polarization efficiently along the wire direction. Furthermore, by tuning the quantum dot (QD) length in the MQDs, optimized polarization condition would be realized. This is why the polymer MQDs seem promising as high-performance EO materials.

7.2 PROCEDURES FOR CALCULATION OF EO EFFECTS BY MOLECULAR ORBITAL METHOD

To predict the performance of EO polymer MQDs, the MO method is used. Figure 7.4 shows models for Hamiltonian in the MO calculation. The system contains N electrons. The whole-system Hamiltonian is expressed as follows:

$$H = \sum_{i=1} \left[-\frac{\hbar^2}{2m} \Delta_i + V(\boldsymbol{q}_i) \right] + \sum_{i<j} \frac{e^2}{|\boldsymbol{q}_i - \boldsymbol{q}_j|} \tag{7.11}$$

Here, i $(i = 1,2,\cdots,N)$ and j $(j = 1,2,\cdots,N)$ represent ith electron and jth electron, respectively. $\boldsymbol{q}_i$ is the position of the ith electron, $\boldsymbol{q}_j$ is the position of the jth electron, Δ_i is the Laplacian for the ith electron, $V(\boldsymbol{q}_i)$ is the interaction energy between the ith electron and nuclei, $\hbar$ is the Planck constant divided by 2π, m is the mass of an electron, and e is the charge of an electron. The second term represents interaction energy arising from the Coulomb force between electrons.

One-electron Hamiltonian for the ith electron is expressed as follows:

$$H_i = -\frac{\hbar^2}{2m} \Delta_i + V(\boldsymbol{q}_i) + \sum_{j\neq i} \int_{-\infty}^{\infty} \varphi_j^*(\boldsymbol{q}_j) \frac{e^2}{|\boldsymbol{q}_i - \boldsymbol{q}_j|} \varphi_j(\boldsymbol{q}_j) d\boldsymbol{q}_j \tag{7.12}$$

(a) Model for Whole-System Hamiltonian

(b) Model for One-Electron Hamiltonian for the ith Electron

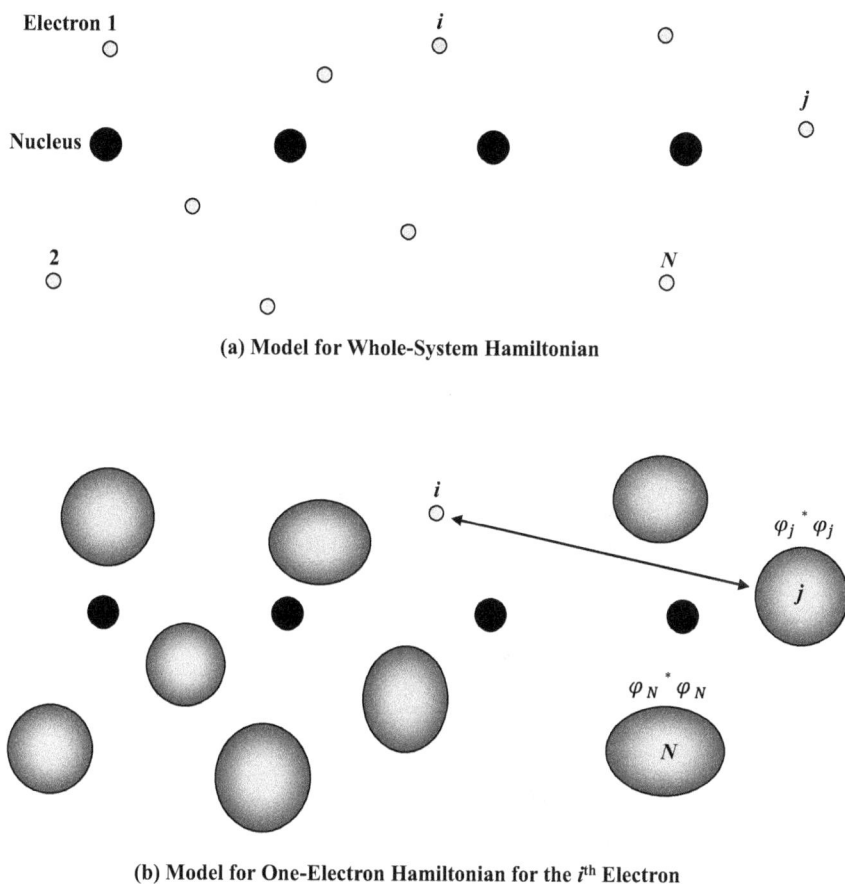

FIGURE 7.4 Models for Hamiltonian in the MO calculation.

Here, $\varphi_j(q_j)$ is a wavefunction of the jth electron. $\varphi_j(q_j)$ is called "molecular orbital." The third term represents the interaction energy between the ith electron and the jth electron, where the jth electron is regarded as an electron cloud. The interaction energy is calculated as an expected value determined by the charge density distribution of the jth electron, $e\varphi_j^*(q_j)\varphi_j(q_j)$.

Schrödinger equations for the whole-system Hamiltonian and the one-electron Hamiltonian are, respectively, given as follows:

$$H\Psi = E\Psi \tag{7.13}$$

$$H_i\varphi_i = \varepsilon_i\varphi_i \tag{7.14}$$

Here, Ψ and E are a wavefunction and energy of the whole system, respectively, and ε_i is energy for φ_i.

$$H \quad \text{---} \quad C \quad \equiv \quad C \quad \text{---} \quad H$$

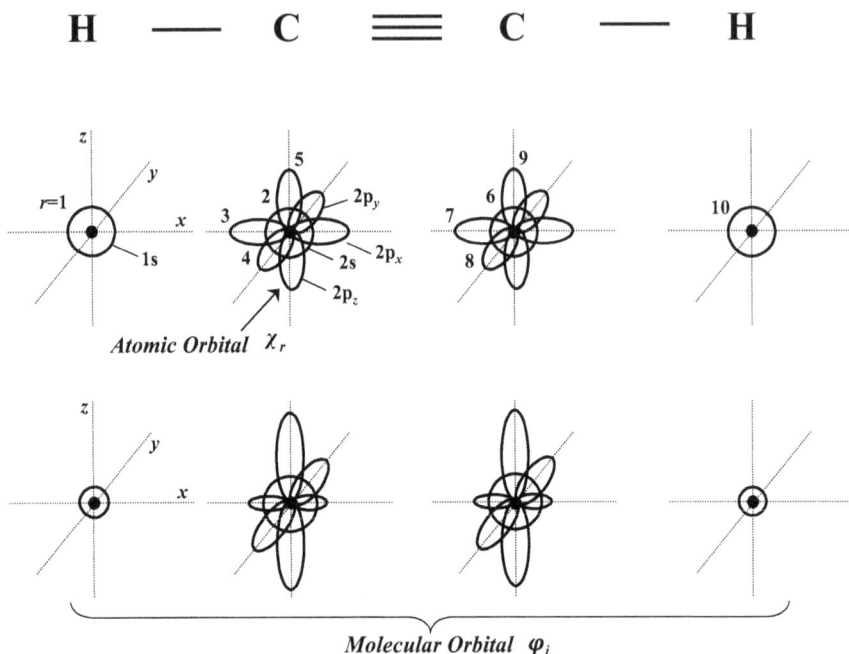

Atomic Orbital χ_r

Molecular Orbital φ_i

FIGURE 7.5 Schematic illustrations for relationship between molecular orbitals φ_i and atomic orbitals χ_r.

The molecular orbitals are expressed as linear combinations of wavefunctions in atoms (atomic orbitals), χ_r.

$$\varphi_i(q_i) = \sum_r c_{ir}\chi_r \qquad i = 1, 2, \cdots, N \qquad (7.15)$$

The relationship between φ_i and χ_r is schematically shown in Figure 7.5 for a case of C_2H_2. φ_i is built up by χ_r with an amplitude of c_{ir}. c_{ir}s are determined by the variation principle, namely, by finding a set of c_{ir}s that minimize ε_i in the following formula:

$$\int_{-\infty}^{\infty} \varphi_i^*(q_i) H_i \varphi_i(q_i) dq_i = \varepsilon_i \int_{-\infty}^{\infty} \varphi_i^*(q_i) \varphi_i(q_i) dq_i \qquad (7.16)$$

The minimized ε_i gives the energy eigenvalue for the molecular orbital of $\varphi_i = \sum_r c_{ir}\chi_r$. In a system with N electrons, N molecular orbitals are derived. However, in real systems, two spin states (α, β) are involved for one molecular orbital, resulting in $2N$ molecular orbitals. So, we should renumber the molecular orbitals including spin states as $\varphi_i (i = 1, 2, \cdots, 2N)$.

Figure 7.6 shows schematic illustrations for energy levels of molecular orbitals and electron occupation in many-electron wavefunctions built-up from the molecular orbitals. The many-electron wavefunctions represent the electronic states of the whole system. The wavefunction for the ground state is expressed as follows:

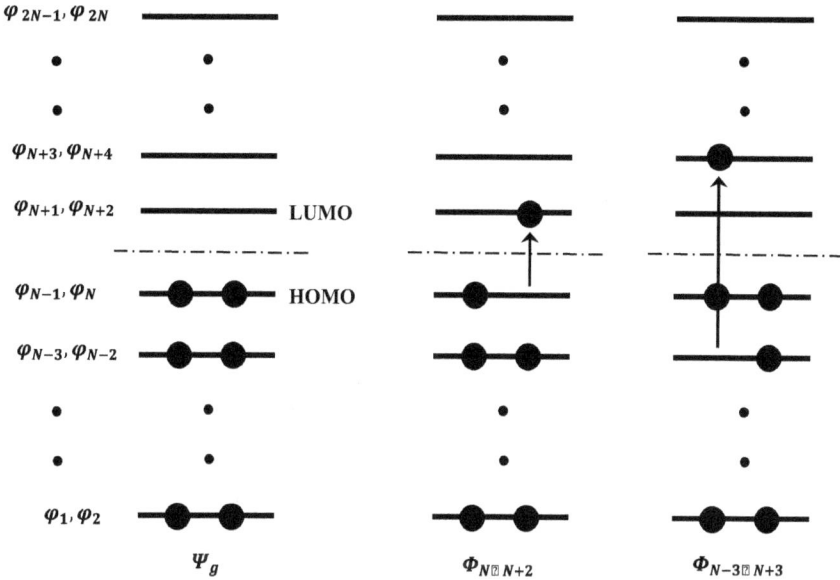

FIGURE 7.6 Schematic illustrations for energy levels of molecular orbitals and electron occupation in many-electron wavefunctions built-up from the molecular orbitals.

$$\Psi_g = \frac{1}{\sqrt{N!}} \sum_p (-1)^p \hat{P} \varphi_1(q_1) \varphi_2(q_2) \cdots \varphi_N(q_N), \qquad (7.17)$$

where, $\hat{P}$ is a permutation operator and p the number of permutations. Electrons occupy molecular orbitals $\varphi_1, \varphi_2, \cdots, \varphi_N$.

In the 0th approximation, many-electron wavefunctions for the excited states are written as follows:

$$\Phi_{i \to j} = \frac{1}{\sqrt{N!}} \sum_p (-1)^p \hat{P} \varphi_1(q_1) \varphi_2(q_2) \cdots \varphi_{i-1}(q_{i-1}) \varphi_{i+1}(q_{i+1}) \cdots \varphi_N(q_N) \varphi_j(q_i) \quad (7.18)$$

where $i \to j$ represents an excited state, where an electron is excited from φ_i to φ_j. $\Phi_{i \to j}$ is called configuration function.

In order to improve accuracy of the molecular orbitals obtained above, the variation principle is again applied to the following equation.

$$\int_{-\infty}^{\infty} \Psi_g^* H \Psi_g dq_1 dq_2 \cdots dq_N = E \int_{-\infty}^{\infty} \Psi_g^* \Psi_g dq_1 dq_2 \cdots dq_N \qquad (7.19)$$

Then, a set of improved c_{ir}s that minimize E is obtained. The minimized E gives the ground state energy E_g for Ψ_g.

In general, the excited states are expressed as linear combinations of $\Phi_{i\to j}$ as follows:

$$\Psi_n = \sum_{i,j} C_{n,i\to j}\Phi_{i\to j} \qquad (7.20)$$

$C_{n,i\to j}$s are determined by the variation principle, namely, by finding a set of $C_{n,i\to j}$s that minimizes E_n in the following formula:

$$\int_{-\infty}^{\infty} \Psi_n{}^* H\Psi_n dq_1 dq_2 \cdots dq_N = E_n \int_{-\infty}^{\infty} \Psi_n{}^*\Psi_n dq_1 dq_2 \cdots dq_N \qquad (7.21)$$

The minimized E_n gives the energy eigenvalues for the excited states of $\Psi_n = \sum_{i,j} C_{n,i\to j}\Phi_{i\to j}$. Thus, a set of wavefunctions and energy of eigenstates are obtained.

Molecular second-order nonlinear optical susceptibility, or molecular hyperpolarizability, along the wire direction is shown below based on Ward's expression [3,4].

For the Pockels effect:

$$\beta = -\frac{e^3}{2\hbar^2}\left[\sum_{\substack{g\neq n' \\ n\neq g \\ n\neq n'}} r_{gn'}r_{n'n}r_{ng} \frac{3\omega_{n'g}\omega_{ng}+\omega^2}{\left(\omega_{n'g}{}^2-\omega^2\right)\left(\omega_{ng}{}^2-\omega^2\right)} + \sum_n r_{gn}{}^2\Delta r_n \frac{3\omega_{ng}{}^2+\omega^2}{\left(\omega_{ng}{}^2-\omega^2\right)^2}\right] \qquad (7.22)$$

For the SHG:

$$\beta = -\frac{e^3}{2\hbar^2}\left[\begin{array}{l} \displaystyle\sum_{\substack{g\neq n' \\ n\neq g \\ n\neq n'}} r_{gn'}r_{n'n}r_{ng} \frac{2\left(\omega_{n'g}\omega_{ng}+2\omega^2\right)}{\left(\omega_{n'g}{}^2-4\omega^2\right)\left(\omega_{ng}{}^2-\omega^2\right)} \\[4ex] +\displaystyle\sum_{\substack{g\neq n' \\ n\neq g \\ n\neq n'}} r_{gn'}r_{n'n}r_{ng} \frac{\omega_{n'g}\omega_{ng}-\omega^2}{\left(\omega_{n'g}{}^2-\omega^2\right)\left(\omega_{ng}{}^2-\omega^2\right)} \\[4ex] +\displaystyle\sum_n r_{gn}{}^2\Delta r_n \frac{3\omega_{ng}{}^2}{\left(\omega_{ng}{}^2-\omega^2\right)\left(\omega_{ng}{}^2-4\omega^2\right)} \end{array}\right] \qquad (7.23)$$

In Equations (7.22) and (7.23), r_{gn} is the transition dipole moment between the ground state and the excited state, and $r_{n'n}$ is that between two excited states. Δr_n is the difference in the dipole moment between the ground state and the excited state. $\hbar\omega_{ng}$ is the excitation energy from the ground state to the excited state. $\hbar\omega$ is the energy of incident photons.

r_{gn}, $r_{n'n}$, and Δr_n are expressed by the following equations:

$$r_{gn} = \int \Psi_g^* r \Psi_n d^3 x, \tag{7.24}$$

$$r_{n'n} = \int \Psi_{n'}^* r \Psi_n d^3 x, \tag{7.25}$$

$$\Delta r_n = \int \Psi_n^* r \Psi_n d^3 x - \int \Psi_g^* r \Psi_g d^3 x, \tag{7.26}$$

where Ψ_g is the wavefunction of the ground state and Ψ_n is that of the excited state. r is the position of an electron, and $d^3 x$ represents $dq_1 dq_2 \cdots dq_N$. $r_n = \int \Psi_n^* r \Psi_n d^3 x$ and $r_g = \int \Psi_g^* r \Psi_g d^3 x$ are, respectively, the dipole moment of the excited state and that of the ground state. Ψ_g and Ψ_n are, respectively, expressed by Equations (7.17) and (7.20), and are constructed by molecular orbitals φ_i. Then, r_{gn}, $r_{n'n}$, and Δr_n are expressed as follows [4]:

$$r_{gn} = \sqrt{2} \sum_{i,j} C_{n,i \to j} m_{ij} \tag{7.27}$$

$$r_{n'n} = \sum_{i,j,k,l}{}' C_{n',i \to j} C_{n,k \to l} \left(\delta_{ik} m_{jl} - \delta_{jl} m_{ik} \right) \tag{7.28}$$

$$\Delta r_n = \sum_{i,j,k,l}{}' C_{n,i \to j} C_{n,k \to l} \left(\delta_{ik} m_{jl} - \delta_{jl} m_{ik} \right) + \sum_{i,j} C_{n,i \to j}^2 \left(m_{jj} - m_{ii} \right) \tag{7.29}$$

$$m_{ij} = \int \varphi_i^* r \varphi_j d^3 x \tag{7.30}$$

Here, the primed summations indicate to exclude terms, in which $i = k$ and $j = l$ simultaneously. r_{gn}, $r_{n'n}$, and Δr_n are calculated by Equations (7.27)–(7.30) once a set of molecular orbitals are obtained by the MO method. By substituting the r_{gn}, $r_{n'n}$, and Δr_n into Equations (7.22) and (7.23), β is obtained.

In the present simulations [1,5,6], the Austin Model 1 (AM1) method, which is a semi-empirical MO method, was used for the calculation of molecular orbitals. AM1 was developed by modifying the core repulsion function in the modified neglect of the diatomic overlap (MNDO) method. Using these molecular orbitals, the ground state wavefunction Ψ_g, a product of the occupied orbitals, and configuration functions $\Phi_{i \to j}$, a product of the orbitals with one-electron excitation from occupied orbital φ_i to unoccupied orbital φ_j, were made. Excited states, Ψ_n were determined by single-excitation configuration interaction (SCI). The CI calculation involved eight unoccupied orbitals above the lowest unoccupied molecular orbital (LUMO) and six occupied orbitals below the highest occupied molecular orbital (HOMO). 2-methyl-4-nitroaniline (MNA) was used as the standard molecule. From the calculation, β of 13×10^{-30} esu for SHG with fundamental wavelength of $1.06\,\mu m$ was obtained. This agrees fairly well with the 12×10^{-30} esu reported by Garito et al. [7].

7.3 GUIDELINES TO ENHANCE OPTICAL NONLINEARITY BY CONTROLLING ELECTRON WAVEFUNCTIONS

As mentioned in Section 7.1, the electronic potential energy vs. P curve in the linear system is parabolic. Deviation of the potential energy curve from the parabola generates a variety of nonlinear optical phenomena. The second-order optical nonlinearity arises from the asymmetric potential energy curve deviation (see Figure 7.7a). The third-order optical nonlinearity arises from the symmetric potential energy curve deviation (see Figure 7.7b). The larger the potential energy curve deviation becomes, the larger the induced nonlinear optical effect becomes.

The driving force for the potential energy curve deformation is explained in terms of the electron wavefunction shapes, as schematically depicted in Figure 7.7 [1,5]. For simplicity, we consider the ground state g and one excited state e, that is, a two-level model. When a perturbation is induced in a molecule by electric fields of incident light or external electric fields, mixing of the excited state wavefunction into the ground state wavefunction occurs, which deforms the potential energy curve. Therefore, the first approach to enhancing the optical nonlinearity lies in "promoting the mixing rate of the ground and excited states, i.e., promoting the oscillator strength, $f \propto r_{ge}^2$, between the ground and excited states." Here, $r_{ge} = \int \Psi_g^* r \Psi_e d^3 x$. f is increased with increasing the wavefunction overlap between the ground and excited states.

However, only promoting the wavefunction mixing rate is not in itself sufficient to enhance the potential energy curve deformation. For example, when the ground and excited state wavefunctions have a similar shape, mixing the wavefunctions results only in a small change in electron distribution. Consequently, there is only a small potential energy curve deformation. Therefore, "promoting the wavefunction difference between the ground and excited states" is the second approach to enhancing the optical nonlinearity.

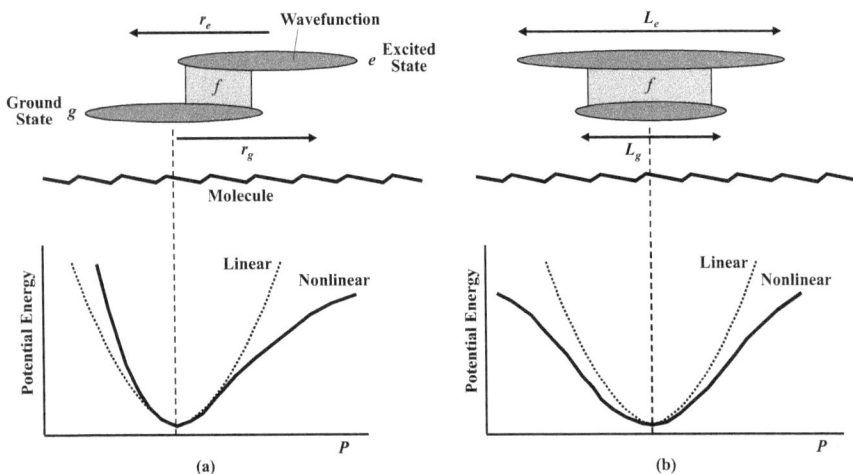

FIGURE 7.7 Microscopic origins for (a) second-order and (b) third-order optical nonlinearity.

From above mentioned arguments, the following qualitative guidelines are derived [1,2,5].

For Second-Order Optical Nonlinearity:
1. Promote wavefunction overlap between the ground and excited states, increasing oscillator strength $f \propto r_{ge}^2$.
2. Promote wavefunction separation between the ground and excited states to extend the wavefunction difference, increasing the difference in the dipole moment between the ground and excited states, $\Delta r_e = r_e - r_g$. Here, $r_e = \int \Psi_e^* r \Psi_e d^3 x$ and $r_g = \int \Psi_g^* r \Psi_g d^3 x$.
3. The second-order optical nonlinearity is proportional to the product of f and Δr_e, namely,

$$\beta \propto r_{ge}^2 \Delta r_e = r_{ge}^2 \left(r_e - r_g \right). \tag{7.31}$$

For Third-Order Optical Nonlinearity:
1. Promote wavefunction overlap between the ground and excited states, increasing the oscillator strength $f \propto r_{ge}^2$.
2. Promote difference in wavefunction spread between the ground and excited states to extend the wavefunction difference, increasing $\Delta L_e = L_e - L_g$. Here $L_e = \int \Psi_e^* r^2 \Psi_e d^3 x = \left(r^2 \right)_e$ and $L_g = \int \Psi_g^* r^2 \Psi_g d^3 x = \left(r^2 \right)_g$. They are the measures of the wavefunction spread.
3. The third-order optical nonlinearity is proportional to the product of f and ΔL_e, namely,

$$\gamma \propto r_{ge}^2 \Delta L_e = r_{ge}^2 \left(\left(r^2 \right)_e - \left(r^2 \right)_g \right). \tag{7.32}$$

For both second-order and third-order optical nonlinearities, note that r_{ge}^2 and $r_e - r_g$, or r_{ge}^2 and $\left(r^2 \right)_e - \left(r^2 \right)_g$, are not independent, but are in a tradeoff relationship. Therefore, it is concluded that "the balance of the wavefunction overlap and the wavefunction separation, or the balance of the wavefunction overlap and the wavefunction spread difference must be considered in optimizing nonlinear optical effects."

The result for β expressed by Equation (7.31) is consistent with two-level model expressions of Equations (7.22) and (7.23) as follows:
For the Pockels effect:

$$\beta = -\frac{e^3}{2\hbar^2} r_{ge}^2 \Delta r_e \frac{3\omega_{eg}^2 + \omega^2}{\left(\omega_{eg}^2 - \omega^2\right)^2} \propto r_{ge}^2 \Delta r_e = r_{ge}^2 \left(r_e - r_g \right) \tag{7.33}$$

For the SHG:

$$\beta = -\frac{e^3}{2\hbar^2} r_{ge}^2 \Delta r_e \frac{3\omega_{eg}^2}{\left(\omega_{eg}^2 - \omega^2\right)\left(\omega_{eg}^2 - 4\omega^2\right)} \propto r_{ge}^2 \Delta r_e = r_{ge}^2 \left(r_e - r_g \right) \tag{7.34}$$

The result for γ expressed by Equation (7.32) can be derived from the following equation for THG.

$$\gamma = -\frac{e^4}{4\hbar^2}\left[\sum_{\substack{n\\n'\\n''}} r_{gn}r_{nn'}\ r_{n'n''}r_{n''g}\ \frac{1}{\left(\omega_{ng}-\omega\right)\left(\omega_{n'g}-2\omega\right)\left(\omega_{n''g}-\omega\right)}\right]$$ (7.35)

Considering one excited state e, which has a major transition dipole moment with the ground state g, and neglecting the energy denominator, γ is roughly expressed as

$$\gamma \propto \sum_{n'} r_{ge}r_{en'}\ r_{n'e}r_{eg} = r_{ge}^2\ (r^2)_e.$$ (7.36)

By assuming $\left(r^2\right)_e \gg \left(r^2\right)_g$, Equation (7.32) is derived.

7.4 ENHANCEMENT OF POCKELS EFFECT IN POLYDIACETYLENE BACKBONES

In order to clarify the relationship between the Pockels effect and the electron wavefunction structures, the Pockels effect was calculated for models with polydiacetylene (PDA) backbones, which have a simple conjugated linear structure [1,2,5,6].

7.4.1 INFLUENCE OF WAVEFUNCTION SHAPES

The influence of wavefunction shapes of the excited state n and the ground state g on r_{gn}, $\Delta r_n = r_n - r_g$, and β is schematically illustrated in Figure 7.8. From left to right, wavefunction separation between the excited and ground states increases while

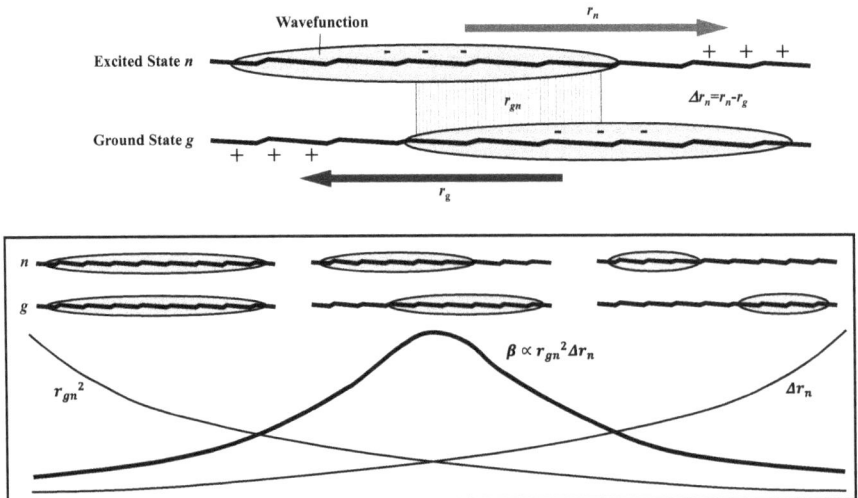

FIGURE 7.8   Influence of wavefunction shapes of the excited state n and the ground state g on r_{gn}, Δr_n, and β. (Reprinted with permission from Yoshimura [5].)

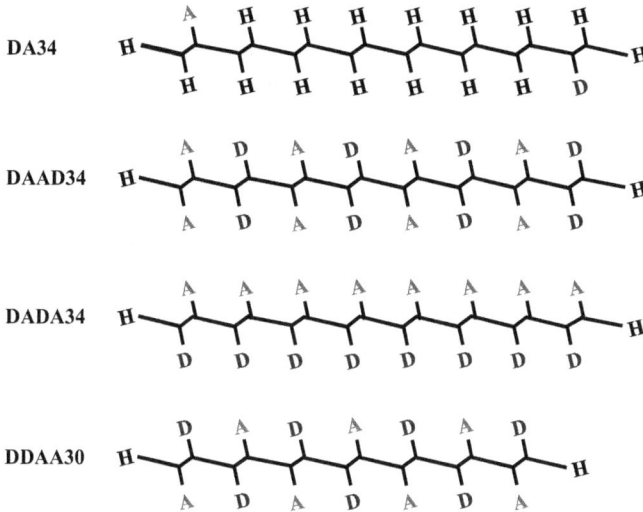

FIGURE 7.9 Four types of conjugated polymer wire models with PDA backbones. (Reprinted with permission from Yoshimura [5].)

wavefunction overlap between them decreases, i.e., Δr_n increases and r_{gn} decreases in a tradeoff relationship. Both the wavefunction separation and overlap must simultaneously be considered to optimize β. According to Equation (7.31) for the two-level model, β is expressed as $\beta \propto r_{gn}^2 \Delta r_n$. So, β becomes the maximum in the middle wavefunction shape. Furthermore, as described in Section 7.4.2, Δr_n and r_{gn} depend on the dimensionality of the conjugated systems. Then, it is expected that β has its maximum in a wavefunction shape of intermediate wavefunction separation with appropriate conjugated system dimensionality existing somewhere between zero and one, i.e., between the QD condition and the quantum wire condition.

Four types of conjugated polymer wire models depicted in Figure 7.9 were considered. The wires have PDA backbones and substituted donor and acceptor groups, respectively, denoted by D and A. DA, DAAD, DADA, and DDAA represent the types of donor/acceptor distribution, and the numbers following them the number of carbon sites in the wire, N_c. As Figure 7.10 shows, the donor and acceptor have

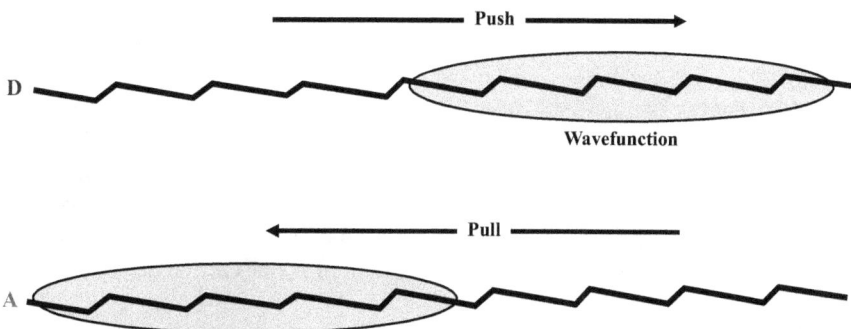

FIGURE 7.10 Push and Pull effects on wavefunction induced by donors and acceptors.

the tendency to push electrons away and pull electrons closer, respectively, when electrons are excited. So, wavefunction shapes can be optimized by adjusting donor/acceptor substitution sites in the wires using the push and pull effects as well as by adjusting wire lengths. In the present model, the donor is -NH_2 and the acceptor is -NO_2. β was calculated by varying N_c from 10 to 34.

Figure 7.11 shows molecular orbitals near the Fermi surface for DA34, DAAD34, DADA34, and DDAA30. Here, GM ($M = 1, 2, 3$) denotes the occupied molecular orbitals and EN ($N = 1, 2, 3$) the unoccupied molecular orbitals. $G1$ is HOMO and $E1$ is LUMO. In DA34, DAAD34, and DADA34, the charge separation in the wire direction appears mainly in HOMO and LUMO. In other molecular orbitals, electrons in the wires tend to be delocalized. In DDAA30, considerable charge separation appears in $E2$ and $E3$ orbitals as in LUMO.

Figure 7.12 shows $C^2_{n,GM \to EN}$ as a function of GM ($M = 1, 2, ..., 6$) and EN ($N = 1, 2, ..., 8$) for the first excited states ($n = 1$) in DA10, DA34, DAAD10, DAAD34, DADA10, DADA34, DDAA14, and DDAA30. $C_{n,GM \to EN}$ represents the fraction of the configuration function $\Phi_{GM \to EN}$ involved in Equation (7.20). The height of the pyramid indicates $C^2_{n,GM \to EN}$. For the short wires with $N_c = 10$ or 14, the first excited states mainly consist of configuration functions with small GM and EN, which means that the configuration interaction can be determined using only a few occupied and unoccupied molecular orbitals. For the long wires with $N_c = 30$ or 34, the pyramids are distributed more widely than for the short wires. The height of the pyramids tends to decrease rapidly with increasing GM and EN. This suggests that six occupied orbitals and eight unoccupied orbitals are enough to determine the configuration interaction for the polymer wires shown in Figure 7.9.

Combining the results of $C^2_{n,GM \to EN}$ shown in Figure 7.12 with shapes of the molecular orbitals shown in Figure 7.11, the wavefunction shapes can qualitatively be estimated. Figure 7.13a and b, respectively, show structures and schematic wavefunction shapes for DA34, DAAD34, and DDAA30. In DA34, the $G1 \to E1$ component, corresponding to the HOMO $\to$ LUMO transition, is extremely small. The wavefunction separation then becomes small, which corresponds to the wavefunction condition on the left side in Figure 7.8. In DAAD34, finite $G1 \to EN$ and $GM \to E1$ components appear, resulting in considerable wavefunction separation, which corresponds to the condition in the center in Figure 7.8. In DDAA30, $G1 \to EN$ components predominate, and the EN orbitals exhibit considerable charge separation, inducing large wavefunction separation like the condition on the right side in Figure 7.8. Thus, it is found that the wavefunction separation increases in the order of DA34, DAAD34, and DDAA30. For DADA34, since the molecular orbital shapes and the $C^2_{n,GM \to EN}$ distribution resemble those for DAAD34, DADA34 seems to have the wavefunction condition in the center in Figure 7.8.

r_{gn}, Δr_n, and β were, respectively, calculated by Equations (7.27), (7.29), and (7.22) for DA34, DAAD34, DADA34, and DDAA30 at a detuning energy of 0.2 eV from the first excited states. The results are shown in Figure 7.14. Here, ρ_β represents β normalized by wire lengths. Although calculation of β was carried out by Equation

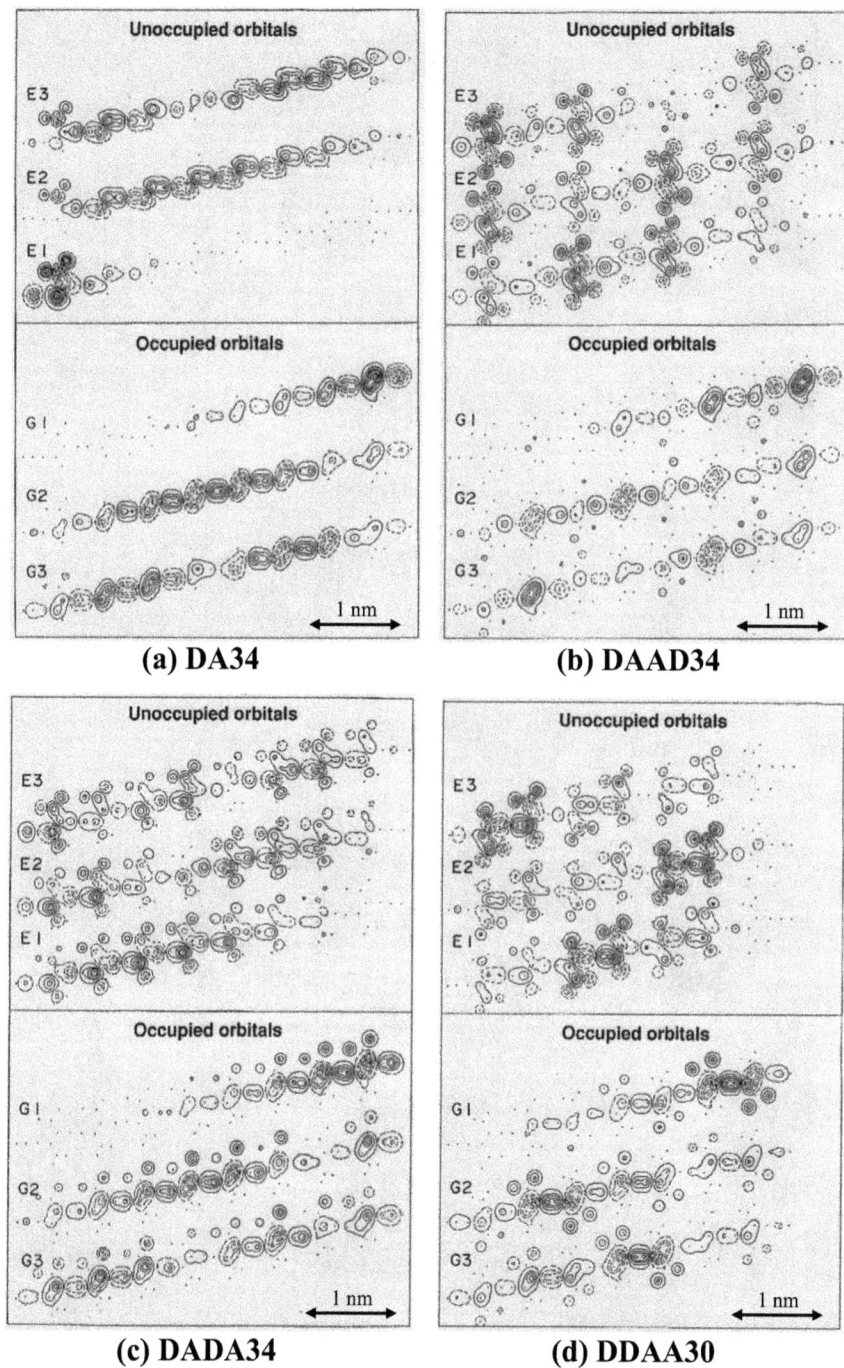

FIGURE 7.11 Contour diagrams of molecular orbitals near the Fermi surface in plane $z = 0.06$ nm for (a) DA34, (b) DAAD34, (c) DADA34, and (d) DDAA30. (Reprinted with permission from Yoshimura [5].)

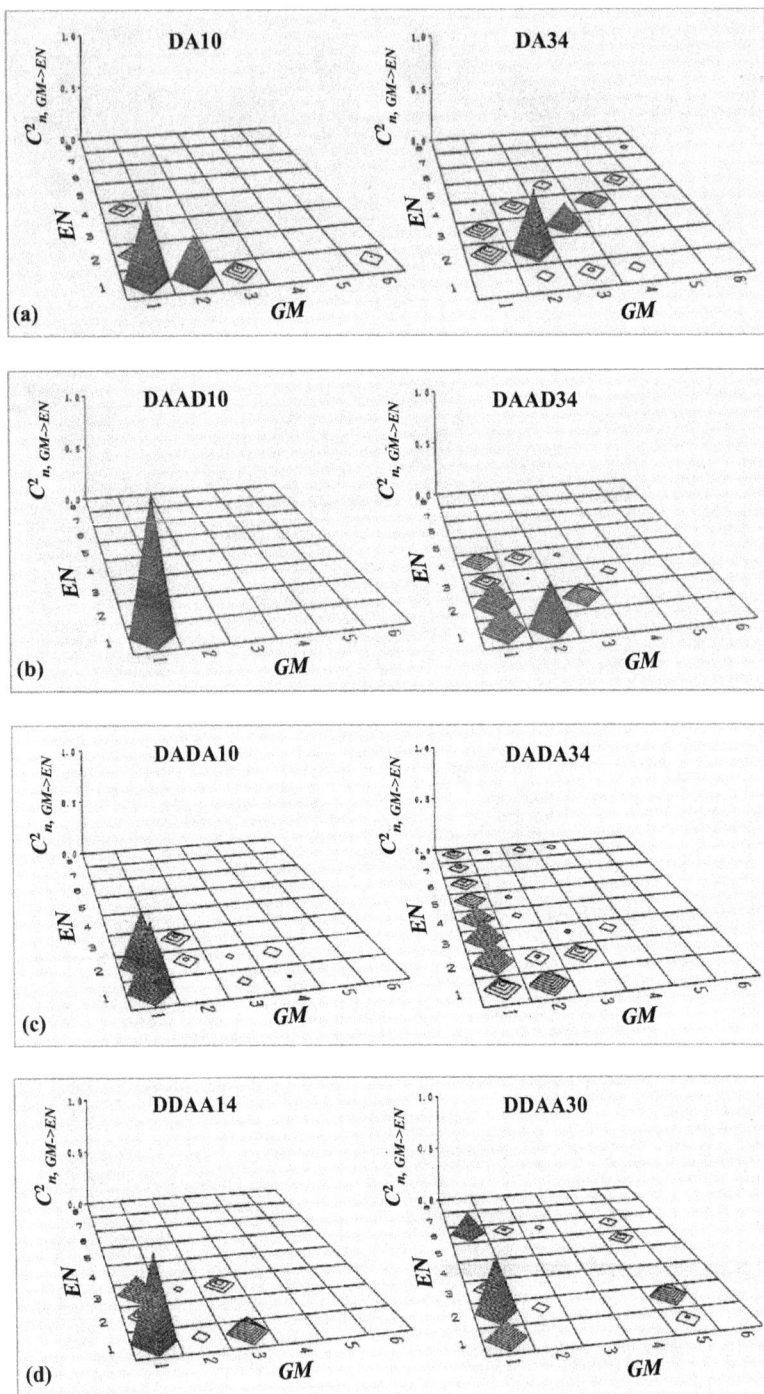

Figure 7.12 $C^2_{n,GM\to EN}$ for the first excited states in (a) DA10 and DA34, (b) DAAD10 and DAAD34, (c) DADA10 and DADA34, and (d) DDAA14 and DDAA30. (Reprinted with permission from Yoshimura [5].)

(a)

(b)

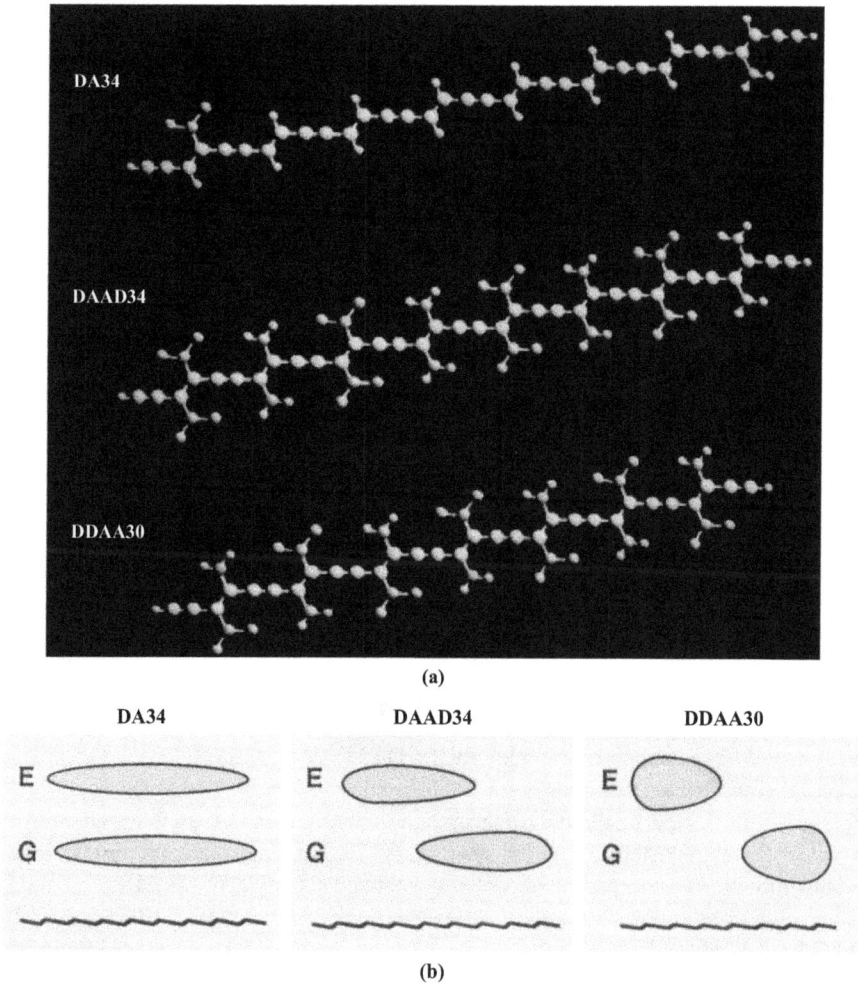

FIGURE 7.13   (a) Structures calculated by the MO method and (b) schematic wavefunction shapes estimated from the results shown in Figures 7.11 and 7.12 for DA34, DAAD34, and DDAA30.

(7.22) involving 48 excited states, the major contribution to ρ_β was from the first excited state. Thus, the overall tendency in Figure 7.14 can be explained in terms of the wavefunction of the first excited state. r_{gn} decreases and Δr_n increases in the order of DA34, DAAD34, DADA34, and DDAA30. ρ_β is small in DA34 and DDAA30, and becomes large between them, in DAAD34 and DADA34, which exhibits a medium wavefunction separation. This parallels the tendency of the qualitative guidelines shown in Figure 7.8 and indicates that a balance between r_{gn} and Δr_n is essential to improve β.

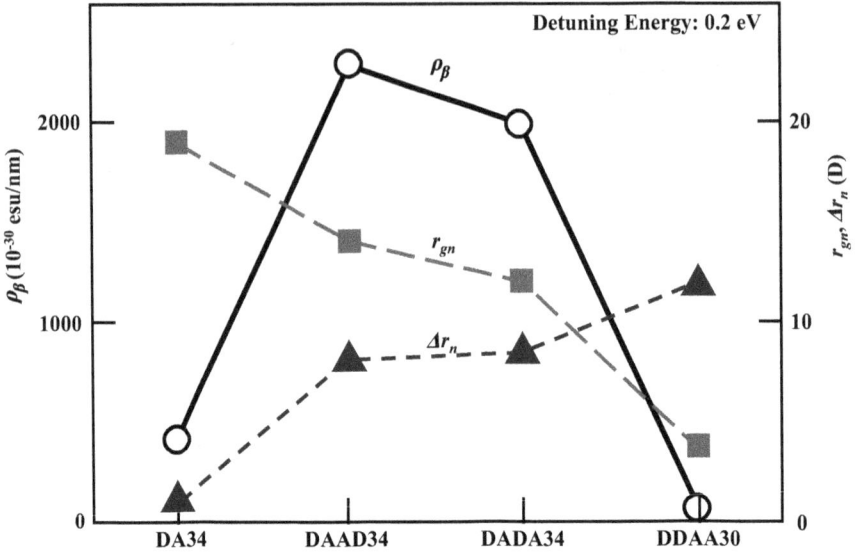

FIGURE 7.14   Calculated r_{gn}, Δr_n, and ρ_β for DA34, DAAD34, DADA34, and DDAA30. ρ_β is β per 1 nm of wire length. (Reprinted with permission from Yoshimura [5].)

7.4.2   INFLUENCE OF WAVEFUNCTION DIMENSIONALITY

In Figure 7.15, calculated ρ_β is plotted for DA, DAAD, DADA, and DDAA as a function of N_c that corresponds to the conjugated wire length. The N_c dependence of ρ_β is greatly affected by the donor/acceptor distribution. In DA, ρ_β is maximized at $N_c = 18$. Morley reported similar behavior in β vs. N_c curves in polyenes and polyphenyls having one donor and one acceptor at opposite sides [8]. In DAAD, ρ_β is maximized at $N_c = 18$ and the large ρ_β is preserved in the whole range of $10 \le N_c \le 34$. In DADA, ρ_β monotonically increases with N_c in the range of $10 \le N_c \le 34$. In DDAA, ρ_β is very small.

DAAD has the largest ρ_β of about 3000×10^{-30} esu/nm at $N_c = 18$. Assuming PDA wire density of 1.3×10^{14} 1/cm², the expected EO coefficient r of DAAD18 is about 3000 pm/V that is 100 times larger than the EO coefficient r_{33} of LN. The wire length of DAAD18 is ~2.2 nm. This indicates that DAAD exhibits enhanced optical nonlinearity in a long QD, namely, a dimensionality of ~0.5 between zero and one. Thus, it is concluded that a large EO effect will be obtained by controlling wavefunction dimensionality as well as the shapes, which can be realized by adjusting the conjugated length and the donor/acceptor substitution sites in polymer wires.

To investigate the N_c dependence of ρ_β precisely, r_{gn} and Δr_n are plotted as a function of N_c in Figure 7.16. In DA, r_{gn} increases and Δr_n decreases with increasing N_c. Consequently, as Figure 7.15 shows, ρ_β becomes maximum at an intermediate conjugated wire length of $N_c = 18$. In DAAD, r_{gn} and Δr_n exhibit similar N_c dependence as DA, resulting in a peak at $N_c = 18$, as shown in Figure 7.15. In DAAD, however, Δr_n is much larger than in DA, keeping the large ρ_β in the whole range of $10 \le N_c \le 34$. In DADA, Δr_n and r_{gn} increase with N_c. This is reflected in the monotonical increase

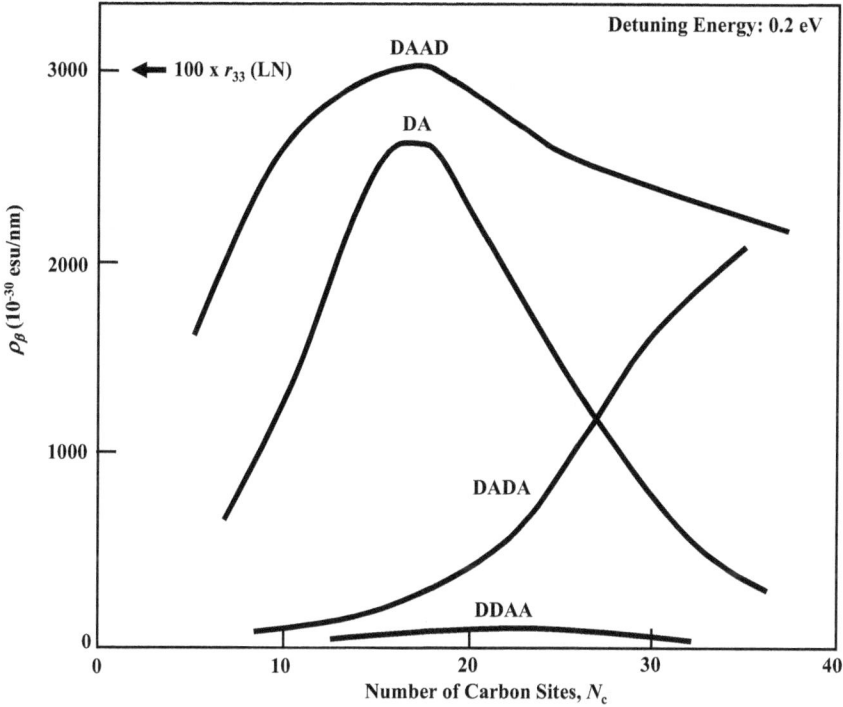

FIGURE 7.15 Calculated ρ_β for DA, DAAD, DADA, and DDAA as a function of N_c that corresponds to the conjugated wire length. (Reprinted with permission from Yoshimura [5].)

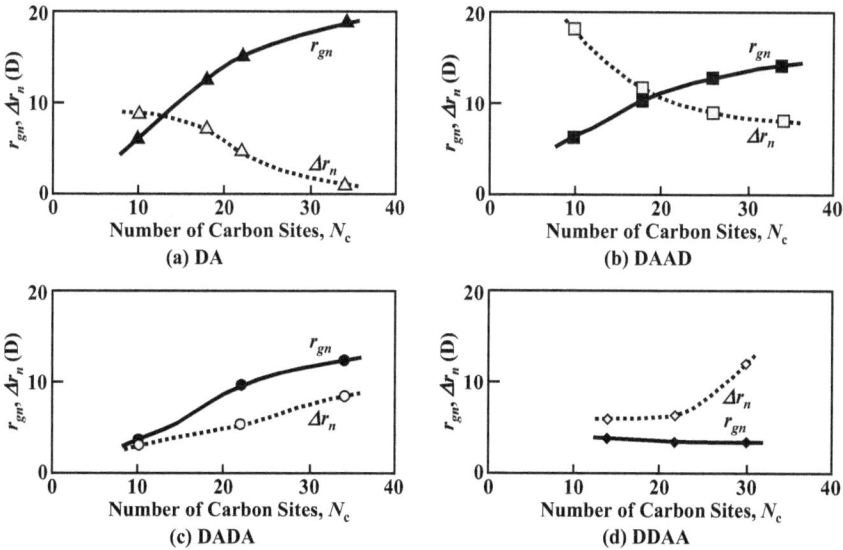

FIGURE 7.16 Influence of N_c on r_{gn} and Δr_n. (Reprinted with permission from Yoshimura [5].)

in ρ_β with N_c. In DDAA, r_{gn} is markedly smaller than in the other three types, which reduces ρ_β, as shown in Figure 7.15.

Because the wavefunction shape affects β via the dipole moments, in order to determine the relationship between wavefunctions and β, we must clarify the relationship between the wavefunctions and the dipole moments. As mentioned for the explanation of Figure 7.11, the charge separation mainly appears in HOMO and LUMO. Therefore, the G1 → EN and GM → E1 compositions in Ψ_n ($n = 1$) mainly determine Δr_n.

In DA, as can be seen in Figure 7.12a, the contribution of G1 → EN and GM → E1 compositions to Ψ_n is reduced drastically with increasing N_c. This reduces the wavefunction separation, decreasing Δr_n. At the same time, the wavefunction overlap between the ground and excited states increases with N_c, increasing r_{gn}. These results are consistent with those presented in Figure 7.16a. In DAAD, as Figure 7.12b shows, although the wavefunction changes in a similar way as in DA with increasing N_c, the change is not so drastic. This is consistent with the results shown in Figure 7.16b. In DA34, G1 → EN and GM → E1 compositions are less than in DAAD34 (see Figure 7.12a and b), which accounts for the fact that, in DA34, Δr_n is smaller and r_{gn} is larger than in DAAD34, as can be seen in Figure 7.16a and b. This is because DA has only one donor and one acceptor. So, it has smaller electronic pull and push power than DAAD, which has many donors and acceptors. In DADA, the contribution of G1 → EN compositions to Ψ_n remains large when N_c increases from 10 to 34 (see Figure 7.12c), causing a monotonical increase in Δr_n as can be seen in Figure 7.16c. For DDAA, because the main compositions in DDAA30 are the G1 → EN series (see Figure 7.12d), and all three orbitals of E1, E2, and E3 exhibit considerable charge separation (see Figure 7.11d), Δr_n becomes large while r_{gn} is considerably suppressed, as can be seen in Figure 7.16d.

7.4.3 INFLUENCE OF SHARPENING ABSORPTION BANDS

Figure 7.17 shows resonant enhancement of ρ_β near the first excited state for DAAD18. The absorption band width of a PDA thin film is known to be about 0.1 eV, so a detuning energy of about 0.2 eV or more might be necessary for low-loss operation. In this case, the EO coefficient is about two orders of magnitude larger than r_{33} of $LiNbO_3$, as indicated in Figure 7.15. If the band width could be reduced one order of magnitude by sharpening the absorption band, the detuning energy required would only be ~0.02 eV, and the EO coefficient would be about 10^4 times larger than r_{33} of $LiNbO_3$. This leads us to believe that sharpening the absorption band [9] is important as well as controlling the wavefunction to improve the nonlinear optical property.

The uniformity of the conjugated length in materials is essential in sharpening. Reduction of the exciton-phonon (or electron-phonon) coupling strength is also important. One way to do this might be to use rigid backbone structures. Bound exciton formation or exciton confinement in QDs in polymer MQDs might be another way to sharpen absorption bands.

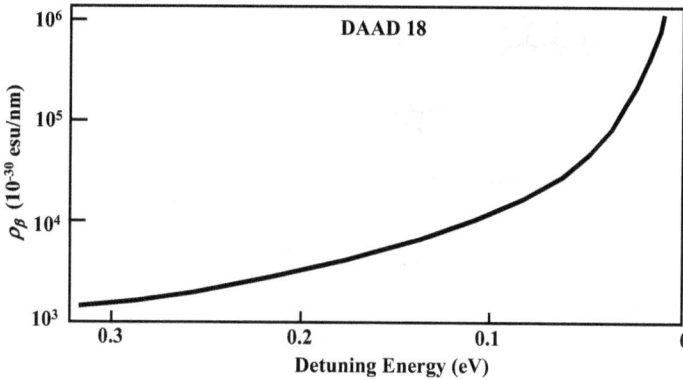

FIGURE 7.17 Dependence of ρ_β on detuning energy in DAAD18. (Reprinted with permission from Yoshimura [5].)

7.4.4 CONTROL OF ENERGY GAPS

As shown in Figure 7.18, the MO calculation revealed that the energy gap of conjugated wires with PDA backbones varies depending on the donor/acceptor distribution. The energy gap decreases from 3.3 eV of intrinsic PDA wires to 1.6 eV by donors/acceptors substitution with the DAAD configuration.

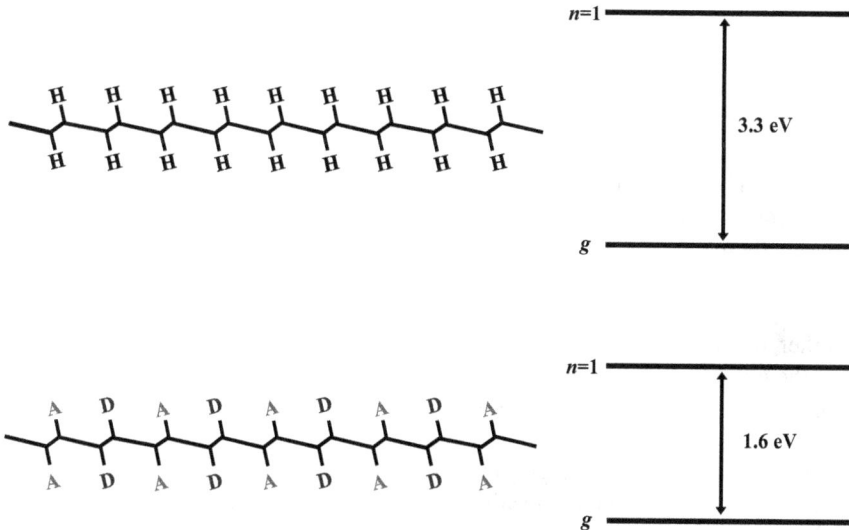

FIGURE 7.18 Energy gap narrowing in conjugated wires with PDA backbones by donor and acceptor substitution.

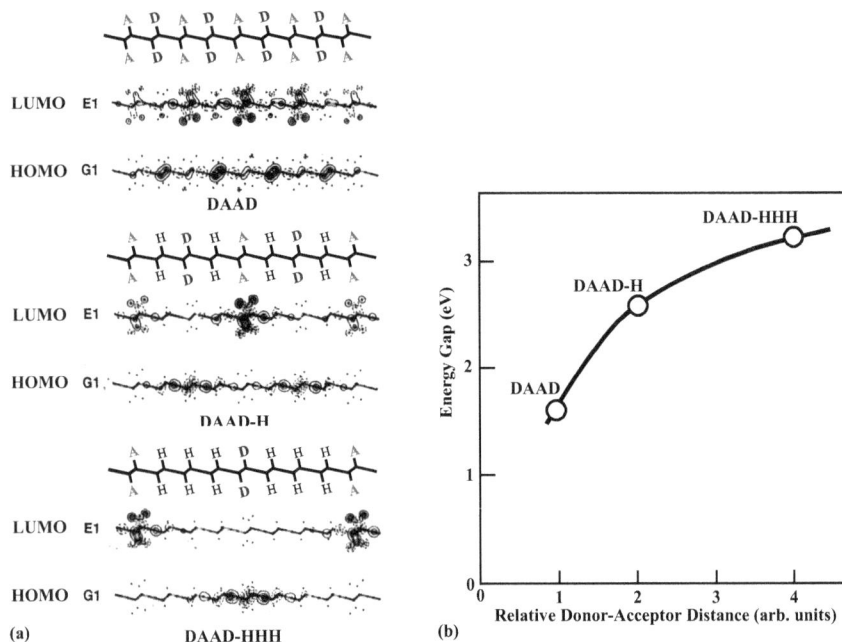

FIGURE 7.19 (a) Three models of conjugated wire structures, in which donors and acceptors are distributed, and the corresponding LUMOs and HOMOs. (b) Calculated energy gaps for the models.

Figure 7.19a presents three models of conjugated wire structures, in which donors and acceptors are distributed, and the corresponding LUMOs and HOMOs. In DAAD, D-D pairs and A-A pairs are placed alternately. In DAAD-H, a H-H pair is inserted between them. In DAAD-HHH, three H-H pairs are inserted between them. In all the three models, electrons are localized near D-D pairs and A-A pairs in HOMO and LUMO, respectively. As shown in Figure 7.19b, the energy gap decreases in the order of DAAD-HHH, DAAD-H, and DAAD. In other words, by decreasing the distance between the D-D pair and the A-A pair, the energy gap becomes narrow. This result implies that the energy gap of the conjugated polymer wires can be controlled by adjusting the donor/acceptor distribution.

The mechanism of the energy gap narrowing by the donor and acceptor substitution is as follows. Electrons are localized around donor sites in HOMO while they are localized around acceptor sites in LUMO (see Figure 7.19a). The electron distribution in HOMO corresponds to the hole distribution in HOMO when the electrons are excited. In the order of DAAD-HHH, DAAD-H, and DAAD, electron-hole distance

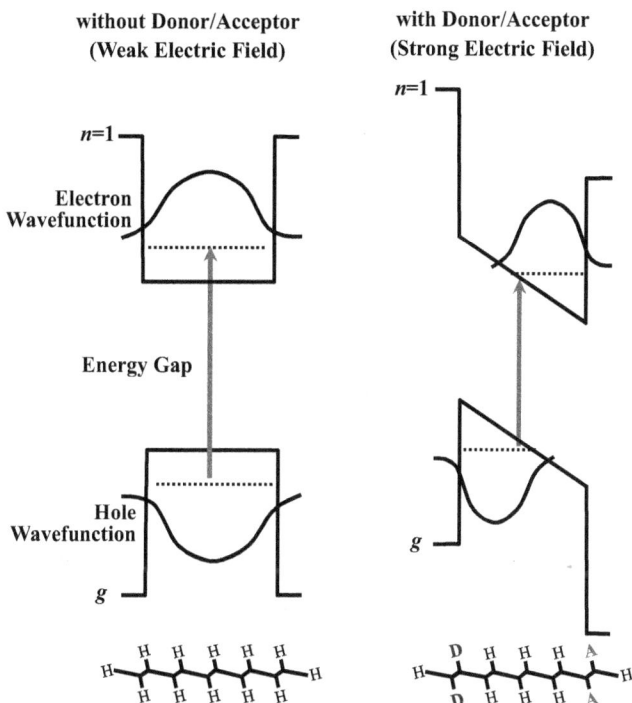

FIGURE 7.20 Mechanism of the energy gap narrowing in conjugated wires by donor and acceptor substitution.

becomes short to increase the electric field in the area in between. The strong electric field causes large potential energy slopes, as depicted in Figure 7.20, narrowing the energy gaps similarly to the case of the multiple quantum well (MQW) light modulators described in Section 2.3.

7.4.5 EO POLYMER MQDs

To apply the tailored materials to practical devices, it is necessary to fabricate the materials in ordered structures. This involves constructing MQD structures in one-dimensional (1-D) conjugated polymer wires. One way to do this is the insertion of disconnecting molecular units to cut the π-conjugation in polymer wires, as described in Section 6.2. Another way is modulation of energy gaps of conjugated polymer wires by adjusting the donor/acceptor distribution.

In Figure 7.21, DAAD18 and a PDA wire with an inserted DAAD18 structure are compared. For the PDA wire with an inserted DAAD18 structure, as mentioned in Section 7.4.4, the energy gap of the DAAD portion narrows comparing to the energy gap of portions without donors and acceptors. Consequently, the DAAD portion

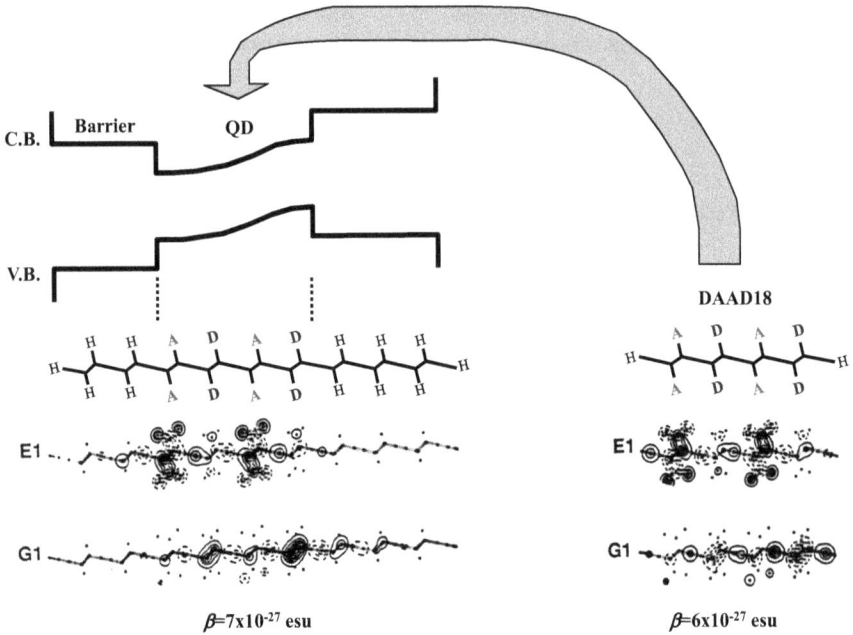

FIGURE 7.21 Insertion of a DAAD18 structure into a conjugated wire with PDA backbone.

becomes a QD while the portions without donors and acceptors become barriers. Although the molecular orbitals of the DAAD QD structure slightly penetrate into the barrier regions by the tunneling, the molecular orbitals and β for the QD structure are almost the same as those for DAAD18.

Using this effect, it might be possible to insert many DAAD structures into a conjugated polymer wire, constructing an EO polymer MQD of a polymer superlattice depicted in Figure 7.22a. This structure enables us to align wavefunctions of the DAAD structures along the wire direction perfectly, which is favorable to attain large Pockels effect. It might also be possible to construct an EO polymer MQD by modulating the donor/acceptor distribution pattern, as depicted in Figure 7.22b. In this example, donors and acceptors are attached to the conjugated wire with the DADA-type distribution, which induces little gap narrowing, in the barrier portions.

As mentioned above, using the energy gap narrowing effect, it will be possible to form EO polymer MQDs. MLD is the most promising method to realize the polymer MQDs. Figure 7.23 shows sandwich-type configuration and in-plane-type configuration of EO waveguides consisting of the EO polymer MQDs fabricated by MLD.

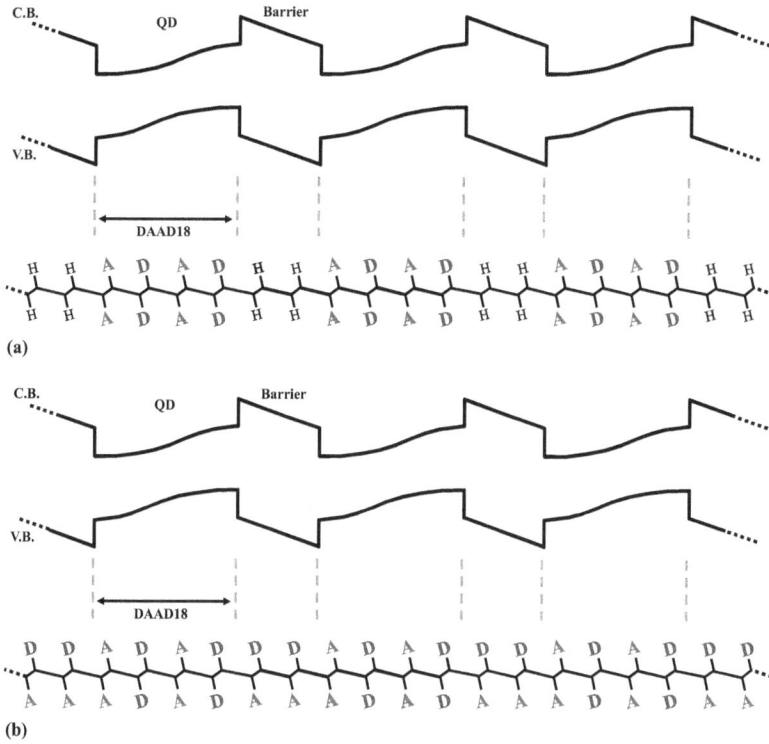

FIGURE 7.22 EO polymer MQD of a polymer superlattice with corresponding electronic potential energy curves. The barrier portions contain (a) no donors/acceptors and (b) donors/acceptors with the DADA-type distribution.

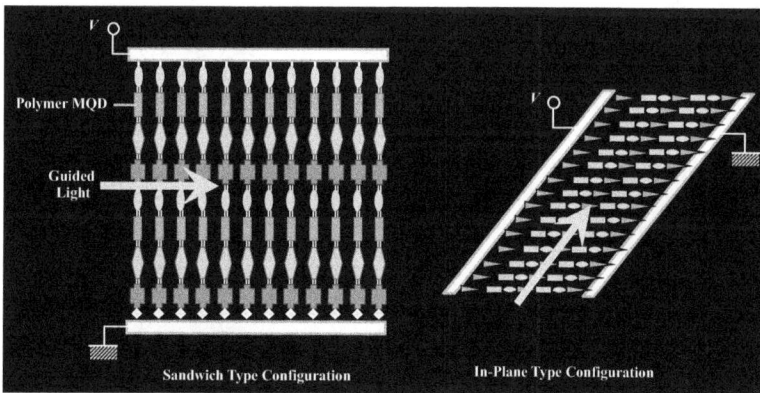

FIGURE 7.23 Possible configuration of EO waveguides consisting of EO polymer MQDs fabricated by MLD.

7.5 ENHANCEMENT OF POCKELS EFFECT IN POLYAZOMETHINE BACKBONES

From the viewpoint of fabrication of EO polymer MQDs, poly-AM is preferable for the backbone because growth of polymer MQDs by MLD has already been demonstrated experimentally for poly-AM, as described in Chapter 6. Matsuura and Hayano theoretically predicted the enhancement of the Pockels effect in the poly-AM conjugated wires utilizing the MO method [10]. In Figure 7.24, models for the calculation and predicted βs of EO conjugated wires with poly-AM backbones at 633 nm are shown. AAAADDDD exhibits the largest β, which corresponds to EO coefficient of 1260 pm/V.

The reason why AAAADDDD has larger β than DDDDAAAA was explained by Matsuura and Hayano as follows [10]. N at the edge of the wire has donor-like characteristics. In AAAADDDD, pull/push strength is increased because the N at the edge and the donor region of DDDD are located on the same side. In DDDDAAAA, on the other hand, pull/push strength is suppressed because the N at the edge and the donor region of DDDD are located in opposite sides.

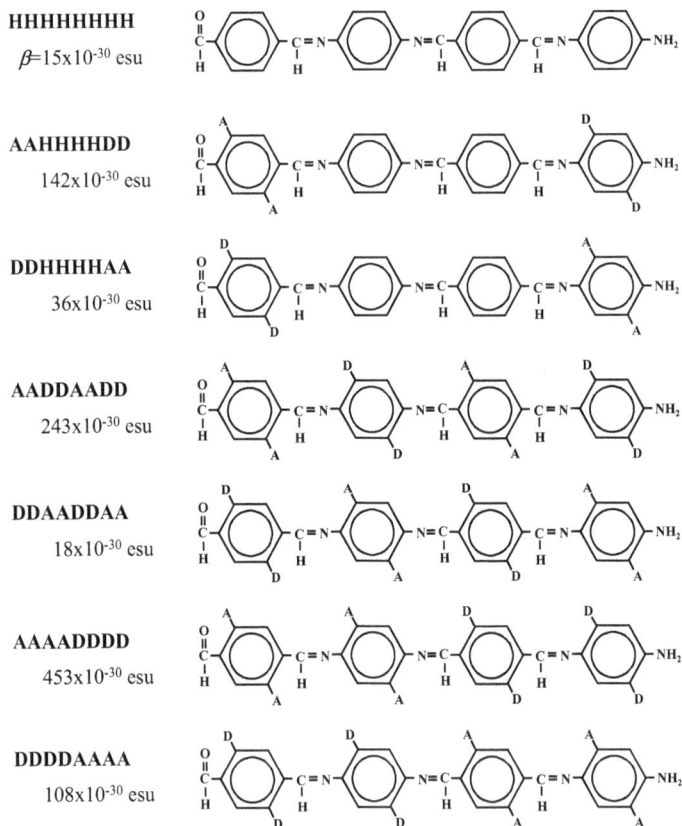

FIGURE 7.24 Models for calculations and predicted βs of EO conjugated wires with poly-AM backbones at 633 nm.

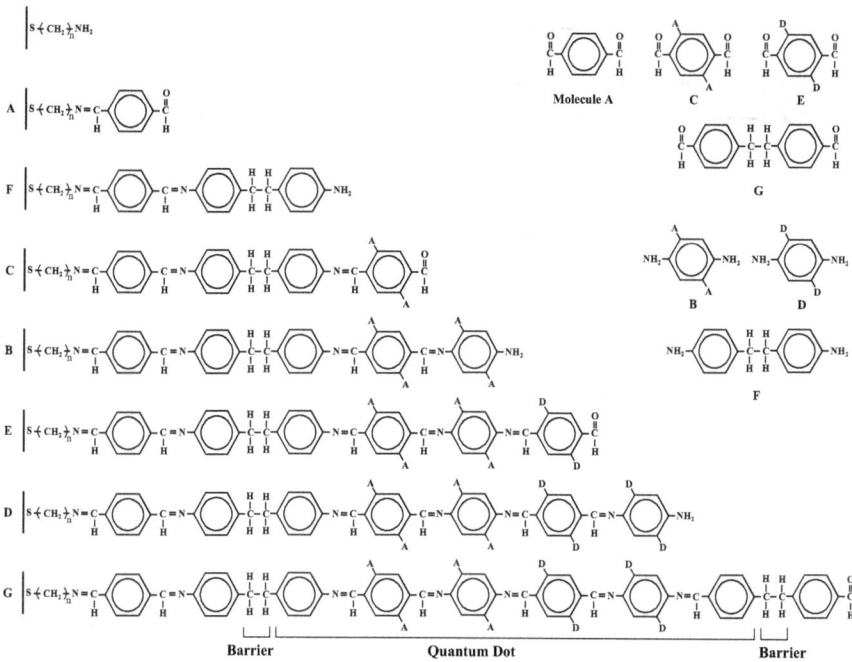

FIGURE 7.25 Possible example of the MLD process to fabricate EO polymer MQDs with poly-AM backbones.

A possible example of the MLD process to fabricate EO polymer MQDs with poly-AM backbones is presented in Figure 7.25. Seven kinds of source molecules, Molecules A-G, are used. Molecules F and G are for terminating conjugated systems with three single bonds in the molecule. By connecting molecules in a sequence of Molecules A, F, C, B, E, D, G, ... using MLD, EO polymer MQDs consisting of QDs having the AAAADDDD-like structure will be constructed.

7.6 ENHANCEMENT OF KERR EFFECT

To improve the Kerr effect, which is one of the third-order nonlinear optical effects, the wavefunction shape control was investigated by constructing QD structures in 1-D conjugated systems [1] according to the qualitative guidelines mentioned in Section 7.3. Figure 7.26a and b gives schematic illustrations of a single QD (SQD) structure and corresponding wavefunctions. By varying the QD length in the conduction band, ML, and that in the valence band, W, the wavefunction spread difference between the ground and excited states is adjusted.

By reducing the QD width D, the system is changed from being two-dimensional (2-D) to 1-D. The third-order nonlinear optical susceptibility γ for the light polarization along the QD length direction was calculated from Equation (7.32). For simplicity, it was assumed that the ground state wavefunction has even parity, the excited state wavefunction has odd parity, and wavefunctions are square waves. The results

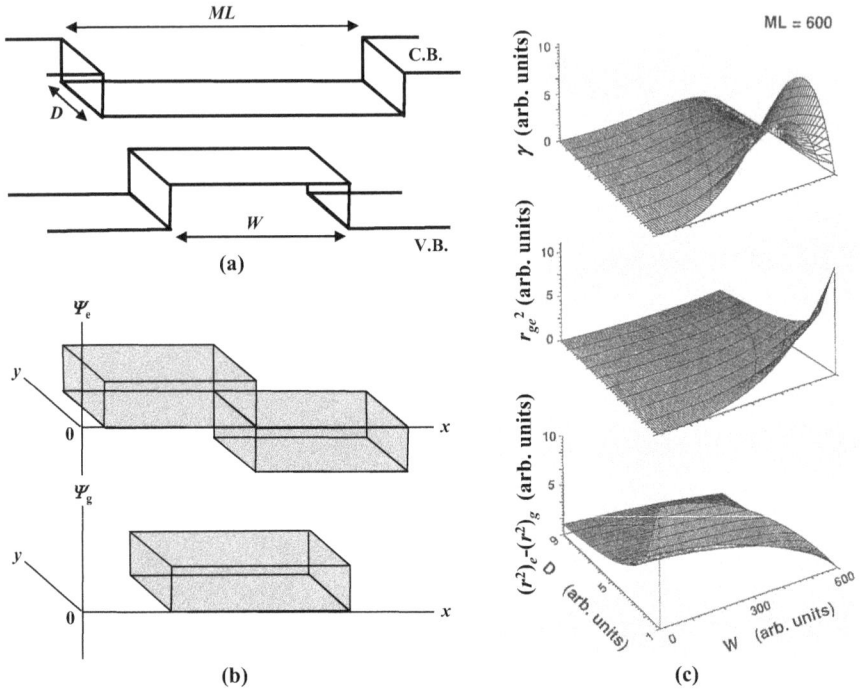

FIGURE 7.26   Enhancement of γ by controlling wavefunction shapes in the SQD structure. (a) SQD structure, (b) schematic illustrations of corresponding wavefunctions, and (c) calculated γ.

are shown in Figure 7.26c. With reduced D, both $\left(r^2\right)_e - \left(r^2\right)_g$ and $r_{ge}{}^2$ increase, then γ increases rapidly. This indicates that the 1-D system is better than the 2-D system to induce large optical nonlinearity, as mentioned in Section 7.1. This is because the wavefunction extent perpendicular to the direction of light polarization contributes little to optical nonlinearity, simply diluting the wavefunction density and reducing the optical nonlinearity.

With increasing W, i.e., with increasing the overlap and decreasing the difference in wavefunction spread between the ground and excited states, r_{ge}^2 increases while $\left(r^2\right)_e - \left(r^2\right)_g$ decreases, with γ peaking at an intermediate region in W.

Figure 7.27a and b gives schematic illustrations of a double quantum dot (DQD) structure with sub-QDs at both sides and the corresponding wavefunctions. The wavefunction spread difference is controlled by the difference in the QD depth between the conduction band and the valence band. By increasing the QD depth, the wavefunction tends to be localized on both sides. Calculated values of γ, $\left(r^2\right)_e - \left(r^2\right)_g$, and r_{ge}^2 are shown in Figure 7.27c as a function of C/D, which is the measure of the wavefunction extent in the valence band, as defined in Figure 7.27b. With increasing

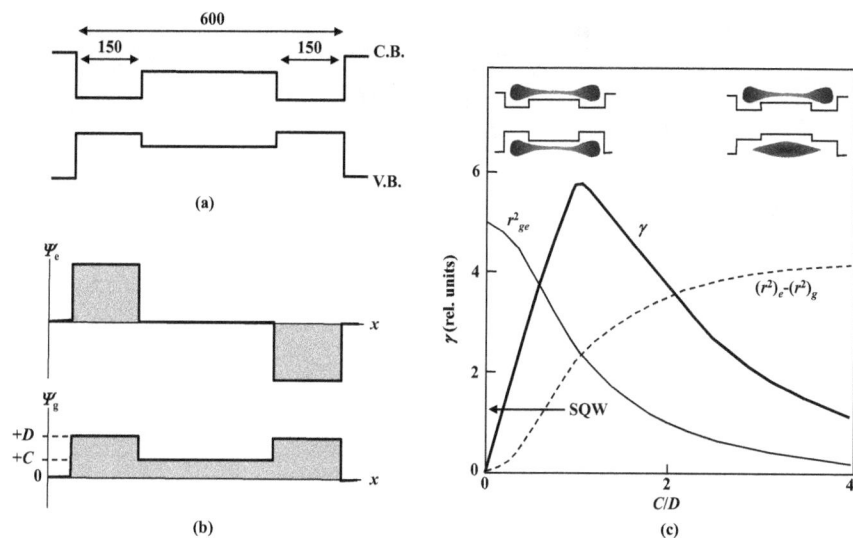

FIGURE 7.27 Enhancement of γ by controlling wavefunction shapes in the DQD structure. (a) DQD structure, (b) schematic illustrations of corresponding wavefunctions, and (c) calculated γ.

C/D, $\left(r^2 \right)_e - \left(r^2 \right)_g$ increases and r_{ge}^2 decreases. γ peaks at an intermediate region in C/D of about one when the balance between the wavefunction overlap and the wavefunction spread difference is optimum. In this case, γ is improved about five times over that for an SQD structure. It was also found that further confinement of electrons in three or more sub-QDs would make it possible to enhance γ more than ten times.

A DQD structure with sub-QDs of 1.4 nm long is shown in Figure 7.28 together with an SQD structure. The sub-QDs are formed in a PDA wire with 38 carbon sites by donor/acceptor substitution. The energy gap in the sub-QD regions is about 1.6 eV, about one half of that in the barrier region. Molecular orbitals reveal that electrons are spread throughout the conjugated wire in the SQD. In the DQD, electrons tend to be confined in the sub-QDs located on both sides of the wire. Here, note that, although electrons are localized on both sides in all $E1$, $E2$, and $E3$ unoccupied orbitals, they are in the barrier region in the $G3$ occupied orbital. This suggests that the wavefunction tends to be delocalized throughout the wire more in the valence band than in the conduction band, reproducing the condition drawn in Figure 7.27b.

γ of DQD and SQD were calculated by the MO method using Equation (7.35) at a detuning energy of 0.2 eV from the first excited state, which induces large resonant enhancement in γ. It was found that γ in the DQD is more than twice that in the SQD. This confirms that it will be possible to improve the third-order optical nonlinearity by adjusting the QD structures.

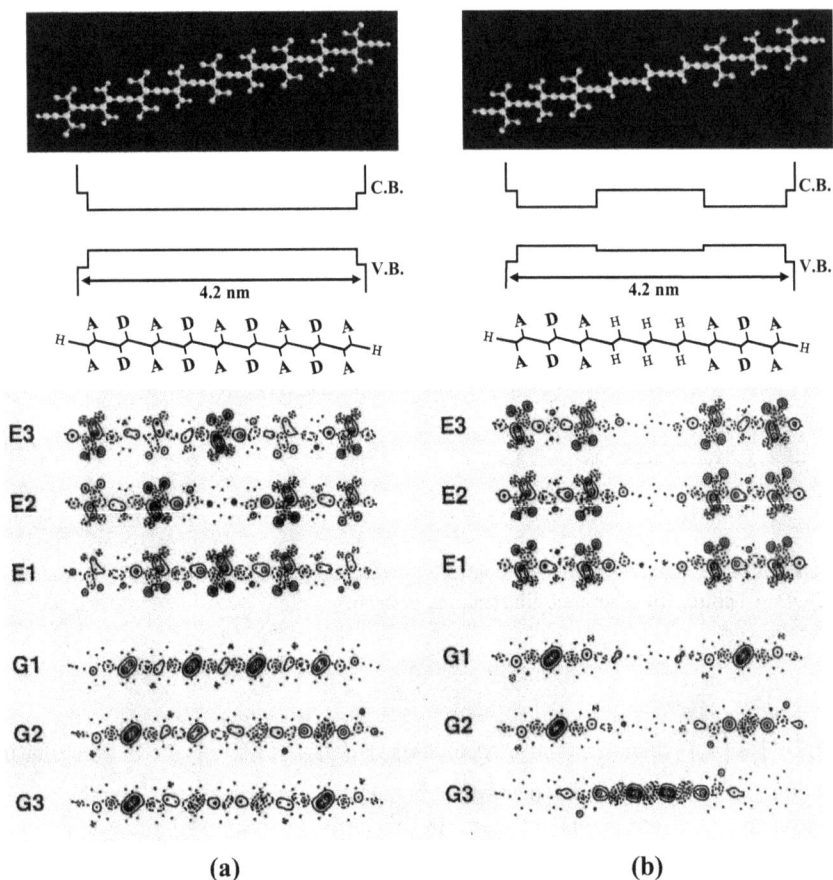

FIGURE 7.28   Structures and corresponding molecular orbitals for (a) SQD and (b) DQD.

REFERENCES

1. T. Yoshimura, "Design of organic nonlinear optical materials for electro-optic and all-optical devices by computer simulation," *Fujitsu Sci. Tech. J.* **27**, 115–131 (1991).
2. T. Yoshimura, *Molecular Nano Systems: Applications to Optoelectronic Computers and Solar Energy Conversion*, Corona Publishing Co., Ltd., Tokyo (2007) [in Japanese].
3. J. Ward, "Calculation of nonlinear optical susceptibilities using diagrammatic perturbation theory," *Rev. Mod. Phys.* **37**, 1–18 (1965).
4. S. J. Lalama and A. F. Garito, "Origin of the non-linear second-order optical susceptibilities of organic systems," *Phys. Rev. A* **20**, 1179–1194 (1979).
5. T. Yoshimura, "Enhancing second-order nonlinear optical properties by controlling the wave function in one-dimensional conjugated molecules," *Phys. Rev. B* **40**, 6292–6298 (1989).
6. T. Yoshimura, "Theoretically predicted influence of donors and acceptors on quadratic hyperpolarizabilities in conjugated long-chain molecules," *Appl. Phys. Lett.* **55**, 534–536 (1989).

7. A. F. Garito, C. C. Teng, K. W. Wong, and O. Zammani-Khamiri, "Molecular optics: Nonlinear optical processes in organic and polymer crystals," *Mol. Cryst. Liq. Cryst.* **106**, 219–258 (1984).
8. J. O. Morley, "Theoretical study of the electronic structure and hyperpolarizabilities of donor-acceptor comulenes and a comparison with the corresponding polyenes and polyynes," *J. Phys. Chem.* **99**, 10166–10174 (1995).
9. T. Yoshimura, "Estimation of enhancement in third-order nonlinear susceptibility induced by a sharpening absorption band" *Opt. Commun.* **70**, 535–537 (1989).
10. A. Matsuura and T. Hayano, "Theoretical prediction of the donor/acceptor site-dependence on first hyperpolarizabilities in conjugated systems containing azomethine bonds," *Mat. Res. Soc. Symp. Proc.* **291**, 503–508 (1993).

8 Organic EO Materials

Electro-optic (EO) devices like light modulators, optical switches, and tunable wavelength filters are the key in optical interconnects and optical communication systems. Lithium niobate (LiNbO$_3$ (LN)) is the most popular commercially-available EO material. However, its optical nonlinearity is insufficient to build *intra*-box optical interconnects and future ultra-high-speed optical communication systems. Organic materials with π-conjugated systems have attracted interest as high-performance EO materials. Their low dielectric constant characteristics are favorable for EO devices, enabling high-speed and low-power operation. As predicted in Chapter 7, the organic multiple quantum dot (MQD) fabricated by molecular layer deposition (MLD) is expected to exhibit large optical nonlinearity and to be the final goal of EO materials. The EO organic MQD, however, is a kind of far-reaching material because it has not been demonstrated experimentally yet. In the present chapter, after various EO materials are compared, some experimental work on already-demonstrated organic EO materials like molecular crystals and poled polymers are presented. Optical switches made of the poled polymers and optical waveguides fabricated by the selective growth are also described.

8.1 COMPARISON OF EO MATERIALS

Typical electro-optic (EO) coefficients r for the Pockels effect and refractive indexes n_0 of organic and inorganic EO materials are compared in Figure 8.1. EO coefficient R for the Kerr effect and n_0 of lead lanthanum zirconate titanate (PLZT) [1] are also included. These values are some of the examples reported previously, and they vary to some extent depending on the material samples and the probe light wavelengths.

The organic EO materials are classified into poled polymers [2], molecular crystals [3,4], and polymer MQDs [5–8]. The most promising EO material seems to be the polymer MQD, which is expected to exhibit a large r of ~1000 pm/V as predicted by simulations based on the molecular orbital (MO) method in Chapter 7. The EO polymer MQD, however, is a kind of far-reaching material because it has not been fabricated experimentally yet. On the other hand, the molecular crystal and the poled polymer are currently available. For the molecular crystal, a thin-film crystal of styrylpyridinium cyanine dye (SPCD) was found to exhibit r of 430 pm/V [9], which is about one order of magnitude larger than r of LN and III-V semiconductor quantum dots (QDs) [10,11]. This suggests the prospect of organic materials for EO devices. For the poled polymer, although r is usually smaller than that of the molecular crystal, it is easily formed by spin-coating into thin films with high optical quality.

In the following sections, details of EO molecular crystals and poled polymers are described as well as their applications to optical waveguides and optical switches.

Inorganic EO Materials

Pockels Effect			Kerr Effect	
	LN	**III-V QD**		**PLZT**
n_0	2.2	3.5	n_0	2.48
r (pm/V)	31	26 [10]	R (m²/V²)	5.7×10^{-16} [1]

Organic EO Materials

Pockels Effect			
	Poled Polymer	**Molecular Crystal**	**Polymer MQD**
n_0	1.55	1.55 *	~2 **
r (pm/V)	3 ~ 25	430 * [9]	~1000 *** [6]

* Measured at 633 nm in wavelength
** Assumed Value   *** Molecular Orbital Calculation

FIGURE 8.1   EO coefficients and refractive indexes of organic and inorganic EO materials.

8.2   MOLECULAR CRYSTALS

8.2.1   AC MODULATION METHOD

Figure 8.2a shows an experimental setup for the ac modulation method, which characterizes the Pockels effect of organic thin films [9]. To apply an electric field to a thin film, a glass substrate, on which slit-type electrodes are formed, is pressed against the thin film. This method makes it possible to form electrodes on a small organic thin-film flake, say, with μm-order sizes. AC voltage is applied between the electrodes. Linearly-polarized light from a He-Ne laser passes through the thin film and the electrode gap, and the light transmitted through an analyzer is detected by a photodetector.

In Figure 8.2b, polarization states at positions (I), (II), and (III) indicated in Figure 8.2a are schematically illustrated. The z and y components of the electric field of the light (E_z and E_y) at positions (I) and (II), and the component along the analyzed polarization direction (E_θ) at position (III) can be written as follows:

Position (I):

$$E_z = E_0 \cos\phi \cos\omega t$$

$$E_y = E_0 \sin\phi \cos\omega t$$

Position (II):

$$E_z = E_0 \cos\phi \cos\omega t$$

$$E_y = E_0 \sin\phi \cos(\omega t - \delta)$$

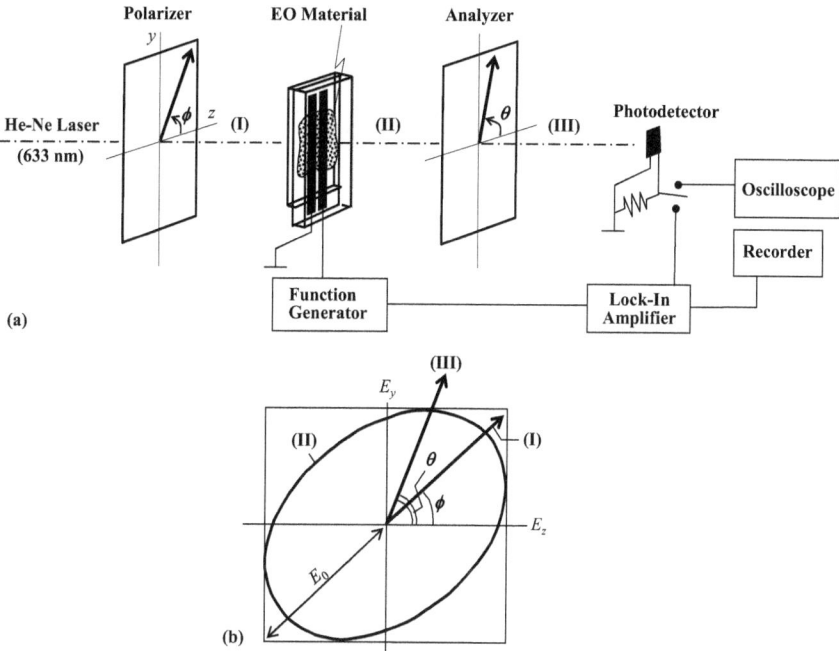

FIGURE 8.2 (a) Experimental setup for the ac modulation method. (b) Polarization states corresponding to positions (I), (II), and (III) indicated in (a). (Reprinted with permission from Yoshimura [9].)

Position (III):

$$E_\theta = E_{\theta 0}(\delta)\cos(\omega t + \alpha)$$

$$E_{\theta 0}(\delta) = E_0\left[(\cos\phi\cos\theta + \sin\phi\sin\theta\cos\delta)^2 + (\sin\phi\sin\theta\sin\delta)^2\right]^{1/2},$$ (8.1)

where ϕ is the angle between the polarization direction of the incident light and the z-axis, θ is the angle between the analyzed polarization direction and the z-axis, and δ is the phase retardation arising from birefringence of the thin film. E_0 and ω are, respectively, amplitude and angular frequency of the incident light. Then, the photodetector output V_{out} can be written as

$$V_{out} = \text{const}[E_{\theta 0}]^2.$$

δ varies depending on the voltage V_A applied between the slit-type electrodes since the thin film has an EO effect. The detector output at $V_A = 0$ is given by

$$V_{dc} = \text{const}[E_{\theta 0}(\delta_0)]^2,$$ (8.2)

where δ_0 is the phase retardation at $V_A = 0$. When V_A is increased from 0 to V, the phase retardation changes from δ_0 to $\delta_0 + \Delta\delta$. $\Delta\delta$ is the EO phase retardation induced by $V_A = V$. Then, the detector output change V_{ac} induced by applying $V_A = V$ is given as follows:

$$V_{ac} = const \left\{ \left[E_{\theta 0}(\delta_0) \right]^2 - \left[E_{\theta 0}(\delta_0 + \Delta\delta) \right]^2 \right\} \tag{8.3}$$

From Equations (8.2) and (8.3),

$$\frac{V_{ac}}{V_{dc}} = \left\{ \left[E_{\theta 0}(\delta_0) \right]^2 - \left[E_{\theta 0}(\delta_0 + \Delta\delta) \right]^2 \right\} / \left[E_{\theta 0}(\delta_0) \right]^2 \tag{8.4}$$

is obtained. Substituting Equation (8.1) into Equation (8.4), an expression for V_{ac}/V_{dc} related to the four parameters, that is, ϕ, θ, δ_0, and $\Delta\delta$ is derived. Here ϕ and θ are known, and δ_0 is estimated from the angular dependence of V_{dc}. Thus, by measuring V_{ac} experimentally in the setup illustrated in Figure 8.2a, the EO phase retardation $\Delta\delta$ is determined from Equation (8.4).

Once $\Delta\delta$ is obtained, the figure of merit (FOM) of the EO phase retardation, F, is estimated according to the following formula:

$$F = (\lambda_0 \Delta\delta D) / (2\pi V_A L) \tag{8.5}$$

where λ_0 is wavelength of the light in vacuum, D is the gap of the slit-type electrodes, and L is the film thickness. The ac modulation method enables high-sensitivity measurement of EO effects in EO thin films since it uses the lock-in technique.

8.2.2 POCKELS EFFECT IN STYRYLPYRIDINIUM CYANINE DYE (SPCD)

Lipscomb et al. studied the Pockels effect in a 2-methyl-4-nitroaniline (MNA) single crystal and estimated the EO coefficient r_{11} to be 67 pm/V [4], which is about two times larger than r_{33} in $LiNbO_3$. SPCD was found to show extremely large second harmonic generation (SHG) activity, which is about one order of magnitude larger than that of MNA [12,13]. This suggests that SPCD exhibits large Pockels effect.

An SPCD thin-film crystal was grown as follows. Two quartz substrates ($55 \times 55 \times 1$ mm^3) were paired in a methanol solution of SPCD (Japanese Research Institute for Photosensitizing Dyes Co. Ltd.). The solution was inserted between the two substrates by the capillary effect. After several days, the methanol evaporated naturally, and SPCD thin-film flakes were grown between the substrates. The two substrates were separated to leave the thin films on one of the substrates. The film size typically ranged from 1 to 10-mm long, <1-mm wide, and 3 to 10-μm thick. Microscopic observation of the SPCD thin films between crossed polarizers revealed that they exhibit birefringence. When the thin films were rotated on a stage, some of them showed uniform extinction, indicating that they are single crystals [9].

In Figure 8.3, schematic appearance and absorption spectra of an SPCD thin-film crystal are shown. Two optical axes are denoted by z and y. Light absorption of the crystal increases rapidly in the photon energy region above 2 eV, and the crystal

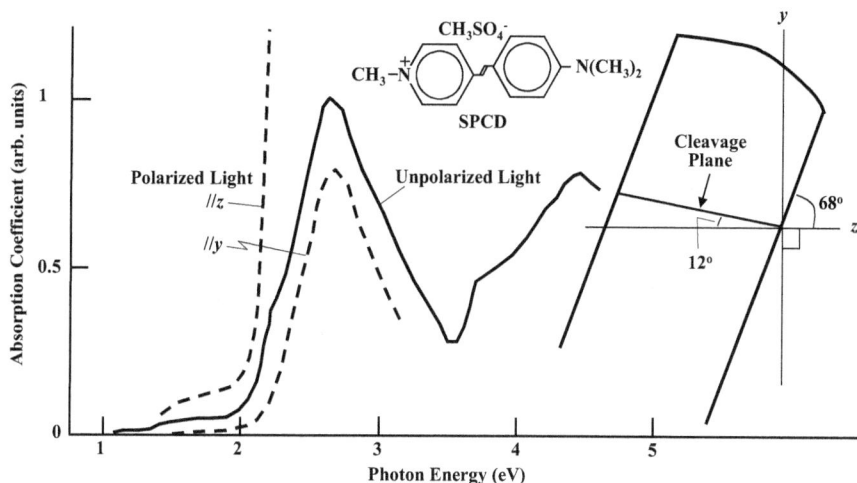

FIGURE 8.3 Schematic appearance and absorption spectra of an SPCD thin-film crystal for polarized light and unpolarized light. (Reprinted with permission from Yoshimura [9].)

shows strong dichroism. The absorption is greater for polarization along the z-axis than along the y-axis. This suggests that the molecular dipole moments of SPCD tend to align along the z-axis in the thin-film crystalline state [9].

Photoluminescence (PL) spectra of an SPCD thin-film crystal, SPCD powder used as the starting material, and water solution of SPCD were measured with an excitation light from He-Cd laser (325 nm) at room temperature. The spectrum of the SPCD powder appeared in the low-energy side (peak position: 1.5 eV) compared to that of the SPCD thin-film crystal (peak position: 2.05 eV), and the PL intensity for the powder was about 1/1000 of that for the crystal. This suggests that the electronic state in the crystal is largely different from the state in the powder. The spectrum of the SPCD crystal was relatively close to that of the water solution. This implies that the electronic state of SPCD molecules in the crystal is like the state in water, that is, the interaction between SPCD molecules is weak.

The refractive index of the SPCD thin-film crystal was estimated to be $n_z = 1.55$ and $n_y = 1.31$ by the prism coupler method [14] and a differential interferometer. The result of $n_z > n_y$ is consistent with the strong dichroism of the crystal and indicates again that the molecular dipole moments align along the z-axis.

The Pockels effect of SPCD thin-film crystals was evaluated using the ac modulation method. To apply an electric field to the crystal, slit-type NiCr electrodes with a 5-μm gap were used. To estimate $\Delta\delta$ and F, V_{dc} and V_{ac} were measured. From the angular dependence of V_{dc}, δ_0 was determined. The result is shown in Figure 8.4a for a 5-μm-thick SPCD thin-film crystal. Figure 8.4b shows typical light modulation induced by the SPCD thin-film crystal, to which 8-kHz sine-wave voltage is applied. V_{ac} is superposed on V_{dc} [9].

V_{ac} vs. θ characteristics of an SPCD thin-film crystal for an electric field application along the z-direction are shown in Figure 8.5 with a solid line. V_{ac} becomes

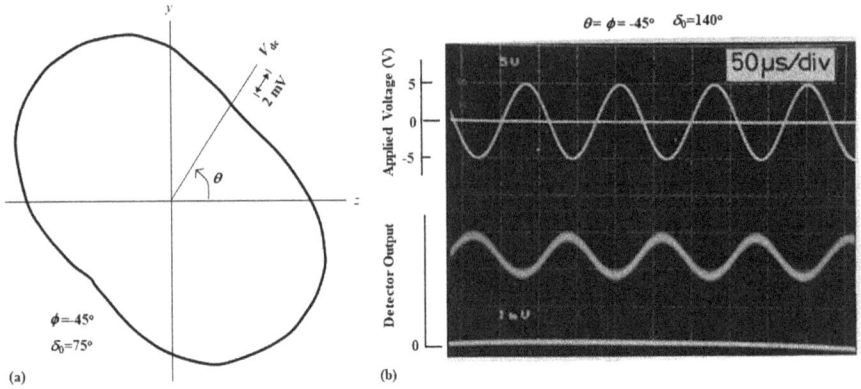

FIGURE 8.4 (a) V_{dc} as a function of θ for a 5-μm-thick SPCD thin-film crystal. (b) Light modulation induced by the SPCD thin-film crystal, to which 8-kHz sine-wave voltage is applied. (Reprinted with permission from Yoshimura [9].)

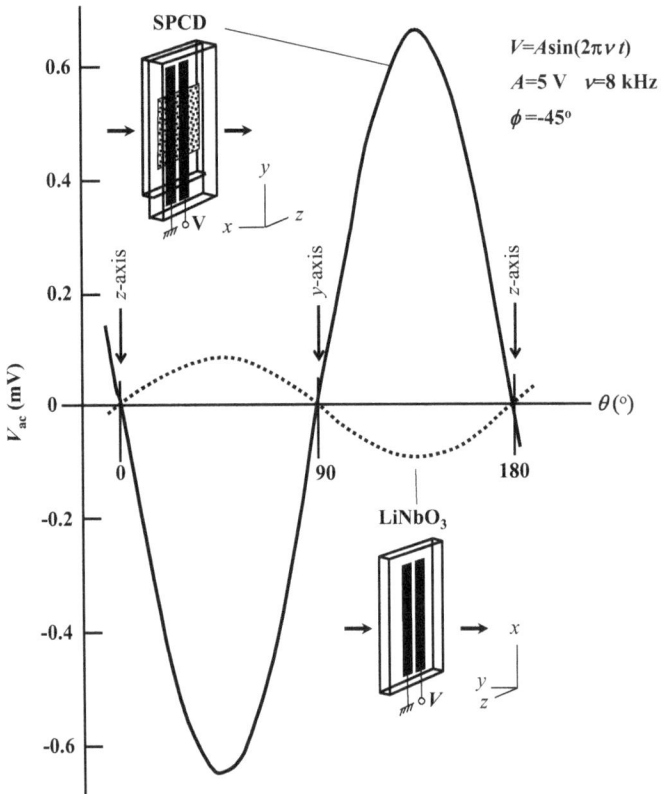

FIGURE 8.5 V_{ac} vs. θ characteristics of an SPCD thin-film crystal (solid line) and an LN crystal (dotted line) for electric field application along the z-direction. (Reprinted with permission from Yoshimura [9].)

zero at $\theta = 0°$, 90°, and 180°, which correspond to the directions of the optical axes (z-axis and y-axis), and has its maximum or minimum at $\theta = 45°$ and 135°. This behavior can be explained by Equations (8.3) and (8.1). For comparison, V_{ac} vs. θ characteristics of an LN crystal are shown with a dotted line in the figure. The SPCD thin-film crystal exhibits V_{ac} about five times larger than the LN crystal. When the electric field was applied along the y-axis of the SPCD thin-film crystal, the maximum value of V_{ac} was reduced to 1/3 of that for the electric field along the z-axis. This means that in the SPCD crystal, the EO effect is larger for the direction along the molecular dipole alignment than for the direction perpendicular to the dipole alignment.

The EO phase retardation $\Delta\delta$ is obtained by substituting the experimental values, that is, $\phi = \theta = -45°, \delta_0 = 75°, V_{dc} = 14$ mV, and $V_{ac} = 0.65$ mV into Equations (8.1) and (8.4). The result is shown in Figure 8.6. The SPCD thin-film crystal shows $\Delta\delta = 7°$ at an applied voltage of 20 V, which corresponds to the half-wave voltage V_π of 500 V, while the LN crystal shows $\Delta\delta = 1.5°$ at 20 V, corresponding to $V_\pi = 2400$ V. Using these results, FOMs of the EO phase retardation were calculated from Equation (8.5). F in the $LiNbO_3$ crystal was 120 pm/V, which is close to the value of 95 pm/V reported by Lipscomb et al. [4]. This confirms the accuracy of the ac modulation method. For the SPCD thin-film crystal, large F of 650 pm/V was obtained. The value is about 5.4 times as large as that of the LN crystal and 2.4 times as large as 270 pm/V of the MNA crystal.

It should be noted that the EO effect in the SPCD thin-film crystals was detected only by applying ac voltage. The reason might be that in SPCD, the EO effect decays rapidly under a constant voltage application due to some mobile ions that cancel the applied voltage between the electrodes.

Since the SPCD crystal has an orthorhombic structure [12], the EO coefficient matrix is written as follows:

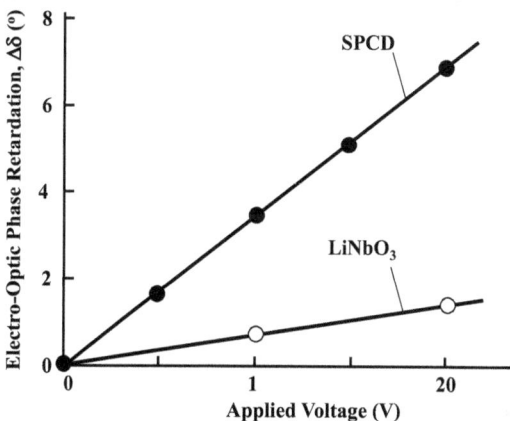

FIGURE 8.6 Dependence of EO phase retardation on applied voltage for the SPCD thin-film crystal (filled circles) and for the $LiNbO_3$ crystal (open circles) with electric field applied along the z-axis. (Reprinted with permission from Yoshimura [9].)

$$
\begin{array}{ccc}
0 & 0 & r_{13} \\
0 & 0 & r_{23} \\
0 & 0 & r_{33} \\
0 & r_{42} & 0 \\
r_{51} & 0 & 0 \\
0 & 0 & 0
\end{array}
$$

According to this matrix, rotation of optical axes does not occur only when the electric field is applied along the principal z-axis of the index ellipsoid. Because no rotation of optical axes was observed when an electric field was applied along the optical z-axis, the optical z-axis and the principal z-axis were found to coincide in the SPCD thin-film crystal [9].

EO coefficient r_{33} in the SPCD crystal and F are related as follows:

$$F = \left|\left(n_z{}^3 r_c\right)/2\right|, \tag{8.6}$$

$$r_c = r_{33} - \left(\frac{n_y}{n_z}\right)^3 r_{23} \tag{8.7}$$

By substituting $F = 6.5 \times 10^{-10}$ m/V and $n_z = 1.55$ into Equation (8.6), r_c is found to be 4.3×10^{-10} m/V. Usually, the coefficient corresponding to the direction along the molecular dipole alignment tends to be the largest [4]. Then, by neglecting r_{23} in Equation (8.7), $r_{33} \approx r_c = 430$ pm/V is obtained. This value is extremely large compared with $r_{33} = 32$ pm/V of LN and $r_{11} = 67$ pm/V of MNA [4].

Finally, the relationship between the EO coefficient r_{33} and SHG efficiency of the SPCD crystal is discussed. In the organic nonlinear optical materials, it has been known that the EO coefficient r and SHG efficiency η are related to the second-harmonic coefficient d and the refractive index n as follows [3,4]:

$$r \propto d/n^4$$

$$\eta \propto d^2/n^3$$

Here, dispersion is ignored for simplicity. From these relationships,

$$r \propto \eta^{0.5}/n^{2.5}$$

is derived, and then we have

$$r_{33}(\text{SPCD})/r_{11}(\text{MNA}) = \left[\eta(\text{SPCD})/\eta(\text{MNA})\right]^{0.5} \times \left[n(\text{MNA})/n(\text{SPCD})\right]^{2.5}. \tag{8.8}$$

Using this expression, the relative ratio of r can be calculated from the SHG efficiency. By substituting $n(\text{SPCD}) = 1.55$ and $n(\text{MNA}) = 2$ at 633 nm [3], and $\eta(\text{SPCD})/\eta(\text{MNA}) = 10$ [13] into Equation (8.8), we have $r_{33}(\text{SPCD})/r_{11}(\text{MNA}) = 6$ as an expected value based on the SHG measurement. The ratio of 6 agrees well with the measured value of $r_{33}(\text{SPCD})/r_{11}(\text{MNA}) = 4.3 \times 10^{-10}/0.67 \times 10^{-10} = 6.4$. This coincidence again supports the accuracy of the ac modulation method.

8.2.3 POCKELS EFFECT IN METHYL-4-NITROANILINE (MNA)

MNA is the most popular molecular crystal for nonlinear optics. The author's research group investigated the effect of growth temperature on the Pockels effect in MNA [15]. As Figure 8.7a shows, paired quartz substrates and MNA powder are set in a test tube and heated. The temperature of MNA is controlled by a computer, as depicted in Figure 8.7b. Near the melting point T_M of MNA, the temperature change rate dT/dt decreases due to an endothermic reaction, and the MNA powder begins to melt. The temperature T_0, at which dT/dt starts to decrease, is detected, and feedback control is employed to maintain the temperature at $T_P = T_0 + \Delta T$. The melt of MNA is introduced into the gap space between the substrates by capillary action, and MNA thin-film crystals grow by slow cooling. The thin films are typically 2–3 mm in length and width, and 5–10 µm in thickness.

In Figure 8.8a, T_P dependence of FOM of the EO phase retardation, measured by the ac modulation method, is shown for MNA thin-film crystals. As T_P decreases toward T_M, F increases, and at $T_P \approx T_M$, the maximum F is obtained. By annealing at 118°C for 120 min, F further increases. One of the following reasons seems to be responsible for the variation in F: (1) a change in the molecular structure, (2) a change in the crystal structure, and (3) a change in the interaction between neighboring

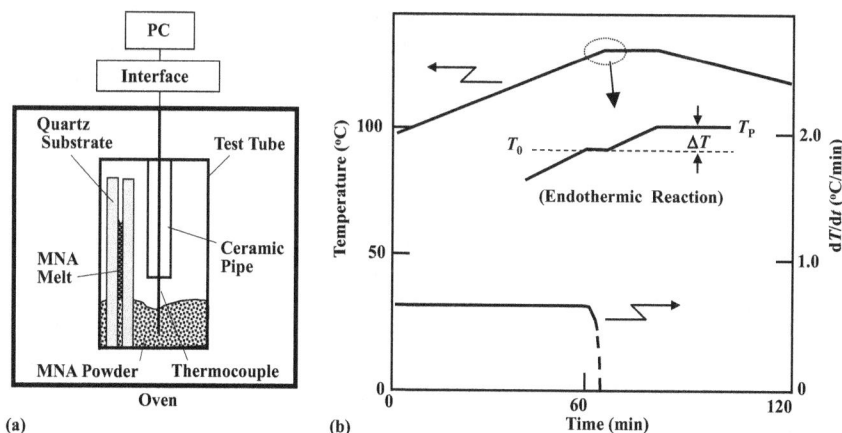

FIGURE 8.7 (a) Growth apparatus for MNA thin-film crystals with automatic temperature control. (b)Temperature-time profile for MNA thin-film crystal growth. (Reprinted with permission from Kubota and Yoshimura [15].)

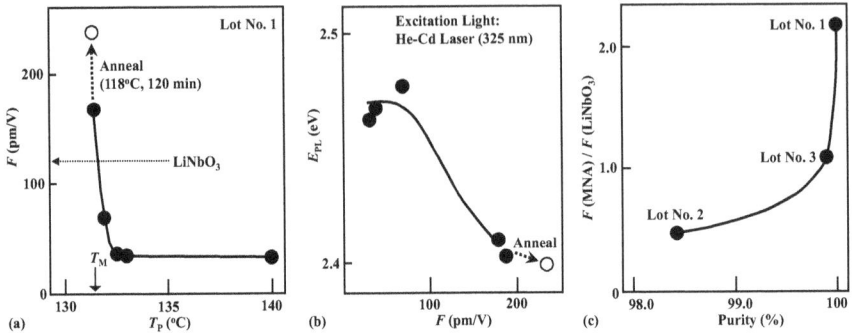

FIGURE 8.8 (a) T_P dependence of F and (b) relationship between F and E_{PL} for MNA thin-film crystals. (c) Dependence of F on purity of the MNA powder used for the crystal growth. (Reprinted with permission from Kubota and Yoshimura [15].)

molecules. Differential thermal analysis and x-ray diffraction analysis revealed that no obvious molecular degeneration and crystal structure change occur for various T_P, and that the change in F might be caused by the reason (3).

It is found from Figure 8.8b that F increases as the peak position of PL spectra, E_{PL}, shifts to the lower-energy side, implying that the change in F is attributed to a change in electronic states of MNA molecules. According to Equation (7.33), the Pockels effect is affected by the energy gap and the dipole moments. Since absorption spectra of the MNA thin-film crystals exhibited little T_P dependence of the energy gap, it is plausible that the change in F is attributed to a change in dipole moments. Figure 8.8c shows the dependence of F on purity of the MNA powder used for the crystal growth. Here, F is normalized with respect to F of the LN crystal. Purity was measured by differential scanning calorimetry (DSC). F decreases with decreasing the purity. From these results, the T_P dependence of F can be explained as follows. During crystal preparation at high temperature, a small number of impurity molecules are produced and they rotate the orientation of a large number of MNA molecules in the crystal. As a result, the interaction between donor and acceptor groups of neighboring molecules changes, decreasing the total dipole moments, and consequently, decreasing F [16].

On the left-hand side in Figure 8.9, electron distributions in MNA molecules are presented for three arrangements: an isolated molecule, three molecules with rotating by $\theta = 0°$, and three molecules with rotating by $\theta = 35°$. The molecular orbitals for each arrangement were calculated by the Austin Model 1 (AM1) method with ANCHOR developed by Fujitsu. Using Equation (7.33), β of the central molecule was calculated for each arrangement, and the results were shown in Figure 8.9 with bars. When θ increases from $0°$ to $35°$, β decreases. With further increases in θ to $65°$, β increases. These results suggest that the interaction between molecules in crystals greatly affects the second-order nonlinearity, and confirm the possibility that the T_P dependence of F in the MNA crystal is attributed to the change in the interaction between MNA molecules. Note that, in the configuration of $\theta = 0°$, F is larger than in the isolated molecule. This indicates that enhancement of β occurs by interaction between neighboring molecules.

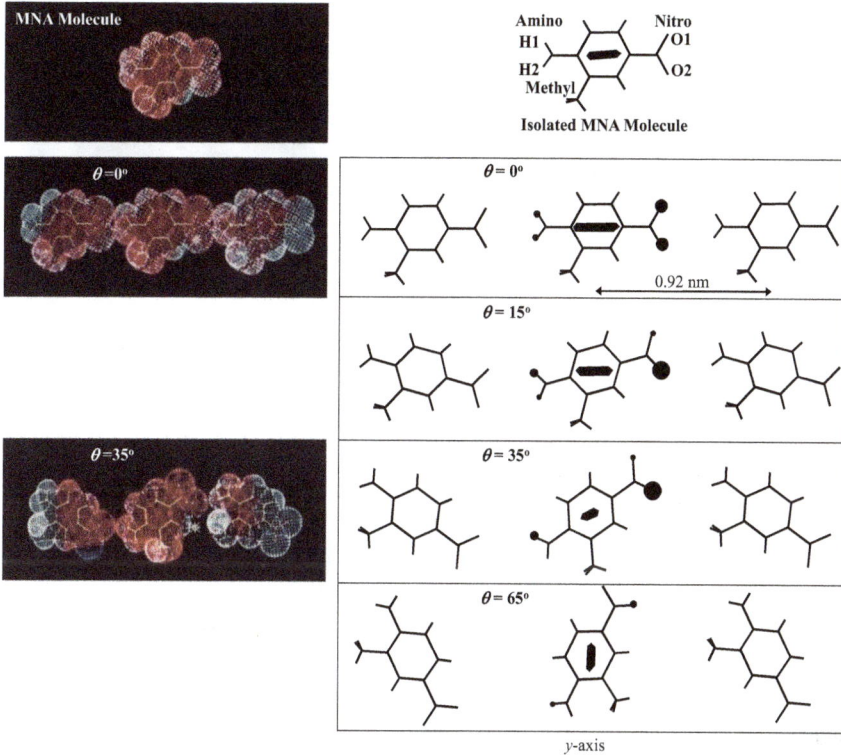

FIGURE 8.9 Influence of arrangements of MNA molecules on electron distributions in MNA molecules, and predicted influence of the MNA-molecule arrangements on β.

8.3 POLED POLYMERS

8.3.1 EO POLED POLYMERS

The poled polymer is the most popular organic EO material [2]. EO poled polymers are suitable for EO devices due to their low dielectric constants and waveguide processability. The typical fabrication process is illustrated in Figure 8.10. After a polymer film containing nonlinear optical molecules is formed by spin-coating, poling is performed by applying an electric field to the polymer film at a poling temperature T_{poling}, above glass transition temperature T_g. The nonlinear optical molecules having electric dipoles rotate to align along the electric field direction in the film. By cooling to room temperature, the orientation of the aligned molecules is fixed to generate the second-order optical nonlinearity.

EO coefficient of poled polymers is expressed by the following formula [17,18]:

$$r = -2Nf^2 f_0^2 \beta\mu E_{poling} / 5n^4 k_B T_{poling} \qquad (8.9)$$

Here, β, μ, and N are molecular second-order nonlinear susceptibility, electric dipole moment, and concentration of nonlinear optical molecules, respectively. E_{poling} is the

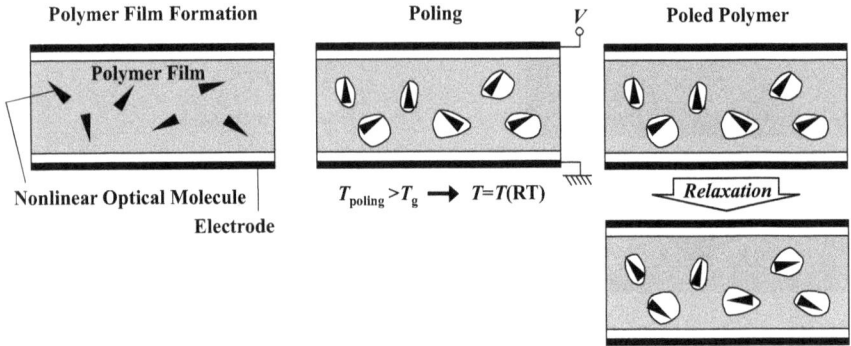

FIGURE 8.10   Fabrication process of EO poled polymers.

poling electric field and k_B the Boltzmann constant. f and f_0 are given by the following expressions using relative dielectric constant ε_s and refractive index n and n_∞ [17,18].

$$f = \left(n^2 + 2\right)/3, \quad f_0 = \varepsilon_s \left(n_\infty^2 + 2\right) / \left(n_\infty^2 + 2\varepsilon_s\right) \qquad (8.10)$$

It is found from Equation (8.9) that r increases with increasing N, β, μ, and E_{poling}, and with decreasing T_{poling}. Figure 8.11 summarizes the influence of β and T_{poling} on r [18]. With decreasing T_{poling}, thermal disturbance during poling is suppressed to improve the molecular alignment along the electric fields, resulting in an increase in r. However,

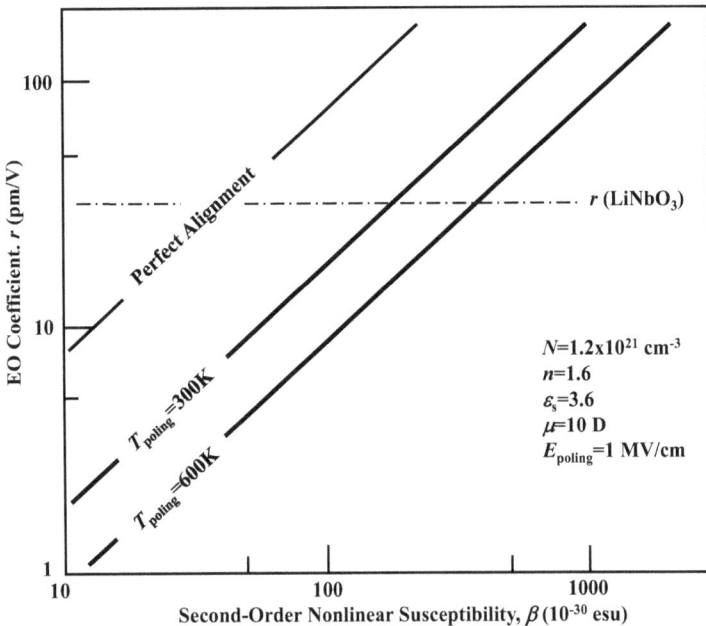

FIGURE 8.11   Influence of β and T_{poling} on r in poled polymers.

low T_{poling} implies low T_g of the polymer, which causes decay of the Pockels effect due to relaxation of the aligned molecules to a random-orientation state (see Figure 8.10). This decreases thermal stability. Thus, in EO poled polymers, to solve the trade-off relationship between r and thermal stability is the most important issue.

8.3.2 POCKELS EFFECT IN SIDE-CHAIN POLYIMIDE

In order to improve the thermal stability after poling, Sotoyama et al. developed EO side-chain polyimide with nonlinear optical molecules of azobenzene dyes attached as side chains [19]. The EO side-chain polyimide is expected to offer high EO coefficients because of the high-density distribution of nonlinear optical molecules and high thermal stability brought by the polyimide backbones.

The chemical reaction for the side-chain polyimide synthesis is shown in Figure 8.12. As shown in Figure 8.13, below 250°C, the absorption peak at 480 nm, which is attributed to the azobenzene dye, is preserved, indicating that obvious thermal decomposition does not occur. The fabrication process for the EO side-chain polyimide is depicted in Figure 8.14. By spin-coating, a polyamic acid film is formed on a glass substrate with an indium tin oxide (ITO) electrode. Next, the solvent is removed by baking at 100°C. The film thickness is a few μm. The polyamic acid film is imidized at 250°C for 2h, and an Au electrode is deposited on it. Then, the polyimide film is poled by applying an electric field of 170 MV/m between the ITO electrode and the Au electrode at 180°C for 1 h to obtain an EO side-chain polyimide.

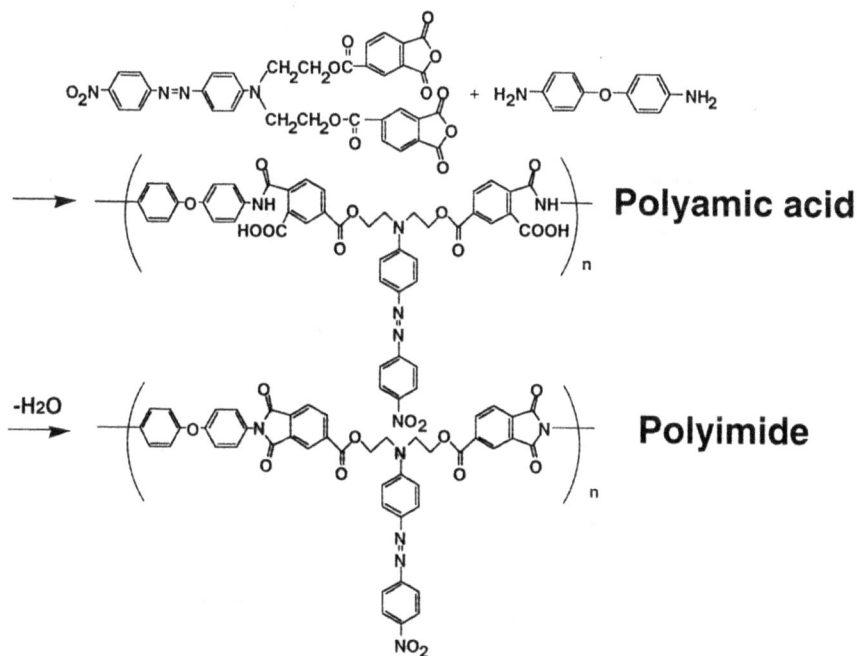

FIGURE 8.12 Chemical reaction for the side-chain polyimide synthesis. (Reprinted with permission from Sotoyama et al. [19].)

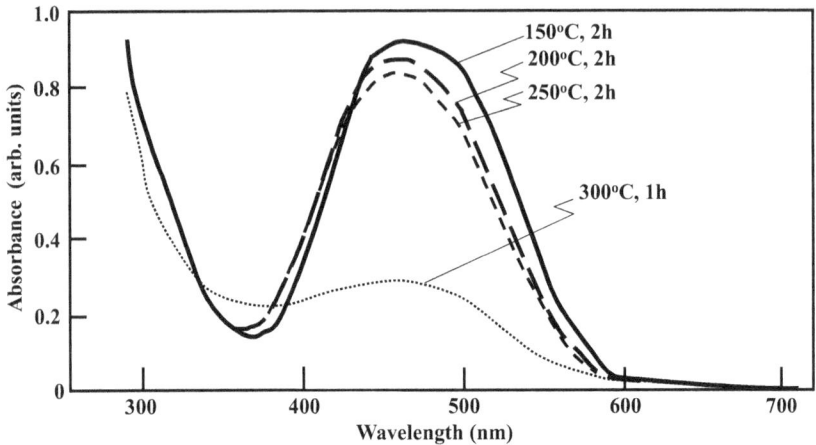

FIGURE 8.13   Absorption spectra of side-chain polyimide after baking at 150°C, 200°C, 250°C, and 300°C. (Reprinted with permission from Sotoyama et al. [19].)

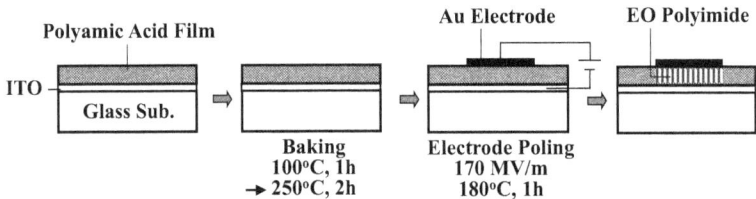

FIGURE 8.14   Fabrication process for the EO side-chain polyimide.

The EO coefficient r_{33} measured by the reflection technique [20] was 10.8 pm/V at 1.3 µm in wavelength. The thermal stability of the EO coefficient of the EO side-chain polyimide was tested by keeping the sample at 120°C. As shown in Figure 8.15, after more than 100 h, the EO coefficient of the sample is retained at nearly 90% of

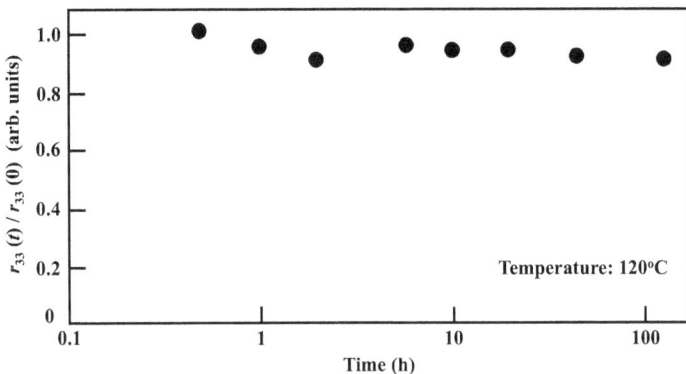

FIGURE 8.15   Relationship between heating time (at 120°C) and the EO coefficient of EO side-chain polyimide. (Reprinted with permission from Sotoyama et al. [19].)

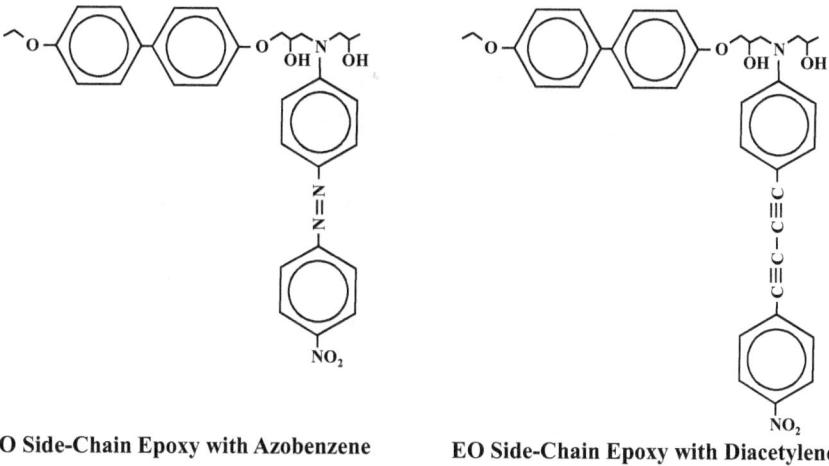

EO Side-Chain Epoxy with Azobenzene   **EO Side-Chain Epoxy with Diacetylene**

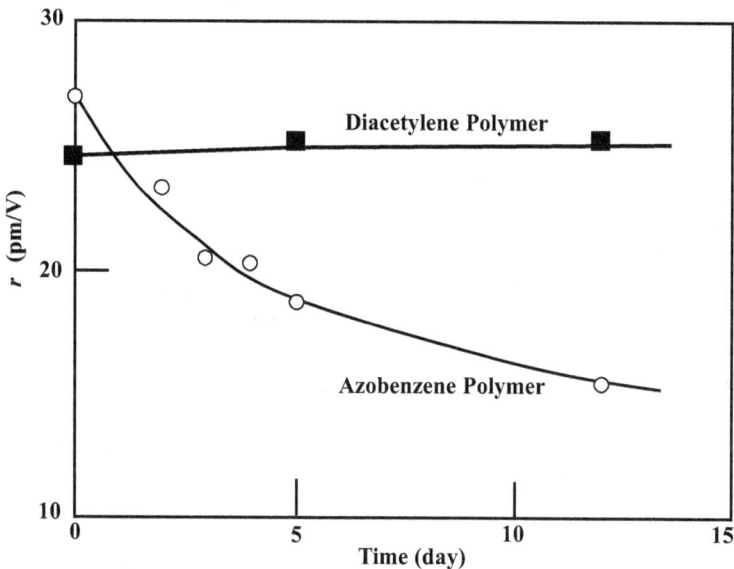

FIGURE 8.16 Influence of nonlinear optical molecule structures on decay of the EO coefficient in EO side-chain epoxy.

the initial value. The thermal stability of the EO side-chain polyimide with azobenzene is much higher than that of EO side-chain epoxy with azobenzene, whose EO coefficient decay is shown with open circles in Figure 8.16. By replacing azobenzene with diacetylene, the thermal stability of the EO polymers is improved.

8.3.3 Optical Switches Utilizing EO Poled Polymers

Sotoyama et al. demonstrated the switching operation of a directional coupler optical switch utilizing the EO side-chain polyimide [19]. Figure 8.17 depicts a photograph

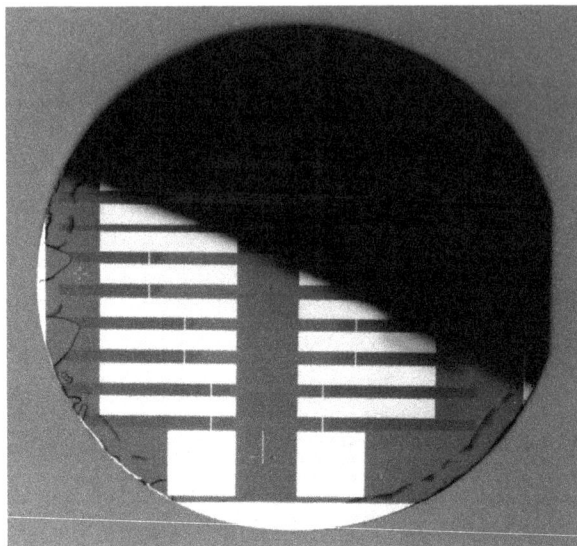

FIGURE 8.17   Photograph of optical switches fabricated on a conductive Si wafer utilizing the EO side-chain polyimide.

of the optical switches fabricated on a conductive Si wafer, which is used as the bottom electrode. As Figure 8.18 shows, the cladding part is made of the EO side-chain polyimide and the core is made of passive fluorinated polyimide. When the applied voltage V is 0 V, a light beam is guided to one of the two branches of the optical switch. When 500 V is applied, the light beam is picked up to another branch, achieving switching operation. The driving voltage is extremely high. This is attributed to the fact that only a small fraction of the light beam feels the refractive index change induced by the EO effect since the EO material does not exist in the core region, where most electric fields of the light beam are confined.

To fabricate integrated optical circuits for high-density optical wiring, small light modulators and optical switches having three-dimensional (3-D) structures are favorable. Sotoyama et al. developed the vertical directional coupler [21], which is a candidate of the 3-D optical switch. In Figure 8.19, a structure of the 3-D optical switch and an experimental setup are shown as well as the structure of p-nitroaniline-bonded epoxy polymer (PNA-epoxy) used for EO core layers. The refractive index of the PNA-epoxy layers is 1.64 before poling. Polyvinyl alcohol (PVA) with a refractive index of 1.53 is used for cladding layers. These layers are stacked on a conductive Si wafer by spin-coating. Two single-mode slab waveguides made from the EO core layers have the same effective index.

When a He-Ne laser beam (633 nm) was coupled into one of the stacked waveguides through the cleaved edge, scattered light from the top surface of the waveguide was observed as bright and dark stripes perpendicular to the guided light direction as

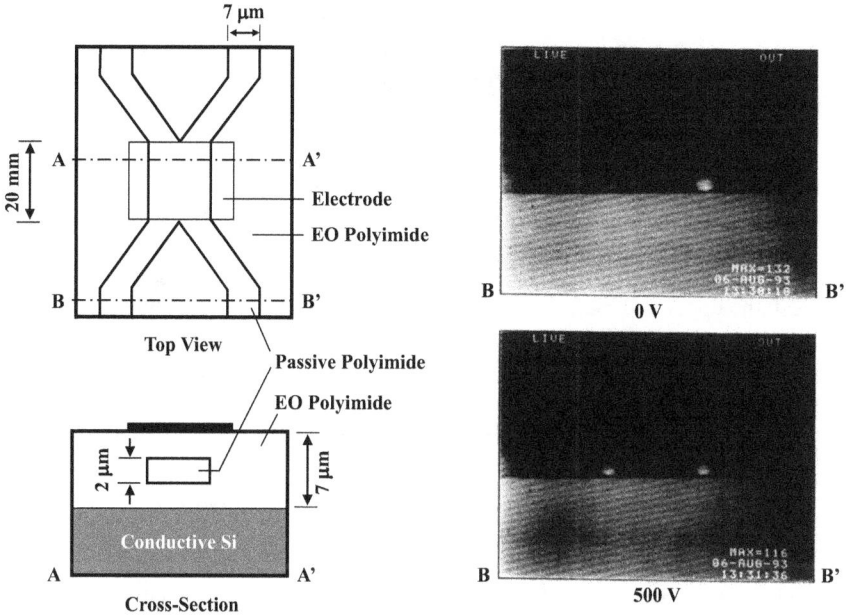

FIGURE 8.18 Schematic illustration and switching operation of an optical switch utilizing the EO side-chain polyimide.

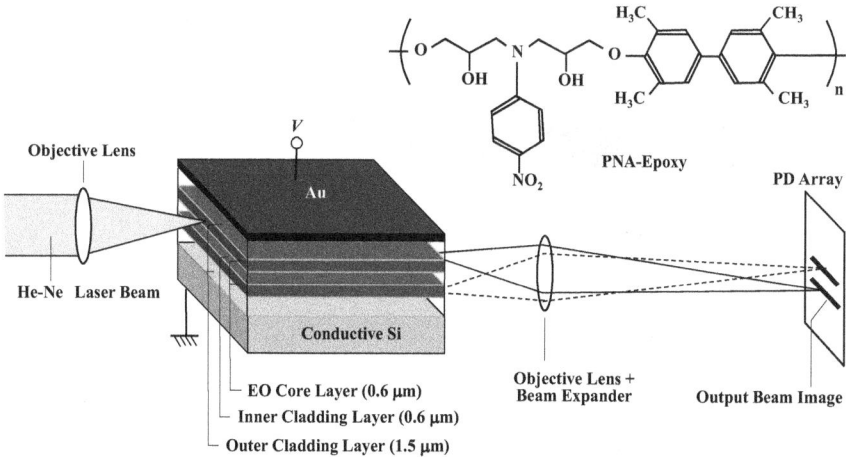

FIGURE 8.19 Structure of the 3-D optical switch, an experimental setup, and structure of paranitroaniline-bonded epoxy polymer (PNA-epoxy). (Reprinted with permission from Sotoyama et al. [21]. Copyright (1992) The Physical Society of Japan and The Japan Society of Applied Physics.)

Bright Stripe

Guided Light

FIGURE 8.20   Schematic drawing of scattered light from the top surface of the waveguide. Bright and dark stripes appear.

schematically drawn in Figure 8.20. This indicates that the guided light is transferred between the upper and the lower waveguides back and forth. The stripe interval was about 0.3 mm for the TE-mode light, being consistent with the calculated coupling length of 0.34 mm in the model shown in Figure 8.19.

After a 100-nm-thick planar Au electrode was deposited on the top of the stacked layers, the PNA-epoxy core layers were poled by applying 200 V between the substrate and the top Au electrode at 60°C for 5 minutes. After poling, the sample was cleaved for a 5-mm waveguide length, and optical switching measurement was performed in the setup shown in Figure 8.19. A He-Ne laser beam focused by a 100×objective lens was coupled to the waveguide in the TM mode. The exiting beam was corrected by a 50×objective lens and a 10×beam expander, and was detected by a photodiode (PD) array.

Figure 8.21a shows the output beam line images from the 3-D optical switch detected by the PD array. Two parallel lines spaced about 1 mm apart appear, corresponding to light beams from the upper waveguide and the lower waveguide. As Figure 8.21b shows, the intensity of these lines varies when voltage is applied between the substrate and the top electrode, indicating that the light beam switching between layered waveguides occurs. The switching voltage is about 30 V.

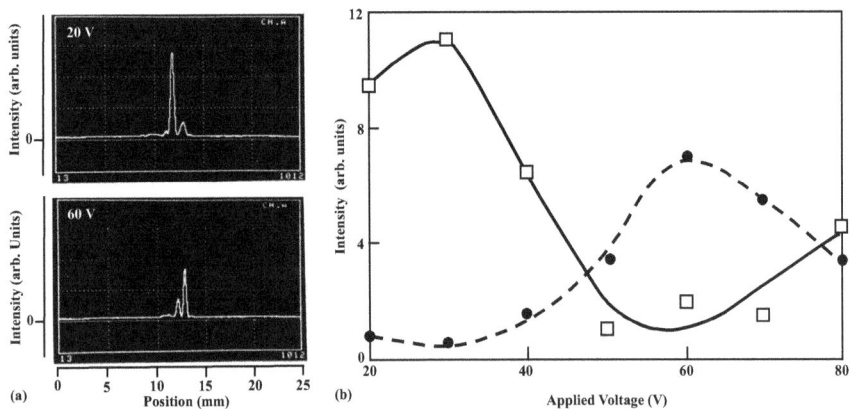

FIGURE 8.21   (a) Output beam line images from the 3-D optical switch detected by the PD array at applied voltage of 20 and 60 V. Horizontal axis represents position on the PD array. (b) Output beam line intensity as a function of applied voltage. Solid circle represents intensity of one line and rectangle the other. (Reprinted with permission from Sotoyama et al. [21]. Copyright (1992) The Physical Society of Japan and The Japan Society of Applied Physics.)

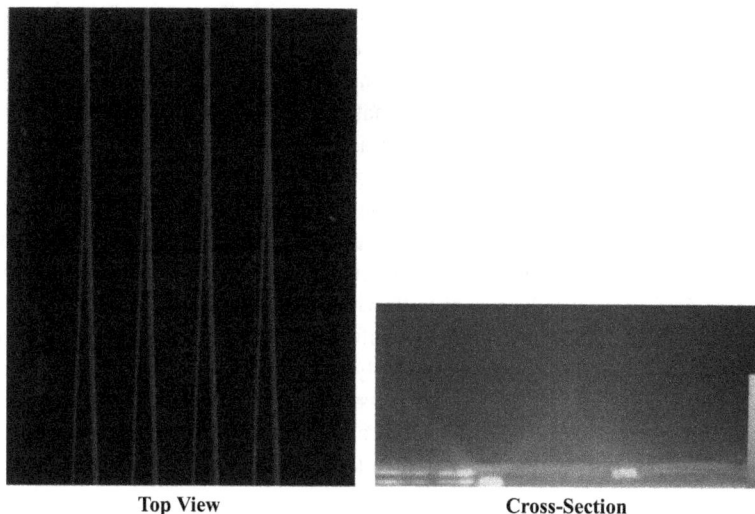

| Top View | Cross-Section |

FIGURE 8.22   An example of stacked channel waveguides.

The switching voltage of 30 V, which is too high for optical interconnect applications, is due to the small EO coefficient of the PNA-epoxy polymer, $r_{33} = 3$ pm/V. The switching voltage will be reduced by implementation of the polymer MQDs fabricated by MLD in future. An optical switch consisting of stacked channel waveguides might be constructed in principle using line-shaped electrodes of a few µm wide. Figure 8.22 shows an example of stacked channel waveguides [22].

8.4   EO WAVEGUIDES FABRICATED BY SELECTIVE GROWTH

As mentioned in Section 8.3.1, in EO poled polymers the trade-off relationship between r and thermal stability is the most serious problem. One of the ways to solve it is to apply the selective growth of polymer wires described in Chapter 4 to EO polymer fabrication. In the present section, several examples of EO waveguide formation by the selective growth are presented.

8.4.1   EO WAVEGUIDES BY ELECTRIC-FIELD-ASSISTED SELECTIVE GROWTH

Currently, r of EO poled polymers is not higher than r of LN due to insufficient alignments of nonlinear optical molecules in the polymer films. To enhance r, the degree of the molecular alignment should be improved. Conventionally, poling is done in a rubber condition at high T_{poling} above T_g after a polymer film is formed by spin-coating. In such cases, the surrounding polymer matrix prevents the nonlinear optical molecules from rotating smoothly and being aligned along the electric field. During the rotation of molecules, stress and free volume are generated in surrounding regions, causing force to relax the oriented molecules back to the initial random state (see Figure 8.10). The poling at high T_{poling} degrades the molecular alignments due to thermal disturbance. Furthermore, spin-coated films tend to contain contaminations, which reduces the electric field in the polymer films.

The electric-field-assisted selective growth described in Section 4.4 solves all these problems because it enables solvent-/contamination-free polymer film growth and smooth molecular alignment in vacuum before the nonlinear optical molecules are inserted into polymer wires, as schematically illustrated in Figure 4.16. When source molecules with sufficiently high reactivity are used, polymer films can be fabricated at low temperature below the glass transition temperature. Thus, nonlinear optical molecules could rotate freely and the disordering due to thermal disturbance could be suppressed to achieve highly-oriented molecular alignment, overcoming the problem of the trade-off relationship between r and thermal stability. By combining the electric-field-assisted selective growth with MLD, precise molecular rotation and film thickness controllability as well as conformability of the films are expected.

EO waveguides of an epoxy-amine polymer [BE/AEANP] were fabricated by Tatsuura et al. utilizing the electric-field-assisted selective growth with source molecules of tetramethyl-biphenylepoxy (BE) and 2(2-aminoethylamino)-5-nitropyridine (AEANP) [23,24]. The reaction between them is shown in Figure 8.23a. AEANP,

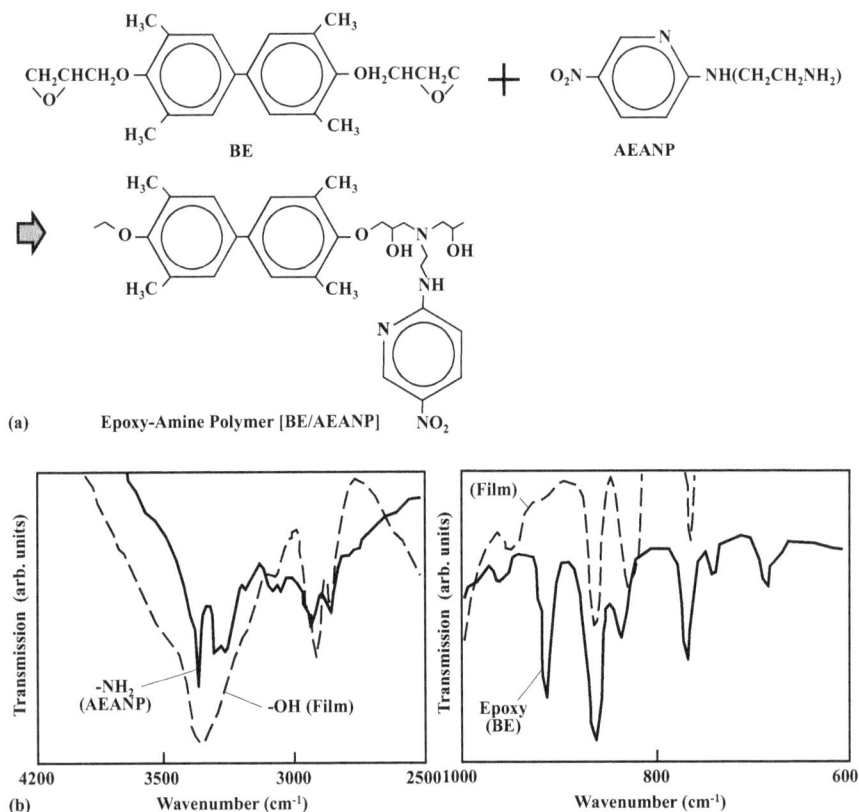

FIGURE 8.23 (a) Reaction for wire growth of epoxy-amine polymer [BE/AEANP]. (b) Infrared spectra of an epoxy-amine polymer [BE/AEANP] film and source molecules. (Reprinted with permission from Tatsuura et al. [23].)

which has a donor of -NH(CH$_2$CH$_2$NH$_2$) and an acceptor of -NO$_2$, is a nonlinear optical molecule with an electric dipole moment.

Epoxy-amine polymer films were grown with a gas pressure of 5×10^{-3} Pa (background pressure of $< 10^{-4}$ Pa) at deposition rates of 0.2–0.5 nm/s. The film grown at substrate temperature T_s of 30°C was yellow and transparent when the AEANP/BE ratio is 1/1. Figure 8.23b shows infrared spectra of the film and the source molecules. In the film spectrum, the band of the epoxy ring (911 cm^{-1}) becomes very small, that of $-$NH$_2$ (3367 cm^{-1}) disappears, and that of $-$OH grows compared to the spectra of AEANP and BE molecules, indicating that polymerization progressed and a polymer film was grown.

To fabricate channel optical waveguides, a thermally oxidized Si wafer with a 1-μm-thick SiO$_2$ layer for cladding was prepared. Al slit-type electrodes with 10- or 20-μm gap were formed on the wafer. Epoxy-amine polymer [BE/AEANP] of 1.3-μm thick was grown under electric field applied in the gap region of the slit-type electrodes. Electric fields were 0.2 MV/cm and 0.1 MV/cm for the gap of 20 and 10 μm, respectively.

A He-Ne laser beam (633 nm) was introduced into the epoxy-amine polymer [BE/AEANP] film on the electrode gap region to guide the beam. As shown in Figure 8.24, a streak of the guided beam is observed in the gap region, demonstrating

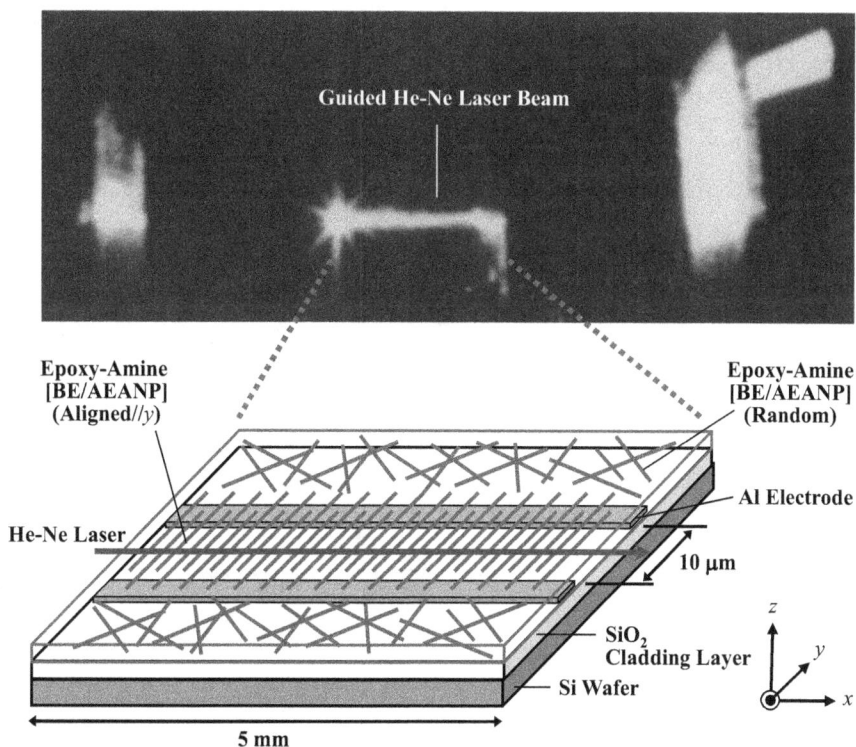

FIGURE 8.24 Guided He-Ne laser beam in a channel waveguide of epoxy-amine polymer [BE/AEANP] fabricated by the electric-field-assisted selective growth.

that a channel waveguide is formed there. Near field pattern observation revealed that a guided beam with polarization parallel to the y-axis is bright while a guided beam with polarization parallel to the z-axis is weak, indicating that the polarization of the guided beam is parallel to the y-axis. This result is attributed to the fact that AEANP molecules are aligned by the electric field along the y-axis. This AEANP alignment increases refractive index for the y-polarization and decreases that for the z-polarization, resulting in polarization-selective confinement of the He-Ne laser beam. The EO coefficient of the channel waveguide evaluated by a Mach–Zehnder interferometer was ~0.1 pm/V at 633 nm.

Polyazomethine (poly-AM) is a candidate for the EO polymer MQD. Tatsuura et al. fabricated EO waveguides of poly-AM [o-PA/MPDA] by the electric-field-assisted selective growth with source molecules of o-phthalaldehyde (o-PA) and 4-methoxy-o-phenylenediamine (MPDA) shown in Figure 8.25a [25]. The methoxy group in MPDA acts as a donor. MO calculations revealed that electric dipole moments are generated along a direction perpendicular to the polymer wire and β of 3–5 times larger than that of p-nitroaniline is expected.

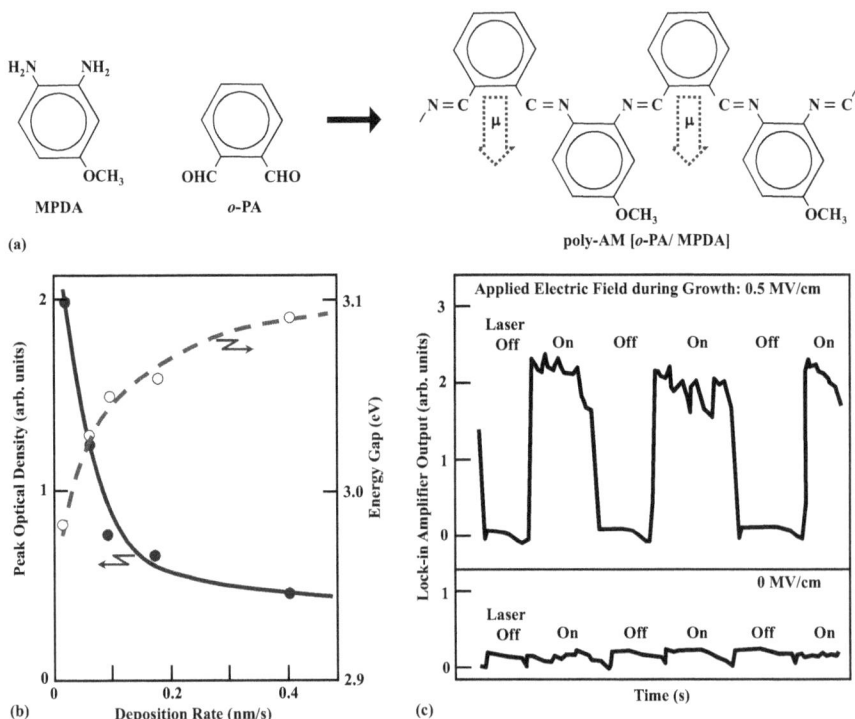

FIGURE 8.25 (a) Reaction for wire growth of poly-AM [o-PA/MPDA]. (b) Deposition rate dependence of the peak optical density and the energy gap obtained from absorption spectra of poly-AM [o-PA/MPDA] films grown by the electric-field-assisted selective growth. (c) EO responses in poly-AM [o-PA/MPDA] films. (Reprinted with permission from Tatsuura et al. [25].)

A thermally oxidized Si wafer, on which Al slit-type electrodes with 5- or 20-μm gap are formed, was used for a substrate. The SiO$_2$ layer thickness was about 1 μm. Yellow-orange poly-AM [o-PA/MPDA] films with thicknesses of 0.8–3.3 μm were grown at molecular gas pressure of 1–10^{-1} Pa under electric fields of 0–0.6 MV/cm in the gap region of the slit-type electrodes.

Deposition rate dependence of the peak optical density and the energy gap obtained from absorption spectra of poly-AM [o-PA/MPDA] films are shown in Figure 8.25b. By decreasing the deposition rate, the peak optical density increases and the energy gap decreases. The results are attributed to an increase in the conjugated length of poly-AM [o-PA/MPDA] by decreasing the deposition rate. Namely, a slow deposition rate prolongs the duration for o-PA and MPDA molecules to migrate, meet each other, and combine with strong bonds on the substrate surface.

The refractive index of the poly-AM [o-PA/MPDA] film was 1.68 at 633 nm. A He-Ne laser beam was directly coupled to the film on the electrode gap region to guide the beam. A channel-like beam propagation over 15-mm long was observed in the electrode gap region. The near field pattern of the exited beam was bright in the TE mode and weak in the TM mode. This is attributed to the fact that the benzene ring planes in the polymer wires were aligned parallel to the substrate surface by the applied electric field.

Figure 8.25c shows EO responses in poly-AM [o-PA/MPDA] films grown with applied electric fields of 0.5 MV/cm and 0 MV/cm. EO effect was measured using a Mach–Zehnder interferometer. The poly-AM [o-PA/MPDA] film was placed in one arm with the film plane perpendicular to a He-Ne laser beam. The polarized direction of the laser beam was along the applied electric field direction in the electrode gap region. For 0.5 MV/cm clear EO responses are observed while no EO responses for 0 MV/cm. The EO response was observed after heating at 80°C for 5 h. This confirms that the conjugated polymer is thermally stable.

These results demonstrate the possibility that conjugated polymer wires can be used for EO waveguides. By combining the electric-field-assisted selective growth with MLD, the EO effect of the conjugated polymer wires will be optimized.

8.4.2 Conjugated Polymer Waveguides by Horizontally-Aligned Selective Growth

Waveguides of poly-AM [TPA/PPDA] were fabricated by the horizontally-aligned selective growth on surfaces with anisotropic atomic-scale structures described in Section 4.3 [26]. Here, TPA and PPDA are, respectively, terephthalaldehyde and p-phenylenediamine. The schematic waveguide structure is shown in Figure 8.26a. A 200-nm-thick SiO under layer was deposited in waveguide core patterns by oblique vacuum evaporation on a glass substrate that was tilted by 45° along the y-axis. A poly-AM [TPA/PPDA] film was grown on the substrate by the carrier-gas-type organic chemical vapor deposition (CVD) using TPA and PPDA. Growth time, carrier gas flow rate, and gas pressure were, respectively, 1 h, 4 NL/min, and 100 Pa or more.

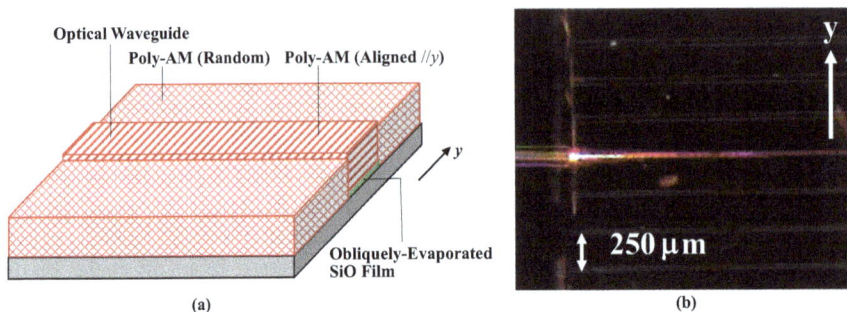

FIGURE 8.26 Poly-AM [TPA/PPDA] waveguides fabricated by the horizontally-aligned selective growth on anisotropic atomic-scale structures. (a) Schematic waveguide structure and (b) microscopic image of fabricated 4-μm-wide waveguides and a 650-nm guided laser beam. (Reprinted with permission from Matsumoto and Yoshimura [26].)

The poly-AM was selectively-aligned along the y-axis on the obliquely-evaporated SiO film region. In the other regions, poly-AM was randomly aligned. Consequently, refractive index of the poly-AM film on the SiO film becomes larger than that of the surrounding poly-AM film for guided light beams with polarization parallel to the y-axis. Figure 8.26b shows a microscopic image of fabricated 4-μm-wide waveguides. A 650-nm guided laser beam is observed, confirming that poly-AM [TPA/PPDA] waveguides are fabricated.

Estimation based on absorption spectra fringes revealed that the refractive index is about 1.65 for polarization perpendicular to the poly-AM wires and 1.95 for polarization parallel to the poly-AM wires. Using the refractive index anisotropicity, optical waveguides and various photonic devices will be fabricated without etching. By combining the process with MLD, EO effect of the conjugated polymer wires will be obtained.

8.5 ACCEPTOR SUBSTITUTION INTO CONJUGATED POLYMER WIRES

As described in Chapter 7, for drastic enhancement of optical nonlinearity of conjugated polymer wires, donor/acceptor substitution into polymer backbones is necessary. A preliminary experiment for the substitution was carried out by using TPA and 2-nitro-1,4-phenylenediamine (NPDA) [8]. As shown in Figure 8.27, NPDA has two -NH$_2$ used for connections with TPA, and one -NO$_2$, an acceptor. In the poly-AM [TPA/NPDA] film, the absorption band of NPDA film at 400–600 nm disappears and a new band appears at 400–450 nm, suggesting that TPA and NPDA combine to form a polymer. Therefore, in principle, donor/acceptor groups can be introduced into conjugated polymer wires by using molecules that contain groups both for connections and for donors/acceptors.

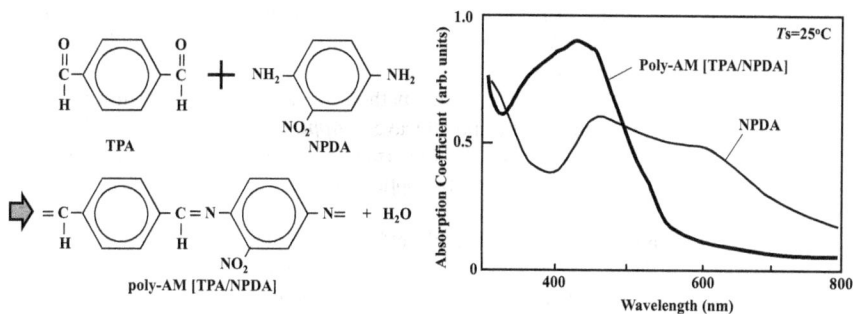

FIGURE 8.27 Molecules and reaction for acceptor-substituted conjugated polymer wire growth, and absorption spectra of poly-AM [TPA/NPDA] film and NPDA film.

REFERENCES

1. T. Aizawa, Furuuchi Chemical Corporation Technical Articles for PLZT Shutter Arrays.
2. K. D. Singer, M. G. Kuzyk, W. R. Holland, J. E. Sohn, S. J. Lalama, R. B. Comizzoli, H. E. Katz, and M. L. Schilling, "Electro-optic phase modulation and optical second-harmonic generation in corona-poled polymer films," *Appl. Phys. Lett.* **53**, 1800–1802 (1988).
3. B. F. Levine, C. G. Bethea, C. D. Thurmond, R. T. Rynth, and J. L. Bernstein, "An organic crystal with an exceptionally large optical second-harmonic coefficient: 2-methyl-4-nitroaniline," *J. Appl. Phys.* **50**, 2523–2527 (1979).
4. G. F. Lipscomb, A. F. Garito, and R. S. Narang, "An exceptionally large electro-optic effect in the organic solid MNA," *J. Chem. Phys.* **75**, 1509–1516 (1981).
5. T. Yoshimura, "Design of organic nonlinear optical materials for electro-optic and all-optical devices by computer simulation," *Fujitsu Sci. Tech. J.* **27**, 115–131 (1991).
6. T. Yoshimura, "Enhancing second-order nonlinear optical properties by controlling the wave function in one-dimensional conjugated molecules," *Phys. Rev. B* **40**, 6292–6298 (1989).
7. T. Yoshimura, "Theoretically predicted influence of donors and acceptors on quadratic hyperpolarizabilities in conjugated long-chain molecules," *Appl. Phys. Lett.* **55**, 534–536 (1989).
8. T. Yoshimura and Y. Kubota, "Quantum well formation in one-dimensional polymer films by molecular layer deposition and chemical vapor deposition," *Mat. Res. Soc. Symp. Proc.* **247**, 829–834 (1992).
9. T. Yoshimura, "Characterization of the EO effect in styrylpyridinium cyanine dye thin-film crystals by an ac modulation method," *J. Appl. Phys.* **62**, 2028–2032 (1987).
10. O. Qasaimeh, K. Kamath, P. Bhattacharya, and J. Phikkips, "Linear and quadratic electro-optic coefficients of self-organized In $_{0.4}$Ga$_{0.6}$As/GaAs quantum dots," *Appl. Phys. Lett.* **72**, 1275–1277 (1998).
11. T. Yoshimura and T. Futatsugi., "Non-linear optical device using quantum dots," U.S. Patent 6,294,794B1 (2001).
12. G. R. Meredith, "Design and characterization of molecular and polymeric nonlinear optical materials: Successes and pitfalls," in *Nonlinear Optical Properties of Organic and Polymeric Materials* (Ed. D. J. Williams), 27–56, American Chemical Society, Washinton, DC (1983).

13. R. J. Twieg and K. Jain, "Organic materials for optical second harmonic generation," in *Nonlinear Optical Properties of Organic and Polymeric Materials* (Ed. D. J. Williams), 57–80, American Chemical Society, Washinton, DC (1983).

14. J. S. Wei and W. D. Westwood, "A new method for determining thin-film refractive index and thickness using guided optical waves," *Appl. Phys. Lett.* **32**, 819–821 (1978).

15. Y. Kubota and T. Yoshimura, "Endothermic reaction aided crystal growth of 2-methyl-4-nitroaniline and their electro-optic properties," *Appl. Phys. Lett.* **53**, 2579–2580 (1988).

16. T. Yoshimura and Y. Kubota, "Pockels effect in organic thin films," in *Springer Proceedings in Physics 36, Nonlinear Optics of Organics and Semiconductors* (Ed. T. Kobayashi), 222–226, Springer-Verlag Berlin, Heidelberg (1989).

17. K. D. Singer, M. G. Kuzyk and J. E. Sohn, "Second-order nonlinear-optical processes in orientationally ordered materials: Relationship between molecular and macroscopic properties," *J. Opt. Soc. Am. B* **4**, 968–976 (1987).

18. T. Yoshimura, S. Tatsuura, and W. Sotoyama, "Organic electro-optic materials and their applications," *Oyo Buturi* **61**, 38–42 (1992) [in Japanese].

19. W. Sotoyama, S. Tatsuura, and T. Yoshimura, "Electro-optic side-chain polyimide system with large optical nonlinearity and high thermal stability," *Appl. Phys. Lett.* **64**, 2197–2199 (1994).

20. C. C. Teng and H. T. Man, "Simple reflection technique for measuring the electro-optic coefficient of poled polymers," *Appl. Phys. Lett.* **56**, 1734–1736 (1990).

21. W. Sotoyama, S. Tatsuura, K. Motoyoshi, and T. Yoshimura, "Directional-coupled optical switch between stacked waveguide layers using electro-optic polymer," *Jpn. J. Appl. Phys.* **31**, L1180–L1181 (1992).

22. W. Sotoyama, S. Tatsuura, and T. Yoshimura, "Fabrication techniques for electro-optic polymer waveguides," *Rev. Laser Eng.* **21**, 1125–1133 (1993) [in Japanese].

23. S. Tatsuura, W. Sotoyama, and T. Yoshimura, "Epoxy-amine polymer waveguide containing nonlinear optical molecules fabricated by chemical vapor deposition," *Appl. Phys. Lett.* **60**, 1158–1160 (1992).

24. S. Tatsuura, W. Sotoyama, and T. Yoshimura, "Electro-optic polymer waveguide fabricated using electric-field-assisted chemical vapor deposition," *Appl. Phys. Lett.* **60**, 1661–1663 (1992).

25. S. Tatsuura, W. Sotoyama, K. Motoyoshi, A. Matsuura, T. Hayano, and T. Yoshimura, "Polyazomethine conjugated polymer film with second order nonlinear optical properties fabricated by electric-field-assisted chemical vapor deposition," *Appl. Phys. Lett.* **62**, 2182–2184 (1993).

26. K. Matsumoto and T. Yoshimura, "Electro-optic waveguides with conjugated polymer films fabricated by the carrier-gas-type organic CVD for chip-scale optical interconnects," *Proc. SPIE* **6899**, 98990E (2008).

9 Integrated Light Modulators and Optical Switches

High-speed/small-size light modulators and optical switches are the key for integrated optical interconnects and optical switching systems. For such large-scale optoelectronic (OE) systems, flux-controlled devices like the total internal reflection (TIR) optical switch, the digital optical switch, and the waveguide prism deflector (WPD) optical switch are suitable because they do not require extremely-strict control of shapes/sizes, refractive indexes, and operation temperature due to their phase-insensitive characteristics. In the present chapter, after various types of light modulators/optical switches are explained, the WPD optical switch is described in detail as a model device for the integrated light modulators/optical switches, and the impact of the polymer multiple quantum dots (MQDs) is discussed. Performance of WPD optical switch is predicted by the beam propagation method for three kinds of electro-optic (EO) materials: styrylpyridinium cyanine dye (SPCD), lead lanthanum zirconate titanate (PLZT), and polymer MQD. The polymer-MQD WPD optical switch is found to be preferable for large-scale OE systems from viewpoints of modulation/switching performance and power consumption as well as robustness. Advanced WPD optical switches with ADD and MUX/DEMUX functions are proposed, and the switching operation is simulated. In addition, a process for integration of WPDs, optical waveguides, and waveguide lenses by the photolithographic packaging with selectively occupied repeated transfer (PL-Pack with SORT) is proposed.

9.1 CLASSIFICATION OF LIGHT MODULATORS AND OPTICAL SWITCHES

Light modulators and optical switches are classified into the phase-controlled devices and the flux-controlled devices, as presented in Figure 9.1. In many cases, the light modulator function and the optical switch function are available in the same device structure. These devices operate by the refractive index change in EO materials or semiconductors. In the EO materials, the refractive index changes by applying electric fields through the Pockels effect or the Kerr effect. In the semiconductors, it changes by carrier injection and depletion.

The Mach–Zehnder light modulator (MZM) and the directional coupler (DC) are phase-controlled devices [1–3]. The former utilizes interference of two light beams propagating in branching waveguide arms, and the latter utilizes interference of two light beams propagating in different modes. The MZM itself operates as a light modulator, and it operates as an optical switch by combining it with a DC. The microring

DOI: 10.1201/9781003094012-9

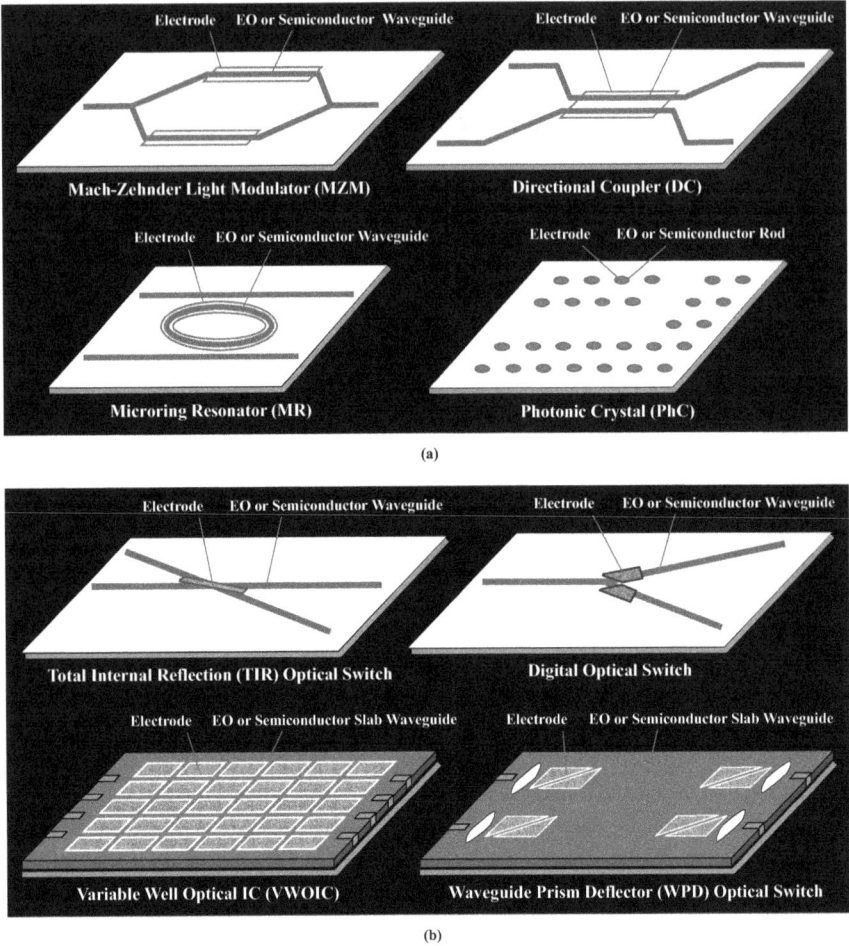

FIGURE 9.1   Various types of light modulators and optical switches. (a) Phase-controlled devices and (b) flux-controlled devices.

resonator (MR) [4] and the photonic crystal (PhC) [5,6] are phase-controlled devices, which utilize accumulated interference. Their device functions appear after super-position of all the light beams is completed to reach the steady state, in which the interference is perfectly built-up.

The phase-controlled devices can work with relatively small refractive index change, enabling low-voltage operation. However, due to their phase-sensitive char-acteristics, extremely-strict control of waveguide shapes/sizes, refractive indexes, and operation temperature is required for stable operation. Furthermore, in the MR and PhC, as described in Section 9.7, building-up duration is necessary to settle the modulation/switching functions, causing delay time. These drawbacks of the phase-controlled devices imply that they are not the final destination in the large-scale OE systems like the integrated optical interconnects and optical switching systems.

As the flux-controlled devices, on the other hand, do not utilize the interference, namely, they are phase-insensitive, they do not require extremely-strict control of shapes/sizes, refractive indexes, and operation temperature, and consequently, they are suitable for the large-scale OE systems. The flux-controlled devices include the TIR optical switch, the digital optical switch, and the variable well optical integrated circuit (VWOIC) [7–10]. These devices work by TIR. The WPD optical switch [8–16] is another flux-controlled device. This works by refraction, and is applicable to wavelength division multiplexing (WDM) in principle, as described in Section 9.6.2.

The flux-controlled devices require relatively large refractive index change. So, to realize high-speed/small-size integrated light modulators and optical switches for the large-scale OE systems, development of materials having enhanced refractive-index-change characteristics is essential.

In Sections 9.2–9.6, as a model device for the integrated light modulators/optical switches, the WPD optical switch is described in detail, and the impact of the polymer MQDs on its performance is discussed.

9.2 WAVEGUIDE PRISM DEFLECTOR (WPD) OPTICAL SWITCHES

The WPD optical switch has been investigated in the author's research group. The conceptual structure is presented in Figure 9.2. An EO slab waveguide consisting of cladding/core/cladding layers is placed between a counter electrode and prism-shaped electrodes. The core layer is made of EO materials, and the refractive index changes by applying voltage between the electrodes. An interface film, in which integrated circuit (IC) dies are embedded, is stacked on the waveguide to drive individual prism-shaped electrodes.

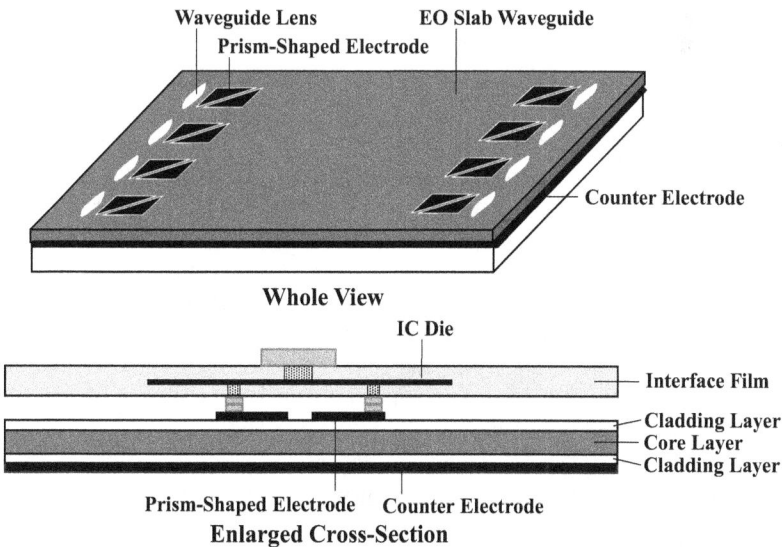

FIGURE 9.2 Conceptual structure of WPD optical switch.

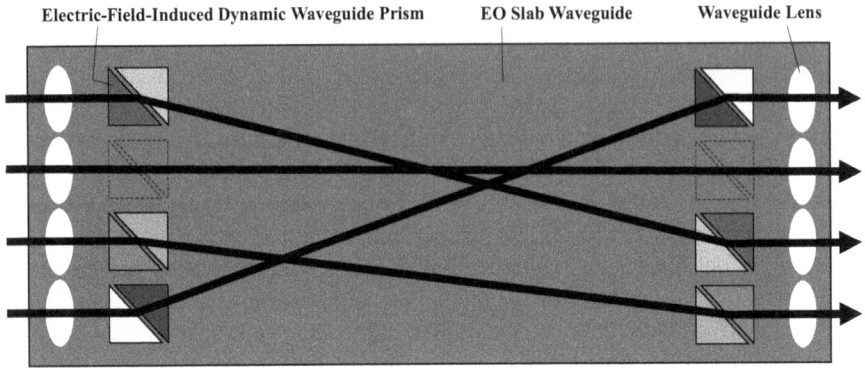

FIGURE 9.3   Operation principle of WPD optical switch.

In the WPD optical switch, as illustrated in Figure 9.3, electric-field-induced dynamic waveguide prisms, which deflect guided light beams, are constructed in the EO slab waveguide by applying voltage to prism-shaped electrodes, enabling the light beam switching through a two-dimensional (2-D) free space like "search light." Waveguide lenses are formed to suppress spreading of the guided light beams in the 2-D free space.

The WPD optical switch is designed considering the following guidelines:

- Reduction of the prism size and the inter-channel pitch.
- Suppression of the beam spreading accompanying the free space propagation in the slab waveguide.

These two points have a reciprocal relationship. When the prism size is reduced, the prism aperture decreases and the diffraction angle increases, resulting in an increase in the beam spreading. So, it is important to find a practical solution for the WPD optical switch structure within this trade-off.

The WPD utilizes the refractive index difference induced by electric fields in a slab waveguide, not the refractive index itself. The refractive index difference is less sensitive to temperature than the refractive index itself. This makes the WPD operation robust against the temperature and refractive index fluctuation. For the processing accuracy, as far as the right prism edge angle is maintained, the size of each part does not largely affect the device operation. Therefore, the WPD optical switch is expected to be more stable against the thermal and structural disturbance than the MZM, DC, MR, and PhC, and to be suited to the large-scale integrated optical interconnects within computers and the 3-D micro optical switching systems (3D-MOSSs) described in Chapter 11.

Candidates of EO materials for the WPD optical switch are presented in Figure 8.1. The electric-field-induced refractive index change, Δn, of EO materials is expressed as follows:

$$\text{Pockels Effect}\quad \Delta n = \frac{1}{2}n_0{}^3 rE \tag{9.1}$$

$$\text{Kerr Effect} \qquad \Delta n = \frac{1}{2} n_0^3 R E^2 \tag{9.2}$$

Here, n_0 and E represent the refractive index of the EO materials with no electric fields and the electric field in the EO materials, respectively. r and R are, respectively, EO coefficients for the Pockels effect and the Kerr effect. Figure 9.4 shows Δn calculated using Equations (9.1) and (9.2) for lithium niobate (LiNbO$_3$ (LN)), III-V quantum dot (QD) [17,18], styrylpyridinium cyanine dye (SPCD) molecular crystal [19], polymer MQD [13,20,21], and lead lanthanum zirconate titanate (PLZT) [22]. The EO material thickness that equals the electrode gap distance d is 1 µm, and the applied voltage is 1 V. PLZT and polymer MQD are expected to give extremely large Δn.

It should be noted that a WPD utilizing the Pockels effect can deflect guided light beams toward both right-hand side and left-hand side directions by inverting the polarity of the applied voltage, as can be seen from Equation (9.1). This enables $n \times n$ (n: 2, 3, 4, ...) optical switches, as depicted in Figure 9.3. For a WPD utilizing the Kerr effect, on the other hand, the switching configuration is limited because the deflection angle control by the polarity inversion does not work as can be seen from Equation (9.2). In Sections 9.3–9.5, design of WPD optical switches and prediction of the switching performance are described for the SPCD molecular crystal and the polymer MQD, which exhibit the Pockels effect, and for PLZT, which exhibits the Kerr effect.

In addition, another flux-controlled device, the VWOIC, shown in Figure 9.5, is briefly explained [8]. An array of rectangular electrodes is formed on an EO slab waveguide. By selectively applying voltage to the electrodes, refractive index changes are induced to form potential wells for guided light beams, constructing

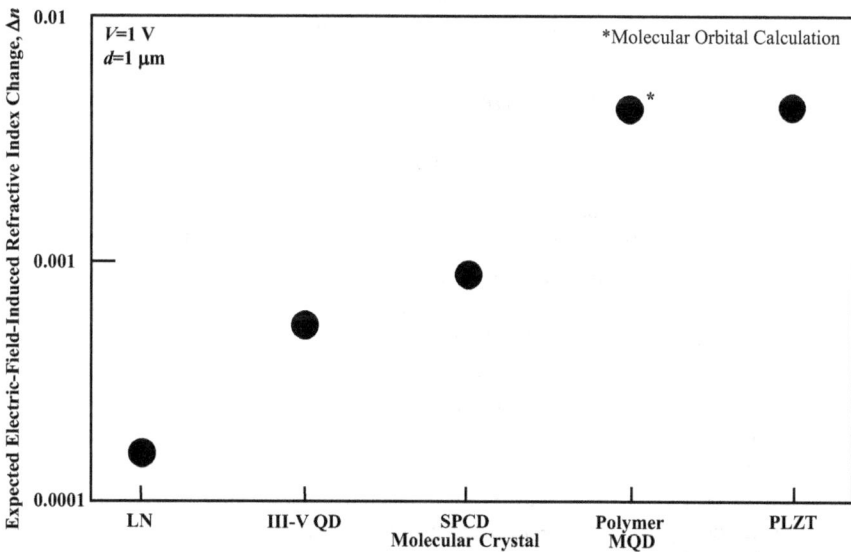

FIGURE 9.4 Expected electric-field-induced refractive index change Δn of various EO materials.

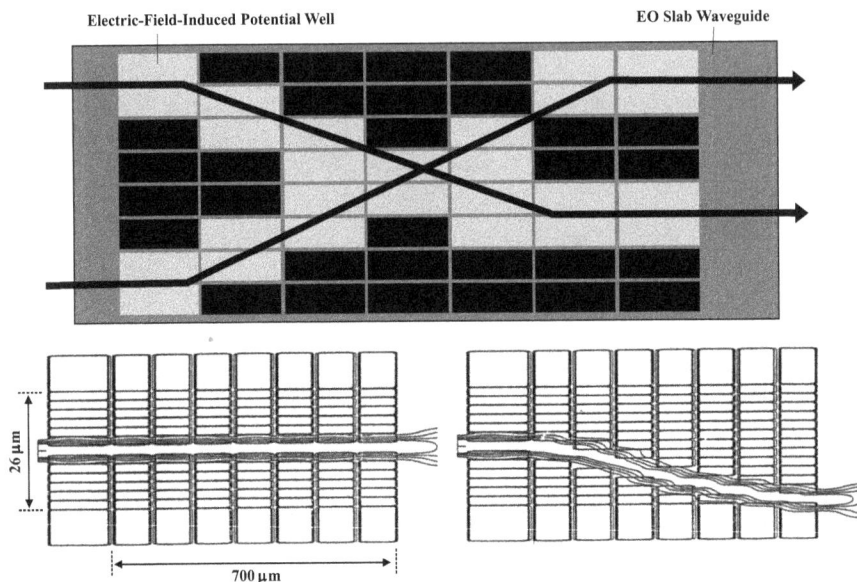

FIGURE 9.5 Operation principle of VWOIC and switching operation simulated by the BPM. (Reprinted with permission from Yoshimura et al. [8].)

electric-field-induced dynamic optical waveguides. Simulations by the beam propagation method (BPM) for a VWOIC with 12×7 electrodes of $2\,\mu m \times 100\,\mu m$ revealed that 1.3-μm light beams are switched in a length of $700\,\mu m$. Here, an electric-field-induced refractive index change of 0.005 is assumed, and two adjacent electrodes are simultaneously selected for the voltage application to construct 4-μm-wide waveguides.

9.3 WPD OPTICAL SWITCHES UTILIZING POCKELS EFFECT

9.3.1 Models and Simulation Procedures

Figure 9.6a shows a model of the WPD utilizing the Pockels effect. Prism-shaped electrodes are formed by a pair on an EO slab waveguide. When $-V$ voltage is applied to one electrode and $+V$ voltage to the other, prism-shaped refractive index changes of $-\Delta n$ and $+\Delta n$ are induced in the slab waveguide just under the electrodes. These waveguide prisms deflect guided light beams by refraction [14].

Figure 9.6b shows a model of the WPD optical switch utilizing the Pockels effect. The EO slab waveguide is an SPCD thin-film molecular crystal. For simplicity, the gap distance between the prism-shaped electrodes and the counter electrode, d, is assumed to equal the SPCD film thickness. Between input/output waveguides and the prism-shaped electrodes, waveguide lenses are placed. The inter-channel pitch is P and the core width of the input and output waveguides D. The prism-shaped electrode has a right-angle triangle shape with a length and an aperture of D_P. By lining up the multiple prism-shaped electrode pairs in a cascade, the deflection angle θ_d increases. The number of the cascade steps is N. The distance between the center

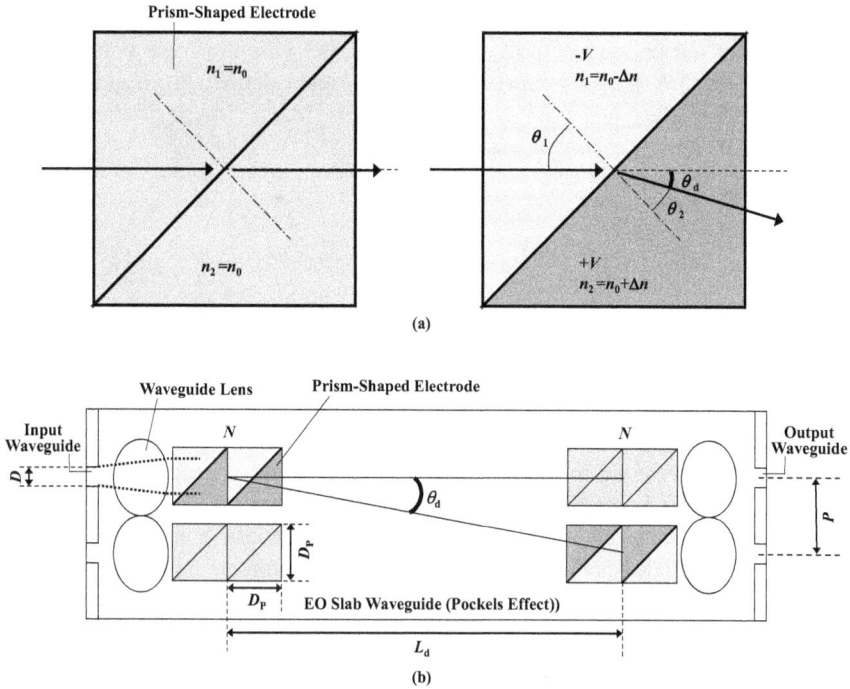

FIGURE 9.6   (a) A model of WPD utilizing the Pockels effect. (b) A model of WPD optical switch utilizing the Pockels effect. (Reprinted with permission from Ninomiya et al. [14].)

of the prism cascade at the input and that at the output is L_d. The polarity in the voltage pattern of the prism-shaped electrodes at the output is opposite to that at the input. This enables the deflected light beams to return in the horizontal direction and smoothly couple to the output waveguide.

The refractive index n_1 in the negative voltage application unit and n_2 in the positive voltage application unit in Figure 9.6a are expressed as follows using the electric-field-induced refractive index change Δn:

$$n_1 = n_0 - \Delta n \tag{9.3}$$

$$n_2 = n_0 + \Delta n \tag{9.4}$$

where n_0 is the refractive index when no voltage is applied. The relationship between the incident angle θ_1 and the transmission angle θ_2 is given by

$$\theta_2 = \sin^{-1}\left(\frac{n_1}{n_2}\sin\theta_1\right). \tag{9.5}$$

The deflection angle θ_d is obtained by the following equation:

$$\theta_d = \theta_1 - \theta_2 \tag{9.6}$$

Here, although refraction also occurs at the left-hand-side and right-hand-side boundaries of the paired-electrode configuration, deflection at these boundaries is neglected. For an N-stage cascade, N iterations of the calculation in Equation (9.5) determine the final θ_d.

L_d in the 0th approximation is expressed in terms of θ_d and P by

$$L_d = \frac{P}{\tan \theta_d}. \tag{9.7}$$

The diffraction angle of the light beam increases with decreasing D_P. Meanwhile, the beam diameter must be less than the inter-channel pitch P. Therefore, restrictions develop in L_d depending on D_P.

The switching operation was simulated by the wide-angle-approximate 2-D BPM using the finite-difference method. The propagating light beam has a wavelength λ of 1.3 μm and is in the TE mode. The Δz step in the light beam propagation direction is 0.5 μm, and the Δx step in the perpendicular direction is $\lambda/30$. The core refractive index of the input and output waveguides is equal to the core refractive index of the slab waveguide, that is, n_0. The difference in the refractive index between the core and the cladding region is 0.03. The electric field distribution of the input light beam is that of the input waveguide fundamental mode. The integrated intensity of the input light beam is 1.0. The EO slab waveguide is assumed to be a 400-nm-thick SPCD thin film with $r = 430$ pm/V and $n_0 = 1.55$.

9.3.2 PREDICTED PERFORMANCE

From preliminary tests, $N = 4$ and $V = 11$ V were extracted as the appropriate conditions for the WPD optical switch. A structure of a 2×2 WPD optical switch is created with these conditions, as schematically illustrated in Figure 9.7. Here, $P = 20$ μm, $D = 4$ μm, and $D_P = 16$ μm. The diameter and the refractive index of the lens are 40 μm and 2.0, respectively. The optical switching length L_{OSW} is 370 μm [14].

Figure 9.8 shows voltage application patterns and switching operation simulated by the BPM for Input 1 → Output 1 and Input 2 → Output 1. Guided light beams are smoothly switched. For Input 1 → Output 1, the crosstalk was −12 dB and the

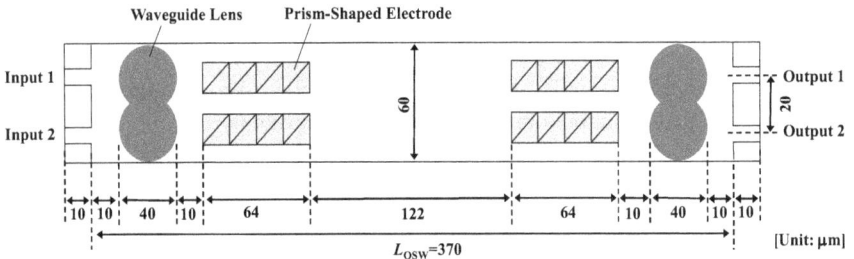

FIGURE 9.7 Schematic structure of 2×2 WPD optical switch with the 400-nm-thick SPCD thin film. (Reprinted with permission from Ninomiya et al. [14].)

Input 1 → Output 1   Input 2 → Output 1

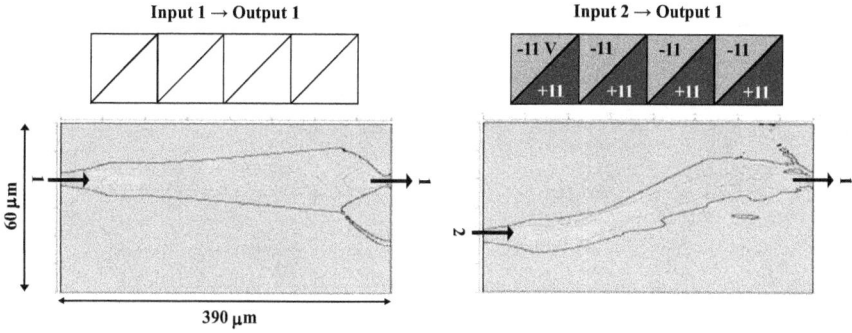

FIGURE 9.8 Voltage application patterns and switching operation simulated by the BPM for 2×2 WPD optical switch with the 400-nm-thick SPCD thin film. (Reprinted with permission from Ninomiya et al. [14].)

insertion loss was 0.44 dB. For Input 2 → Output 1, the crosstalk was −14 dB and the insertion loss was 0.96 dB.

A WPD is regarded as a capacitor with a prism-shaped electrode, and the switching time τ, the energy consumption per bit P_b, and the power consumption *Power* can be estimated based on the charge–discharge model for the prism-shaped capacitors as the 0th approximation. The capacitance of the one prism-shaped capacitor, C, is given by

$$C = \frac{\varepsilon_s \varepsilon_0 S}{d},\tag{9.8}$$

where S represents the prism-shaped electrode area and d is the electrode gap distance. ε_0 and ε_s are, respectively, the dielectric constant in vacuum and the relative dielectric constant. Then, the switching time is expressed by

$$\tau = R_{ON}C,\tag{9.9}$$

where R_{ON} is the ON resistance of the driver circuit. The energy consumption per bit for the charge and discharge, and the power consumption at a switching rate of F are, respectively, given by the following expressions:

$$P_b = CV^2\tag{9.10}$$

$$Power = FCV^2\tag{9.11}$$

By substituting $S = (1/2) \times (16 \times 10^{-6}) \times (16 \times 10^{-6}) = 128 \times 10^{-12} \text{m}^2$, $d = 4 \times 10^{-7} \text{m}$, and an assumed relative dielectric constant of an SPCD crystal $\varepsilon_s \approx 4$ into Equation (9.8), $C = 11$ fF is obtained. Then, from Equation (9.9), τ is calculated to be 1.1 ps for R_{ON} of 100 Ω. When $F = 5 \times 10^9$ 1/s and $V = 11$ V, *Power* is calculated to be 6.9 mW from Equation (9.11).

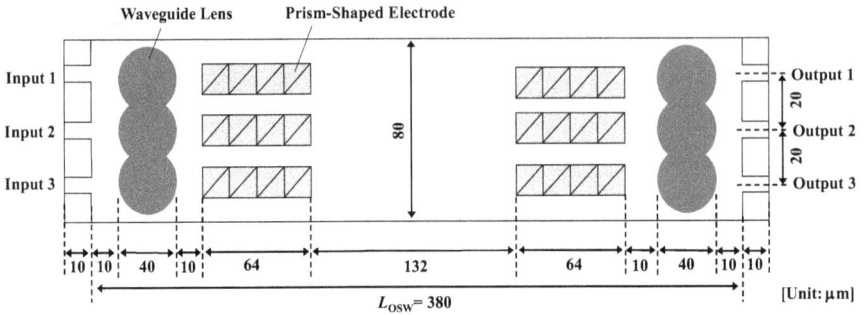

FIGURE 9.9 Schematic structure of 3×3 WPD optical switch with the 400-nm-thick SPCD thin film. (Reprinted with permission from Ninomiya et al. [14].)

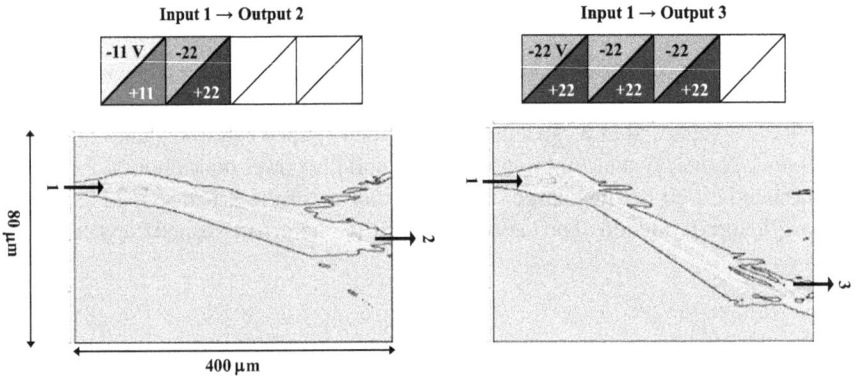

FIGURE 9.10 Voltage application patterns and switching operation simulated by the BPM for 3×3 WPD optical switch with the 400-nm-thick SPCD thin film. (Reprinted with permission from Ninomiya et al. [14].)

Figure 9.9 schematically depicts a 3×3 WPD optical switch structure. In Figure 9.10, voltage application patterns and the simulated switching operation for Input 1 → Output 2 and Input 1 → Output 3 are presented. Guided light beams are switched in L_{OSW} of 380 μm. The crosstalk was $\leq -10\,\text{dB}$ and the insertion loss $\leq 1.2\,\text{dB}$.

9.4 WPD OPTICAL SWITCHES UTILIZING KERR EFFECT

9.4.1 MODELS AND SIMULATION PROCEDURES

Figure 9.11 shows a model of the WPD and a preliminary model of the 2×2 WPD optical switch utilizing the Kerr effect. Between input waveguides and output waveguides, an EO slab waveguide, on which prism-shaped electrodes are formed, is placed. Passive slab waveguides containing waveguide lenses are placed between the input/output waveguides and the EO slab waveguide. When voltage is applied to the

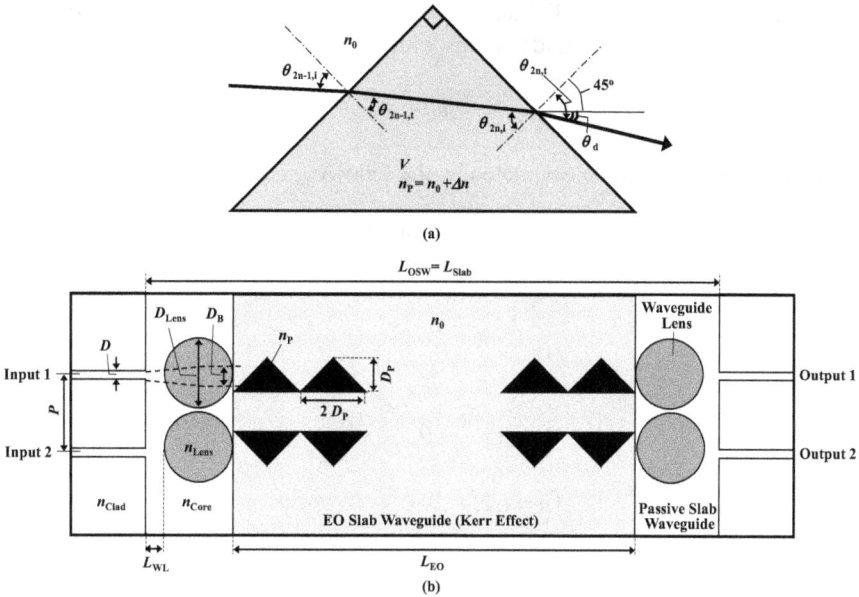

FIGURE 9.11   (a) A model of WPD utilizing the Kerr effect. (b) A preliminary model of WPD optical switch utilizing the Kerr effect. (Reprinted with permission from Yoshimura et al. [15]. Copyright (2004) The Optical Society.)

prism-shaped electrodes, electric-field-induced dynamic waveguide prisms deflect optical signals from the input waveguides to switch them through the slab waveguides into the output waveguides [11,15].

For the input/output waveguides, core width D is 4 µm, core refractive index n_{Core} is 1.52, and cladding refractive index n_{Clad} is 1.50, which permit light beams to propagate in the fundamental mode of $m = 0$ and the first excited mode of $m = 1$. The inter-channel pitch P is 20 µm. The total length of the slab waveguide region including the EO and passive slab waveguides is L_{Slab}. The length of the EO slab waveguide is L_{EO}. The EO slab waveguide is a 400-nm-thick PLZT thin film. The gap distance between the prism-shaped electrodes and the counter electrode is $d = 400$ nm The aperture D_P and the length of the right-angle prism-shaped electrodes are, respectively, 16 and 32 µm. Two-step prism cascading configuration is considered. By applying voltage to the slab waveguide, waveguide prisms with a refractive index of n_P are formed. For waveguide lenses, the diameter D_{Lens} is 40 µm and the refractive index is n_{Lens}. The refractive index of the surrounding passive slab waveguides is n_{Core} of 1.52. The lens expands the light beam size to D_B. The distance between the lens and the input/output waveguides is L_{WL}. The guided beam wavelength λ is 1.3 µm. Reflection from the integrated components is not taken into account. Anti-reflection (AR) coating on the component edge surfaces using the atomic layer deposition (ALD) or the molecular layer deposition (MLD) might make the model viable.

In order to determine an optimized model of the 2×2 WPD optical switch, proper n_{Lens}, L_{WL}, L_{Slab}, and L_{EO} were adjusted. As shown in Figure 9.11a, refractive index change Δn is induced in the EO slab waveguide by driving voltage V according to Equation (9.2) to form waveguide prisms with a refractive index of n_P given by

$$n_P = n_0 + \Delta n. \tag{9.12}$$

Deflection angle, θ_d, can be calculated by the following equations:

$$n_0 \sin \theta_{2n-1,i} = n_P \sin \theta_{2n-1,t}$$

$$n_P \sin \theta_{2n,i} = n_0 \sin \theta_{2n,t}$$

$$\theta_{2n,i} = \frac{\pi}{2} - \theta_{2n-1,t}$$

$$\theta_{2n+1,i} = \frac{\pi}{2} - \theta_{2n,t} \tag{9.13}$$

$$(n = 1, 2, \cdots, N)$$

$$\theta_d = \theta_{2N,t} - \pi / 4$$

Here, N is the prism count in the multi-step prism cascading configurations. Final deflection angle is obtained by performing calculations sequentially for $n = 1, 2, ...,$ and N. As the 0th approximation, neglecting the sizes of prism-shaped electrodes and passive slab waveguides containing waveguide lenses, L_{Slab} necessary to achieve the switching operation is given by

$$L_{Slab} \approx \frac{P}{\tan \theta_d}. \tag{9.14}$$

The spread angle of a light beam propagating in the slab waveguide is approximately estimated by $2\theta_S = \lambda/(\pi n_0 D_B)$. The light beam width should be less than P at the end of the slab waveguide after propagation. Then, L_{Slab} should satisfy

$$L_{Slab} < \frac{1}{2} \frac{P}{\tan \theta_S}. \tag{9.15}$$

In Figure 9.12, driving voltage dependence of L_{Slab} derived from Equation (9.14) using parameters for PLZT in Figure 8.1 is shown for a two-step prism cascading configuration. The diffraction limit boundary derived from Equation (9.15) is also shown for D_B of 5 µm. It can be seen that the lowest possible driving voltage is around 1 V, predicting proper L_{Slab} of around 500 µm. According to this preliminary estimation, parameters listed in Table 9.1 were adopted in the present calculation, namely, driving voltage of $V = 1.2$ V, a waveguide prism refractive index of $n_{P(On)} = 2.52$ generated by 1.2-V application, and a waveguide prism refractive index for 0-V application, $n_{P(Off)}$, which is equal to n_0. In this condition, it is found from Figure 9.12 that the proper L_{Slab} is around 310 µm.

Based on the parameters mentioned above, n_{Lens}, L_{WL}, and L_{EO} were optimized. Figure 9.13 shows the BPM simulation results for n_{Lens} of 1.8, 2.0, and 2.5. For n_{Lens}

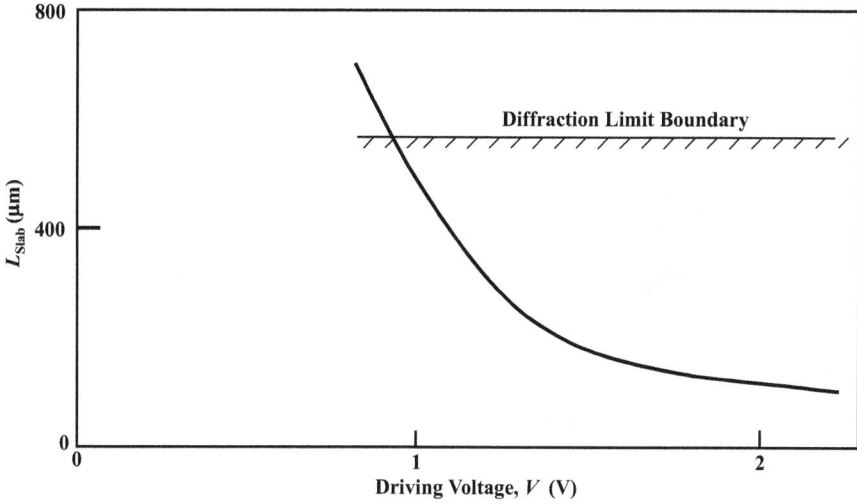

FIGURE 9.12 Driving voltage dependence of L_{Slab} and diffraction limit boundary for D_B of 5 μm. (Reprinted with permission from Yoshimura et al. [15]. Copyright (2004) The Optical Society.)

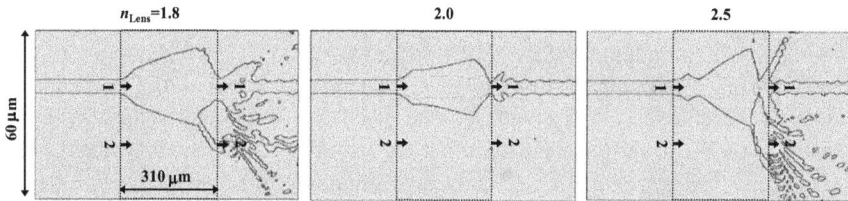

FIGURE 9.13 BPM simulation results of the guided light beam collimation in the EO slab waveguide for n_{Lens} of 1.8, 2.0, and 2.5. (Reprinted with permission from Yoshimura et al. [15]. Copyright (2004) The Optical Society.)

TABLE 9.1

Driving Voltage and Prism Refractive Indexes in 2×2 WPD Optical Switches with the 400-nm-Thick PLZT Thin Film and the 400-nm-Thick Polymer MQD

	400-nm-Thick PLZT Thin Film	400-nm-Thick Polymer MQD
Driving voltage, V (V)	1.2	4
Prism refractive index (Off), $n_{P(Off)}$	2.48	2
Prism refractive index (On), $n_{P(On)}$	2.52	2.04

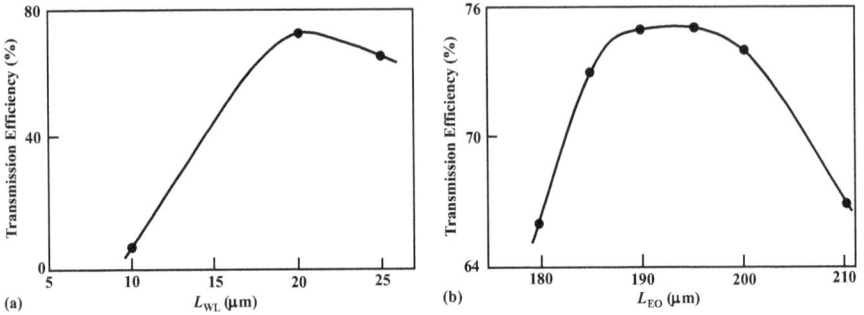

FIGURE 9.14   (a) L_{WL} dependence and (b) L_{EO} dependence of transmission efficiency from the input waveguide into the output waveguide. (Reprinted with permission from Yoshimura et al. [15]. Copyright (2004) The Optical Society.)

of 1.8, collimation of the guided light beam in the EO slab waveguide is insufficient while, for n_{Lens} of 2.5, focusing effect is too strong so that large beam size expansion occurs. n_{Lens} of 2.0 is a proper condition to produce collimated light beams, resulting in smooth coupling to the output waveguide in a distance of 310 µm from the input waveguide. Figure 9.14a shows L_{WL} dependence of transmission efficiency from the input waveguide into the output waveguide. The optimum L_{WL} is found to be 20 µm. For L_{EO}, as shown in Figure 9.14b, the optimum is 190–200 µm. From a viewpoint of miniaturization of optical switches, 190 µm should be the optimum L_{EO}. Thus, the optical switching length of the WPD optical switch, L_{OSW}, which equals L_{Slab}, is calculated to be 310 µm by adding $2(D_{Lens}+L_{WL})$ to L_{EO}.

Design parameters determined for the 2×2 WPD optical switch with the 400-nm-thick PLZT thin film are summarized in Table 9.2 and the optimized structure is shown in Figure 9.15. The area of the right-angle prism-shaped electrodes, S, is $(1/2)\times32\times16=256\,\mu m^2$. The waveguide lens refractive index of 2.0 is available by using Ti_xSi_yO, SiO_xN_y, and so on. The size of the main part of the optical switch is 310 µm × 20 µm. When the channel pitch is widened to 60 µm by bending waveguides of 2.5-mm radius at the input and output ports, the total switch length L_{SW} becomes 1190 µm, and the width W_{SW} becomes ~100 µm including side margins.

9.4.2   PREDICTED PERFORMANCE

Figure 9.16 shows BPM simulation results of switching operation for the 2×2 WPD optical switch with the 400-nm-thick PLZT thin film shown in Figure 9.15. When no voltage is applied to the prism-shaped electrodes, a 1.3-µm guided beam propagates from Input 1 to Output 1, achieving the "Bar" condition. By 1.2-V application to the electrodes, the guided beam propagates from Input 1 to Output 2, achieving "Cross" condition. For both "Bar" and "Cross," crosstalk was less than −20 dB. Transmission efficiency was 80% and 69% for "Bar" and "Cross," respectively. Maximum insertion loss within the WPD optical switch was 1.7 dB including losses in bending waveguides.

As mentioned in Section 9.3, since a WPD is regarded as a prism-shaped capacitor with an electrode area of S, electrode gap distance of d, and dielectric constant

TABLE 9.2
Design Parameters for 2×2 WPD Optical Switch with the 400-nm-Thick PLZT Thin Film

EO Slab Waveguide

Length, L_{EO}	190 μm
Thickness (electrode gap distance), d	400 nm

Input/Output Channel Waveguide

Core width, D	4 μm
Channel separation, P	20 μm
Core refractive index, n_{Core}	1.52
Cladding refractive index, n_{Clad}	1.50

Waveguide Lens

Lens input/output waveguide distance, L_{WL}	20 μm
Diameter, D_{Lens}	40 μm
Refractive index, n_{Lens}	2.0

Right-Angle Prism-Shaped Electrode

Aperture, D_P	16 μm
Length, $2D_P$	32 μm
Cascaded prism count, N	2
Optical switching length, L_{OSW}	310 μm
Guided light wavelength in vacuum, λ	1.3 μm

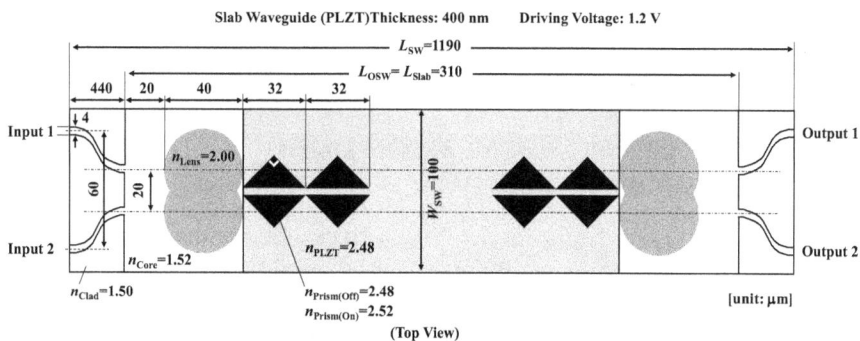

FIGURE 9.15 Optimized structure of 2×2 WPD optical switch with the 400-nm-thick PLZT thin film. (Reprinted with permission from Yoshimura et al. [11].)

of $\varepsilon_s \varepsilon_0$, the capacitance, the switching time, the energy consumption per bit, and the power consumption can be obtained by Equations (9.8)–(9.11). It is found from these equations that low dielectric constant characteristics are preferable for high-speed and low power operation. The predicted electrical performance is summarized in Table 9.3. In the model shown in Figure 9.15, where $S = 256 \times 10^{-12}\,\text{m}^2$, $d = 4 \times 10^{-7}\,\text{m}$, and $\varepsilon_s = 1000$ for PLZT, C is 5.67 pF, τ is 567 ps for R_{ON} of 100 Ω, P_b is 8.17 pJ/bit for $V = 1.2\,\text{V}$, and *Power* is 327 mW for $F = 4 \times 10^{10}$ 1/s.

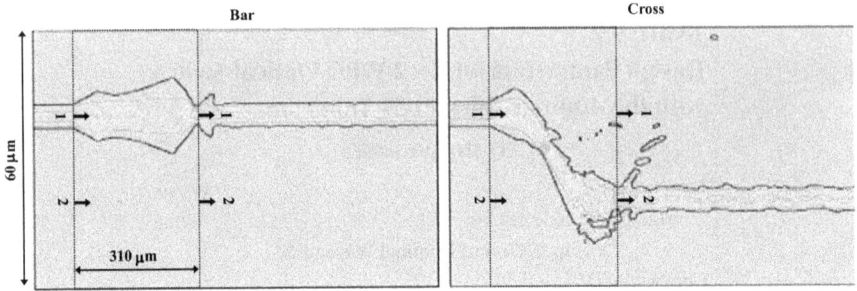

FIGURE 9.16 Switching operation simulated by the BPM for 2×2 WPD optical switch with the 400-nm-thick PLZT thin film. (Reprinted with permission from Yoshimura et al. [11].)

TABLE 9.3
Predicted Electrical Performance in 2×2 WPD Optical Switches with the 400-nm-Thick PLZT Thin Film and the 400-nm-Thick Polymer MQD

	400-nm-Thick PLZT Thin Film	400-nm-Thick Polymer MQD
Relative dielectric constant, ε_s	1000	4[a]
Capacitance, C [pF]	5.67	0.0227
Switching time ($R_{on} = 100 \, \Omega$), τ [ps]	567	2.27
Energy consumption per bit, P_b [pJ/bit]	8.17	0.363
Power consumption ($F = 4 \times 10^{10}$ 1/s), *Power* [mW]	327	14.5

[a] Assumed value.

9.5 IMPACTS OF POLYMER MQDs ON LIGHT MODULATOR/OPTICAL SWITCH PERFORMANCE

In Table 9.1, V, $n_{P(Off)}$, and $n_{P(On)}$ for the WPD with the 400-nm-thick polymer MQD are listed. When predicted r of 1000 pm/V is assumed for the polymer MQD as presented in Figure 8.1, Δn of 0.04 is expected to be induced by applying V of 4 V to the 400-nm-thick polymer MQD. Δn of 0.04 is the same as Δn induced in the 400-nm-thick PLZT by 1.2-V application. Therefore, if the effect of the difference in n_0 on the switching operation is neglected, by setting the driving voltage to 4 V, switching operation equivalent to that presented in Figure 9.16 is expected in the 2×2 WPD optical switch with the 400-nm-thick polymer MQD having the optimized structure shown in Figure 9.15.

In Table 9.3, predicted electrical performance is compared between the 2×2 WPD optical switch with the 400-nm-thick polymer MQD and that with the 400-nm-thick PLZT thin film. For the 400-nm-thick polymer MQD with assumed ε_s of 4, C is 0.0227 pF. τ is 2.27 ps for R_{ON} of 100 Ω, P_b is 0.363 pJ/bit for $V = 4$ V, and *Power* is 14.5 mW for $F = 4 \times 10^{10}$ 1/s. It is found that, by replacing PLZT with polymer MQD, switching time and energy consumption are reduced due to the low dielectric constant characteristics of the polymer.

In Figure 9.17, (Driving Voltage) $\times$ (Optical Switching Length), $V \times L_{OSW}$, is compared for various devices. For the WPD optical switch with the 400-nm-thick PLZT thin film shown in Figure 9.15, L_{OSW} is 310 μm and V is 1.2 V, resulting in $V \times L_{OSW} = (1.2\,V) \times (310 \times 10^{-4}\,cm) = 0.0372$ Vcm. This value is the lowest in the figure. By replacing the PLZT thin film with the 400-nm-thick polymer MQD, $V \times L_{OSW}$ becomes $(4\,V) \times (310 \times 10^{-4}\,cm) = 0.124$ Vcm.

In Figure 9.18, the energy consumption/bit P_b is compared for various devices. Here, the data for the MZM with InGaAsP/Si photonics is missing. The energy consumption of the WPD optical switch with the 400-nm-thick PLZT thin film is high. By replacing the PLZT thin film with the 400-nm-thick polymer MQD, the energy consumption of the WPD optical switch is drastically reduced.

It should be noted that in WPD optical switches with polymer MQDs, prism-shaped electrodes formed by a pair can be employed as shown in Figure 9.6 because polymer MQDs exhibit the Pockels effect. The paired-electrode configuration enables the driving voltage to be reduced from 4 to 2 V. Consequently, as indicated by a star in Figure 9.17, $V \times L_{OSW}$ in the WPD optical switch with the polymer MQD is reduced and becomes comparable to $V \times L_{OSW}$ in the WPD optical switch with PLZT. It is smaller than $V \times L_{OSW}$ in the MZMs with Si photonics [1], InGaAsP/Si photonics [2], and LN [3]. For P_b, by employing the paired-electrode configuration in the WPD optical switch with the polymer MQD, the electrode area is doubled while V^2 becomes 1/4, resulting in P_b reduction from 0.363 pJ/bit to 0.182 pJ/bit. This value is the lowest, as indicated by a star in Figure 9.18. These results indicate that the WPD optical switch with the polymer MQD is preferable for high-speed/small-size integrated light modulators/optical switches.

Figure 9.19 shows EO coefficient dependence of the deflection angle in WPD optical switches simulated by the BPM. When the DAAD-type polymer MQD described in Chapter 7 is employed, much larger deflection angles induced by low driving voltage are expected comparing to the case employing LN. Furthermore, the dielectric constant for electrical signals is expected to be much smaller in the polymer MQD

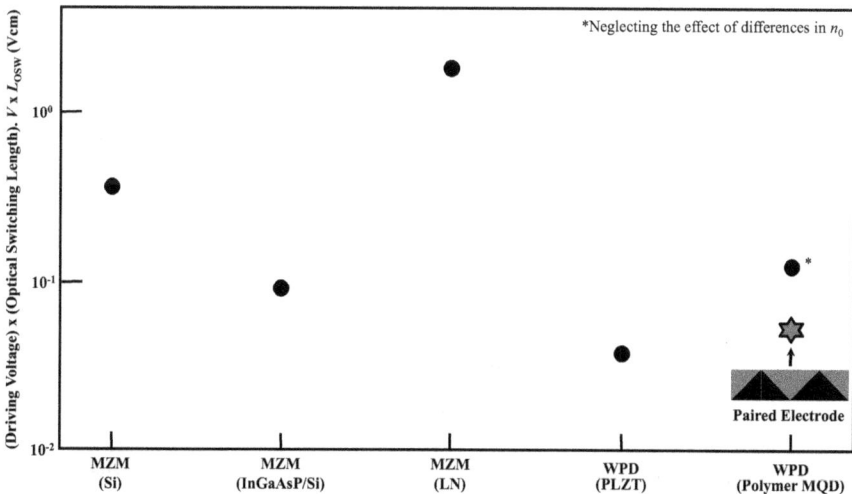

FIGURE 9.17 (Driving Voltage) $\times$ (Optical Switching Length), $V \times L_{OSW}$, for various devices.

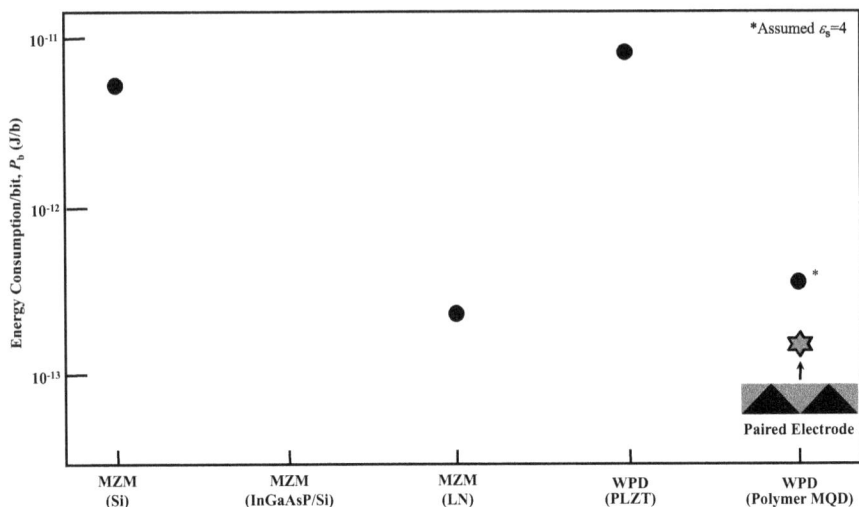

FIGURE 9.18   Energy consumption/bit, P_b, for various devices.

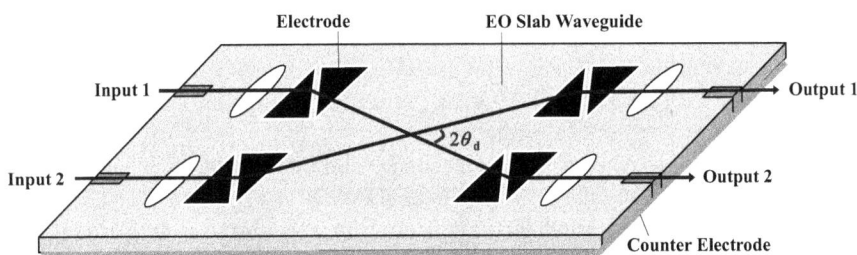

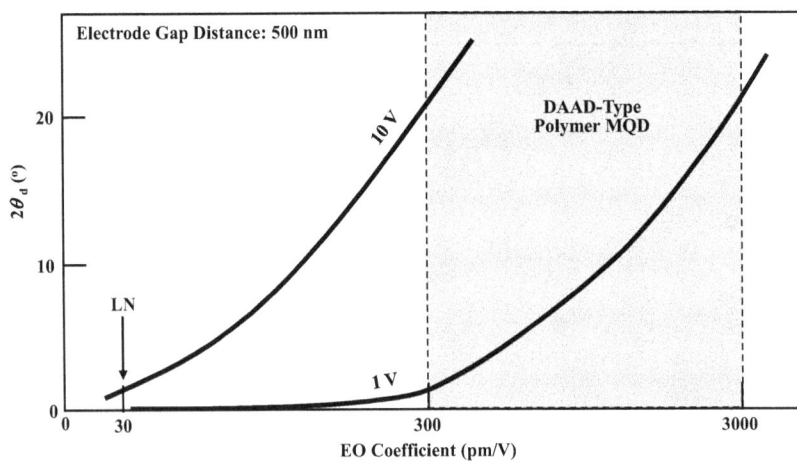

FIGURE 9.19   EO coefficient dependence of deflection angle in WPD optical switches simulated by the BPM.

than in PLZT and LN because the EO effect in polymer wires mainly arises from electronic polarization. This feature of the polymer MQD is desirable to achieve high-speed and low power operation. The above-described discussions again suggest that the WPD optical switch with the polymer MQD is a promising candidate of high-speed/small-size integrated light modulators/optical switches due to the large EO effect and low dielectric constant characteristics.

9.6 ADVANCED WPD OPTICAL SWITCHES

9.6.1 WPD OPTICAL SWITCHES WITH ADD FUNCTIONS

For optical switching systems, optical switches with ADD functions are sometimes required. Hirotaka Tomita invented the WPD optical switch with ADD functions and simulated the switching performance [23]. The structure is presented in Figure 9.20. To give the ADD functions, "ADD Waveguides" with a curvature of 5-mm radius are inserted. Single-mode optical filters [15], which extract the fundamental mode from the optical signals, are placed after the Output.

By applying driving voltage to four prism-shaped electrodes in the upper-half region, optical signals from Input 1 are connected to Output 1, and at the same time, optical signals from Input 2 are also connected to Output 1, accomplishing the ADD operation. Similarly, by applying driving voltage to four electrodes in the lower-half region, optical signals from Input 1 and Input 2 are connected to Output 2 simultaneously. When driving voltage is applied to all electrodes, optical signals from Input 1 are connected to Output 1, and optical signals from Input 2 are connected to Output 2. When no driving voltage is applied to the electrodes, optical signals from Input 1 are connected to Output 2, and optical signals from Input 2 are connected to Output 1.

Figure 9.21 shows the switching operation of the WPD optical switch with ADD functions simulated by the BPM. Smooth adding can be seen. The insertion loss and crosstalk are, respectively, 1.4 dB and −22 dB. After passing through the single-mode optical filter, they, respectively, become 5.4 dB and −21 dB. These results suggest a possibility that the WPD optical switches will realize various kinds of functional optical circuits.

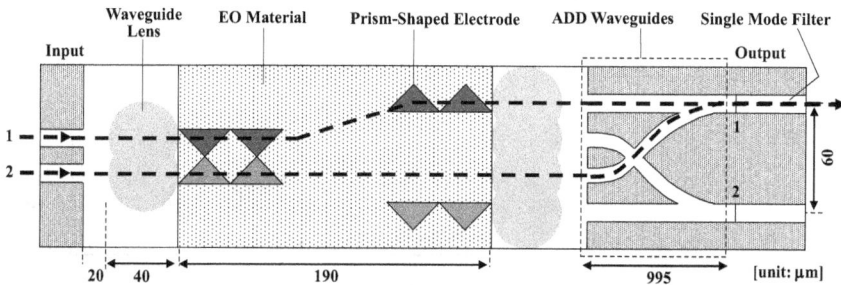

FIGURE 9.20 WPD optical switch with ADD functions, which was invented and simulated by Hirotaka Tomita.

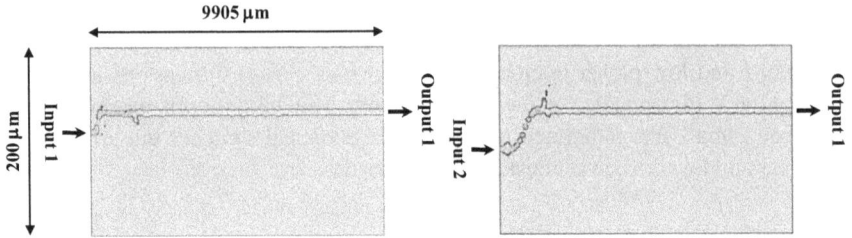

FIGURE 9.21 Switching operation simulated by the BPM for the WPD optical switch with ADD functions.

9.6.2 WPD Optical Switches with MUX/DEMUX Functions

WPD optical switches are expected to be used in WDM. Figure 9.22a shows an example of MUX consisting of WPDs and slab waveguides made of materials with large wavelength dispersion. Due to the dispersion, the angle of refraction becomes

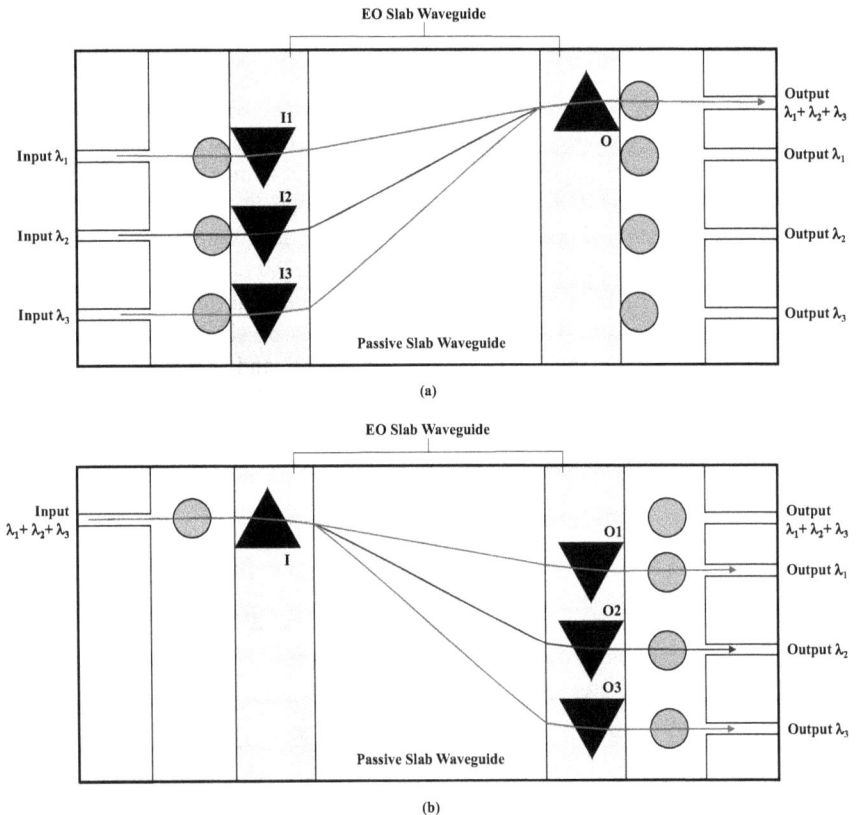

FIGURE 9.22 (a) MUX and (b) DEMUX consisting of WPDs and slab waveguides made of materials with large wavelength dispersion.

large in the order of λ_1, λ_2, and λ_3 in the present model. When no voltage is applied to the prism-shaped electrodes, signal light beams of λ_1, λ_2, and λ_3 emitted from Input λ_1, Input λ_2, and Input λ_3 are, respectively, guided into Output λ_1, Output λ_2, and Output λ_3. By applying driving voltage to electrodes of I2 and O, a signal light beam of λ_2 is selectively guided into Output $\lambda_1 + \lambda_2 + \lambda_3$. By applying driving voltage to all of the electrodes, signal light beams of λ_1, λ_2, and λ_3 are guided into Output $\lambda_1 + \lambda_2 + \lambda_3$.

Figure 9.22b shows an example of DEMUX. When no voltage is applied to the prism-shaped electrodes, signal light beams of λ_1, λ_2, and λ_3 emitted from Input $\lambda_1 + \lambda_2 + \lambda_3$ are guided into Output $\lambda_1 + \lambda_2 + \lambda_3$. By applying driving voltage to all of the electrodes, signal light beams of λ_1, λ_2, and λ_3 are, respectively, guided into Output λ_1, Output λ_2, and Output λ_3.

As described, by introducing electrical signals to individual prism-shaped electrodes, light modulation and optical switching accompanied by MUX/DEMUX operation are performed for individual wavelengths. Such MUX/DEMUX functions of WPD optical switches will enable WDM integrated optical interconnects.

9.7 NANOSCALE OPTICAL SWITCHES

MRs and PhCs are expected to work as ultra-small light modulators/optical switches. These devices, however, as mentioned in Section 9.1, have a drawback of delayed responses, causing a band width limit in ultra-high-speed systems. In the present section, the transient response is simulated by the 2-D finite-difference time-domain (FDTD) method.

Hiroshi Kaburagi simulated the operation of an MR, whose model is depicted in Figure 9.23. An EO ring waveguide with a diameter of 10 μm is inserted between two straight waveguides with a gap of 100-nm. The core is PLZT with a refractive index of 2.48. The core width is 500 nm. The cladding region is air with a refractive

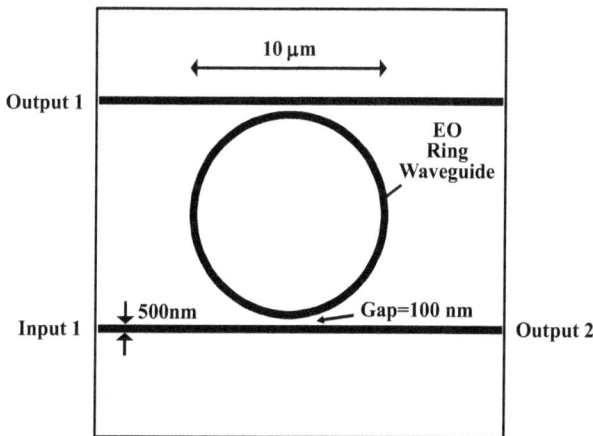

FIGURE 9.23 Model of an MR. (This work was performed by Hiroshi Kaburagi in Tokyo University of Technology.)

$\lambda=1.558\ \mu m$ **Pulse Wave Form** $E(t) = E_0 e^{-(t-t_0)^2} \cos \omega_c (t - t_0)$ $0 \leq t \leq 2t_0$
$E(t) = 0$ $t < 0, 2t_0 < t$

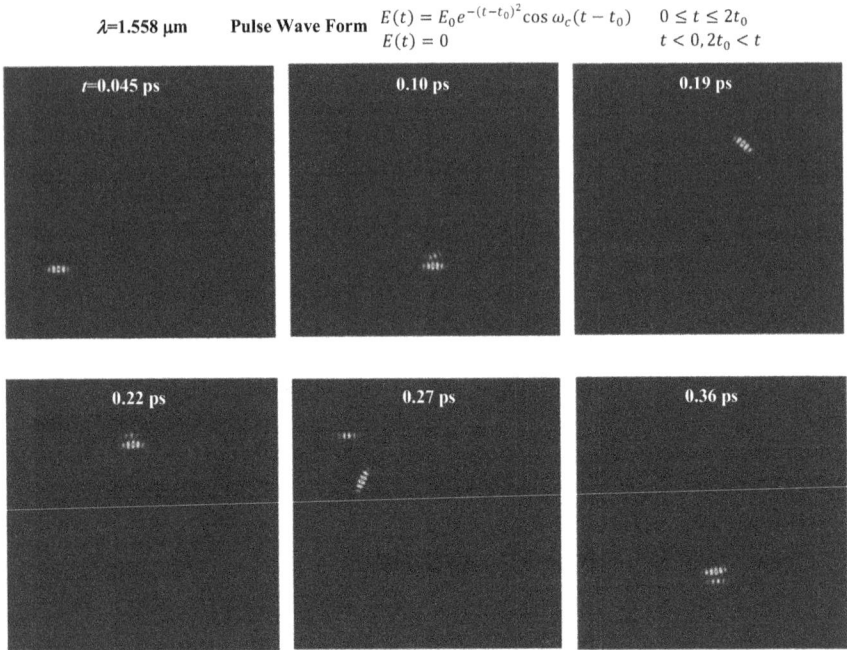

FIGURE 9.24 FDTD simulation of light pulse propagation in the MR. (This work was performed by Hiroshi Kaburagi in Tokyo University of Technology.)

index of 1.0. As Figure 9.24 shows, when a light pulse of 1.558 μm in wavelength is introduced from Input 1, it partially transfers from the straight waveguides to the ring waveguide. Here, the brightness of the light pulse image is intensified at 0.19 ps. By tuning the configuration and refractive index to a resonance condition, cw light beams introduced from Input 1 are gradually accumulated into the ring waveguide, as depicted in Figure 9.25. At the same time, the light beams are switched from Output 2 to Output 1. When voltage is applied to the ring waveguide, the resonance condition is broken to switch back the light beams from Output 1 to Output 2, as depicted in Figure 9.26. Because the resonance condition depends on the wavelength of the guided light beams, the resonance wavelength is adjusted by the driving voltage to achieve the tunable wavelength filter operation.

It is found from Figure 9.25 that it takes about 4.5 ps to complete the accumulated interference in the MR. The data rate limit is estimated to be ~110 Gbps.

The PhC has periodic variations of refractive indexes like dielectric multilayer wavelength filters. As in the case of energy bands for electrons in crystals having periodic potential energy variations, energy bands for photons appear in PhCs, as depicted in Figure 9.27 [5,6]. When the photon energy is in the photonic band gap (PBG), light beams cannot exist in the PhCs in steady states. When there are imperfections in a PhC, light beams with photon energy within the PBG can exist in the imperfection parts. By creating line-shaped imperfections for PhC waveguides, light beams are confined there.

λ=1.558 μm
n_{Core}=2.48
n_{Clad}=1.0

Pulse Wave Form

$$E(t) = \frac{1}{2}\left(1 - \cos\left(\frac{\pi t}{T_\omega}\right)\right)E_0 \sin \omega_0 t \qquad 0 \leq t \leq T_\omega$$

$$E(t) = E_0 \sin \omega_0 t \qquad\qquad T_\omega < t$$

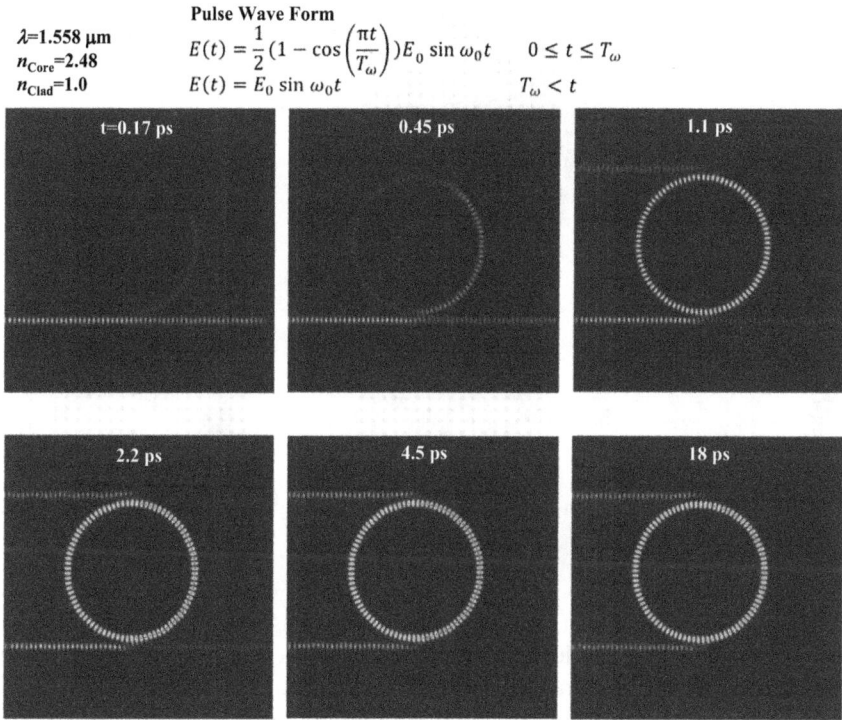

FIGURE 9.25 FDTD simulation of light beam resonance in the MR. (This work was performed by Hiroshi Kaburagi in Tokyo University of Technology.)

λ=1.558 μm, t=4.5 ps

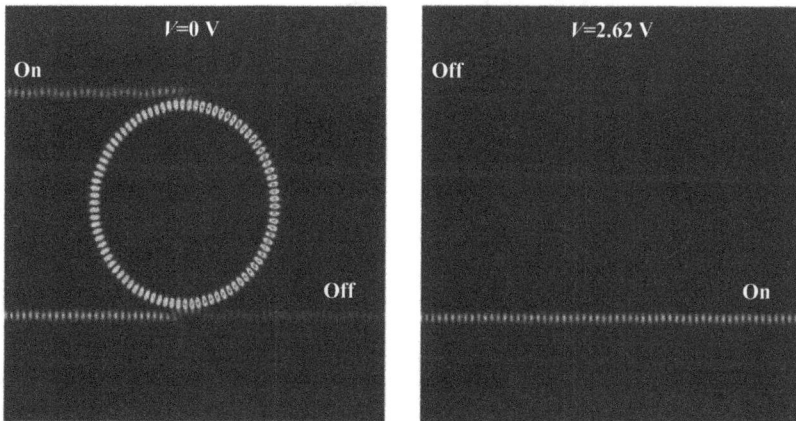

FIGURE 9.26 Simulated switching operation of the MR. (This work was performed by Hiroshi Kaburagi in Tokyo University of Technology.)

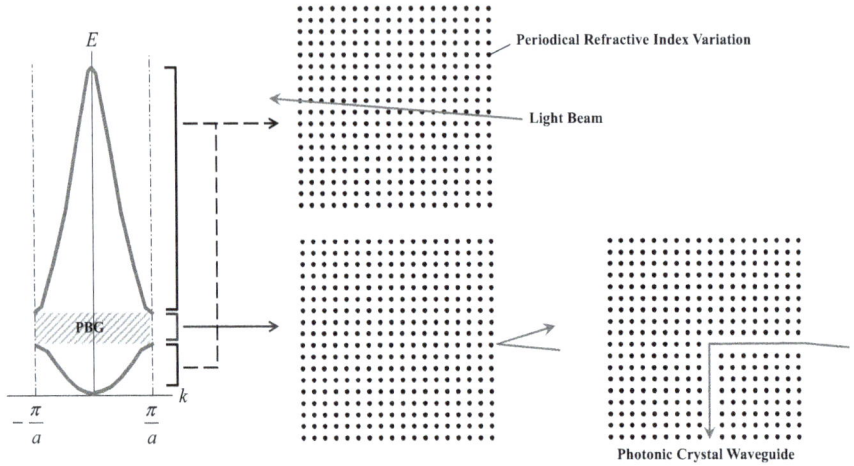

FIGURE 9.27 Band diagram for a photonic crystal and light beam propagation in a PhC waveguide.

The FDTD simulation of transient light beam propagation in a two-dimensional PhC waveguide was carried out by Keiichi Okada [24] for a model, where 104-nm-radius holes are arranged with a 521-nm pitch in a medium having a refractive index of 3.38. Light polarization is parallel to the plane for calculation. The results are shown in Figure 9.28. For a wavelength of $\lambda = 1.19\,\mu m$, which corresponds to photon energy out of PBG, light beams leak from the optical waveguide. For $\lambda = 1.30\,\mu m$, which corresponds to photon energy near the center of PBG, light

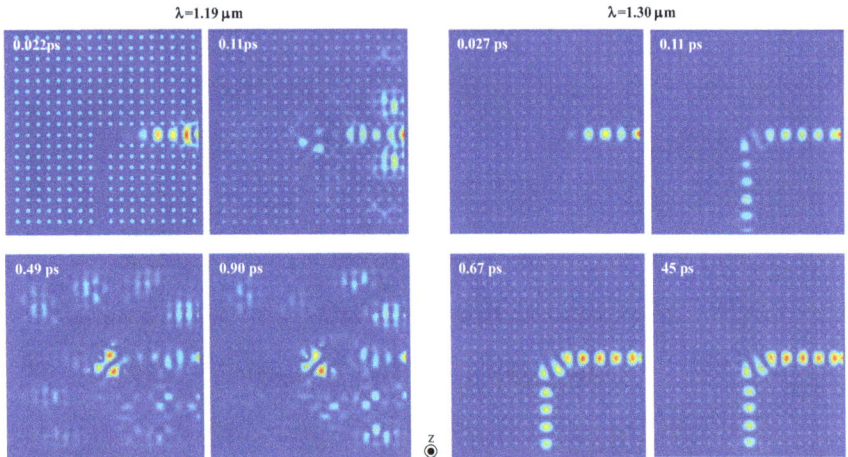

FIGURE 9.28 FDTD simulation of transient light beam propagation in a PhC waveguide for wavelengths out of and within PBG.

beams are strongly confined in the optical waveguide. However, confined light intensity is weak at the initial stage, and with time it increases to reach a saturated state. Figure 9.29 shows the transient light beam propagation, where the drawing contrast is adjusted to enhance leaked light beams in the surrounding perfect PhC region. At $t=0.05$ and 0.1 ps, considerable leakage of light beams is observed. After 0.6 ps, the leakage is suppressed and strong light beam confinement is achieved. Such delay before settling the waveguide function arises from the fact that PhC devices work by accumulated interference at a steady state. At $t=0.05$ and 0.1 ps, only a part of reflected light beams exists in the PhC, namely, light beam interference is not completed yet. At $t=0.6$ ps, superposition of all the reflected light beams is completed in whole of the PhC, namely, the interference is built-up to reach the steady state, resulting in the perfect light beam confinement in the optical waveguide. 0.6 ps corresponds to a light beam propagation distance of ~150 μm that is 17 times of the PhC size in the present model.

Transient responses of the light beam confinement in PhC waveguides are shown in Figure 9.30. For $\lambda=1.20$ μm, which corresponds to photon energy at the edge of PBG, a long delay of ~0.5 ps is observed. For $\lambda=1.30$ μm located near the center of PBG, on the other hand, the confined light intensity rapidly increases to its saturated value with a short delay less than 0.2 ps. As can be seen in Figure 9.31, the light beam distribution after reaching the steady state is wider for $\lambda=1.20$ μm than

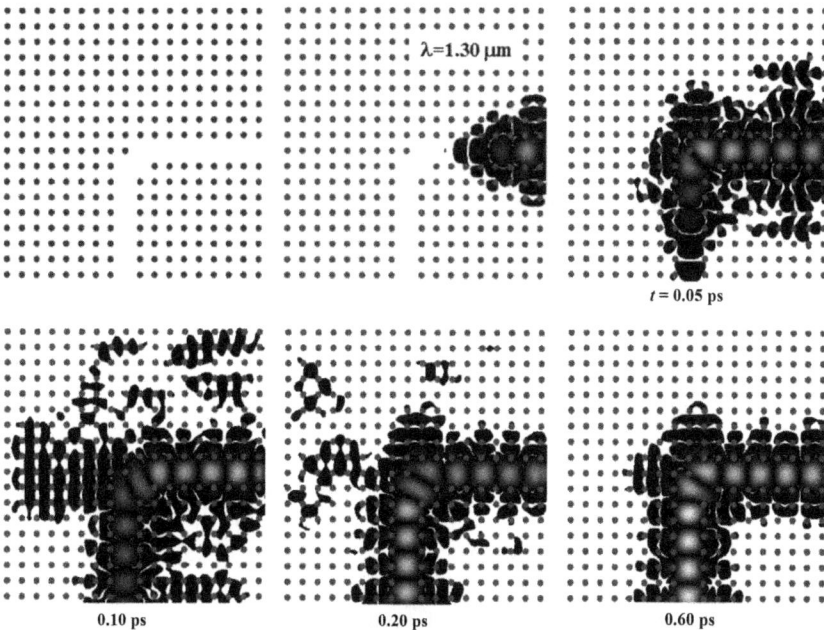

FIGURE 9.29 FDTD simulation of transient light beam propagation in a PhC waveguide with enhanced drawing contrast. (Reprinted with permission from Yoshimura et al. [24].)

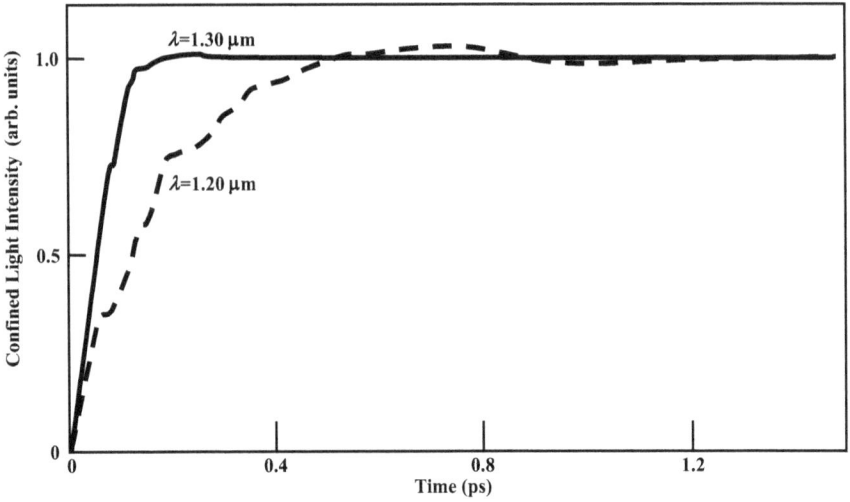

FIGURE 9.30   FDTD simulation of transient responses of light beam confinement in PhC waveguides.

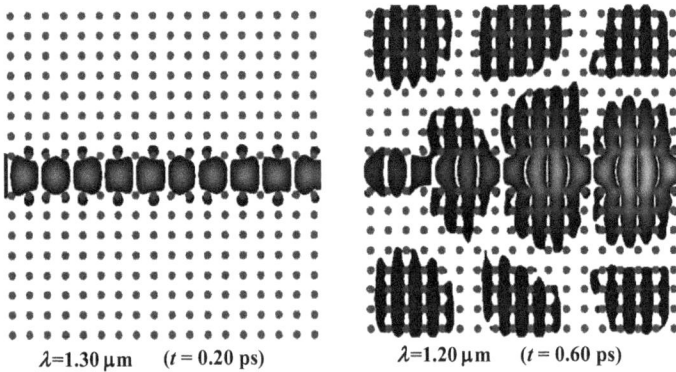

FIGURE 9.31   FDTD simulation of light beam distribution in PhC waveguides after reaching steady state.

for $\lambda = 1.30\,\mu$m. This indicates that for $\lambda = 1.20\,\mu$m, the light beams penetrate into the deep part of the surrounding perfect region, and, consequently, the interference occurs over a wide area comparing with the case of $\lambda = 1.30\,\mu$m.

Figure 9.32 presents schematic illustrations of the MR and PhC with the EO polymer MQD fabricated by MLD using the vertically-aligned selective growth from seed cores described in Section 4.5. First, seed cores with a pattern of an MR waveguide and a PhC waveguide are formed on a substrate. Then, by MLD, polymer wires are grown from the seed cores to form EO waveguides and EO rods of the polymer MQD. By applying voltage to the waveguides and the rods, optical switching operation will be available.

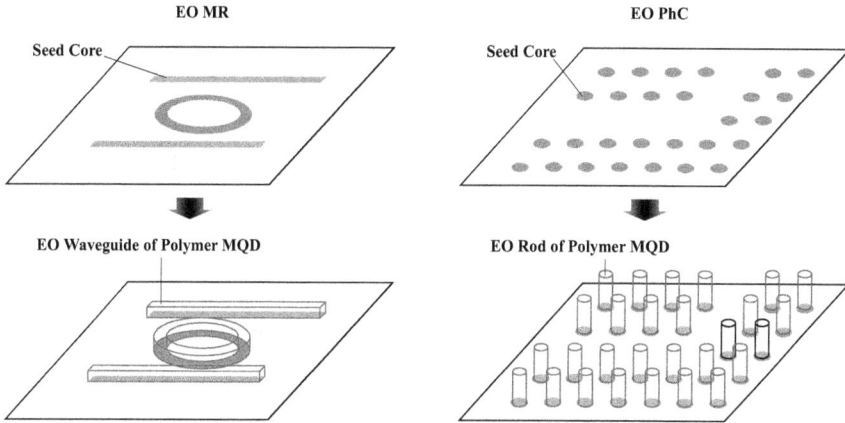

FIGURE 9.32 Schematic illustrations of the MR and PhC with the EO polymer MQD fabricated by MLD using the vertically-aligned selective growth from seed cores.

9.8 HETEROGENEOUS INTEGRATION OF LIGHT MODULATORS/OPTICAL SWITCHES

In order to build actual systems, a process for integrating light modulators/optical switches with optical components like optical waveguides and waveguide lenses/filters/mirrors as well as electronic components like integrated circuits (ICs) for drivers and large-scale integrated circuits (LSIs) is necessary. Such heterogeneous integration is expected to be achieved by the photolithographic packaging with selectively occupied repeated transfer (PL-Pack with SORT) [11,25–28], which transplants different kinds of small/thin dies onto a substrate. For integration of EO polymer MQDs, the molecular nanoduplication (MND) [24,29] described in Section 5.4 is applicable.

Figure 9.33 shows a process flow of the PL-Pack with SORT. For simplicity, two kinds of devices are considered although, in general, many kinds of devices are integrated. A thin film for a Device-I die array and a thin film for a Device-II die array are, respectively, grown on Growth Wafer-I and Growth Wafer-II. Small/thin dies are obtained by dividing the thin films with boundary grooves. The die size is $d \times d$. Using the epitaxial lift-off (ELO) process [30–34], all the Device-I dies are picked up on a pick-up substrate at one time, and are selectively pre-transferred from the pick-up substrate to Supporting Substrate I. The dies left on the pick-up substrate are provided to another supporting substrate. Similarly, the Device-II dies are selectively pre-transferred to Supporting Substrate II, and the dies left on the pick-up substrate are provided to another supporting substrate. The distribution patterns of the small/thin dies on Supporting Substrates I and II are, respectively, designed to match the pad distribution patterns for Device-I and Device-II on final substrates [11,25–28].

For the final transfer, Supporting Substrate I is attached to one of the final substrates, where catch-up sites are formed with a pitch of p_p, to place Devices-I dies at sites for Device-I on the final substrate selectively. Then, Supporting Substrate II is attached to the final substrate to place Device-II dies at sites for Device-II. The Supporting Substrates I and II are further attached to other final substrates or other

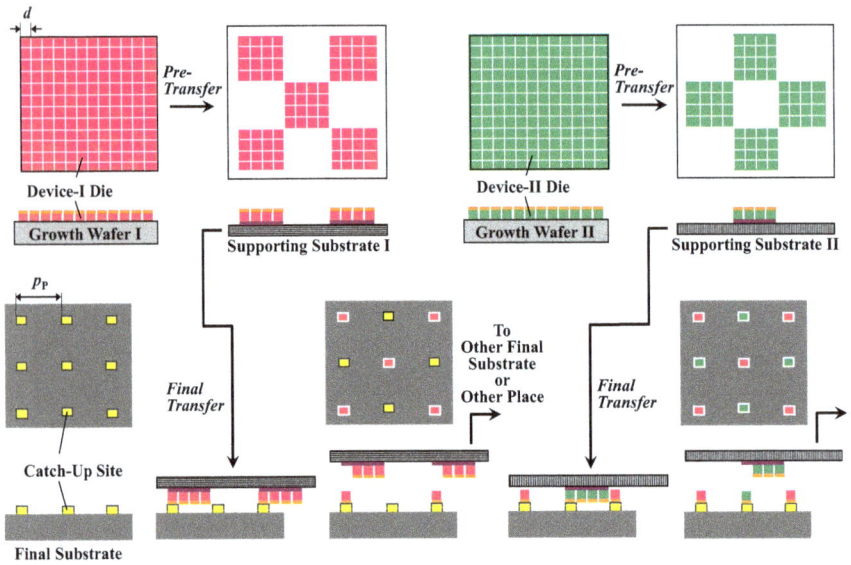

FIGURE 9.33   Process flow of the PL-Pack with SORT.

places of the final substrate to provide Device-I dies and Device-II dies selectively. By repeating the procedure, all the dies on the supporting substrates are transferred to final substrates.

In the present model, there are $4 \times 4 = 16$ dies for one catch-up site. This implies that one supporting substrate can provide dies to 16 final substrates. Because two supporting substrates are available for Device-I and Device-II, respectively, 32 final substrates, on which Device-I dies and Device-II dies are distributed with a designated configuration, can be obtained from the Growth Wafer-I and Growth Wafer-II. Device-I and Device-II could be light modulators, optical switches, wavelength filters, vertical cavity surface-emitting lasers (VCSELs), photodetectors (PDs), optical waveguides, waveguide lenses, mirrors, and so on. PL-Pack with SORT is superior to the conventional flip-chip-bonding-based packaging from the viewpoints of material consumption and assembling process cost.

Figure 9.34 shows an example of the PL-Pack with SORT process for integration of small/thin WPD optical switches, optical waveguides, and waveguide lenses [11,25–28].

1. WPD optical switch dies fabricated on a growth wafer are pre-transferred on Supporting Substrate I, and waveguide lens dies fabricated on another growth wafer are pre-transferred on Supporting Substrate II. Here, when the WPD optical switches are made of EO polymer MQDs, the MND process can be used instead of the ELO process.
2. Anti-reflection (AR) coating is applied to the edges of the dies by the selective MLD (or ALD).

3. Meanwhile, waveguide films are prepared on final substrates.
4. AR coating is applied to the edges of the waveguide cores by MLD. Catch-up sites are formed on the waveguide films.
5. The WPD optical switch dies and the waveguide lens dies are selectively transferred onto the final substrates.
6. Slab waveguide coating is performed to complete the integrated optical switches.

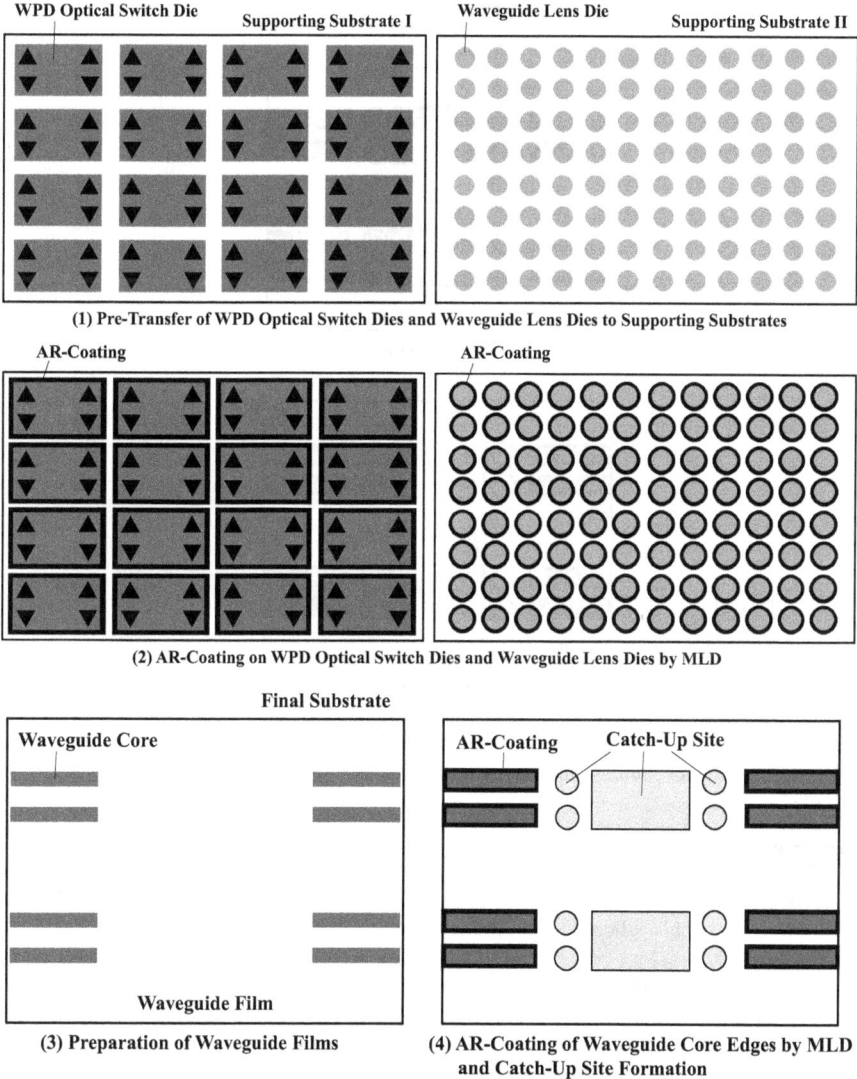

(1) Pre-Transfer of WPD Optical Switch Dies and Waveguide Lens Dies to Supporting Substrates

(2) AR-Coating on WPD Optical Switch Dies and Waveguide Lens Dies by MLD

(3) Preparation of Waveguide Films

(4) AR-Coating of Waveguide Core Edges by MLD and Catch-Up Site Formation

FIGURE 9.34 PL-Pack with SORT process for integration of WPD optical switches, optical waveguides, and waveguide lenses.

(Continued)

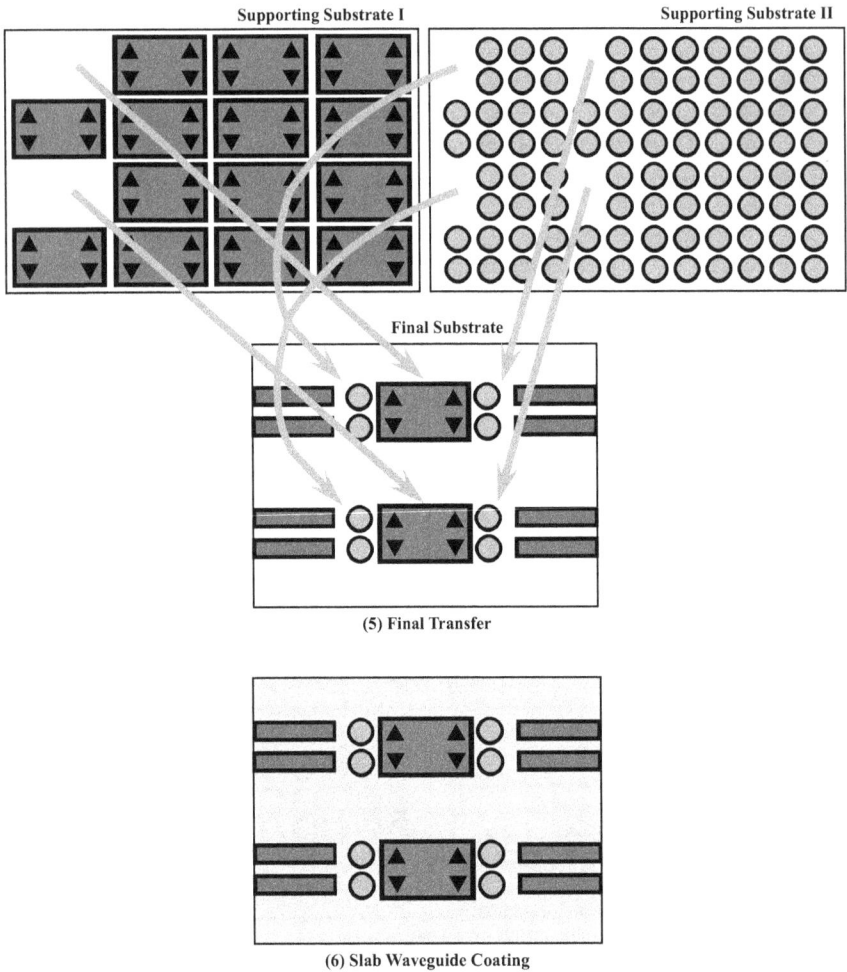

FIGURE 9.34 (*Continued*)   PL-Pack with SORT process for integration of WPD optical switches, optical waveguides, and waveguide lenses.

REFERENCES

1. W. M. J. Green, M. J. Rooks, L. Sekaric, and Y. A. Vlasov, "Ultra-compact, low RF power, 10 Gb/s silicon Mach–Zehnder modulator," *Opt. Express* **15**, 17106–17113 (2007).
2. T. Hiraki, T. Aihara, K. Hasebe, K. Takeda, T. Fujii, T. Kakitsuka, T. Tsuchizawa, H. Fukuda, and S. Matsuo, "Heterogeneously integrated III–V/Si MOS capacitor Mach–Zehnder modulator," *Nature Photonics* **11**, 482–485 (2017).
3. C. Wang, M. Hang, B. Stern, M. Lipson, and M. Loncar, "Nanophotonic lithium niobate electro-optic modulators," *Opt. Express* **26**, 1547–1555 (2018).
4. R. M. Kubacki, "Micro-optic enhancement and fabrication through variable in-plane index of refraction (VIPIR) engineered silicon nanocomposite technology," *Proc. SPIE* **5347**, 233–246 (2004).

5. E. Yablonovitch, "Inhibited spontaneous emission in solid state physics and electronics," *Phys. Rev. Lett.* **58**, 2059–2062 (1987).

6. E. Yablonovitch, T. J. Gmitter, and K. M. Leung, "Photonic band structure: The face-centered-cubic case employing nonspherical atoms," *Phys. Rev. Lett.* **67**, 2295–2298 (1991).

7. T. Yoshimura, S. Tsukada, S. Kawakami, Y. Arai, H. Kurokawa, and K. Asama, "3D micro optical switching system (3D-MOSS) –packaging design-," *2001 IEEE LEOS Annual Meeting Conference Proceedings*, San Diego, California, 499 (2001).

8. T. Yoshimura, S. Tsukada, S. Kawakami, Y. Arai, H. Kurokawa, and K. Asama, "3D micro optical switching system (3D-MOSS) architecture," *Proc. SPIE* **4653**, 62–70 (2002).

9. T. Yoshimura, "Waveguide-type electro-optic devices," Japanese Patent, Tokukai Hei 4–181231 (1992).

10. T. Yoshimura, "Optical Integrated Circuit Devices," Japanese Patent, Tokukai Hei 4–204633 (1992).

11. T. Yoshimura, M. Ojima, Y. Arai, and K. Asama, "Three-dimensional self-organized micro optoelectronic systems for board-level reconfigurable optical interconnects --performance modeling and simulation--," *IEEE J. Select. Top. Quantum Electron.* **9**, 492–511 (2003).

12. T. Yoshimura, S. Tsukada, S. Kawakami, M. Ninomiya, Y. Arai, H. Kurokawa, and K. Asama, "Three-dimensional micro-optical switching system architecture using slab-waveguide-based micro-optical switches," *Opt. Eng.* **42**, 439–446 (2003).

13. T. Yoshimura, "Design of organic nonlinear optical materials for electro-optic and all-optical devices by computer simulation," *FUJITSU Sc. Tech. J.* **27**, 115–131 (1991).

14. M. Ninomiya, Y. Arai, T. Yoshimura, H. Kurokawa, and K. Asama, "Characteristics evaluation for waveguide-prism-deflector type micro optical switches (WPD-MOS)," *Electron. Commun. Japan, Part 2* **86**, 38–48 (2003) (Translated from Denshi Joho Tsushin Gakkai Ronbunshi J85-C, 1192–1201 (2002)).

15. T. Yoshimura, M. Ojima, Y. Arai, N. Fujimoto, and K. Asama, "Simulation on crosstalk reduction in multi-mode-waveguide-based micro optical switching systems using single mode filters," *Appl. Opt.* **43**, 1390–1395 (2004).

16. T. Yoshimura and Y. Arai, "3D optoelectronic micro system," U.S. Patent 7,387,913 B2 (2008).

17. O. Qasaimeh, K. Kamath, P. Bhattacharya, and J. Phikkips, "Linear and quadratic electro-optic coefficients of self-organized $In_{0.4}Ga_{0.6}As/GaAs$ quantum dots," *Appl. Phys. Lett.* **72**, 1275–1277 (1998).

18. T. Yoshimura and T. Futatsugi., "Non-linear optical device using quantum dots," U.S. Patent 6,294,794 (2001).

19. T. Yoshimura, "Characterization of the electro-optic effect in styrylpyridinium cyanine dye thin-film crystals by an ac modulation method," *J. Appl. Phys.* **62**, 2028–2032 (1987).

20. T. Yoshimura, "Enhancing second-order nonlinear optical properties by controlling the wave function in one-dimensional conjugated molecules," *Phys. Rev. B* **40**, 6292–6298 (1989).

21. A. Matsuura and T. Hayano, "Theoretical prediction of the donor/acceptor site-dependence on first hyperpolarizabilities in conjugated systems containing azomethine bonds," *Mat. Res. Soc. Symp. Proc.* **291**, 503–508 (1993).

22. T. Aizawa, Furuuchi Chemical Corporation Technical Articles for PLZT Shutter Arrays.

23. N. Fujimoto, H. Tomita, and T. Yoshimura, "A path-independent-loss type optical cross-connect system using high-speed waveguide prism micro-optical switches in WDM networks," Research Reports of the School of Engineering, Kinki University, No. 40, 81–85 (2006).

24. T. Yoshimura, Y. Suzuki, N. Shimoda, T. Kofudo, K. Okada, Y. Arai, and K. Asama, "Three-dimensional chip-scale optical interconnects and switches with self-organized wiring based on device-embedded waveguide films and molecular nanotechnologies," *Proc. SPIE* **6126**, 612609 (2006).

25. T. Yoshimura, K. Kumai, T. Mikawa, and O. Ibaragi, "Photolithographic packaging with selectively occupied repeated transfer (PL-pack with SORT) for scalable optical link multi-chip-module (S-FOLM)," *IEEE Trans. Electron. Packag. Manufact.* **25**, 19–25 (2002).

26. T. Yoshimura, *Self-Organized 3D Integrated Optical Interconnects: With All-Photolithographic Heterogeneous Integration*, Jenny Stanford Publishing, Singapore (2021).

27. T. Yoshimura and Y. Arai, "3D optoelectronic micro system," U.S. Patent 20060261432A1 (2006).

28. T. Yoshimura, J. Roman, W-C. Wang, M. Inao, and M. T. McCormack, "Device transfer method," U.S. Patent 6,669,801 (2003).

29. T. Yoshimura, H. Tomita, T. Inoguchi, T. Yamamoto, S. Moriya, Y. Suzuki, and K. Asama, "Micro/nano devices/systems, self-organized lightwave networks, and molecular nano duplication method," Japanese Patent, Tokukai 2005–258376 (2005) [in Japanese].

30. E. Yablonovitch, T. Gmitter, J. P. Harbison, and R. Bhat, "Extreme selectivity in the lift-off of epitaxial GaAs films," *Appl. Phys. Lett.* **51**, 2222–2224 (1987).

31. W. Chang, G. Pike, C. Kao, E. Yablonobitch, *MRS 1997 Fall Meeting*, Abstracts, H10.1, 210 (1997).

32. R. Pu, C. Duan, and C. Wilmsen, "Hybrid integration of VCSEL's to CMOS integrated circuits," *IEEE J. Select. Top. Quantum Electron.* **5**, 201–208 (1999).

33. J. J. Chang, S. Jung, M. Vrazel, K. Jung, M. Lee, M. A. Brooke, N. M. Jokerst, and S. Wills, "A hybrid optically interconnected microprocessor: An InP I-MSM integrated onto a mixed signal CMOS analog optical receiver with a digital CMOS microporcessor," *Proc. SPIE* **4089**, 708–714 (2000).

34. C. B. Morrison, R. L. Strijek, and J. H. Bechtel, "New processing approach for grafting optoelectronic devices and applications to multichip modules," *Proc. SPIE* **3005**, 120–127 (1997).

10 Self-Organized Lightwave Network (SOLNET)

Self-organized lightwave network (SOLNET) enables (1) self-aligned optical coupling between misaligned optical devices with different core diameters, providing optical solder functions, (2) as-desired drawing of optical waveguides in a free space, and (3) targeting light waves onto specific objects. SOLNET reduces the efforts to assemble optoelectronic (OE) devices with strict positional/angular control, and simplifies the fabrication process of vertical optical waveguides. SOLNET is the key to build the three-dimensional (3-D) integrated optical interconnects and optical switching systems described in Chapter 11, and is applicable to the solar energy conversion systems described in Chapter 12. SOLNET is also useful for photo-assisted cancer therapy described in Chapter 13, where SOLNET is combined with liquid-phase molecular layer deposition (LP-MLD) to perform laser surgery and photodynamic therapy. Details of SOLNET are described in a book, T. Yoshimura, *Self-Organized Lightwave Networks: Self-Aligned Coupling Optical Waveguides*, CRC Press/Taylor & Francis, 2018. In the present chapter, after the concept of SOLNET is explained, optical coupling performance is predicted by the finite-difference time-domain (FDTD) method. And then, experimental demonstrations are presented for one-photon SOLNETs and two-photon SOLNETs.

10.1 CONCEPT OF SOLNET

Three capabilities of SOLNET are schematically shown in Figure 10.1 [1–4].

1. Self-aligned optical coupling between misaligned optical devices with different core diameters.
2. As-desired drawing of optical waveguides in a free space.
3. Targeting light waves onto specific objects.

The inherent light wave nature of "Go Straight" causes a problem that strict positional/angular control is necessary to guide a light wave from Device A to Device B. By the capability (1) of SOLNET, as depicted in Figure 10.1a, the light wave can propagate from Device A to Device B automatically even if there are lateral/angular misalignments between the two devices. This provides optical solder functions with the self-aligned optical coupling effect. By the capability (2), 3-D optical wiring can be fabricated in a free space, as depicted in Figure 10.1b. Thus, SOLNET reduces the efforts to assemble OE devices with strict positional/angular control, and simplifies the fabrication process of vertical optical waveguides. SOLNET is the key to build the 3-D integrated optical interconnects and optical switching systems described in

DOI: 10.1201/9781003094012-10

(a) Self-Aligned Optical Coupling

(b) As-Desired Drawing of Optical Waveguides

(c) Targeting Light Waves onto Specific Objects

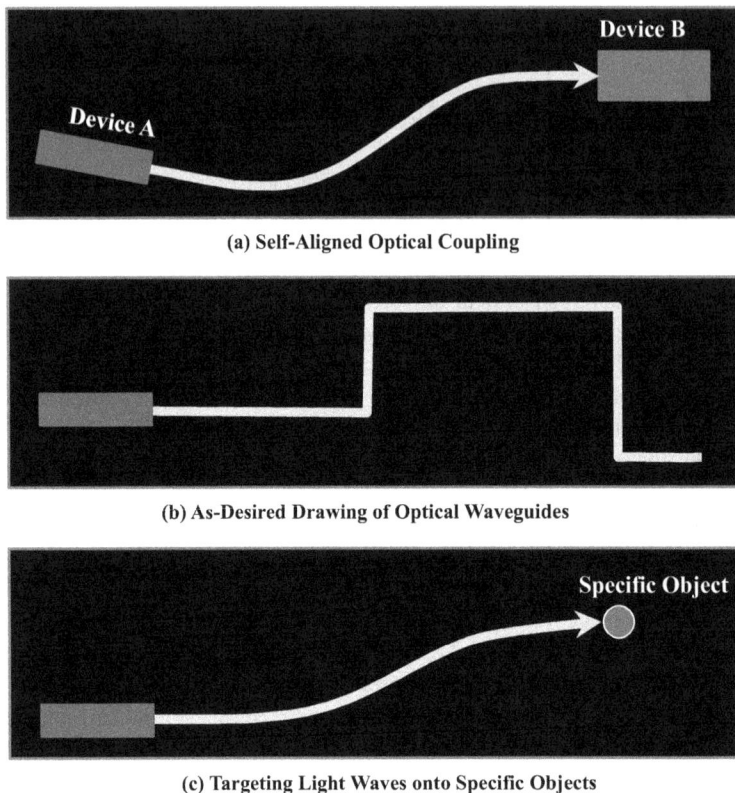

FIGURE 10.1 Three capabilities of SOLNET. (a) Self-aligned optical coupling, (b) as-desired drawing of optical waveguides, and (c) targeting light waves onto specific objects.

Chapter 11, and is applicable to the solar energy conversion systems described in Chapter 12.

The capability (3) of SOLNET is useful for photo-assisted cancer therapy described in Chapter 13, where SOLNET is combined with LP-MLD to perform laser surgery and photodynamic therapy.

10.1.1 TYPES OF SOLNET

Figure 10.2 presents four types of SOLNETs: two-beam-writing SOLNET (TB-SOLNET) [1–8], phosphor SOLNET (P-SOLNET) [1,7,8], reflective SOLNET (R-SOLNET) [1,2,7–16], and luminescence-assisted SOLNET (LA-SOLNET) [1,7,8]. OE devices, such as optical waveguides, light modulators, optical switches, laser diodes (LDs), and photodiodes (PDs), are placed in counter-direction arrangements with a photo-induced refractive-index increase (PRI) material filling in between. The refractive index of the PRI material increases upon write beam exposure. SOLNETs are formed by the interaction between a plurality of the write beams.

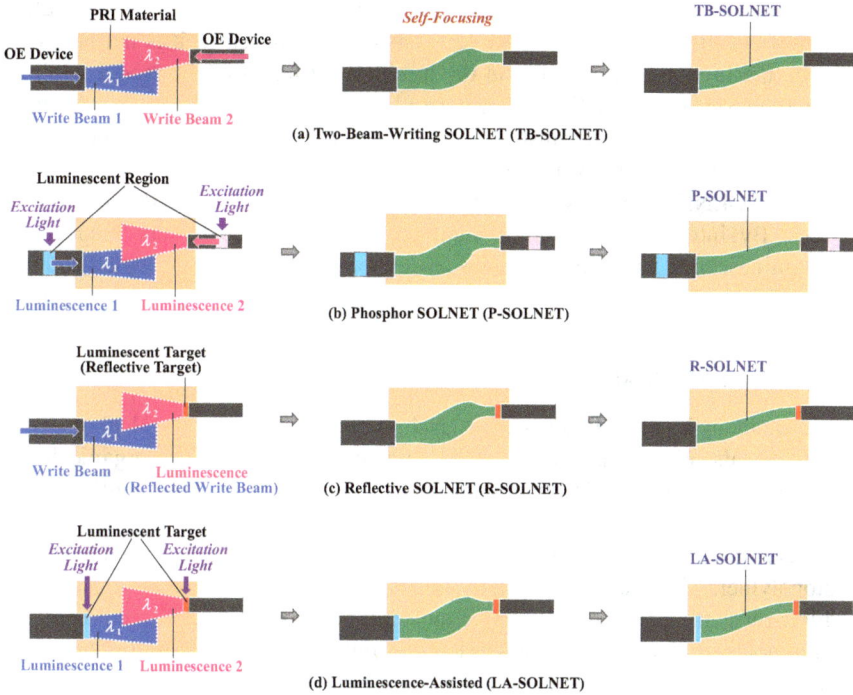

FIGURE 10.2 Types of SOLNETs. (a) Two-beam-writing SOLNET (TB-SOLNET), (b) phosphor SOLNET (P-SOLNET), (c) reflective SOLNET (R-SOLNET), and (d) luminescence-assisted (LA-SOLNET). (Reprinted with permission from Yoshimura et al. [8].)

In TB-SOLNET, a first write beam (write beam 1) with a wavelength of λ_1 is introduced into a PRI material from an OE device while a second write beam (write beam 2) with a wavelength of λ_2 is introduced from another OE device. The refractive index in the region, where the two write beams overlap, increases rapidly compared with that in the surrounding region. Then, because the two write beams tend to be confined in the higher-refractive-index region, they merge through self-focusing to form a self-aligned coupling waveguide of TB-SOLNET between the OE devices, even if misalignments and core size mismatching exist between them.

In P-SOLNET, luminescent regions are prepared in OE device cores to generate luminescence by excitation lights from outside. Luminescence 1 of λ_1 and luminescence 2 of λ_2, respectively, correspond to write beam 1 and write beam 2 in the TB-SOLNET. In the case that the cores are emissive intrinsically, the luminescent region preparation is not necessary, and the luminescent regions in the model imply core regions, which are exposed to the excitation lights.

In R-SOLNET, a luminescent target is placed on a front core edge of one of the OE devices. When a write beam of λ_1 is introduced from the other OE device into the PRI material, the target absorbs the write beam and emits luminescence of λ_2, which corresponds to the second write beam. The luminescence acts as a trigger to let the incident write beam flow toward the emitting point. It paves the way of a self-organized waveguide toward the target. This phenomenon is called the "pulling water" effect. In the case that the core is emissive intrinsically, it is not necessary to place the luminescent target, and the luminescent target in the model implies a luminescent surface of the OE device core edge. It is also possible to use a reflective target like a wavelength filter instead of a luminescent target, where a reflected write beam induces the "pulling water" effect.

In LA-SOLNET, luminescent targets are placed on front core edges of both OE devices. By introducing excitation lights from outside, the targets, respectively, emit luminescence 1 of λ_1 and luminescence 2 of λ_2 to form a self-aligned coupling waveguide.

10.1.2 PRI Material

There are many kinds of PRI materials such as photopolymers [17–23], photo-definable materials, photosensitive glass [24], photosensitive organic/inorganic hybrid materials [25,26], photorefractive materials [27,28], and third-order nonlinear optical materials [29].

Figure 10.3a shows a typical mechanism of the refractive-index increase in a photopolymer. High-refractive-index photoreactive molecules (high-n molecules) and low-refractive-index photoreactive molecules (low-n molecules) are mixed. The high-n molecules have higher photoreactivity to the write beam than the low-n molecules. When the photopolymer is exposed to a write beam, the high-n molecules are combined to make dimers, oligomers, and polymers. Then, high-n molecules diffuse into the exposed region from the surrounding region to compensate for the reduction of the high-n molecule concentration, and are combined to make dimers, oligomers, and polymers. This repeated process increases the refractive index of the exposed region. Curing by blanket light exposure or heating fixes the refractive-index distributions to produce permanent SOLNETs.

Figure 10.3b shows a refractive-index change mechanism in photorefractive materials such as lithium niobate (LiNbO$_3$ (LN)) crystals and photorefractive polymers. When a photorefractive material is exposed to a write beam, electrons excited by the write beam are captured in electron traps, generating local electric fields. The electric fields give rise to refractive-index change due to the electro-optic (EO) effect to induce inhomogeneous refractive-index distribution. Blanket exposure of erasing beams releases the trapped electrons to clear the refractive-index distribution. This makes the material ready for rewriting to produce dynamic SOLNETs.

Figure 10.3c shows a case, in which a third-order nonlinear optical material such as polydiacetylene and compound semiconductors are used for the PRI material. When the material is exposed to a write beam, inhomogeneous refractive-index distribution is induced due to the optical Kerr effect. The refractive-index distribution disappears just after the write beam exposure stops to reset the material. This enables to produce dynamic SOLNETs.

The dynamic SOLNETs is expected to provide optical switches. For example, by selectively introducing excitation lights from outside as depicted in Figure 10.4a, the switching between the bar-state and the cross-state is performed. Multi-channel switching is also possible, as illustrated in Figure 10.4b.

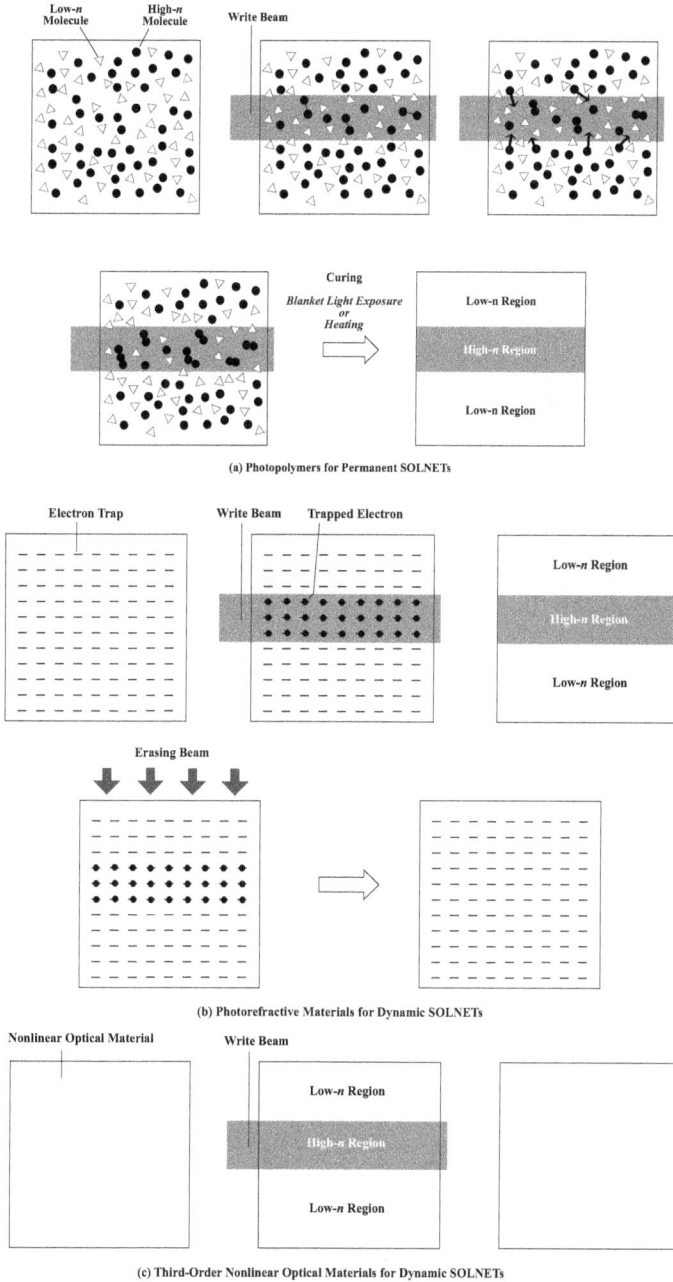

FIGURE 10.3 Typical mechanisms of the refractive-index increase in PRI materials. (a) Photopolymers for permanent SOLNETs, (b) photorefractive materials for dynamic SOLNETs, and (c) third-order nonlinear optical materials for dynamic SOLNETs.

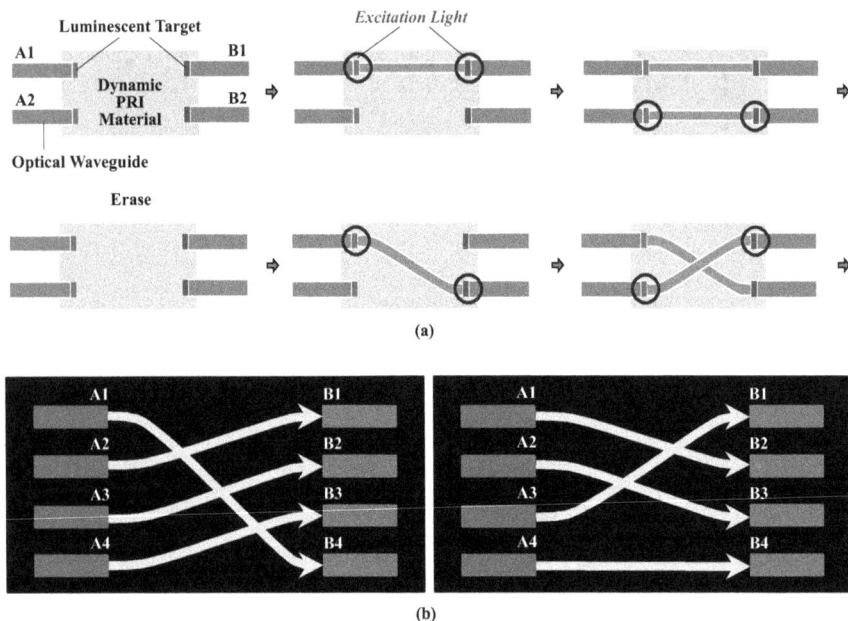

FIGURE 10.4 Concept of optical switches employing dynamic SOLNETs. (a) Switching between the bar-state and the cross-state and (b) multi-channel switching.

10.1.3 ONE-PHOTON AND TWO-PHOTON SOLNETS

In the one-photon SOLNET, a PRI material exhibiting one-photon photochemistry is used. As depicted in Figure 10.5a, a λ_1-write beam excites an electron from S_0 to S_n state in sensitizing molecules to induce chemical reaction, specifically an increase in the refractive index of the PRI material. In parallel, a λ_2-write beam excites an electron from S_0 to $S_{n'}$ state to induce an increase in the refractive index. Therefore, the refractive-index increase rate R is expressed as

$$R = \gamma_1 I_1 + \gamma_2 I_2, \tag{10.1}$$

where I_1 and I_2 are, respectively, the intensity of the λ_1-write beam and the λ_2-write beam, and γ_1 and γ_2 are, respectively, the PRI material sensitivity to the λ_1-write beam and to the λ_2-write beam. In the case that $\gamma_1 \approx \gamma_2 = \gamma$, Equation (10.1) is simplified as

$$R = \gamma (I_1 + I_2). \tag{10.2}$$

In the two-photon SOLNET [1,2,6–8], a PRI material exhibiting two-photon photochemistry is used. Brauchle et al. reported two-photon sensitizing molecules such as biacetyl (BA) and camphorquinone (CQ) for holography [30]. They incorporated these molecules in materials for hologram formation. These two-photon sensitizing molecules are applicable to the two-photon SOLNET formation. As depicted in in Figure 10.5b, an electron excited from S_0 to S_n state by a λ_1-write beam transfers to T_1 state, and is further excited to T_n state by a λ_2-write beam, resulting in chemical

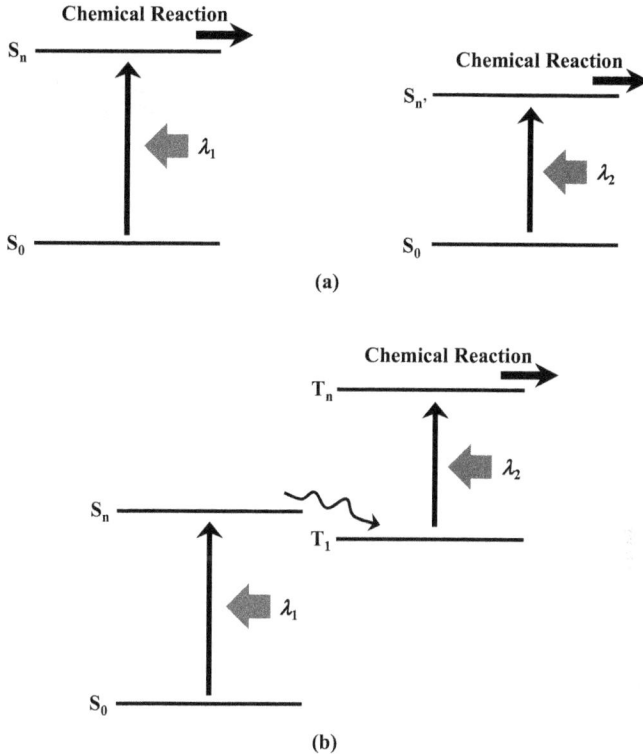

FIGURE 10.5 Energy-level schemes for sensitization with (a) one-photon photochemistry and (b) two-photon photochemistry [7].

reaction to increase the refractive index. In this case, because two-step electron excitation occurs in series, the refractive-index increase rate is expressed as

$$R \propto I_1 I_2. \tag{10.3}$$

The above-described difference in the write beam intensity dependence of R causes differences in formation dynamics between one-photon and two-photon SOLNETs. As illustrated in Figure 10.6a, for the one-photon SOLNET, because I_1 and I_2 are large near front core edges of the OE devices, the refractive index increases rapidly near the EO device front edges. This prevents the write beams from expanding into the write-beam-overlapping region in the middle, resulting in two separated optical waveguides stretching from individual OE devices. This means that the one-photon photochemistry cannot form SOLNETs when the misalignment is extremely large.

For the two-photon SOLNET, on the other hand, the refractive index increases with a rate proportional to $I_1 I_2$, which implies that it increases only if the λ_1-write beam and the λ_2-write beam coexist. Then, as shown in Figure 10.6b, the refractive-index increase near the OE device front edges becomes less notable because I_2 is very small near the left-hand-side device and I_1 is very small near the right-hand-side device. In the write-beam-overlapping region in the middle, where both of I_1 and I_2

(a)

(b)

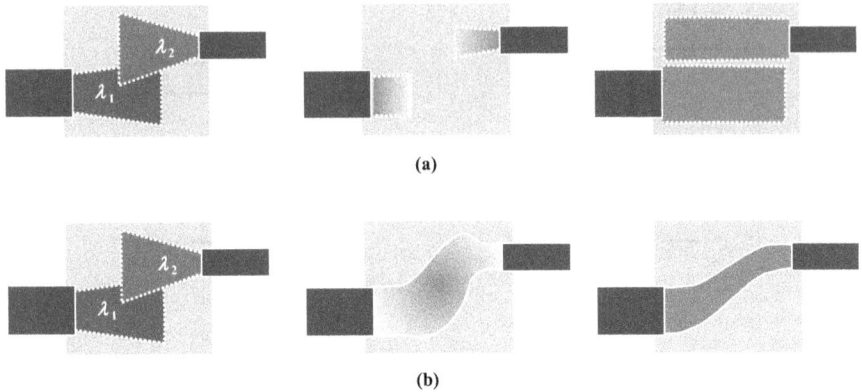

FIGURE 10.6 Schematic illustrations of waveguide formation dynamics for cases with extremely large misalignments in PRI materials (a) with one-photon photochemistry and (b) with two-photon photochemistry.

are not so small, the refractive-index increase develops to make the two write beams merge to form a coupling optical waveguide of SOLNET. Thus, the two-photon photochemistry enhances the write-beam-overlapping effect to extend tolerance to misalignments. The two-photon SOLNET also has an advantage that it suppresses the waveguide broadening caused by overexposure.

10.1.4 Fabrication Processes of Luminescent Targets

To form R-SOLNETs and LA-SOLNETs, luminescent targets should be deposited just on the core regions of OE device edges [16]. Figure 10.7a shows the light curing process. First, an edge of an OE device is coated with a luminescent material like a phosphor-doped photopolymer. Then, to cure the region just on the core, the luminescent material is exposed to curing light that propagates in the core. The luminescent target is obtained by etching the noncured region.

Figure 10.7b shows the selective deposition process. Different surface properties are given to the core surface and the cladding surface of an OE device edge. For example, the core and cladding surfaces are, respectively, hydrophobic and hydrophilic, which is the case of a Si waveguide with SiO_2 cladding. When a hydrophobic luminescent material is introduced on the surface, it is deposited on the hydrophobic core surface selectively to form a luminescent target. The selective deposition process can also be performed by surface treatments such as self-assembled monolayer (SAM) formation with designated patterns, as shown in Figure 10.7c.

The luminescent material can be organic or inorganic thin films, which are grown by vacuum evaporation [14], chemical vapor deposition (CVD) [31,32], atomic layer deposition (ALD) [33], or molecular layer deposition (MLD) [31,34–36]. MLD enables selective conformal growth of organic thin films with designated molecular arrangements as mentioned in Chapter 4. Poly-azomethine (poly-AM) thin films grown by MLD emit blue-green luminescence, as described in Section 6.4.5 [34], and they can selectively be grown on hydrophilic surfaces, as described in Section 4.2 [31,32,34,36]. When hydrophilic and hydrophobic treatments are, respectively,

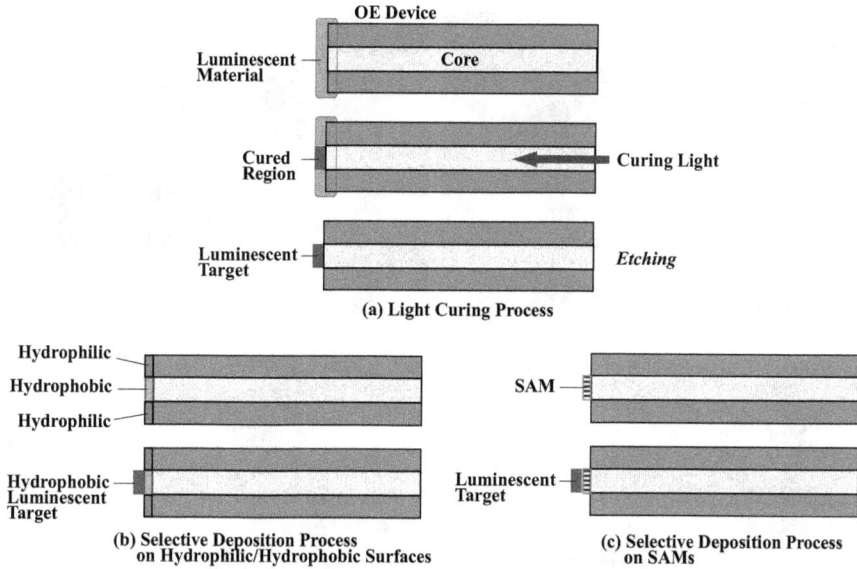

FIGURE 10.7 Fabrication processes of luminescent targets. (a) Light curing process, (b) selective deposition process on hydrophilic/hydrophobic surfaces, and (c) selective deposition process on SAMs. (Reprinted with permission from Yoshimura [16].)

applied on the core and cladding surfaces, a poly-AM thin film is deposited on the core surface selectively to form a luminescent target.

10.2 PREDICTED PERFORMANCE OF SOLNETs

10.2.1 MODELS AND SIMULATION PROCEDURES

Models for computer simulation of SOLNETs in the nano–nano coupling are shown in Figure 10.8a. Input and output waveguides with a core width of 600 nm are placed with a gap distance of 32 μm. A PRI material is placed in the gap region. The refractive index of the cores is 2.0 and that of the cladding region is 1.5. The refractive index of the PRI material increases from 1.5 to 1.7 upon write beam exposure. Lateral misalignment between input and output waveguides is expressed by d [1,6–8].

In TB-SOLNET, 30 point light sources for the λ_1-write beam are distributed across the core along the y-direction with a pitch of 20 nm at $x=1.4$ μm in the input waveguide, and 30 point light sources for the λ_2-write beam are distributed at $x=47.4$ μm in the output waveguide. Here, the origin of the x–y coordinate is set at the left-hand-side bottom corner of the model. The electric field of the point light source is expressed as follows:

$$
\left.
\begin{aligned}
E(t) &= \frac{1}{2}\left(1-\cos\left(\frac{\pi t}{T_\omega}\right)\right)E_0 \sin \omega_0 t \quad 0 \le t \le T_\omega \\
E(t) &= E_0 \sin \omega_0 t \qquad\qquad\qquad\quad T_\omega \le t
\end{aligned}
\right\}
\tag{10.4}
$$

(a) Nano-Nano Coupling

(b) Micro-Nano Coupling

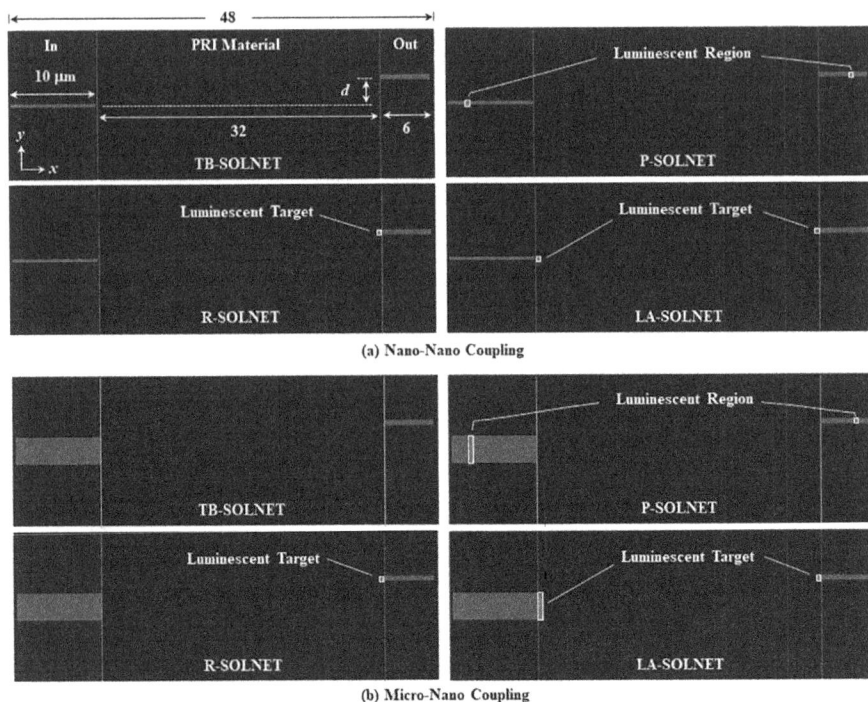

FIGURE 10.8   Models for computer simulation of SOLNETs (a) between nanoscale wave-guides and (b) between a microscale waveguide and a nanoscale waveguide [7].

ω_0 is the angular frequency of the light wave and $\omega_0 T_\omega = 4\pi$. The electric-field ampli-tude of the write beam at the point light source location is $E_0 = 6$ mV/m.

In P-SOLNET, the locations and arrangements of the point light sources in the luminescent regions are the same as those for the write beams in TB-SOLNET. This implies that the model for P-SOLNET is exactly the same as that for TB-SOLNET, and consequently, the results for TB-SOLNET are regarded as the results for both TB-SOLNET and P-SOLNET.

In R-SOLNET and LA-SOLNET, the point light source arrangements in the lumi-nescent target are the same as those for the write beam in TB-SOLNET. For the R-SOLNET, the electric-field amplitude of the emitted luminescence is determined by assuming that it is proportional to the temporary time average of absolute values of the λ_1-write-beam electric field at the point light source location. Efficiency for the luminescence generation is set to 0.7.

Brauchle et al. reported that the absorption spectra of a two-photon sensitizing mol-ecule CQ for λ_1 and λ_2 are, respectively, located in wavelength regions of 400–520 and 580–1100nm (see Figure 10.20) [30]. Based on these data, simulations were performed for a case of $[\lambda_1 = 500\text{nm}, \lambda_2 = 600 \text{ nm}]$. To assess the coupling efficiency between the input and output waveguides, a 650-nm probe beam is used. The point light source loca-tions and arrangements for the probe beam are the same as those for the λ_1-write beam.

Models for computer simulation of SOLNETs in the micro–nano coupling are shown in Figure 10.8b. Core widths of input and output waveguides are, respectively, 3 μm and 600 nm. The basic structures of the models are the same as those in the nano–nano coupling except for the following parameters. Seventy-five point light sources are distributed across the core with a pitch of 40 nm in the write beam generation region, the luminescent region, and the luminescent target of the 3-μm-wide input waveguide.

The computer simulation was performed by the two-dimensional FDTD (2-D FDTD) method under the small-absorption-limit condition, in which the light absorption term in PRI materials is neglected. For the simulation of SOLNET formation, after inputting the initial refractive-index distribution, electric and magnetic fields are calculated using the absorption boundary conditions. Because the refractive index of the PRI material changes with time, its distribution is rewritten at each time step n [1,6–8].

The energy density of the λ_1-write beam exposure and that of the λ_2-write beam exposure on the PRI material over a time interval for one step, Δt, are, respectively, $(1/2)\varepsilon E_1^2 v \Delta t$ and $(1/2)\varepsilon E_2^2 v \Delta t$, where ε is the dielectric constant and v is the light wave velocity. E_1 and E_2 denote the respective electric-field amplitudes for the λ_1-write beam and the λ_2-write beam. Then, for the one-photon SOLNET, the refractive-index change Δn during Δt is derived from Equation (10.1) as

$$\Delta n = \frac{1}{2} \varepsilon v \left(\gamma_1 E_1^2 + \gamma_2 E_2^2 \right) \Delta t. \tag{10.5}$$

In the present calculation, it is assumed that the refractive-index changes in proportion to write beam exposure until it reaches a saturation limit Δn_{sat}, and γ_1 and γ_2 are set to the same value, γ. In this case, Equation (10.5) is simplified with a constant of proportionality:

$$\Delta n = C_{1Photon} (E_1^2 + E_2^2)\Delta t, \tag{10.6}$$

$$C_{1\,Photon} = \frac{1}{2} \gamma \varepsilon v. \tag{10.7}$$

For the two-photon SOLNET, since R is proportional to $I_1 I_2$, Δn during Δt can be expressed as

$$\Delta n = C_{2Photon} (E_1^2 E_2^2)\Delta t. \tag{10.8}$$

The values of $C_{1Photon}$ and $C_{2Photon}$ are adjusted so that duration of two-photon SOLNET formation is comparable to that of one-photon SOLNET formation. This means adjusting the PRI material sensitivity. In the present simulation, constants of proportionality shown in Table 10.1 are used.

TABLE 10.1

Constants of Proportionality in 2-D FDTD Simulation for SOLNETs

Simulation Model	$C_{1Photon}$ [$\times 10^{14}$ (m/V)^2/s]	$C_{2Photon}$ [$\times 10^{18}$ (m/V)^4/s]
TB-SOLNET/P-SOLNET	1.5	1.5
R-SOLNET	0.83	1.5
LA-SOLNET	2.6	9.1

The mesh sizes are $\Delta x = \Delta y = 20$ nm, and $\Delta t = 0.0134$ fs. The polarization direction is perpendicular to the calculated plane. In real systems, the typical PRI material response time for SOLNET formation is over 1 s as can be seen, for example, in Figures 10.16, 10.21, and 10.24. In the 2-D FDTD method, however, it is difficult to perform calculation with such long time spans because Δt should be on a sub-fs scale. To solve the problem, the time parameter in the results was rescaled as follows: $\Delta t' = \Delta t \times 10^{15}$ and $\gamma' = \gamma \times 10^{-15}$. This corresponds to the FDTD calculation with $\Delta t = 0.0134$ fs referring to a PRI material with very high sensitivity.

10.2.2 PREDICTED PERFORMANCE

As mentioned in Section 10.2.1, because TB-SOLNET and P-SOLNET are calculated by the same model, the results for TB-SOLNETs are, at the same time, those for P-SOLNETs [1,6–8].

Figure 10.9a shows one-photon TB-SOLNET formation between 600-nm-wide waveguides by write beams of $\lambda_1 = 500$ nm and $\lambda_2 = 600$ nm for $d = 3000$ nm. The left-hand-side, middle, and right-hand-side columns, respectively, represent the relative dielectric constant, i.e., the square of the refractive index, n^2, the intensity of the write beams, E^2 (Write Beam), and the intensity of the probe beams, E^2 (Probe Beam). At writing time of $t_w = 0.81$ s, traces of optical waveguides begin to grow from the input and output waveguide edges. With writing time, the optical waveguides extend to construct completely-separated two optical waveguides, and at the same time, considerable waveguide broadening occurs. In two-photon TB-SOLNET, on the other hand, as can be seen in Figure 10.9b, the two write beams begin to merge at 2.2 s and a self-aligned coupling waveguide is rapidly constructed. A 650-nm probe beam introduced from the input waveguide is smoothly guided toward the output waveguide at 2.7 s.

In Figure 10.10a, one-photon TB-SOLNET for various lateral misalignments is summarized. For $d = 600$ nm, a self-aligned coupling waveguide of SOLNET is formed, and the probe beam is guided to the output waveguide. For $d \geq 1800$ nm, normal SOLNET cannot be formed. In two-photon TB-SOLNETs, as shown in Figure 10.10b, normal SOLNET is formed even when d is increased up to 4200 nm. In Figure 10.11, dependence of maximum coupling efficiency on lateral misalignments is shown for TB-SOLNETs and butt joints. By introducing SOLNET, the lateral misalignment tolerance greatly increases from that in butt joints. In accordance with the results shown in Figure 10.10, the two-photon TB-SOLNET exhibits wider tolerance than the one-photon TB-SOLNET as predicted in Section 10.1 (see Figure 10.6).

(a) One-Photon TB-SOLNET/P-SOLNET (Nano-Nano): Failed

(b) Two-Photon TB-SOLNET/P-SOLNET (Nano-Nano): Succeeded

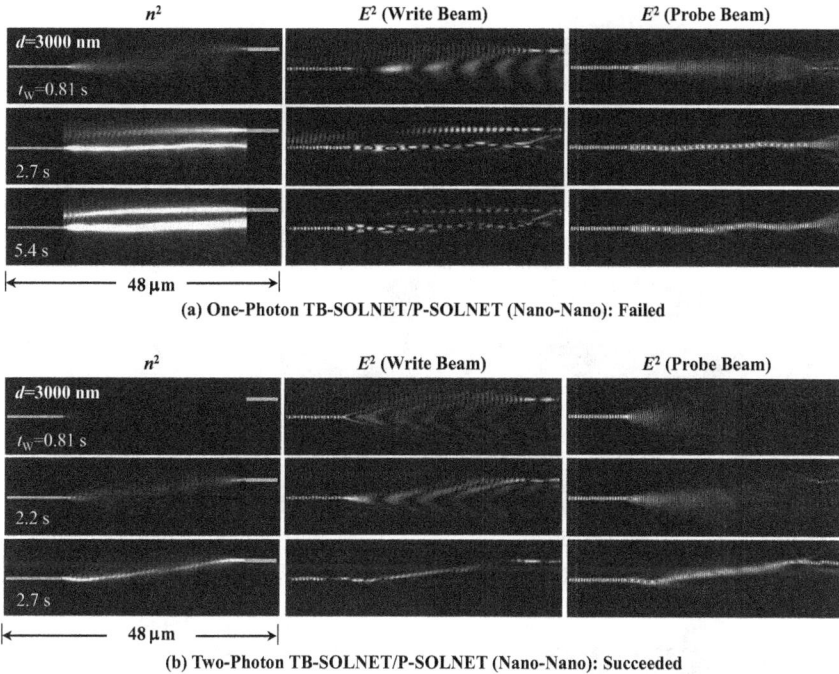

FIGURE 10.9 (a) One-photon TB-SOLNET/P-SOLNET formation and (b) two-photon TB-SOLNET/P-SOLNET formation between 600-nm-wide waveguides with a lateral misalignment of 3000 nm [7].

Formation of two-photon R-SOLNET and LA-SOLNET between 600-nm-wide waveguides is shown in Figure 10.12. Here, the middle columns represent the intensity of the write beam E^2 (Write Beam) and the luminescence E^2 (Luminescence) in R-SOLNET, and E^2 (Luminescence) in LA-SOLNET. For two-photon R-SOLNET with $d = 1800$ nm, when a 500-nm write beam is introduced from the input waveguide, a luminescent target on the output waveguide edge is excited by the write beam and emits 600-nm luminescence. The write beam and the luminescence merge to construct a self-aligned coupling waveguide at 4.0 s. For two-photon LA-SOLNET with $d = 3000$ nm, 500-nm luminescence 1 and 600-nm luminescence 2 are, respectively, emitted from luminescent targets on the input and output waveguide edges to construct a self-aligned coupling waveguide. As can be seen in Figure 10.13, two-photon R-SOLNET and LA-SOLNET are normally formed even when d is 3000 nm, exhibiting wide misalignment tolerance.

In Figure 10.14, two-photon TB-SOLNETs formed between 3-μm-wide and 600-nm-wide waveguides are summarized for various lateral misalignments. A self-aligned coupling waveguide is formed for $d = 600$, 1800, and 3000 nm. The probe beam is guided into the output waveguide for $d = 600$ and 1800 nm, indicating that the two-photon TB-SOLNET achieves wide lateral misalignment tolerance. The probe beam leakage occurring near the output waveguide for $d = 3000$ nm is due to the small bending radius of the SOLNET. It was found that two-photon R-SOLNET is

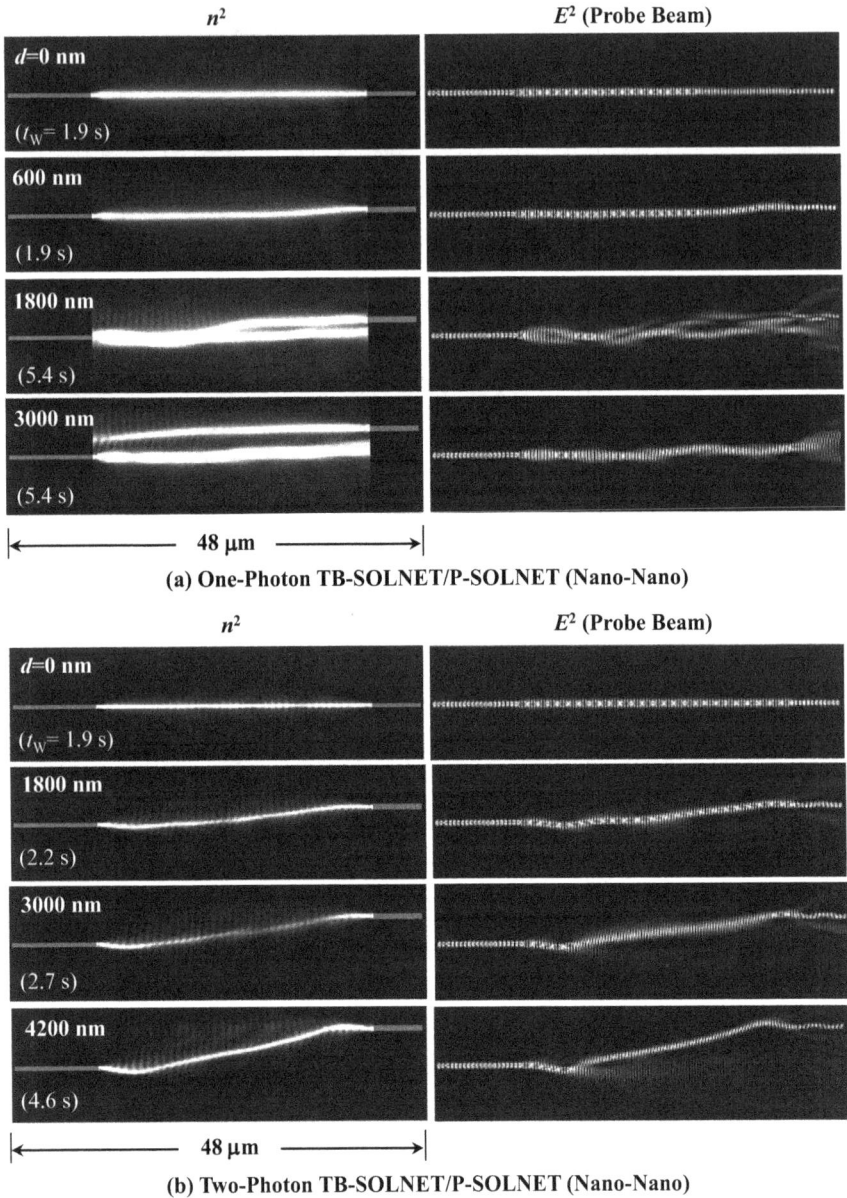

FIGURE 10.10 (a) One-photon TB-SOLNETs/P-SOLNETs and (b) two-photon TB-SOLNETs/P-SOLNETs between 600-nm-wide waveguides at optimum writing time indicated in brackets for various lateral misalignments [7].

formed for $d = 600$ and $1800\,\text{nm}$, and LA-SOLNET for $d = 600$, 1800, and $3000\,\text{nm}$. These results suggest that the SOLNETs enable self-aligned optical couplings with a mode size conversion function [1,6–8].

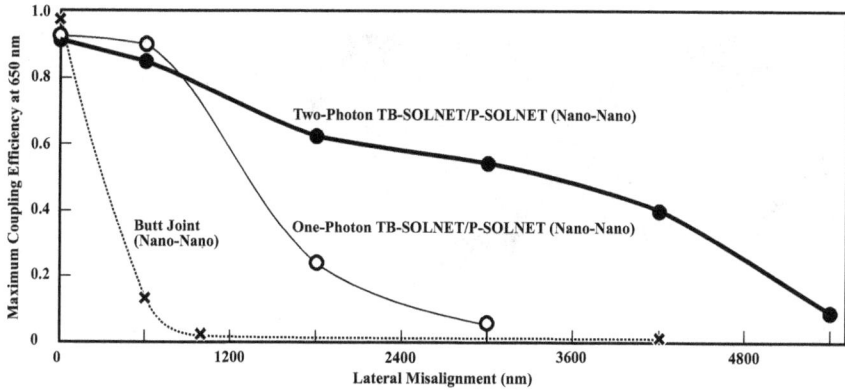

FIGURE 10.11 Dependence of maximum coupling efficiency on lateral misalignments in TB-SOLNETs/P-SOLNETs formed between 600-nm-wide waveguides [7].

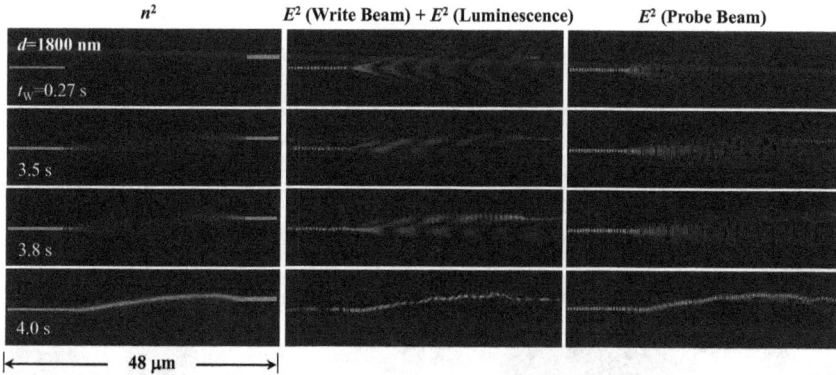

(a) Two-Photon R-SOLNET (Nano-Nano)

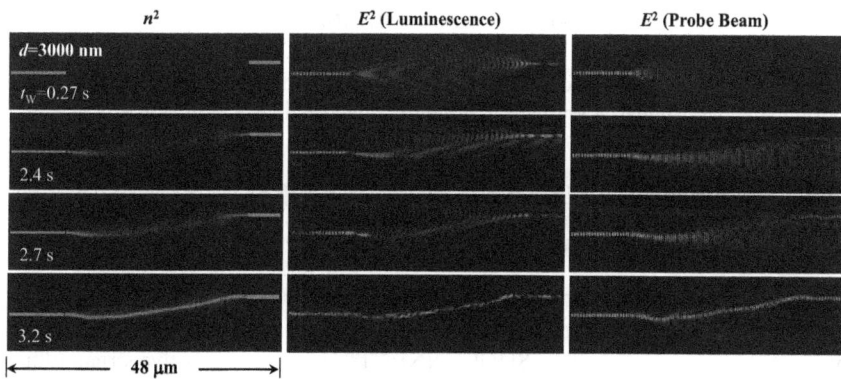

(b) Two-Photon LA-SOLNET (Nano-Nano)

FIGURE 10.12 (a) Two-photon R-SOLNET formation between 600-nm-wide waveguides with a lateral misalignment of 1800 nm and (b) two-photon LA-SOLNET formation between 600-nm-wide waveguides with a lateral misalignment of 3000 nm [7].

(a) Two-Photon R-SOLNET (Nano-Nano)

(b) Two-Photon LA-SOLNET (Nano-Nano)

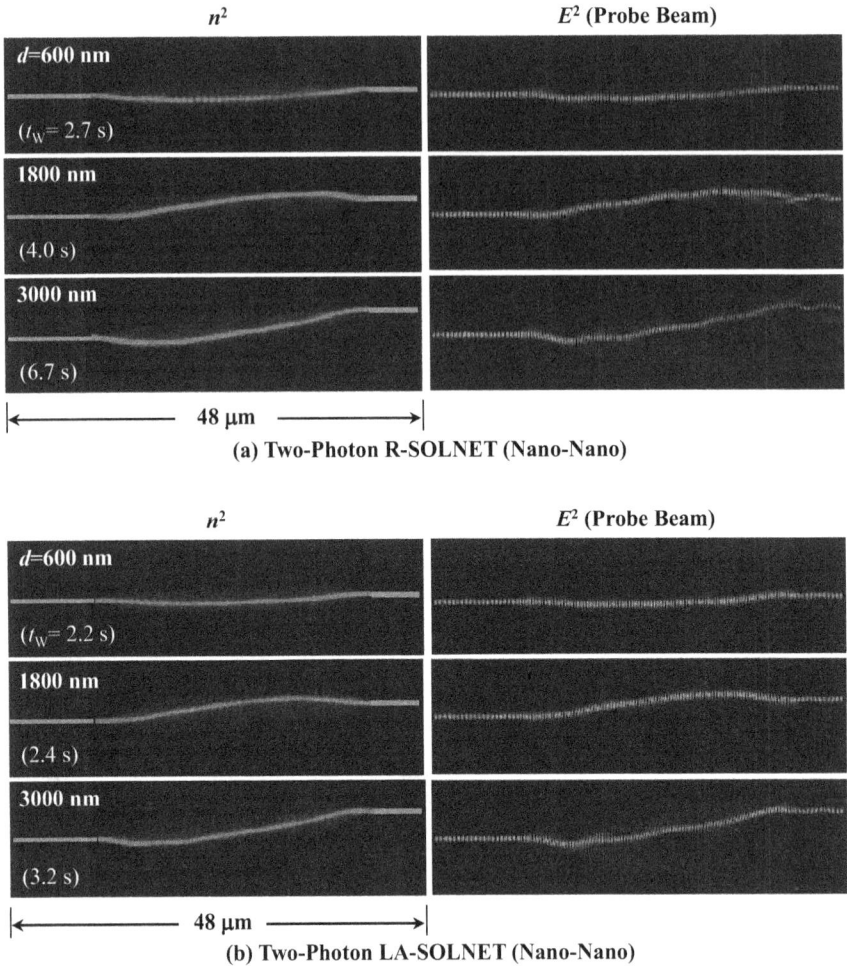

FIGURE 10.13 (a) Two-photon R-SOLNETs and (b) two-photon LA-SOLNETs between 600-nm-wide waveguides at optimum writing time indicated in brackets for various lateral misalignments [7].

10.3 EXPERIMENTAL DEMONSTRATIONS OF SOLNETs

10.3.1 ONE-PHOTON SOLNETs

Iida et al. demonstrated formation of one-photon R-SOLNETs using luminescent targets made of a laser dye, coumarin 481, dissolved in SUNCONNECT® (coumarin 481/SUNCONNECT®). SUNCONNECT® is a photosensitive organic/inorganic hybrid material developed by Nissan Chemical Industries [25,26], which is also used as the PRI material in the present experiment. To enhance the targeting effect, the sensitivity of the PRI material to the luminescence should be comparable with that to the write beam. As Figure 10.15 shows, the absorption spectrum of SUNCONNECT® PRI material has a peak in UV region, and the absorption

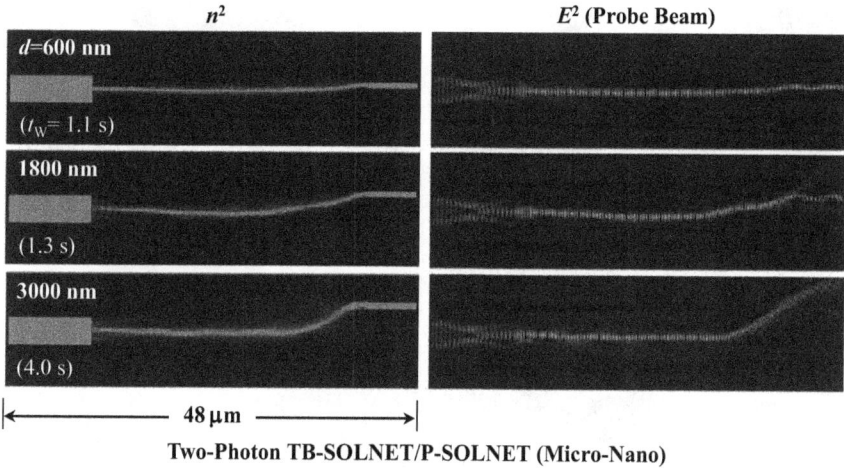

Two-Photon TB-SOLNET/P-SOLNET (Micro-Nano)

FIGURE 10.14   Two-photon TB-SOLNETs/P-SOLNETs between 3-μm-wide and 600-nm-wide waveguides at optimum writing time indicated in brackets for various lateral misalignments [7].

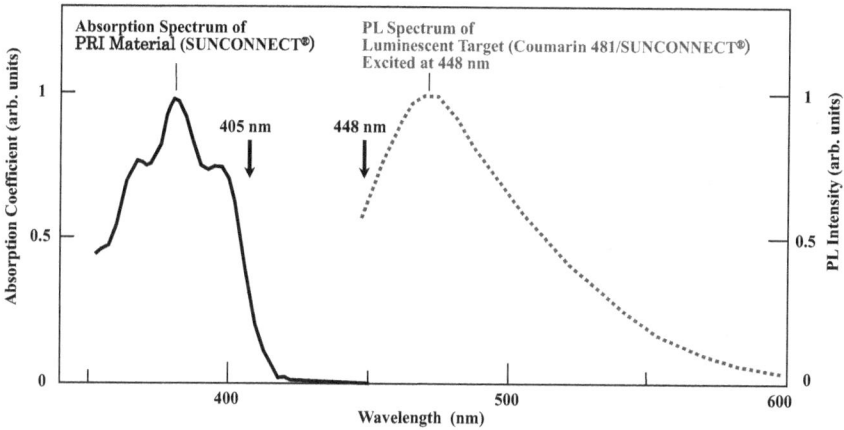

FIGURE 10.15   Absorption spectrum of SUNCONNECT® PRI material and PL spectrum of a coumarin 481/SUNCONNECT® luminescent target excited by a write beam at 448 nm. (Reprinted with permission from Yoshimura et al. [15]. Copyright (2014) The Optical Society.)

coefficient decreases at longer wavelengths, indicating that the sensitivity decreases with increasing wavelength. The photoluminescence (PL) spectrum of the coumarin 481/SUNCONNECT® luminescent target excided by 448-nm light has a peak around 470 nm. The wavelength separation between the 448-nm write beam and the PL peak is 22 nm, being preferable for reducing the sensitivity unbalance between the write beam and the luminescence [15].

To deposit a luminescent target on the core region of an optical fiber edge, the light curing process shown in Figure 10.7a was used. The fiber edge was coated with coumarin 481 dissolved in SUNCONNECT® at a concentration of ~0.1 wt%. The

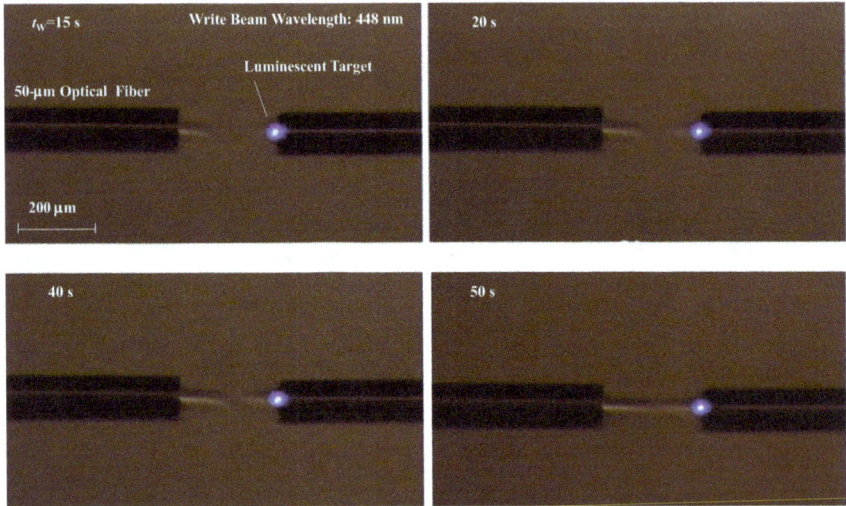

FIGURE 10.16 One-photon R-SOLNET formation between a 50-μm optical fiber and a luminescent target on the core edge of another 50-μm optical fiber in SUNCONNECT® by a 448-nm write beam with power of 20 μW. The lateral misalignment is 20 μm. (Reprinted with permission from Yoshimura et al. [15]. Copyright (2014) The Optical Society.)

coating was exposed to UV light from the fiber core, curing the region just on the core. The luminescent target was obtained by wet-etching the coating to remove the noncured part using a mixture of isopropyl alcohol/4-methyl-2-pentanone.

Figure 10.16 shows one-photon R-SOLNET formation between an optical fiber with a core diameter of 50 μm (50-μm optical fiber) and a luminescent target on the core edge of another 50-μm optical fiber in SUNCONNECT®. The gap distance and the lateral misalignment between the two optical fibers are 250 and 20 μm, respectively. A 448-nm write beam with power of 20 μW is introduced from the optical fiber on the left. Bright luminescence from the luminescent target is observed. As the write beam irradiation progresses, a trace of waveguide growth from the core of the optical fiber on the left is observed. At the same time, a trace of waveguide growth by luminescence emitted from the luminescent target is observed. The two traces finally merge to construct a self-aligned coupling waveguide of one-photon R-SOLNET.

Figure 10.17 shows maximum coupling efficiency between 50-μm optical fibers connected by one-photon R-SOLNETs for a gap distance of 300 μm. The probe beam wavelength is 850 nm. The coupling efficiency decreases with increasing the lateral misalignment between the optical fibers. By forming R-SOLNET, the coupling efficiency is increased by 10%–30%, and the lateral misalignment tolerance is widened. Coupling efficiency of ~80% (coupling loss: ~1 dB) is maintained even when the lateral misalignment increases to 20 μm.

Seki et al. succeeded in the formation of R-SOLNETs using tris(8-hydroxyquinolinato) aluminum (Alq3), which is known as phosphor for green emission in organic

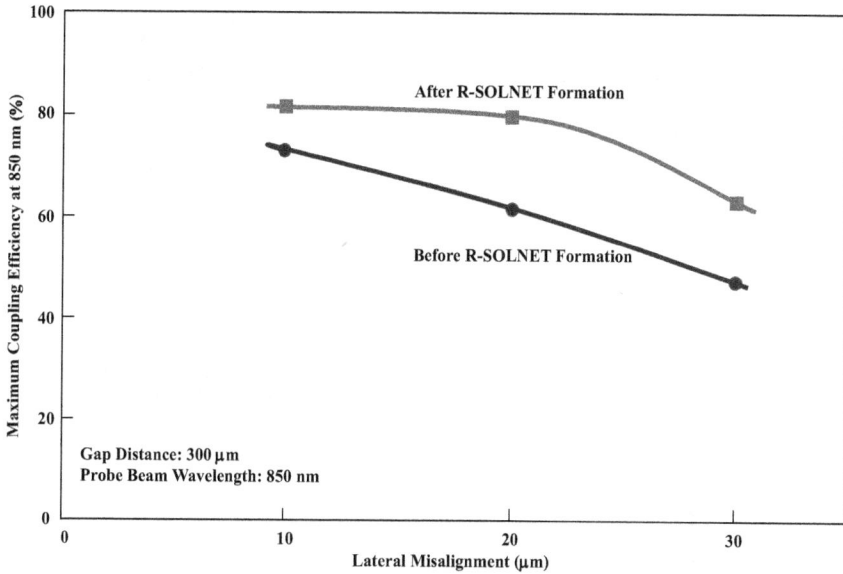

FIGURE 10.17   Dependence of maximum coupling efficiency on lateral misalignments in one-photon R-SOLNETs with luminescent targets between 50-μm optical fibers. (Reprinted with permission from Yoshimura et al. [15]. Copyright (2014) The Optical Society.)

light emitting diodes. A luminescent target of Alq3-dispersed polyvinyl alcohol (Alq3/PVA) was placed on an optical fiber edge. The PRI material was NOA81/NOA65, which is a mixture of acrylic materials of NOA81 (NORLAND) with refractive index of 1.56 and NOA65 with refractive index of 1.52. The mixing weight ratio of NOA81:NOA65 was 2:1. On a slide glass, a 50-μm optical fiber and an optical fiber with the Alq3/PVA luminescent target were placed with a gap distance of ~4 mm in the NOA81/NOA65 PRI material [13,14].

As shown in Figure 10.18a, when a 405-nm write beam is introduced from the left-hand-side 50-μm optical fiber, the luminescent target emits green/blue luminescence to pull the write beam to the target and form a one-photon R-SOLNET that connects the fiber core and the luminescent target. It is found from Figure 10.18b that the R-SOLNET width gradually expands from the fiber core diameter to the width of the luminescent target, realizing a tapered R-SOLNET. This indicates that the one-photon R-SOLNET acts as optical solder with a mode size conversion function.

Figure 10.19 shows a one-photon R-SOLNET formation using an Alq3 thin-film luminescent target deposited on an optical fiber edge by vacuum evaporation. By introducing a 405-nm write beam from the left-hand-side 50-μm optical fiber into the NOA81/NOA65 PRI material, green/blue luminescence is generated from the Alq3 thin film, as presented in Figure 10.19b. The write beam is pulled toward the Alq3 thin-film luminescent target to form a one-photon R-SOLNET shown in Figure 10.19c.

(a) Write Beam Exposure

(b) Tapered R-SOLNET with Luminescent Targets

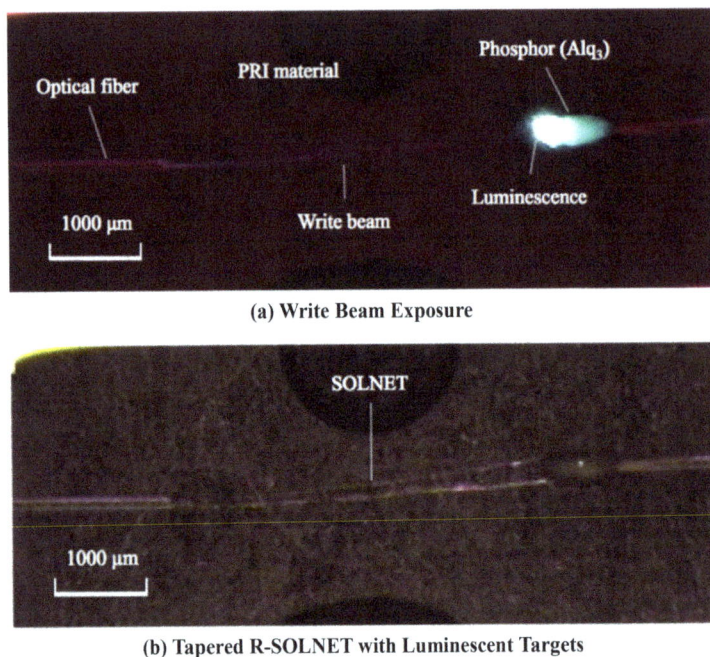

FIGURE 10.18   One-photon R-SOLNET formation between a 50-μm optical fiber and a large Alq3/PVA luminescent target. The R-SOLNET has a mode size conversion function. (a) Write beam exposure and (b) tapered R-SOLNET with luminescent targets. (Reprinted with permission from Seki and Yoshimura [13].)

10.3.2   TWO-PHOTON SOLNETS

A two-photon TB-SOLNET was formed in a two-photon PRI material made of a host material of New-SUNCONNECT® and two-photon sensitizing molecules of CQ. New-SUNCONNECT® is a photosensitive organic/inorganic hybrid material, whose light absorption is smaller than that of SUNCONNECT® [25,26] in blue wavelength regions. The small light absorption of the host material is preferable to feature the effect of the two-photon sensitizing molecules [6].

Absorption spectra of CQ and BA are presented in Figure 10.20 for the $S_0 \rightarrow S_n$ transition and the $T_1 \rightarrow T_n$ transition [30]. Brauchle et al. reported that the $T_1 \rightarrow T_n$ absorption spectrum of CQ is almost the same as that of BA. For the λ_1-write beam that induces the $S_0 \rightarrow S_n$ transition, a wavelength of 448 nm was chosen because CQ absorbs the 448-nm write beam. For the λ_2-write beam, a wavelength of 850 nm, which is within the wavelength region for the $T_1 \rightarrow T_n$ transition, was chosen.

Figure 10.21 shows the two-photon TB-SOLNET formation between 50-μm optical fibers placed with a gap distance of 500 μm in a two-photon PRI material containing 20wt%-CQ (20wt%-CQ/New-SUNCONNECT®). A 448-nm write beam from the left-hand-side optical fiber and an 850-nm write beam from the right-hand-side optical fiber are introduced into the two-photon PRI material. CQ is excited by the 448-nm write beam to emit green luminescence, which is probably accompanied

(a) Before Write Beam Exposure

(b) Luminescence from Alq3 Thin Film

(c) Write Beam Exposure and R-SOLNET with Alq3 Thin-Film Luminescent Targets

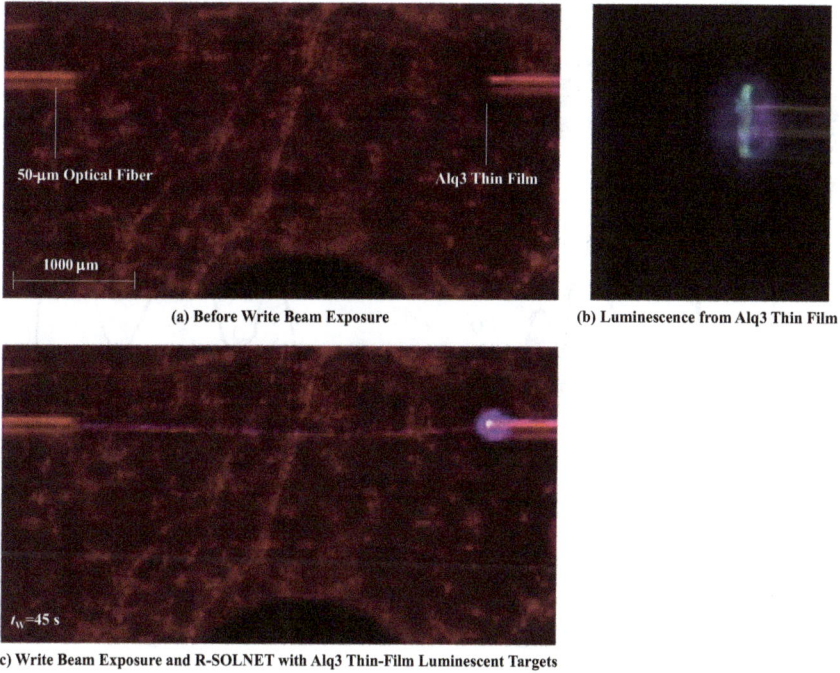

FIGURE 10.19 One-photon R-SOLNET formation between a 50-μm optical fiber and an Alq3 thin-film luminescent target deposited on an optical fiber edge by vacuum evaporation. (a) Before write beam exposure, (b) luminescence from Alq3 thin film, and (c) write beam exposure and R-SOLNET with Alq3 thin-film luminescent targets. (Reprinted with permission from Yoshimura and Seki [14]. Copyright (2013) The Optical Society.)

by the $S_n \rightarrow S_0$ transition or the $T_1 \rightarrow S_0$ transition. Since the intensity of the green luminescence increases with increasing the intensity of the 448-nm write beam, the green luminescence is enhanced with an increase in the 448-nm write beam confinement. Therefore, the green luminescence can be used to monitor SOLNET formation dynamics. A green luminescence line, which corresponds to a trace of two-photon TB-SOLNET, stretches from the left-hand side fiber core, and is initially deviated from the right-hand-side fiber core. With writing time, the luminescence line is gradually stretched toward the right-hand-side fiber core location. Finally, at 120 s, the green luminescence line connects the left-hand-side and right-hand-side fiber cores, indicating the two-photon TB-SOLNET formation. In the photograph, which was taken after the write beam was turned off, a trace of the two-photon TB-SOLNET connecting the two optical fiber cores is observed. The two-photon TB-SOLNET keeps connecting the two optical fiber cores even after the left-hand-side optical fiber moved as Figure 10.22 shows.

Two-photon R-SOLNET was formed using luminescent targets of SUNCONNECT® containing a laser dye called DCM. In Figure 10.23, an absorption spectrum and a PL spectrum of DCM in methanol [37] are shown together with the $S_0 \rightarrow S_n$ absorption spectrum of CQ and the $T_1 \rightarrow T_n$ absorption spectrum of BA, which is almost the same

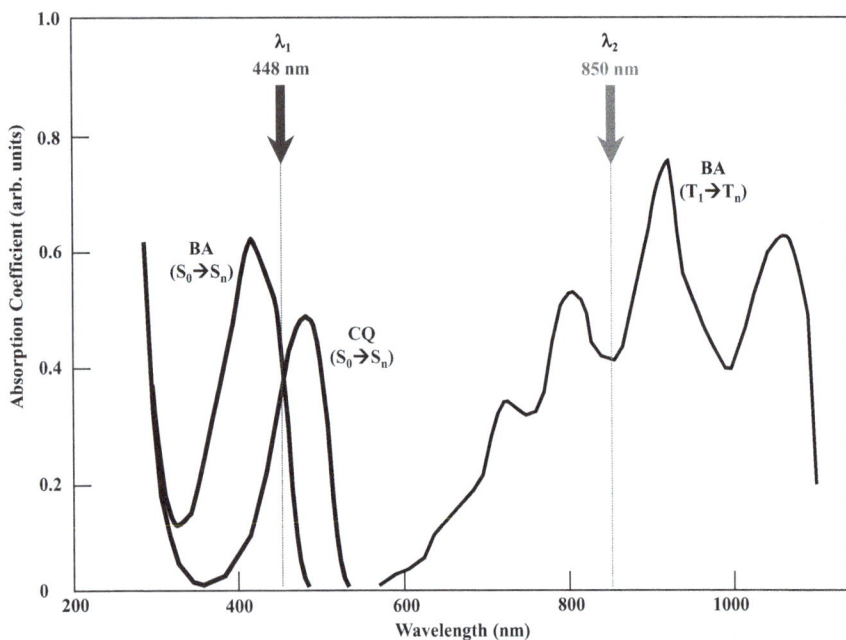

FIGURE 10.20 Absorption spectra of two-photon sensitizing molecules, BA and CQ. (Reprinted with permission from Yoshimura et al. [8].)

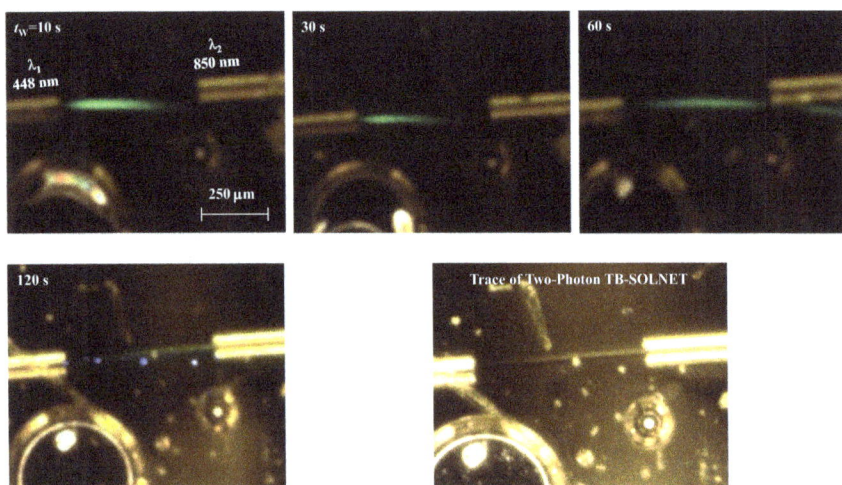

FIGURE 10.21 Two-photon TB-SOLNET formation between 50-μm optical fibers, and a trace of the two-photon TB-SOLNET in 20wt%-CQ/New-SUNCONNECT®. (Reprinted with permission from Yoshimura et al. [8].)

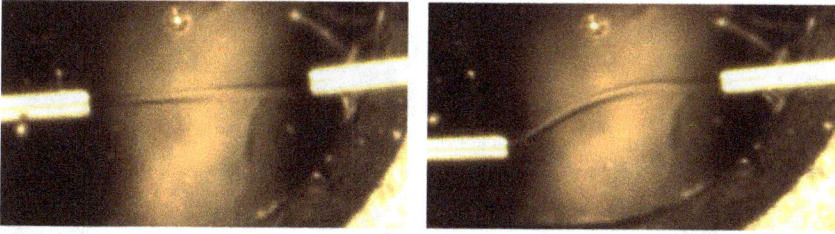

FIGURE 10.22 Photographs of the two-photon TB-SOLNET before and after the left-hand-side optical fiber moved. (Reprinted with permission from Yoshimura et al. [8].)

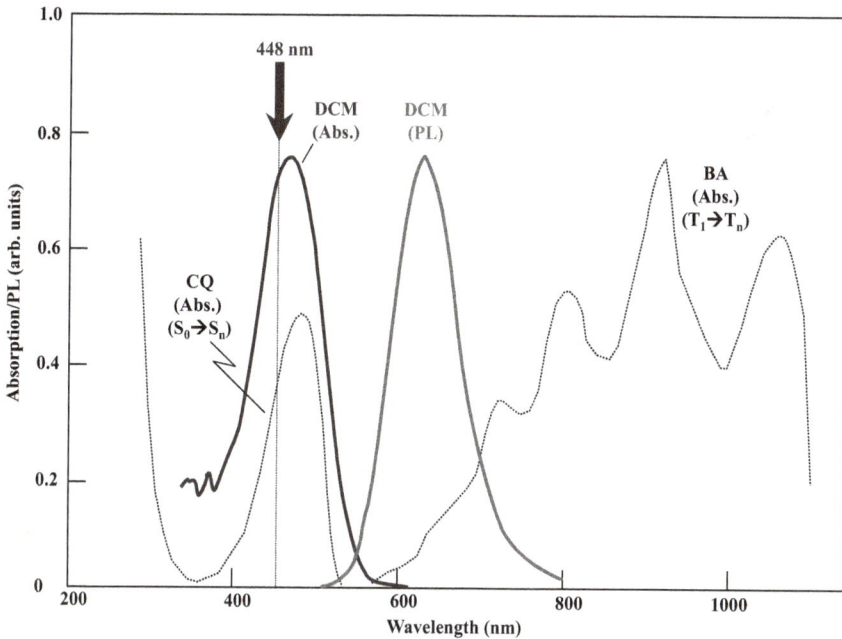

FIGURE 10.23 Absorption and PL spectra of DCM in methanol, $S_0 \rightarrow S_n$ absorption spectrum of CQ, and $T_1 \rightarrow T_n$ absorption spectrum of BA, which is almost the same as that of CQ. (Reprinted with permission from Yoshimura et al. [8].)

as that of CQ. DCM absorbs the 448-nm write beam and emits luminescence in a wavelength region of 600–700 nm, which can be absorbed in the $T_1 \rightarrow T_n$ transition by CQ. Therefore, DCM is a suitable material for luminescent targets in the two-photon R-SOLNET [8].

The light curing process was used to deposit a luminescent target on the core region of a 50-μm fiber edge. After the fiber edge was coated with a luminescent material of DCM dissolved in SUNCONNECT®, the coating was exposed to a 448-nm light from the fiber core, curing the region just on the core. The luminescent target was obtained by wet-etching using a mixture of isopropyl alcohol/4-methyl-2-pentanone.

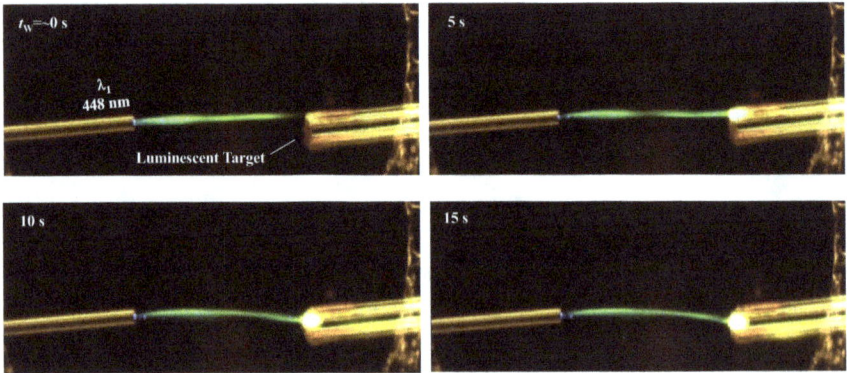

FIGURE 10.24 Two-photon R-SOLNET formation between a 50-μm optical fiber and a 50-μm optical fiber having a DCM/SUNCONNECT® luminescent target in CQ/New-SUNCONNECT® PRI material. There are a horizontal misalignment and a vertical misalignment between the two optical fibers. (Reprinted with permission from Yoshimura et al. [8].)

Figure 10.24 shows formation of a two-photon R-SOLNET with a luminescent target of 1 wt%-DCM/SUNCONNECT® between 50-μm optical fibers in the CQ/New-SUNCONNECT® PRI material. The diameter of the right-hand-side optical fiber seems much larger than that of the left-hand-side optical fiber. This is because the right-hand-side optical fiber is located at a position closer to the objective lens. In other words, the two optical fibers are arranged with a misalignment along both horizontal and vertical directions. When a 448-nm write beam is introduced from the left-hand-side optical fiber, a green luminescence line is gradually pulled toward the luminescent target on the core edge of the right-hand-side optical fiber with writing time to connect the two optical fiber cores. This confirms that a two-photon R-SOLNET with luminescent targets is formed even when such a large misalignment exists.

REFERENCES

1. T. Yoshimura, *Self-Organized Lightwave Networks: Self-Aligned Coupling Optical Waveguides*, CRC/Taylor & Francis, Boca Raton, Florida (2018).
2. T. Yoshimura, W. Sotoyama, K. Motoyoshi, T. Ishitsuka, K. Tsukamoto, S. Tatsuura, H. Soda, and T. Yamamoto, "Method of producing optical waveguide system, optical device and optical coupler employing the same, optical network and optical circuit board," U.S. Patent 6,081,632 (2000).
3. T. Yoshimura, J. Roman, Y. Takahashi, W. V. Wang, M. Inao, T. Ishitsuka, K. Tsukamoto, K. Motoyoshi, and W. Sotoyama, "Self-organizing waveguide coupling method "SOLNET" and its application to film optical circuit substrates," Proc. 50th Electronic Components & Technology Conference (ECTC), Las Vegas, Nevada, 962–969 (2000).
4. T. Yoshimura, J. Roman, Y. Takahashi, W. V. Wang, M. Inao, T. Ishituka, K. Tsukamoto, K. Motoyoshi, and W. Sotoyama, "Self-organizing lightwave network (SOLNET) and its application to film optical circuit substrates," *IEEE Trans. Comp., Packag. Technol.* **24**, 500–509 (2001).

5. T. Yoshimura, A. Hori, Y. Yoshida, Y. Arai, H. Kurokawa, T. Namiki, and K. Asama, "Coupling efficiencies in reflective self-organized lightwave network (R-SOLNET) simulated by the beam propagation method," *IEEE Photon. Technol. Lett.* **17**, 1653–1655 (2005).

6. T. Yoshimura, D. Takeda, T. Sato, Y. Kinugasa, and H. Nawata, "Polymer waveguides self-organized by two-photon photochemistry for self-aligned optical couplings with wide misalignment tolerances," *Opt. Commun.* **362**, 81–86 (2016).

7. T. Yoshimura and H. Nawata, "Micro/nanoscale self-aligned optical couplings of the self-organized lightwave network (SOLNET) formed by excitation lights from outside," *Opt. Commun.* **383**, 119–131 (2017).

8. T. Yoshimura., S. Yasuda, H. Yamaura, Y. Yamada, M. Takashima, R. Ito, and T. Hamazaki, "Self-aligned coupling waveguides experimentally formed by two-photon photochemistry for 3-D integrated optical interconnects," *Opt. Commun.* **430** 284–292 (2019).

9. T. Yoshimura, T. Inoguchi, T. Yamamoto, S. Moriya, Y. Teramoto, Y. Arai, T. Namiki, and K. Asama, "Self-organized lightwave network based on waveguide films for three-dimensional optical wiring within boxes," *J. Lightwave Technol.* **22**, 2091–2100 (2004).

10. H. Kaburagi and T. Yoshimura, "Simulation of micro/nanometer-scale self-organized lightwave network (SOLNET) using the finite difference time domain method," *Opt. Commun.* **281**, 4019–4022 (2008).

11. T. Yoshimura and H. Kaburagi, "Self-organization of optical waveguides between misaligned devices induced by write-beam reflection," *Appl. Phys. Express* **1**, 062007 (2008).

12. T. Yoshimura, K. Wakabayashi, and S. Ono, "Analysis of reflective self-organized light-wave network (R-SOLNET) for Z-connections in three-dimensional optical circuits by the finite difference time domain method," *IEEE J. Select. Top. Quantum Electron.* **17**, 566–570 (2011).

13. M. Seki and T. Yoshimura, "Reflective self-organizing lightwave network using a phosphor," *Opt. Eng.* **51**, 074601 (2012).

14. T. Yoshimura and M. Seki, "Simulation of self-organized parallel waveguides targeting nanoscale luminescent objects," *J. Opt. Soc. Am. B* **30**, 1643–1650 (2013).

15. T. Yoshimura, M. Iida, and H. Nawata, "Self-aligned optical couplings by self-organized waveguides toward luminescent targets in organic/inorganic hybrid materials," *Opt. Lett.* **39**, 3496–3499 (2014).

16. T. Yoshimura, "Simulation of self-aligned optical coupling between micro- and nanoscale devices using self-organized waveguides," *J. Lightwave Technol.* **33**, 849–856 (2015).

17. S. J. Frisken, "Light-induced optical waveguide uptapers," *Opt. Lett.* **18**, 1035–1037 (1993).

18. A. S. Kewitsch and A. Yariv, "Self-focusing and self-trapping of optical beams upon photopolymerization," *Opt. Lett.* **21**, 24–26 (1996).

19. M. Kagami, T. Yamashita, and H. Ito, "Light-induced self-written three-dimensional optical waveguide," *Appl. Phys. Lett.* **79**, 1079–1081 (2001).

20. K. Dorkenoo, O. Crégut, L. Mager, and F. Gillot, "Quasi-solitonic behavior of self-written waveguides created by photopolymerization," *Opt. Lett.* **27**, 1782–1784 (2002).

21. F. Huang, H. Takase, Y. Eriyama, and T. Ukachi, "Single-mode optical fiber interconnection by using self-written waveguides", *Proc. 12th Int. POF. Conf.*, Seattle, USA, 199–202 (2003).

22. N. Hirose and O. Ibaragi, "Optical component coupling using self-written waveguides," *Proc. SPIE* **5355**, 206–214 (2004).

23. S. Jradi, O. Soppera, and D. J. Lougnot, "Fabrication of polymer waveguides between two optical fibers using spatially controlled light-induced polymerization," *Appl. Opt.* **47**, 3987–3993 (2008).

24. T. M. Monro, C. M. de Sterke and L. Poladian, "Investigation of waveguide growth in photosensitive germanosilicate glass," *J. Opt. Soc. Am. B* **13**, 2824–2832 (1996).

25. T. Sato, "Novel organic-inorganic hybrid materials for optical interconnects," *Proc. SPIE* **7944**, 79440M (2011).

26. H. Nawata, "Organic-inorganic hybrid material for on-board optical interconnection and its applications in optical coupling," *IEEE CPMT Symposium Japan*, 121–124 (2013).

27. E. Fazio, M. Alonzo, F. Devaux, A. Toncelli, N. Argiolas, M. Bazzan, C. Sada, and M. Chauvet, "Luminescence-induced photorefractive spatial solitons," *Appl. Phys. Lett.* **96**, 091107 (2010).

28. E. Fazio, A. Zaltron, A. Belardini, N. Argiolas, and C. Sada, "Influence of iron doping on spatial soliton formation and fixing in lithium niobate crystals," *Opt. Mater.* **37**, 175–180 (2014).

29. Y. R. Shen, *The Principles of Nonlinear Optics*, John Wiley & Sons, New York (1984).

30. C. Brauchle, U.P. Wild, D.M. Burland, G.C. Bjorkund, and D.C. Alvares, "Two-photon holographic recording with continuous-wave lasers in the 750–1100-nm range," *Opt. Lett.* **7**, 177–179 (1982).

31. T. Yoshimura, S. Ito, T. Nakayama, and K. Matsumoto, "Orientation-controlled molecule-by-molecule polymer wire growth by the carrier-gas-type organic chemical vapor deposition and the molecular layer deposition," *Appl. Phys. Lett.* **91**, 033103 (2007).

32. T. Yoshimura, N. Terasawa, H. Kazama, Y. Naito, Y. Suzuki, and K. Asama, "Selective growth of conjugated polymer thin films by the vapor deposition polymerization," *Thin Solid Films* **497**, 182–184 (2006).

33. M. Pessa, R. Makela, and T. Suntola, "Characterization of surface exchange reactions used to grow compound films," *Appl. Phys. Lett.* **31**, 131–133 (1981).

34. T. Yoshimura and S. Ishii, "Effect of quantum dot length on the degree of electron localization in polymer wires grown by molecular layer deposition," *J. Vac. Sci. Technol. A* **31**, 031501 (2013).

35. T. Yoshimura, S. Tatsuura, and W. Sotoyama, "Polymer films formed with monolayer growth steps by molecular layer deposition," *Appl. Phys. Lett.* **59**, 482–484 (1991).

36. T. Yoshimura, "Molecular layer deposition (MLD): Monomolecular-step growth of polymers with designated sequences," *Macromol. Symp.* **361**, 141–148 (2016).

37. Catalogue of Laserdyes, Acros Organics.

11 Integrated Optical Interconnects and Optical Switching Systems

Currently, light wave technologies are ready to come into boxes of computers.. To realize the *intra*-box optical interconnects, the author proposed the self-organized three-dimensional (3-D) integrated optical interconnects and the 3-D micro-optical switching systems (3D-MOSSs) that provide reconfigurable networking functions in the optical interconnects. These large-scale optoelectronic (OE) systems are expected to be constructed using the scalable film optical link modules (S-FOLMs) with 3-D optical wiring backbones, where the self-organized lightwave networks (SOLNETs) are implemented. In the present chapter, after concepts of the OE systems and S-FOLMs are explained, the system performance is theoretically predicted for models involving the waveguide prism deflector (WPD) optical switches described in Chapter 9. They are made of lead lanthanum zirconate titanate (PLZT) and polymer multiple quantum dot (MQD) described in Chapters 6–9. The impacts of the polymer MQD on the OE systems are discussed, and it is revealed that, by replacing PLZT with polymer MQD, the data rate to transmit optical signals will increase and the power dissipation will decrease.

11.1 *INTRA*-BOX OPTICAL INTERCONNECTS

As summarized in Figure 11.1, light waves have sequentially penetrated into shorter-distance systems in the order of long-haul communication systems, metro networks, local-area networks (LANs), and *inter*-box optical interconnects in high-performance computing systems such as data centers and supr computers. Currently, light wave technologies are ready to come into boxes of computers to construct *intra*-box optical interconnects.

In the *intra*-box optical interconnects, *optics excess* becomes a serious problem. It includes excess components/materials, spaces for components, fabrication efforts for packaging, and design efforts. To solve the problem, the author proposed a concept of the self-organized 3-D integrated optical interconnects [1–3], which are made from the S-FOLMs [1,4–12]. S-FOLMs provide 3-D-stacked OE multichip modules (MCMs), OE printed circuit boards (PCBs), and OE backplanes (BPs) within the boxes of computers, as depicted in Figure 11.1. S-FOLMs also provide the 3D-MOSSs [1,10–12], which are required to implement reconfigurable networking functions into the *intra*-box optical interconnects.

DOI: 10.1201/9781003094012-11

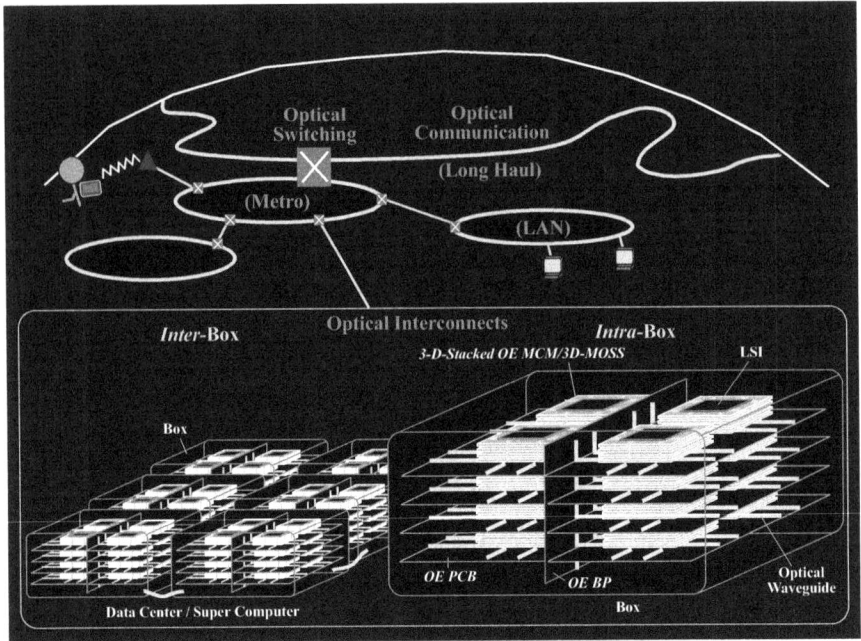

FIGURE 11.1 Penetration of light waves into electronic systems.

Details of S-FOLMs are presented in Section 11.2, and details of self-organized 3-D integrated optical interconnects and 3D-MOSSs are, respectively, described in Sections 11.3 and 11.4, where the impacts of the polymer MQDs on these OE systems are discussed.

11.2   3-D OPTOELECTRONIC (OE) PLATFORMS BASED ON SCALABLE FILM OPTICAL LINK MODULES (S-FOLMs)

The concept of the S-FOLM is shown in Figure 11.2 [1,4–12]. Firstly, packaging elements such as electrical substrates and OE-Films are prepared. The packaging elements are denoted as follows:

S: Electrical substrates.
W: Optical waveguide films.
D: Films, in which small/thin OE dies for light modulators, optical switches (SWs), wavelength filters, vertical-cavity surface-emitting lasers (VCSELs), photodetectors (PDs), and so on are embedded.
DW: Optical waveguide films, in which small/thin OE dies are embedded.
C: Films, in which small/thin electronic dies, such as integrated circuits (ICs), large-scale ICs (LSIs), and capacitors, are embedded.

Then, the packaging elements are combined by the Film/Z-connection technology [13] to build products of the S-FOLM. Z-connections act as both electrical

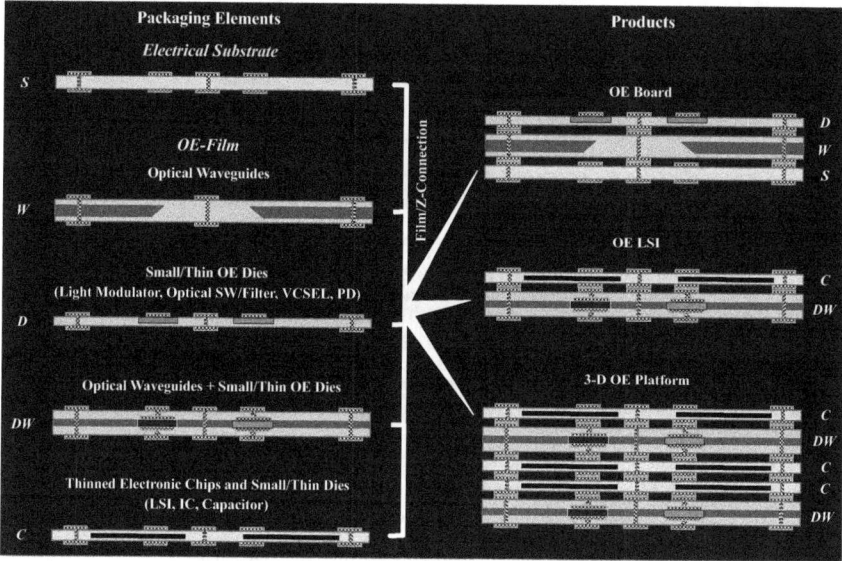

FIGURE 11.2   Concept of S-FOLM.

and mechanical joints between films. The S-FOLM realizes maximized variety of products built from minimized types of packaging elements, providing all levels of optical networking units. For example, an OE board is constructed by combining electrical substrate (S), OE-Film (W), and OE-Film (D). A combination of OE-Film (DW) and OE-Film (C) with thinned LSIs produces an OE LSI. By stacking a plurality of OE-Films (DW) and OE-Films (C), a 3-D OE platform is obtained. In the S-FOLM, since small/thin divided OE dies are embedded just under the individual pad positions of LSIs/ICs by the photolithographic packaging with selectively occupied repeated transfer (PL-Pack with SORT) [1,5,10,14,15] described in Section 9.8, the *optics excess* can be reduced, resulting in the system size/cost reduction.

Figure 11.3 shows examples of 3-D OE platforms [1,4–12]. The multilayer OE MCM consists of a plurality of OE-Films (W), OE-Film (DW) with embedded small/thin light modulator/PD dies, and OE-Film (C) with small/thin IC dies. Vertical optical signal routing is carried out by optical Z-connections consisting of vertical waveguides and vertical mirrors. The second layer (OE-Film (DW)) is for the E-O/O-E signal conversion. Electrical signals from LSI outputs are converted into optical signals by light modulators, and optical signals are converted into electrical signals for LSI inputs by PDs. The first layer (OE-Film (C)) is an interface film having functions like driver, amplifier, and error correction.

In the 3-D-stacked OE MCM, OE-Films (DW) are stacked together with OE-Films (C) containing thinned LSIs. Optical signals generated by the light modulators at E-O signal conversion sites propagate to O-E signal conversion sites to be converted into electrical signals by PDs. In some cases, the optical signals are coupled to OE PCB by backside connections to be transmitted to other OE units, and vice versa. By placing light sources outside of the stacked structure, the heat generation in the OE MCM is suppressed.

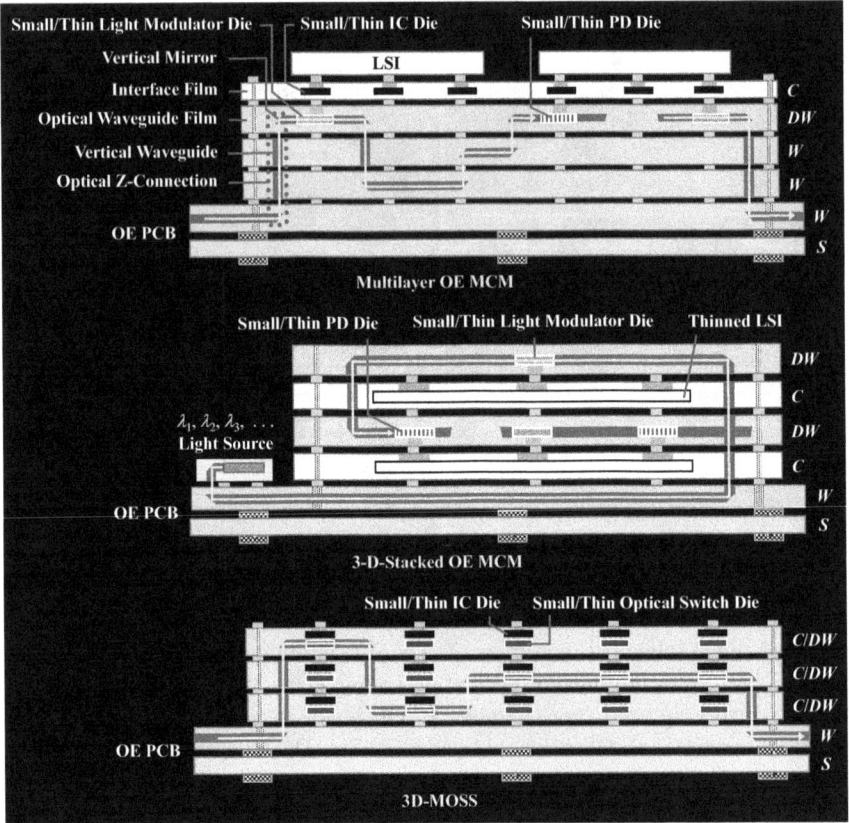

FIGURE 11.3   Examples of 3-D OE platform based on S-FOLM.

In the 3D-MOSS, OE-Films (*DW*) with small/thin optical switch die arrays are stacked with coupled OE-Films (*C*) with small/thin IC dies for optical switch drivers.

11.3 SELF-ORGANIZED 3-D INTEGRATED OPTICAL INTERCONNECTS WITHIN BOXES OF COMPUTERS

11.3.1 CONCEPT

Figure 11.4 depicts a model for a 3-D-stacked OE MCM in the self-organized 3-D integrated optical interconnects [1–3]. First, a 3-D pre-form shown in Figure 11.4a is prepared. OE-Films, which contain small/thin OE dies for light modulators and PDs as well as nanoscale waveguides with vertical mirrors, are stacked together with OE-Films containing thinned LSIs. To form SOLNETs between waveguides and between waveguides and OE dies, luminescent targets, luminescent regions, and photo-induced refractive-index increase (PRI) materials are distributed.

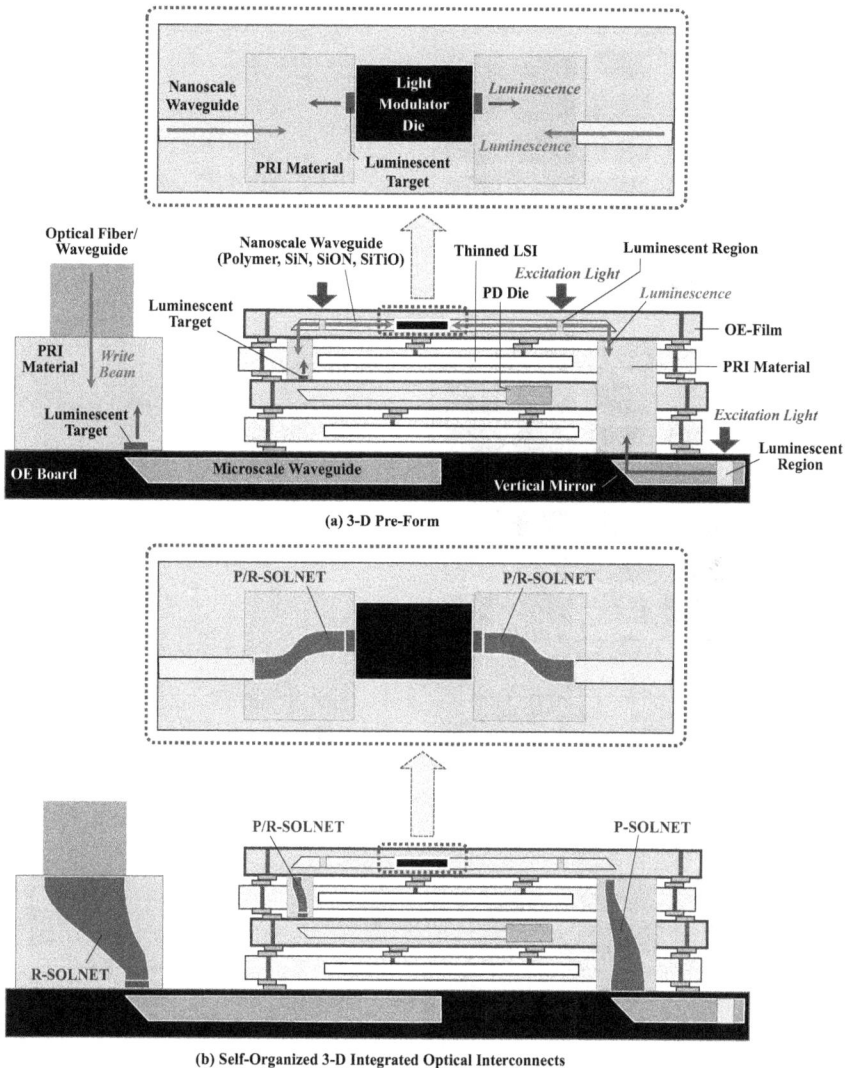

FIGURE 11.4 A model for 3-D-stacked OE MCM in the self-organized 3-D integrated optical interconnects. (Reprinted with permission from Yoshimura et al. [3].)

Then, external excitation lights are introduced into the 3-D pre-form by focusing optics or optical fibers. As a result, as depicted in Figure 11.4b, self-aligned waveguides of reflective SOLNETs (R-SOLNETs), phosphor SOLNETs (P-SOLNETs), and P/R-SOLNETs are formed. Here, P/R-SOLNETs mean R-SOLNETs formed by write beams from the luminescent regions. The detail of SOLNET is in Chapter 10.

Figure 11.5 shows an example of a PL-Pack with SORT process [1,5,10,14,15] for integration of light modulators and nanoscale waveguides with vertical mirrors.

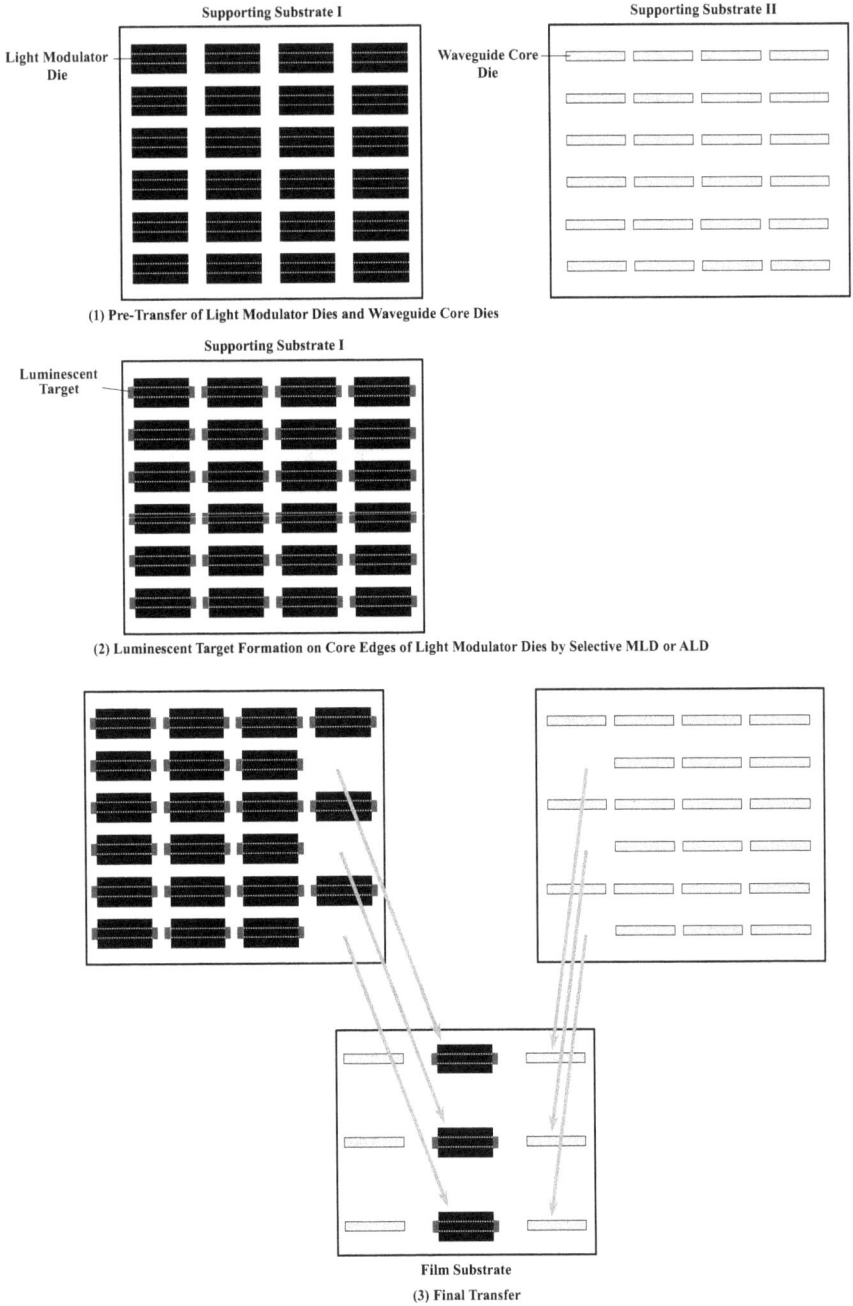

FIGURE 11.5 PL-Pack with SORT process for the integration of light modulators and nanoscale waveguides with vertical mirrors.

1. Light modulator dies and waveguide core dies are, respectively, picked up on Supporting Substrates I and II from growth substrates by the epitaxial lift-off (ELO).
2. Luminescent targets are deposited on the core edges of the light modulator dies by selective molecular layer deposition (MLD) or atomic layer deposition (ALD) using the process described in Section 10.1 (see Figure 10.7).
3. The light modulator dies and waveguide core dies are selectively transferred to final substrates to complete the heterogeneous integration.

11.3.2 IMPACTS OF POLYMER MQDs

Figure 11.6 depicts the concept of the OE amplifier/driver-less substrate (OE-ADLES) for inter-LSI connections. In the conventional optical interconnects, bulk-chip-based OE modules are used. In this case, chips of VCSELs, PDs, VCSEL drivers, and PD amplifiers as well as spaces for them, which are unnecessary in electrical interconnects, are necessary. Packaging with precise OE device alignments and design considering optics and electronics simultaneously to fabricate the OE modules are also necessary, causing *optics excess*. Furthermore, signal delay/skew in drivers and amplifiers for the E-O/O-E signal conversion, and noise due to relatively long metal lines for LSI-driver-VCSEL and PD-amplifier-LSI connections are increased [1,16,17].

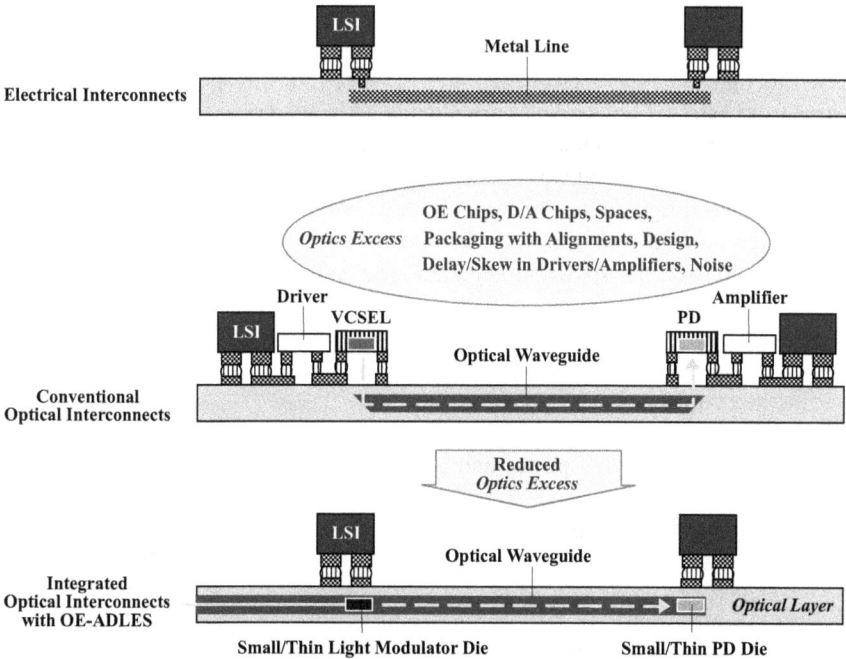

FIGURE 11.6 Concept of OE-ADLES for inter-LSI connection.

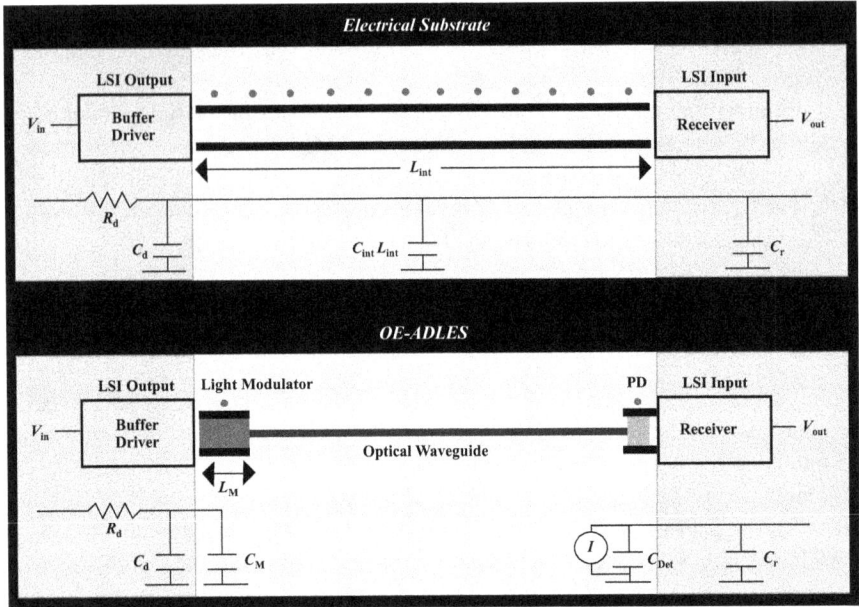

FIGURE 11.7 Models for the electrical substrate and OE-ADLES. (From Yoshimura et al., *Proc. SPIE* **6126**, 612609 (2006).)

In integrated optical interconnects with OE-ADLES, the *optics excess* is reduced. The E-O/O-E signal conversion is performed by small/thin light modulator dies and PD dies that are buried in an optical layer. LSI outputs and inputs are directly connected to light modulator electrodes and PD electrodes, respectively. Strong light beams from external light sources are modulated by electrical signals from LSI outputs through the light modulators without drivers. The generated optical signals are transmitted through optical waveguides to the PDs, where the signals are converted back to electrical signals for LSI inputs without amplifiers. Thus, a simple interconnect configuration just like the electrical interconnects can be built. The OE-ADLES is expected to have "low-power-dissipation" and "high-data-rate" capabilities due to the following reasons:

- Light sources are put outside, reducing power dissipation within the interconnect units.
- Light modulators are driven directly by LSI output electrical signals, reducing the driving power and the rising/falling time for optical signal generation.
- Strong light beams are available from the external light sources, typically high-power laser diodes (LDs) and mode-locked LDs, achieving fast PD response.

In order to estimate power dissipation and *RC* delay in the OE-ADLES, models shown in Figure 11.7 are considered. For the electrical substrate, data transmission is treated as charge/discharge processes for capacitance in a buffer driver, a transmission line, and a receiver through gate resistance of the buffer driver. For the OE-ADLES, data transmission is treated as charge/discharge processes for capacitance in a buffer driver and a light modulator through gate resistance of the buffer driver, and charge/discharge processes for capacitance in a PD and a receiver by currents generated in the PD [1,16,17].

In Table 11.1, expressions for the RC delay and power dissipation are summarized as well as parameters for the calculation. Attenuation along the transmission lines is neglected. External power dissipation at external light sources is not included. The WPD optical switch described in Chapter 9 is assumed as the light modulator. PLZT and polymer MQD described in Chapters 6–9 are assumed as the electro-optic (EO) materials.

In Figure 11.8, RC delay and power dissipation of an electrical substrate and an OE-ADLES at a data rate of 5 Gbps are compared. For RC delay, the OE-ADLES

TABLE 11.1

Expressions for RC Delay and Power Dissipation, and Parameters for the Calculation

	Electrical Substrate	**OE-ADLES**
RC delay	$\tau = R_d\left(C_{int}\, L_{int} + C_d + C_r\right)$	$\tau = R_d\left(C_M + C_d\right) + V_{out}\left(C_{Det} + C_r\right)/\left(\eta_c\, P_{Det}\right) + \tau_P$
Power dissipation	$P = f\left(C_{int}\, L_{int} + C_d + C_r\right)V_{out}^2$	$P = f\left[\left(C_M + C_d\right)V_{in}^2 + \left(C_{Det} + C_r\right)V_{out}^2\right]$
	$C_{int} = \varepsilon_0 \varepsilon_s w/d$	$C_M = \varepsilon_0 \varepsilon_s A_M/d$

	Electrical Substrate	**OE-ADLES**	
		PLZT	**Polymer MQD (calc.)**
Relative dielectric constant, ε_s	3	300	4
Refractive index, n		2.48	2
EO coefficient		$R = 5.7 \times 10^{-16}\,\mathrm{m^2/V^2}$	$r = 1.35 \times 10^{-9}\,\mathrm{m/V}$

Source: Reprinted with permission from Yoshimura et al. [17].

A_M, Light modulator area: $16 \times 16\,\mu\mathrm{m^2}$; C_d, Driver capacitance: 0.02 pF; C_{Det}, Photodetector capacitance: 0.01 pF; C_{int}, Transmission line capacitance per cm; C_M, Light modulator capacitance; C_r, Receiver capacitance: 0.02 pF; d, Gap between electrodes: 1 μm; f, Data rate; L_{int}, Transmission line length; P_{Det}, Incident power to photodetector; R_d, Driver gate resistance: 50 Ω; V_{in}, Power supply voltage: 3 V; V_{out}, Received detector voltage: 3 V; w, Electrode width: 4 μm; η_c, Responsibility of photodetector: 0.8 A/W; λ, Wavelength: 1.3 μm; τ_p, Photodetector internal propagation delay: 25 ps.

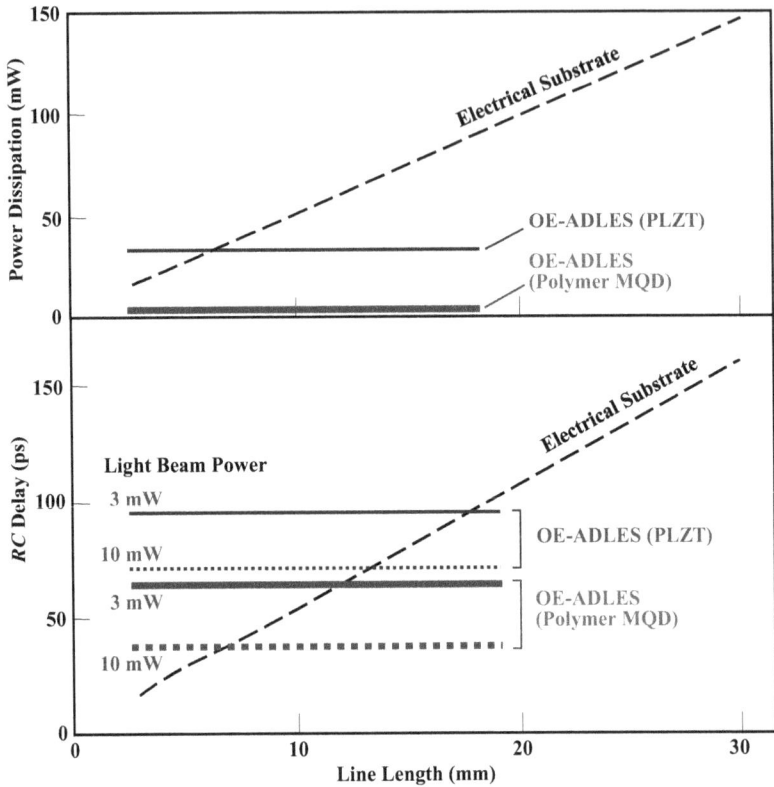

FIGURE 11.8   Results of calculation for RC delay and power dissipation at a data rate of 5 Gbps in an electrical substrate and an OE-ADLES. (From Yoshimura et al., *Proc. SPIE* **6126**, 612609 (2006).)

with polymer MQD has an advantage over the electrical substrate in a line length range longer than 12 mm for light beam power of 3 mW. By increasing the light beam power to 10 mW, RC delay becomes shorter, moving the break-even point toward a shorter line length of a few mm. For power dissipation, drastic reduction can be seen in the OE-ADLES with polymer MQD. The break-even point is located at a line length of sub mm. In the OE-ADLES with PLZT, the break-even point is located at longer line length regions for both RC delay and power dissipation, comparing to the OE-ADLES with polymer MQD. This comes from the large dielectric constant of PLZT.

The superiority of the OE-ADLES in the calculated RC delay and power dissipation is attributed to the capacitance reduction. Namely, in the electrical substrate, a large amount of charge is required to charge up the electrical transmission line due to its large line capacitance while in the OE-ADLES only a small amount of charge is required to charge up the electrodes of the light modulator and PD having small capacitance.

A key issue in the OE-ADLES is availability of EO materials with a large EO coefficient and a small relative dielectric constant. The polymer MQD seems promising for the EO material. Size reduction of the light modulators and PDs is another important issue. When they are miniaturized by one order of magnitude for each dimension, that is, the device thickness is reduced by one order of magnitude to 100 nm and the device area by two orders of magnitude to $1.6 \times 1.6\,\mu m^2$, their capacitance and the PD internal propagation delay will be reduced by one order of magnitude in principle. This will move the break-even points to shorter line length regions.

11.4 SELF-ORGANIZED 3-D MICRO-OPTICAL SWITCHING SYSTEMS (3D-MOSSs)

11.4.1 Concept

Figure 11.9a presents a conventional 32×32 Banyan network consisting of five optical switch arrays. Each array contains sixteen 2×2 optical switches. Adjacent arrays are connected by optical waveguides in connection areas (CAs). In CA 1, an optical waveguide experiences 15 waveguide crossing points, seven in CA 2, three in CA 3, and one in CA 4. To construct a 3D-MOSS, the conventional Banyan network is divided into four subnetwork blocks of BL 1, 2, 3, and 4. Optical waveguides corresponding to *inter-* and *intra-*block connections are, respectively, represented by thick and thin lines. As presented in Figure 11.9b, the blocks are separated and stacked to each other to construct a 32×32 3D-MOSS with a four-layer structure of BL 1, 2, 3, and 4. The long optical waveguides in CAs 1 and 2 are replaced with short-distance optical Z-connections, which experience no waveguide crossing points. Thus, the 3D-MOSS drastically reduces the waveguide crossing point count and the wiring distance. This is a kind of warping. The advantage of 3D-MOSS becomes remarkable by increasing the channel count. For a 1024×1024 Banyan network, an optical waveguide in CA 1 experiences 511 waveguide crossing points in the conventional planar configuration. The 3D-MOSS reduces the crossing point count to 0 [1,10–12,18,19].

CAs consisting of in-plane waveguides and corner-turning mirrors are denoted by "in-plane CAs," and those containing optical Z-connections by "vertical CAs." In the present model, CAs 3 and 4 are in-plane CAs, and CAs 1 and 2 are vertical CAs.

An example of a procedure for self-organization of 3D-MOSSs is shown in Figure 11.10. First, luminescent targets are deposited on core edges of optical switch dies using selective MLD or ALD with processes mentioned in Sections 10.1 and 11.3.1 (see Figures 10.7 and 11.5). Next, OE-Films are fabricated by embedding the optical switch dies into optical waveguide films using PL-Pack with SORT with a process similar to that presented in Figure 11.5. Then, as depicted in Figure 11.10a, a 3-D pre-form is prepared by stacking the OE-Films by inserting PRI materials in parts, where SOLNETs will be formed. By illuminating luminescent regions of optical waveguides with excitation lights, luminescence 1 is generated from the luminescent regions, and propagates in the optical waveguides. The luminescence 1 generates luminescence 2 from luminescent targets. As a result, as depicted in Figure 11.10b, the luminescence 1 and the luminescence 2 form P/R-SOLNETs for connections between the optical switches and the optical waveguides. For the vertical waveguide construction, P-SOLNETs are formed by two write beams of luminescence 1 [1,10–12,18,19].

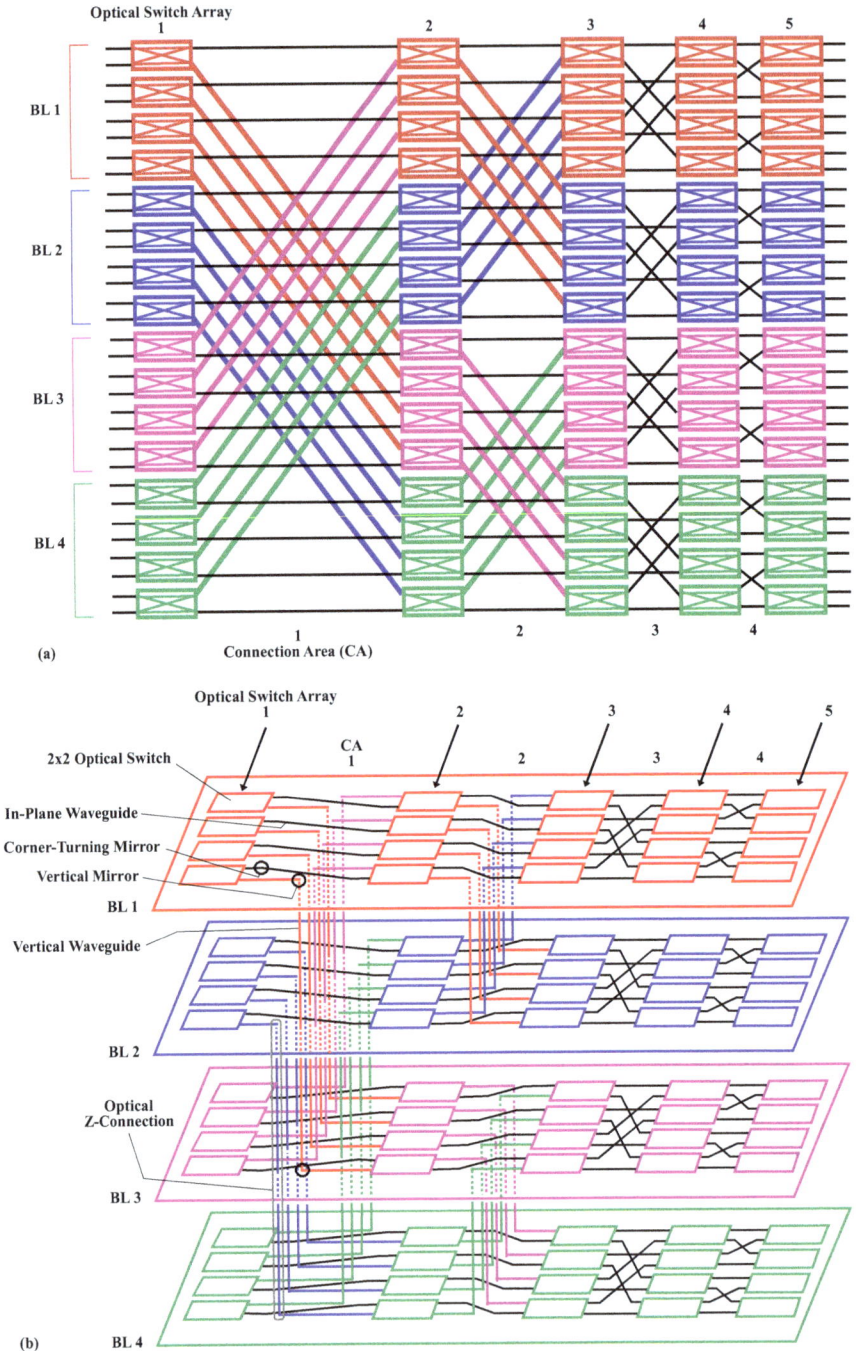

FIGURE 11.9 (a) Conventional 32×32 Banyan network consisting of five optical switch arrays. (b) 3D-MOSS for 32×32 Banyan network. (Reprinted with permission from Yoshimura et al. [10].)

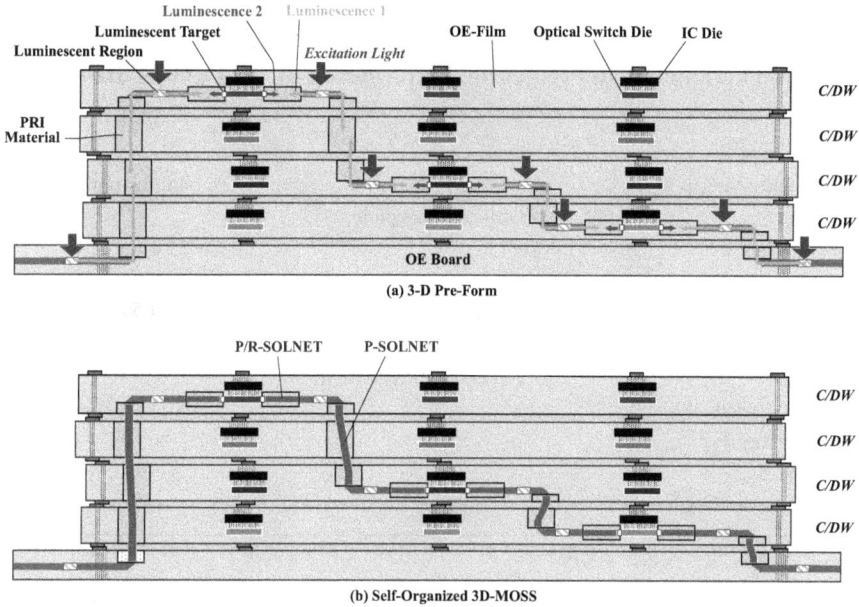

FIGURE 11.10 Procedure for self-organization of 3D-MOSSs. (Reprinted with permission from Yoshimura et al. [10].)

11.4.2 Impacts of Polymer MQDs

Table 11.2 summarizes structural parameters to predict the performance of 1024×1024 3D-MOSSs. The 3-D optical wiring backbone consists of optical waveguides distributed with a pitch P_{WG} of 20 μm. Probe beams of 1.3 μm in wavelength is used to calculate coupling efficiency (or insertion loss). OE-Film thickness T_F is 46 μm, and the gap between the OE-Films, T_G, is 4 μm. So, thickness per one layer, $T_F + T_G$, is 50 μm [1,10–12,18,19].

2×2 WPD optical switches shown in Figure 9.15 are assumed to be located with a pitch P_{SW} of 200 μm in optical switch arrays. The optical switch array count, that is, step count S_{SW} is 10. One array contains $N_C/2$ optical switches with length L_{SW} of 1190 μm and width W_{SW} of 100 μm. Although, in the model shown in Figure 9.15, 400-nm-thick PLZT is used for the EO material, 4-μm-thick PLZT is assumed in the present model. In this case, switching voltage of 12 V is required to induce refractive-index change equivalent to the change induced by 1.2 V in the 400-nm-thick PLZT.

Figure 11.11 shows top views of 1024×1024 Banyan network models for a conventional planar structure and a 3D-MOSS structure. In the 3D-MOSS, the network is divided into N_L blocks, which are stacked to make a N_L-layer 3-D structure. The area of the network is drastically reduced in the 3D-MOSS structure with $N_L = 16$, comparing with the conventional planar structure. The sizes inserted in Figure 11.11, as well as in Figures 11.12 and 11.13, are calculated using Equations (11.2)–(11.5) described later.

TABLE 11.2

Structural Parameters to Predict the Performance of 1024×1024 3D-MOSSs

Channel count, N_C		1024
Layer count, N_L		1, 2^2, 2^4, 2^6, 2^9
OE-film thickness/gap, T_F/T_G		46 μm/4 μm
Optical Waveguide		
Width, w		4 μm
Refractive index	Core, n_{Core}	1.52
	Cladding, n_{Clad}	1.50
Pitch, P_{WG}		20 μm
2×2 WPD Optical Switch		
Length, L_{SW}, width, W_{SW}		1190 μm, 100 μm
Pitch in a array, P_{SW}		200 μm
Step count, S_{SW}		10
Probe beam wavelength		1.3 μm

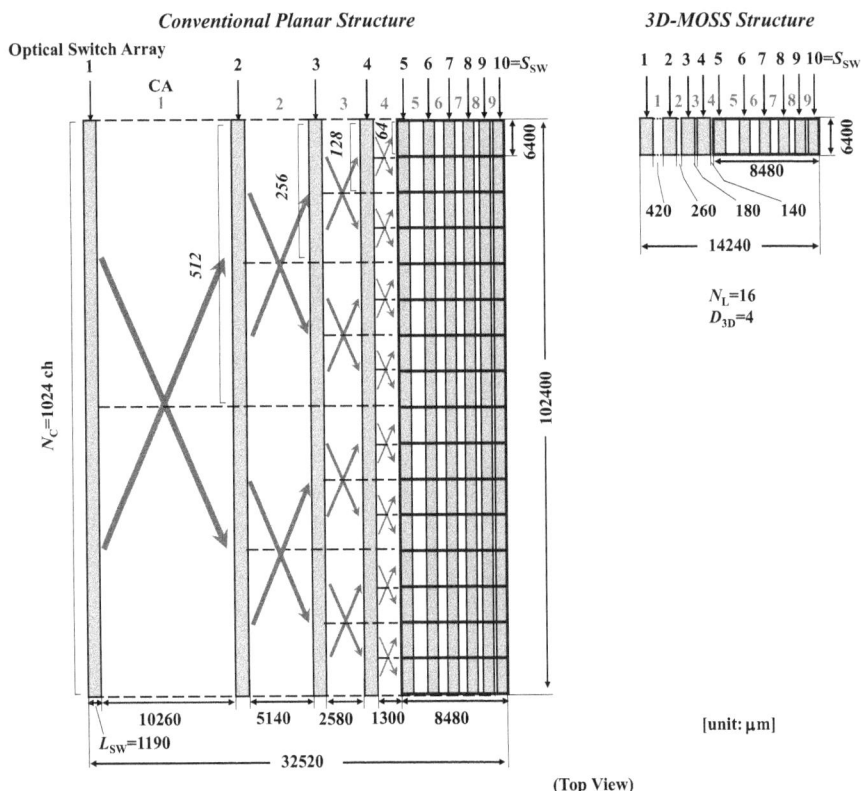

FIGURE 11.11   Top views of 1024×1024 Banyan network models for a conventional planar structure and a 3D-MOSS structure. (Reprinted with permission from Yoshimura et al. [10].)

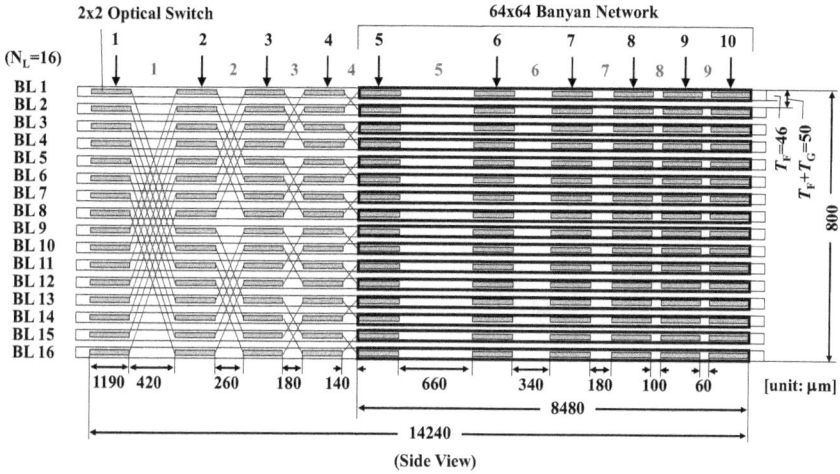

FIGURE 11.12 Side view of 1024×1024 3D-MOSS with N_L of 16. (Reprinted with permission from Yoshimura et al. [10].)

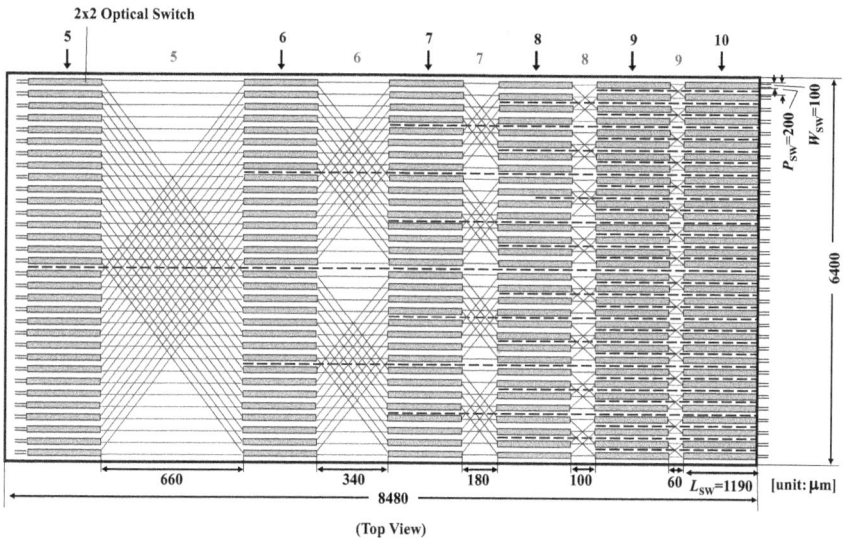

FIGURE 11.13 Top view of 64×64 Banyan network with a conventional planar structure. (Reprinted with permission from Yoshimura et al. [10].)

In the N_L-layer 3-D structure, the degree of 3-D characteristics of the networks is expressed as follows:

$$D_{3D} = \log_2 N_L \tag{11.1}$$

D_{3D} indicates the number of vertical CAs in a 3D-MOSS. Namely, CAs 1 to D_{3D} are vertical CAs, and CAs D_{3D} + 1 to S_{SW} − 1 are in-plane CAs. In the case of N_L = 16, D_{3D} = 4. As can be seen in a side view of the 3D-MOSS with N_L of 16 illustrated in Figure 11.12, optical Z-connections exist in CAs 1–4, while there are no optical Z-connections in CAs 5–9. The region involving CAs 5–9 corresponds to a 64×64 Banyan network with a conventional planar structure shown in Figure 11.13.

Length $L(IP)_i$ of in-plane CA i and length $L(V)_i$ of vertical CA i are expressed as follows:

$$L(IP)_i = \left(N_C / 2^i + 1 \right) P_{WG} \tag{11.2}$$

$$L(V)_i = \left(N_L / 2^{i-1} + 5 \right) P_{WG} \tag{11.3}$$

Width W and height H of 3D-MOSS are expressed as follows:

$$W = \left((N_C / 2) / 2^{D_{3D}} \right) P_{SW} \tag{11.4}$$

$$H = N_L \left(T_F + T_G \right) \tag{11.5}$$

The process to derive these equations is described in detail in references [1,10]. By substituting parameters listed in Table 11.2 into Equations (11.2) and (11.4), the sizes for i = 5~9 in the model with N_L = 16 can be calculated, as shown in Figures 11.12 and 11.13. Equations (11.3) and (11.4) give the sizes for i = 1–4, and Equation (11.5) gives the height of the 3D-MOSS, as shown in Figures 11.11 and 11.12.

Then, total 3D-MOSS length $L_{3D\text{-MOSS}}$, width $W_{3D\text{-MOSS}}$, and height $H_{3D\text{-MOSS}}$, including switch array length are given by

$$L_{3D-MOSS} = L_{SW} S_{SW} + \left(\sum_{i=1}^{D_{3D}} \left(N_L / 2^{i-1} + 5 \right) \right) P_{WG}$$

$$+ \left(\sum_{i=D_{3D}+1}^{S_{SW}-1} \left(N_C / 2^i + 1 \right) \right) P_{WG}, \tag{11.6}$$

$$W_{3D-MOSS} = \left((N_C / 2) / 2^{D_{3D}} \right) P_{SW}, \tag{11.7}$$

$$H_{3D-MOSS} = N_L \left(T_F + T_G \right). \tag{11.8}$$

In Figure 11.14, major losses caused in 3D-MOSS are indicated by "γ." $\gamma_{P(IP)}$ is propagation loss of in-plane waveguides, $\gamma_{P(V)}$ propagation loss of vertical waveguides, γ_{CR} loss induced at a right-angle waveguide–waveguide cross point, γ_{CT} loss of a pair of corner-turning mirrors in a crank-shaped waveguide, γ_Z loss of a pair of vertical

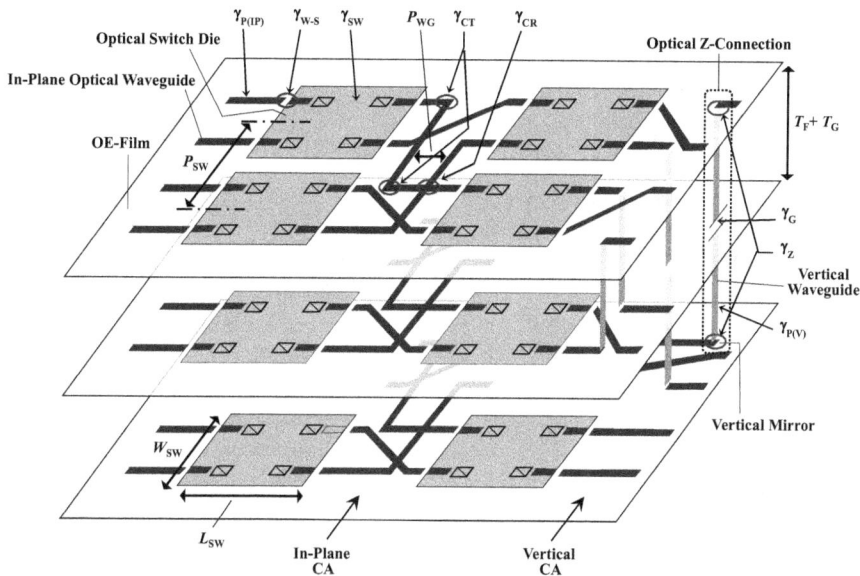

FIGURE 11.14 Schematic illustration of a part of 3D-MOSS. Major losses caused in 3D-MOSS are indicated by "γ." (Reprinted with permission from Yoshimura et al. [18].)

mirrors in an optical Z-connection, γ_G loss at a gap between OE-Films, γ_{SW} loss within a 2×2 optical switch, and γ_{W-S} waveguide–switch coupling loss. The estimated major losses are listed in Table 11.3. The detail of the estimation is described in references [1,10]. The simulation by the beam propagation method (BPM) revealed that γ_G increases to 0.89 dB when a 25% misalignment exists between the OE-Films. Here, "25% misalignment" means that the misalignment is 25% of the waveguide core width. For a core width of 4 μm, the 25% misalignment is 1 μm. γ_G with the 25% misalignment decreases to 0.14 dB by implementing SOLNETs. The BPM simulation also revealed that γ_{W-S} is 0.83 dB for a 25% misalignment. When SOLNETs are applied to the couplings, as in the case of γ_G, γ_{W-S} decreases to 0.14 dB [1,10–12,18,19].

As described in references [1,10], general expressions for the eight kinds of total losses in 3D-MOSSs are derived in terms of the losses listed in Table 11.3. The results are summarized in Table 11.4. The insertion loss of 3D-MOSSs is obtained by summation of these eight losses, and the occupation area by $L_{3D-MOSS} \times W_{3D-MOSS}$. In Figure 11.15, the results are presented as a function of D_{3D}. $D_{3D} = 0$ ($N_L = 1$) implies a conventional planar structure, and $D_{3D} = 9$ ($N_L = 2^9$) implies that all the CAs are vertical CAs. The occupation area decreases monotonically with D_{3D}. The insertion loss has a minimum at D_{3D} of ~4 when no misalignments exist in the 3D-MOSS. This result indicates that a certain optimum layer count exists to minimize the insertion loss. The dashed curve represents the insertion loss for a 25%-misalignment case, where the misalignment exists between adjacent layers and between in-plane waveguides and optical switch waveguides (note that the dashed curve is drawn with 1/24 reduction). The misalignment causes a monotonic increase in insertion loss with

TABLE 11.3
Estimated Major Losses in 3D-MOSS

Propagation in optical waveguide: γ_P		0.5 dB/cm
Waveguide–waveguide cross point: γ_{CR}		0.0088 dB
Pair of corner-turning mirrors in crank-shaped waveguide: γ_{CT}		0.97 dB
Pair of vertical mirrors in optical Z-connection: γ_Z		0.97 dB
Gap between OE-Films : γ_G	with no misalignment	0.0088 dB
	with 25% misalignment	0.89 dB
	with 25% misalignment using SOLNET	0.14 dB
2×2 optical switch: γ_{SW}		1.7 dB
Waveguide–switch coupling: $\gamma_{W\text{-}S}$	with no misalignment	0 dB
	with 25% misalignment	0.83 dB
	with 25% misalignment using SOLNET	0.14 dB

Source: Reprinted with permission from Yoshimura et al. [10].

TABLE 11.4
General Expressions for the Eight Kinds of Total Losses in 3D-MOSSs

In-Plane Optical Connections

In-plane propagation: $\Gamma_{P(IP)}$

$$\left[\left(\sum_{i=1}^{D_{3D}} (N_L/2^{i-1} + 5) \right) P_{WG} + \left(\sum_{i=D_{3D}+1}^{S_{SW}-1} (N_C/2^i + 1) \right) P_{WG} + \left(\sum_{i=D_{3D}+1}^{S_{SW}-1} (N_C/2)/2^i \right) P_{SW} \right] \gamma_P$$

Waveguide–waveguide cross point: Γ_{CR}

$$\left[\sum_{i=D_{3D}+1}^{S_{SW}-1} N_C/2^i \right] \gamma_{CR}$$

Pair of corner-turning mirrors: Γ_{CT}

$$[S_{SW} - 1 - D_{3D}] \gamma_{CT}$$

Optical switch: Γ_{SW}

$$[S_{SW}] \gamma_{SW}$$

Waveguide–switch coupling: $\Gamma_{W\text{-}S}$

$$[2S_{SW}] \gamma_{W\text{-}S}$$

Vertical Optical Connections

Vertical propagation: $\Gamma_{P(V)}$

$$[2N_L (T_F + T_G)] \gamma_P$$

Pair of vertical mirrors: Γ_Z

$$[2 + D_{3D}] \gamma_Z$$

Gap between OE-Films: Γ_G

$$[2N_L] \gamma_G \quad [2 + D_{3D}] \gamma_G \ \text{ for SOLNET}$$

Source: Reprinted with permission from Yoshimura et al. [10].

increasing D_{3D}. At $D_{3D}=4$, the loss is 73 dB. By implementing SOLNETs into the 3D-MOSS with a 25% misalignment, the insertion loss decreases down to 32 dB at $D_{3D}=4$, reaching a level comparable to that of the 3D-MOSS with no misalignments.

In the present calculation, the PLZT thickness, namely, the electrode gap distance d, for the WPD is increased from 400 nm to 4 μm, and the operation voltage is increased from 1.2 to 12 V comparing to the WPD depicted in Figure 9.15. The capacitance C, switching time τ, and power consumption *Power* are given by Equations (9.8), (9.9), and (9.11) in Chapter 9 for one prism-shaped capacitor. In the present 3D-MOSS model with the 4-μm-thick PLZT, C is calculated to be 0.57 pF [1,10–12,18,19].

When the number of the prism-shaped capacitors in a WPD optical switch is N_{Prism}, τ for a WPD, optical switch is given by

$$\tau = N_{Prism} \, R_{ON} C. \tag{11.9}$$

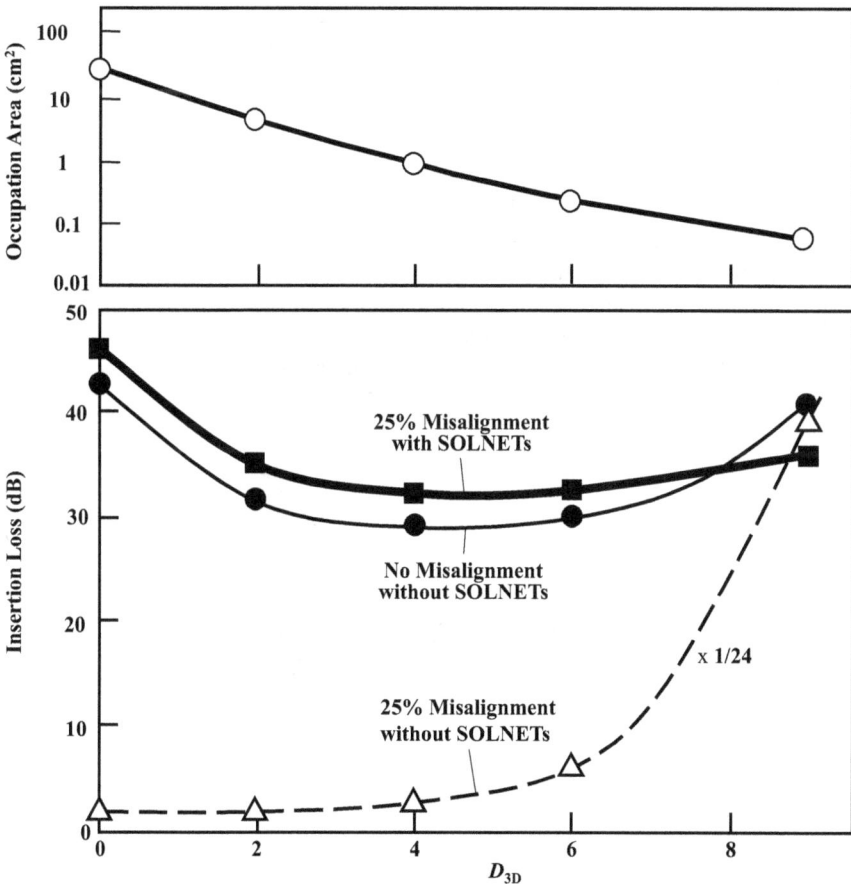

FIGURE 11.15 Calculated occupation areas and insertion losses of 3D-MOSSs plotted as a function of D_{3D}. The dashed curve is drawn with 1/24 reduction. (Reprinted with permission from Yoshimura et al. [18].)

Because the total WPD optical switch count in a 3D-MOSS is $(N_C/2)S_{SW}$, *Power* in a 3D-MOSS for a switching rate of F is expressed as follows:

$$Power = F \, CV^2 \, N_{Prism} \left(N_C/2 \right) S_{SW} \tag{11.10}$$

By substituting $C = 0.57$ pF, $V = 12$ V, $N_{Prism} = 8$, $N_c = 1024$, and $S_{SW} = 10$ into Equations (11.9) and (11.10), $\tau = 454$ ps for $R_{ON} = 100$ Ω and *Power* $= 33$ W for $F = 1 \times 10^7$ 1/s are obtained for the 1024×1024 3D-MOSS. *Power* of 33 W corresponds to power consumption density of 36 W/cm^2 since the area of the 1024×1024 3D-MOSS model is 1.42×0.64 cm^2. This value is comparable to acceptable heat generation of 30 W/cm^2 reported by Takahashi [20] for stacked LSI of 1-cm^2 area.

Figure 11.16 shows the calculated dependence of the power consumption density on the switching rate in 1024×1024 3D-MOSSs, where WPD optical switches with 4-μm-thick PLZT, 400-nm-thick PLZT, or 400-nm-thick polymer MQD are assumed. The power consumption density increases with the switching rate. The achievable switching rate is limited by the heat-releasing capability of the system. Therefore, for a given heat-releasing rate, the lower the heat generation rate becomes, the higher the achievable switching rate becomes as far as the switching time of the WPD optical switch is sufficiently short.

For 4-μm-thick PLZT, the switching rate is limited below 1×10^7 1/s in a case that the limit of the heat releasing is ~30 W/cm^2. When the PLZT thickness is decreased to 400 nm, the operation voltage decreases to 1.2 V (see Table 9.1) and the allowable switching rate increases to 1×10^8 1/s. This indicates that 3-D optical wiring backbones consisting of high-index-contrast (HIC) waveguides is favorable for high-speed/low-power operation. By replacing the 400-nm-thick PLZT with 400-nm-thick polymer MQD, whose properties are listed in Tables 9.1 and 9.3, further switching rate improvement is expected, suggesting a possibility of switching rate more than 10^9 1/s.

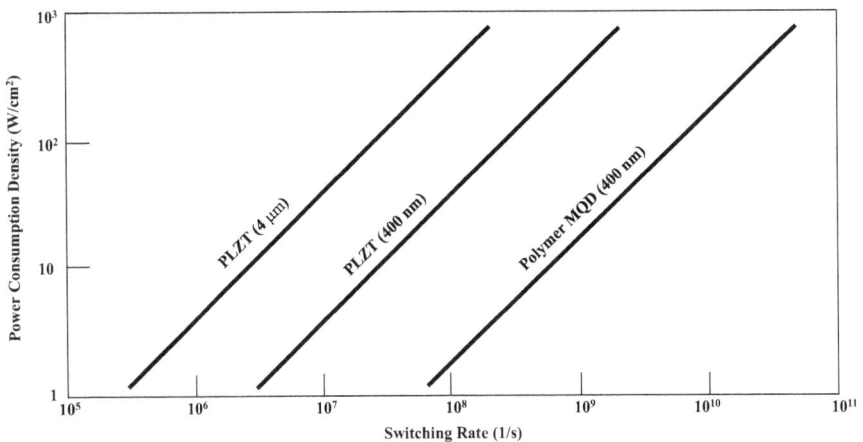

FIGURE 11.16 Calculated dependence of the power consumption density on the switching rate in 1024×1024 3D-MOSSs, where WPD optical switches with 4-μm-thick PLZT, 400-nm-thick PLZT, or 400-nm-thick polymer MQD are used.

REFERENCES

1. T. Yoshimura, *Self-Organized 3D Integrated Optical Interconnects: with All-Photolithographic Heterogeneous Integration*, Jenny Stanford Publishing, Singapore (2021).
2. T. Yoshimura, S. Yasuda, H. Yamaura, and Y. Yamada, "Self-organized lightwave network (SOLNET) formed by two-photon photochemistry for 3-D integrated optical interconnects," *2018 IEEE Optical Interconnects Conference (OI)*, Santa Fe, New Mexico, 33–34 (2018).
3. T. Yoshimura, S. Yasuda, H. Yamaura, Y. Yamada, M. Takashima, R. Ito, and T. Hamazaki, "Self-aligned coupling waveguides experimentally formed by two-photon photochemistry for 3-D integrated optical interconnects," *Opt. Commun.* **430**, 284–292 (2019).
4. T. Yoshimura, J. Roman, Y. Takahashi, M. Lee, B. Chou, S. I. Beilin, W. V. Wang, and M. Inao, "Proposal of optoelectronic substrate with film/Z-connection based on OE-film," *Proc. 3rd IEMT/IMC Symposium*, 140–145 (1999).
5. T. Yoshimura, J. Roman, Y. Takahashi, M. Lee, B. Chou, S. Beilin, W. Wang, and M. Inao, "Optoelectronic scalable substrates based on film/Z-connection and its application to film optical link module (FOLM)," *Proc. SPIE* **3952**, 202–213 (2000).
6. T. Yoshimura, Y. Takahashi, M. Inao, M. Lee, W. Chou, S. Beilin, W-C. Wang, J. Roman, and T. Massingill, "Three-dimensional opto-electronic modules with electrical and optical interconnections and methods for making," U.S. Patent 6,690,845 B1 (2004).
7. T. Yoshimura, Y. Takahashi, M. Inao, M. Lee, W. Chou, S. Beilin, W-C. Wang, J. Roman, and T. Massingill, "Systems based on opto-electronic substrates with electrical and optical interconnections and methods for making," U.S. Patent 6, **343**, 171 B1 (2002).
8. T. Yoshimura, J. Roman, Y. Takahashi, W. V. Wang, M. Inao, T. Ishitsuka, K. Tsukamoto, K. Motoyoshi, and W. Sotoyama, "Self-organizing waveguide coupling method "SOLNET" and its application to film optical circuit substrates," *Proc. 50th Electronic Components & Technology Conference (ECTC)*, Las Vegas, Nevada, 962–969 (2000).
9. T. Yoshimura, T. Inoguchi, T. Yamamoto, S. Moriya, Y. Teramoto, Y. Arai, T. Namiki, and K. Asama, "Self-organized lightwave network based on waveguide films for three-dimensional optical wiring within boxes," *J. Lightwave Technol.* **22**, 2091–2100 (2004).
10. T. Yoshimura, M. Ojima, Y. Arai, and K. Asama, "Three-Dimensional self-organized micro optoelectronic systems for board-level reconfigurable optical interconnects–performance modeling and simulation–," *IEEE J. Select. Top. Quantum Electron.* **9**, 492–511 (2003).
11. T. Yoshimura and Y. Arai, "3D optoelectronic micro system," U.S. Patent 20060261432A1 (2006).
12. T. Yoshimura, *Self-Organized Lightwave Networks: Self-Aligned Coupling Optical Waveguides*, CRC/Taylor & Francis, Boca Raton, Florida, 2018.
13. B. Chou, S. Beilin, H. Jiang, D. Kudzuma, M. Lee, M. McCormack, T. Massingill, M. Peters, J. Roman, Y. Takahashi, and W. V. Wang, "Multilayer high density flex technology," *Proc. 49th Electronic Components & Technology Conference (ECTC)*, San Diego, California, 1181–1189 (1999).
14. T. Yoshimura, K. Kumai, T. Mikawa, O. Ibaragi, and O. Bonkohara, "Photolithographic packaging with selectively occupied repeated transfer (PL-pack with SORT) for scalable optical link multi-chip-module (S-FOLM)," *IEEE Trans. Electron. Packag. Manufact.* 25, 19–25 (2002).
15. T. Yoshimura, J. Roman, W-C. Wang, M. Inao, and M. T. McCormack, "Device transfer method," U.S. Patent 6,669,801 B2 (2003).

16. T. Yoshimura, J. Roman, Y. Takahashi, S. I. Beilin, W. V. Wang, and M. Inao, "Optoelectronic amplifier/driver-less substrate <OE-ADLES> for polymer-waveguide-based board-level interconnection -calculation of delay and power dissipation-," *Nonlinear Opt.* **22**, 453–456 (1999).

17. T. Yoshimura, Y. Suzuki, N. Shimoda, T. Kofudo, K. Okada, Y. Arai, and K. Asama, "Three-dimensional chip-scale optical interconnects and switches with self-organized wiring based on device-embedded waveguide films and molecular nanotechnologies," *Proc. SPIE* 6126, 612609 (2006).

18. T. Yoshimura, Y. Arai, H. Kurokawa, and K. Asama, "Predicted insertion loss reductions achieved by implementing three-dimensional micro optical network in chip-scale optical interconnects," *IEEE Photonics Technol. Lett.* **16**, 647–649 (2004).

19. T. Yoshimura, S. Tsukada, S. Kawakami, M. Ninomiya, Y. Arai, H. Kurokawa, and K. Asama, "Three-dimensional micro-optical switching system architecture using slab-waveguide-based micro-optical switches," *Opt. Eng.* **42**, 439–446 (2003).

20. K. Takahashi, "Research and Development on ultra-high-density 3-dimensional LSI-chip-stack packaging technologies," *The 3rd Annual Meeting on Electronics System Integration Technologies Digest* (edited and published by the Electronic System Integration Technology Research Department, Association of Super-Advanced Electronics Technologies (ASET)), 43–94 (2002).

12 Solar Energy Conversion Systems

Organic multiple quantum dots (MQDs) grown by molecular layer deposition (MLD) are expected to be applied to spectral sensitization of photovoltaic devices and artificial photosynthesis devices for solar energy conversion systems. In the present chapter, the waveguide-type thin-film photovoltaic (WTP) device with the organic-MQD sensitization, that is, molecular-MQD and polymer-MQD sensitization, is proposed. The organic-MQD sensitization is expected to serve a process with multi-step excitation imitating the photosynthesis in plants. Proof of concept of the molecular-MQD sensitization for the WTP devices is demonstrated by growing stacked structures of n-type and p-type dye molecules on ZnO thin films by liquid-phase MLD (LP-MLD) and measuring photocurrents in the guided light configuration. A possibility of the polymer-MQD sensitization is also discussed. Three-dye molecular-MQD sensitization of ZnO is described to give another proof of concept. Proposals of film-based integrated solar energy conversion systems, thin-film artificial photosynthesis devices, and molecular sensing are presented as future MLD applications.

12.1 WAVEGUIDE-TYPE THIN-FILM PHOTOVOLTAIC (WTP) DEVICES WITH ORGANIC-MQD SENSITIZATION

12.1.1 CLASSIFICATION OF WTP DEVICES

In Figure 12.1, the conventional dye-sensitized solar cell (DSC) and the proposed WTP devices are compared. The DSC depicted in Figure 12.1a has received much attention as a cost-effective device [1]. It uses porous semiconductors, which increase the number of adsorbed sensitizing dye molecules on the semiconductor surfaces to enhance light absorption. The porous structures, however, raise the internal resistivity because they narrow the carrier transport channels and decrease the carrier mobility, degrading the photovoltaic response.

To avoid the problem, it is preferable to use thin-film semiconductors with flat surfaces and high crystalline quality for enhanced carrier mobility. This reduces the internal resistivity and improves the photovoltaic response. In this case, however, normally-incident light passes through only a monomolecular dye layer, resulting in small light absorption.

The WTP devices, which employ the guided light configuration, solve the problem [2–10]. They use thin-film semiconductors, and are classified into three types as presented in Figure 12.1b. In the WTP device with dye sensitization, light beams propagate in a thin-film semiconductor with adsorbed dye molecules and pass through a

DOI: 10.1201/9781003094012-12

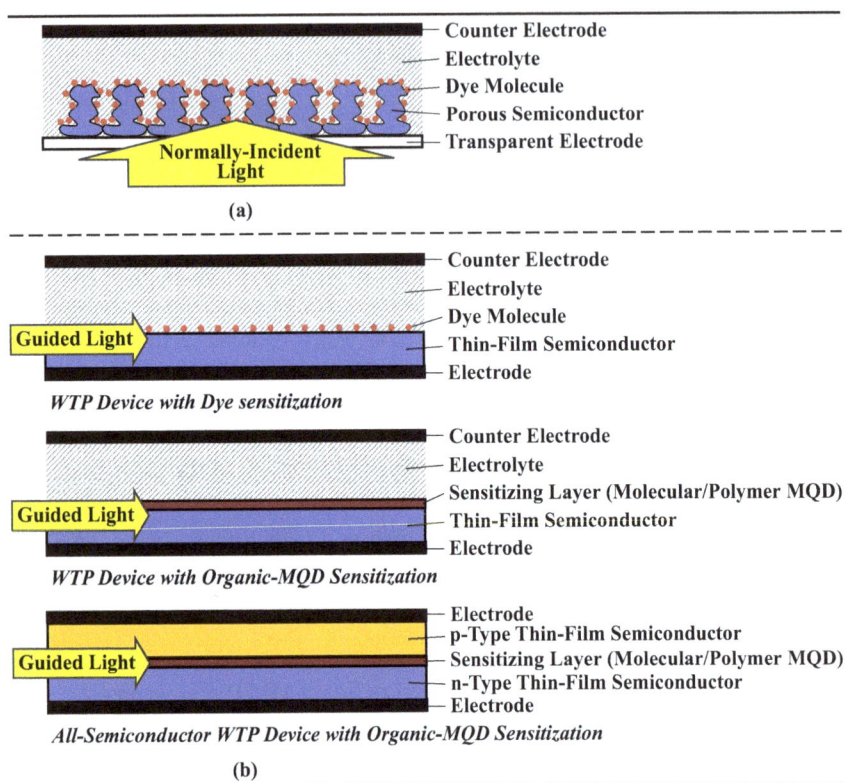

FIGURE 12.1 Comparison of (a) the conventional dye-sensitized solar cell (DSC) and (b) the proposed waveguide-type thin-film photovoltaic (WTP) devices: WTP device with dye sensitization, WTP device with organic-MQD sensitization, and all-semiconductor WTP device with organic-MQD sensitization.

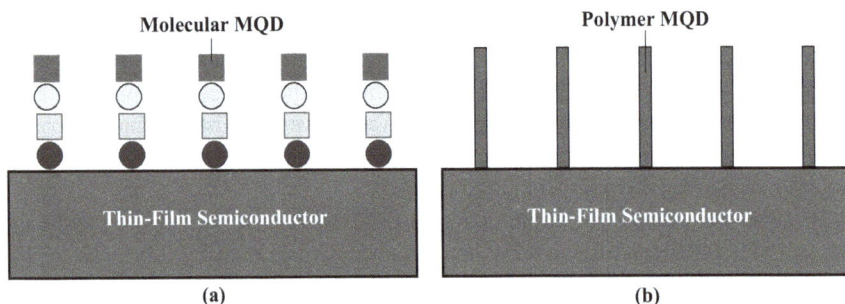

FIGURE 12.2 Schematic structures for the organic-MQD sensitization. (a) Molecular-MQD sensitization and (b) polymer-MQD sensitization.

lot of dye molecules to enhance light absorption. An electrolyte is located between the semiconductor and the counter electrode. A solid-state electrolyte is preferable for robustness and stability. In addition, the guided light configuration enables us to replace expensive and electrically-resistive transparent electrodes with low-cost and low-resistivity metal electrodes. In the WTP device with organic MQD sensitization, a sensitizing layer of organic MQDs, that is, molecular MQDs or polymer MQDs, is grown on a thin-film semiconductor. In the all-semiconductor WTP device with organic-MQD sensitization, a p-type thin-film semiconductor and an n-type thin-film semiconductor are stacked and a sensitizing layer of organic MQDs is inserted between them. Light beams propagate in the layered structure.

Schematic structures for the organic-MQD sensitization are presented in Figure 12.2. Molecular MQDs or polymer MQDs are grown by MLD on thin-film semiconductors [2–12].

12.1.2 ORGANIC-MQD SENSITIZATION

The Si solar cell is the most popular photovoltaic device. Nevertheless, as illustrated in Figure 12.3a, for photons, whose energy is much higher than the band gap of Si, excess energy of the photons is lost as heat to reduce the light energy conversion efficiency. The organic-MQD sensitization is expected to suppress the excess energy loss in the light absorption process [5–11].

Figure 12.3b shows an example of the molecular-MQD sensitization for ZnO. Four kinds of molecules M1–4 having different energy gaps are connected to form a molecular MQD. In the present model, p-type and n-type molecules are connected in a p/n/p/n sequence on n-type ZnO. The wide absorption wavelength region from near infrared to visible is divided into a plurality of narrow wavelength regions corresponding to absorption bands of individual molecules of M1, 2, 3, and 4. Each molecule absorbs light with wavelengths matched to its own energy gap, and injects the excited electrons into ZnO to generate photocurrents. This might suppress the

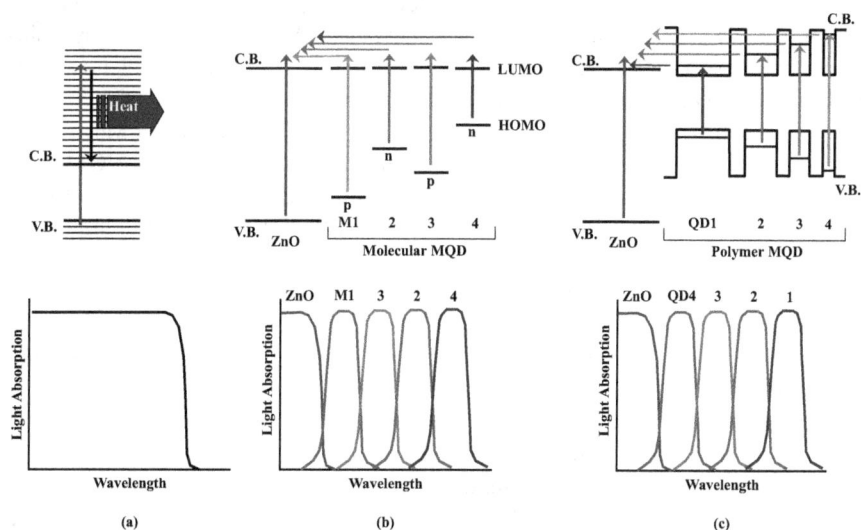

FIGURE 12.3 Schematic illustrations of energy levels and absorption spectra in (a) Si, (b) ZnO with molecular-MQD sensitization, and (c) ZnO with polymer-MQD sensitization.

energy loss arising from the excess photon energy. The structure is regarded as a molecular tandem structure. Figure 12.3c shows an example of the polymer-MQD sensitization. A polymer MQD contains four kinds of quantum dots (QDs) QD1–4 having different lengths. Absorption bands of individual QDs are narrow because of their zero-dimensional characteristics, and the peak positions are determined by the dot lengths. The superposition of the absorption bands of the QDs covers a wide wavelength range. In order to improve the carrier transport performance in the polymer MQDs, the tunneling effect should be adjusted by choosing appropriate barriers. Furthermore, to expand the wavelength region for sensitization to the longer wavelength side, backbone polymers with narrow band gaps are required.

12.1.3 ORGANIC-MQD SENSITIZATION WITH MULTI-STEP EXCITATION

Figure 12.4 shows the molecular-MQD sensitization with multi-step excitation imitating the photosynthesis process in plants [5–13]. In the process shown in Figure 12.4a, gradual energy steps are formed in molecular wires, just like the light-harvesting antennas in the photosynthesis, to perform smooth electron (or energy) transfer to ZnO. In the model drawn at the left, the electron transfer from the redox system occurs at the top of the molecular wires. In the model at the right, it occurs at the inner part of the molecular wires.

In the process shown in Figure 12.4b, two-step excitation, which is similar to the z-scheme process for the photosynthesis in plants, is employed. In the model at the left, an electron excited by a photon with a wavelength of λ_1 in M1 is transferred into ZnO to generate photocurrents. An electron excited by a photon with λ_2 to the lowest unoccupied molecular orbital (LUMO) in M2 is transferred to the highest occupied molecular orbital (HOMO) of M1. The hole left in M2 is compensated by the redox system. In the model at the right, an electron excited by a photon with λ_2 in M2 is transferred into ZnO via M1. An electron excited by a photon with λ_4 in M4 is transferred to HOMO of M2 via M3. The same electron flow occurs when an electron is excited by a photon with λ_1 in M1 instead of in M2, and when an electron is excited by a photon with λ_3 in M3 instead of in M4. For the z-scheme-like molecular arrangements, a molecule with high-energy HOMO and a molecule with low-energy LUMO should be combined. For example, the former could be crystal violet (CV), pinacyanol, etc. and the latter could be phthalocyanine, C_{60}, etc.

Similar z-scheme-like sensitization might be possible by two-photon sensitizing molecules with four-level two-photon absorption characteristics. As shown in Figure 12.4c, a photon with λ_2 excites an electron from singlet S_0 state to S_n state in the molecule, and the excited electron in S_n state transfers to triplet T_1 state. Then, a photon with λ_1 further excites the electron to T_n state to complete the four-level two-photon absorption [14,15]. The transfer of the excited electron to the conduction band of ZnO generates photocurrents. Protoporphyrin, Zn(II) tetraphenylporphyrin, biacetyl, camphorquinone, benzyl, and so on are known as the two-photon sensitizing materials.

The sensitization mechanism employing the two-step excitation in the z-scheme process has the following advantages:

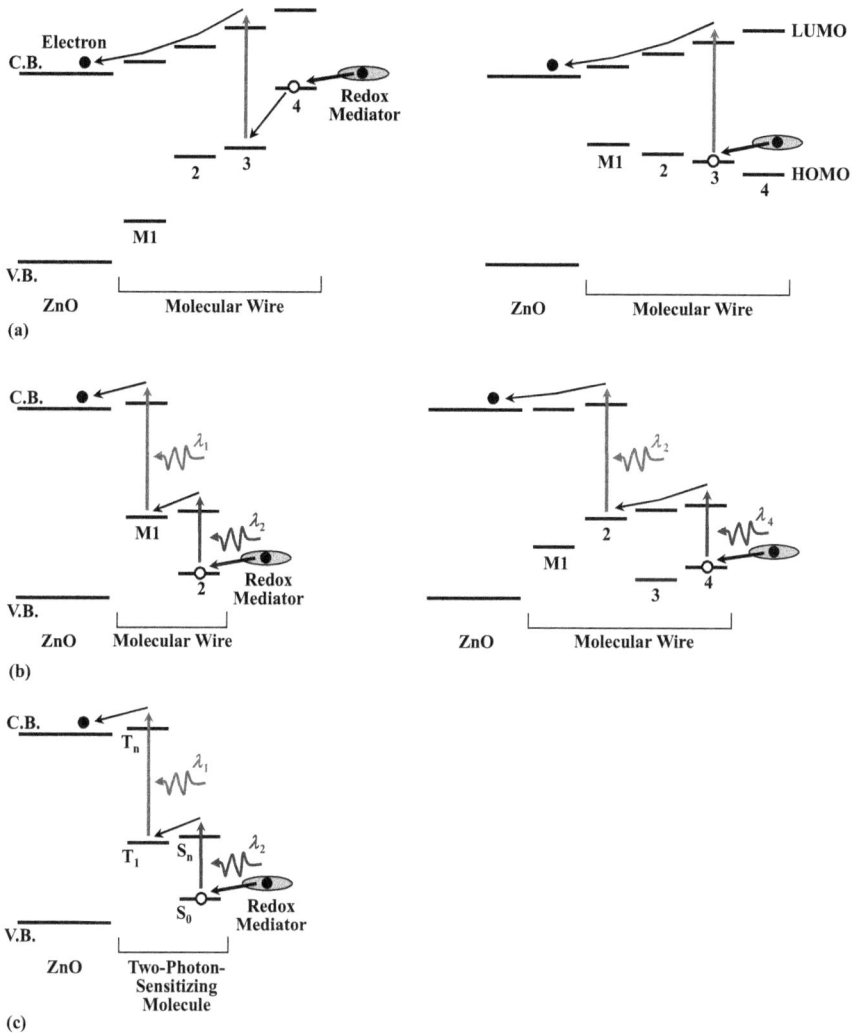

FIGURE 12.4 Molecular-MQD sensitization imitating the photosynthesis process in plants. (a) Electron (or energy) transfer via gradual energy steps, (b) z-scheme-like electron transfer via two-step excitation, and (c) z-scheme-like electron transfer performed by sensitizing molecules with four-level two-photon absorption characteristics.

- It suppresses the energy loss arising from the excess photon energy, which occurs in Si photovoltaic devices and DSCs with black dyes in the light absorption process, to improve the light energy conversion efficiency.
- It increases the difference in energy between the Fermi level of the semiconductor and the HOMO of the sensitizing molecule to increase the generated voltage in the photovoltaic devices.

Multi-step excitation, where more than three steps are involved, will enhance these advantages. To make the multi-step excitation effective for the sensitization, the balance of the excitation rates at individual steps should be adjusted. The optimum condition is that the rates are identical for all steps.

Energy levels for the all-semiconductor WTP device with organic-MQD sensitization via the two-step excitation are schematically illustrated in Figure 12.5 [5–10,12]. In the molecular-MQD sensitization, as shown in Figure 12.5a, an electron excited

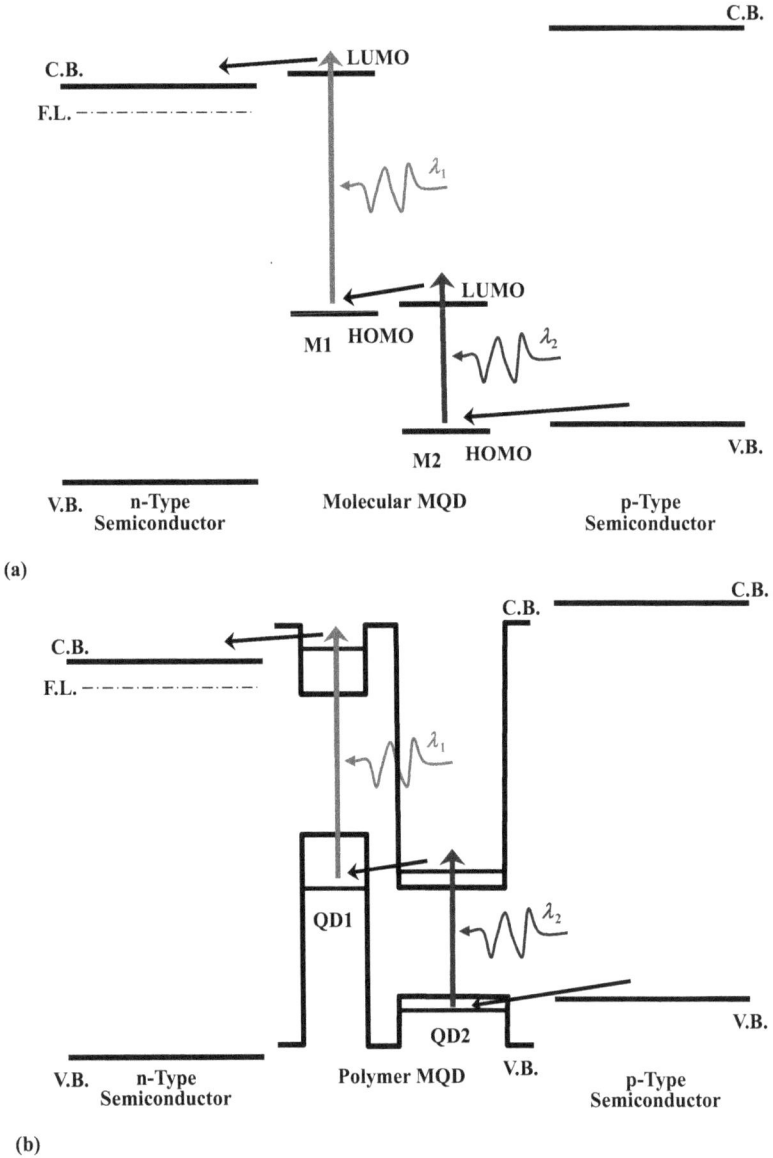

(a)

(b)

FIGURE 12.5 Schematic illustrations of energy levels for the all-semiconductor WTP device with organic-MQD sensitization via the two-step excitation: (a) with molecular-MQD sensitization and (b) with polymer-MQD sensitization.

by a photon with λ_1 in M1 is injected into the n-type semiconductor. An electron excited by a photon with λ_2 in M2 is transferred to HOMO of M1. The hole left in HOMO of M2 is compensated by an electron from the p-type semiconductor. Thus, electrons travel cyclically as ---[p-type semiconductor] → [HOMO/M2] → [LUMO/M2] → [HOMO/M1] → [LUMO/M1] → [n-type semiconductor] → [external circuits] → [p-type semiconductor] →---. This sensitization mechanism suppresses the energy loss arising from the excess photon energy, and at the same time, increases the generated voltage. In the polymer-MQD sensitization, the similar two-step excitation of electrons might be possible by adjusting the band structures, as shown in Figure 12.5b. An electron excited by a photon with λ_1 in QD1 is injected into the n-type semiconductor. An electron excited by a photon with λ_2 in QD2 is transferred to the valence band of QD1. The hole left in the valence band of QD2 is compensated by an electron from the p-type semiconductor. Thus, electrons travel cyclically as ---[p-type semiconductor] → [V.B./QD2] → [C.V./QD2] → [V.B./QD1] → [C.V./QD1] → [n-type semiconductor] → [external circuits] → [p-type semiconductor] →---.

12.2 PROOF OF CONCEPT OF ORGANIC-MQD SENSITIZATION FOR WTP DEVICES

In the present section, molecular-MQD sensitization of ZnO in the guided light configuration for WTP devices is demonstrated. A stacked structure of n-type and p-type dye molecules, which is regarded as a short molecular wire, is grown on ZnO by LP-MLD. Preliminary results of molecular-MQD sensitization performed by LP-MLD with three kinds of dye molecules are presented. A possibility of polymer-MQD sensitization is also discussed.

12.2.1 TWO-DYE SENSITIZATION BY P/N-DYE-STACKED STRUCTURES

Prior to the demonstration of the organic-MQD sensitization utilizing MLD, the two-dye sensitization by p/n-dye-stacked structures is explained. It was invented and developed by Kohei Kiyota, and was used for ZnO photoreceptors in the electrophotography [16]. The sensitization mechanism is applied to the molecular-MQD sensitization.

The two-dye sensitization was investigated by photocurrent and surface potential measurement using fluorescein (FL), eosine (EO), and rose bengal (RB) for p-type dyes, and CV, malachite green (MG), and brilliant green (BG) for n-type dyes. The dye molecule structures are shown in Figure 3.24. The photocurrent measurement was performed for 20-μm-thick ZnO powder layers formed on glass substrates with slit-type SnO_2 electrodes (width: 10 mm, gap distance: 0.5 mm) with applying 30 V to the slit gap. The surface potential measurement and the dye adsorption on ZnO powder layers were carried out by the procedures described in Section 3.5.2. Figure 12.6 shows reflection spectra of the dyed ZnO powder layers. Absorption bands of p-type dyes appear in shorter wavelength regions comparing to those of n-type dyes [17–19].

Photocurrents in the dyed ZnO layers at the maximum absorption wavelengths of adsorbed dye molecules for the same light intensity are listed in Table 12.1. The photocurrents are much larger for the p-type dye than for the n-type dye while the peak

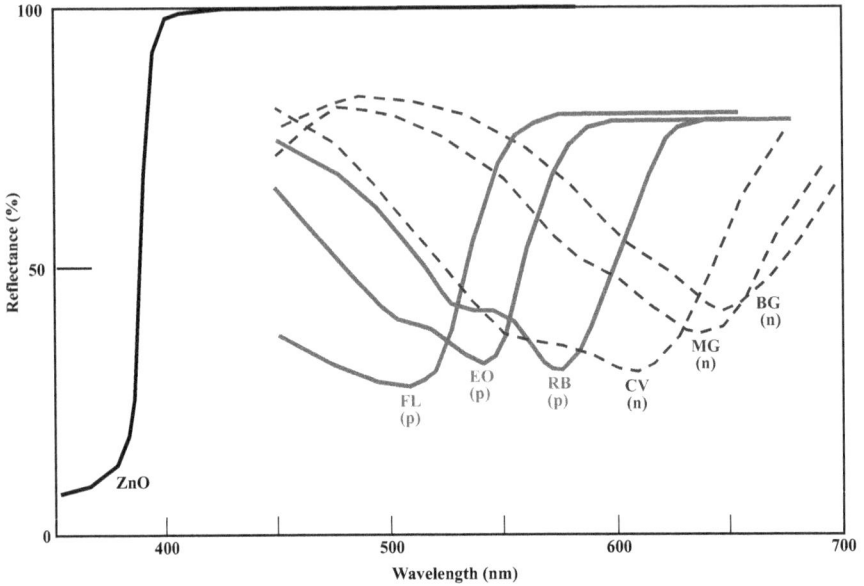

FIGURE 12.6   Reflection spectra of dyed ZnO powder layers.

TABLE 12.1
Photocurrents in Dyed ZnO Layers at the Maximum Absorption Wavelengths of Adsorbed Dye Molecules

Dye	Type	Photocurrent (nA)
FL	p	120
EO	p	190
RB	p	180
CV	n	3
MG	n	0.02
BG	n	0.01

Source: Reprinted with permission from Yoshimura et al. [17]. Copyright (1981) The Physical Society of Japan and The Japan Society of Applied Physics.

values of light absorption are almost the same for both p-type and n-type dyes. This confirms that the p-type dye molecule injects photocarriers into ZnO more efficiently than the n-type dye molecule as Meier reported [20].

A model for the two-dye sensitization is presented in Figure 12.7. Because ZnO is n-type, a p-type dye molecule has strong bond strength to ZnO, exhibiting high electron injection efficiency. An n-type dye molecule has weak bond strength to ZnO, exhibiting low electron injection efficiency. By inserting a p-type dye molecule between an n-type dye molecule and n-type ZnO to construct a p/n-dye-stacked

	p-Type Dye	n-Type Dye
Bond Strength to ZnO:	Strong	Weak
Electron Injection Efficiency:	High	Low
Wavelength Region of Light Absorption Band:	450-600 nm	500-700 nm

Two-Dye Sensitization

p/n-Dye-Stacked Structure

ZnO (n) / p-Type Dye / n-Type Dye

Widening Wavelength Region for Spectral Sensitization

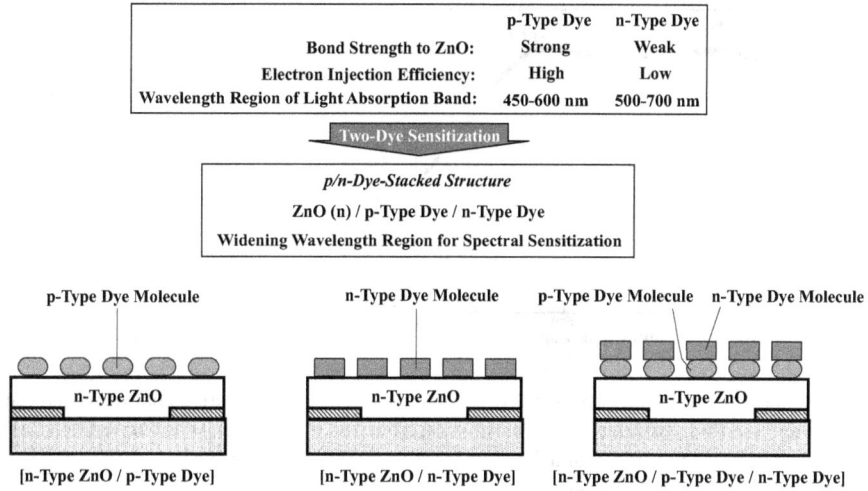

FIGURE 12.7 Model for the two-dye sensitization by p-/n-dye-stacked structures.

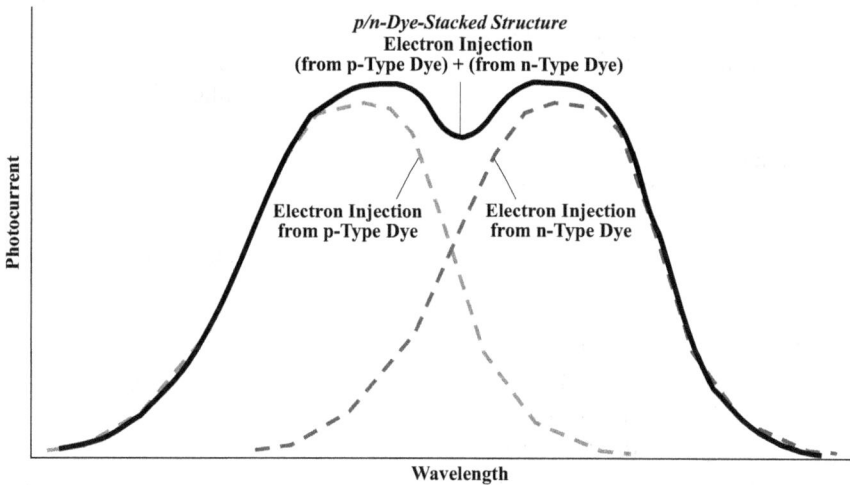

FIGURE 12.8 Expected photocurrent spectrum widening achieved by the p-/n-dye-stacked structures.

structure, strong bonds are formed between the n-type ZnO and p-type dye molecule and between the p-type and n-type dye molecules. Then, an excited electron in the n-type dye molecule can be injected into ZnO via the p-type dye molecule efficiently. As a result, as illustrated in Figure 12.8, photocurrents generated by the electron injection from the n-type dye is superposed on photocurrents generated by the electron injection from the p-type dye to widen the wavelength region for spectral sensitization in the p/n-dye-stacked structure.

A typical two-dye sensitization of ZnO is shown in Figure 12.9a for a combination of RB and CV. Photocurrents are much smaller in CV-adsorbed ZnO (ZnO/CV)

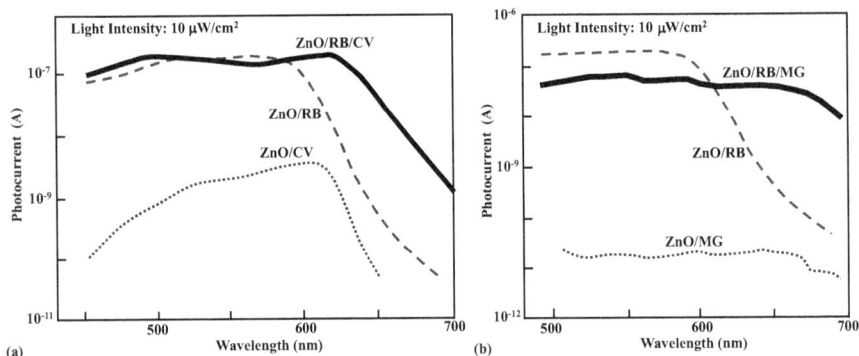

FIGURE 12.9   Typical two-dye sensitization of ZnO for dye combinations of (a) RB and CV and (b) RB and MG. (Reprinted with permission from Yoshimura et al. [17]. Copyright (1981) The Physical Society of Japan and The Japan Society of Applied Physics.)

than in RB-adsorbed ZnO (ZnO/RB). In the p/n-dye-stacked structure of ZnO/RB/CV, the photocurrents generated by the electron injection from CV to ZnO are drastically raised to extend the photocurrent spectrum to the longer wavelength region. Similar two-dye sensitization is observed in a combination of RB and MG, as shown in Figure 12.9b. The photocurrents of ZnO/RB/MG are smaller than in ZnO/RB/CV. This indicates that the efficiency of the two-dye sensitization depends on the dye combination.

As mentioned in Section 3.5.2, dye molecules and oxygen molecules adsorbed on a ZnO surface induce a depletion layer near the surface, generating an electric double layer. When visible light falls on a dyed ZnO layer, the surface potential varies as shown in Figure 12.10 along the following process [18,19].

1. Dye molecules absorb light to generate electron-hole pairs in them.
2. The excited electrons in the dye molecules are injected into ZnO and the holes left in the molecules are compensated by electrons trapped in the ionized oxygen on the ZnO surface (see Figure 12.14).
3. Consequently, the strength of the electric double layer at the ZnO surface decreases, resulting in the surface potential variation.

The saturated values of the surface potential are determined by the electron injection rate from the dye molecules into ZnO and the electron trapping rate by adsorbed oxygen. This implies that the larger the electron injection rate becomes, the larger the saturated value becomes. The results shown in Figure 12.10 indicate that the electron injection rate increases in the order of MG, CV, and RB.

The initial slope of the surface potential curve, defined by $(dV/dt)_{t=0} = (1/C)(dQ/dt)_{t=0}$, is proportional to the electron injection rate. Here, Q and C are the electric charge and the capacitance of the electric double layer at the ZnO surface, respectively. As Figure 12.11 shows, in ZnO/RB/CV, the electron injection rate is much larger than in ZnO/CV in the 600–700 nm wavelength region. This result is parallel to that obtained by the photocurrent measurement shown in Figure 12.9a.

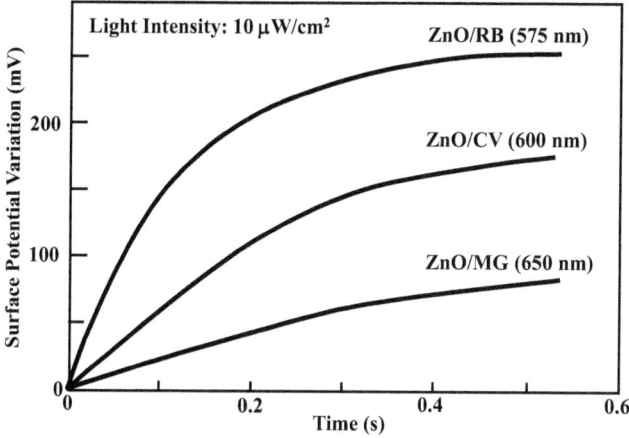

FIGURE 12.10 Surface potential variation of dyed ZnO powder layers illuminated by light at the maximum absorption wavelengths of the absorbed dye molecules. (Reprinted with permission from Yoshimura et al. [19]. Copyright (1980) The Physical Society of Japan and The Japan Society of Applied Physics.)

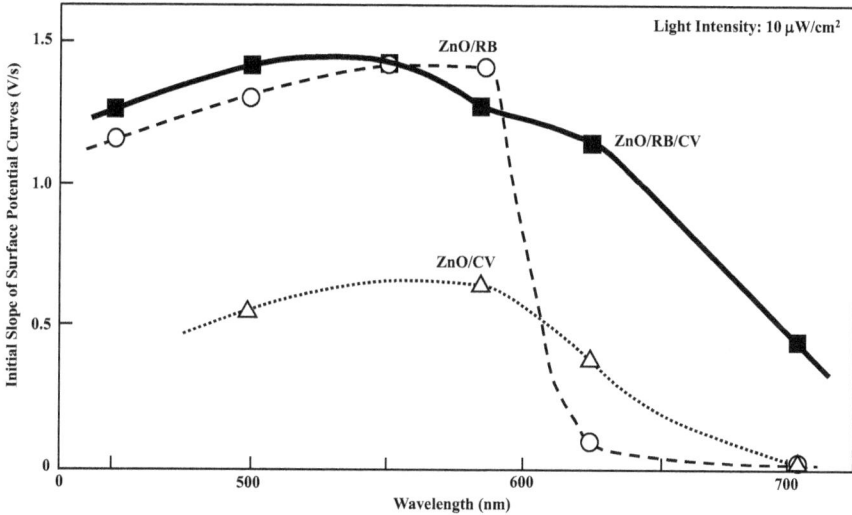

FIGURE 12.11 Wavelength dependence of the initial slope of the surface potential curve of the dyed ZnO powder layers. (Reprinted with permission from Yoshimura et al. [19]. Copyright (1980) The Physical Society of Japan and The Japan Society of Applied Physics.)

The mechanism of the two-dye sensitization is shown in Figure 12.12. K_p is the efficiency, with which the electrons are injected from the p-type dye molecule into ZnO. K_{pn} is the efficiency, with which the electrons excited in the n-type dye molecule are transferred into the p-type dye molecule. Then, the efficiency, with which the electrons excited in the n-type dye molecule are injected into the ZnO, is given as follows [17]:

$$\eta = K_p K_{pn} \qquad (12.1)$$

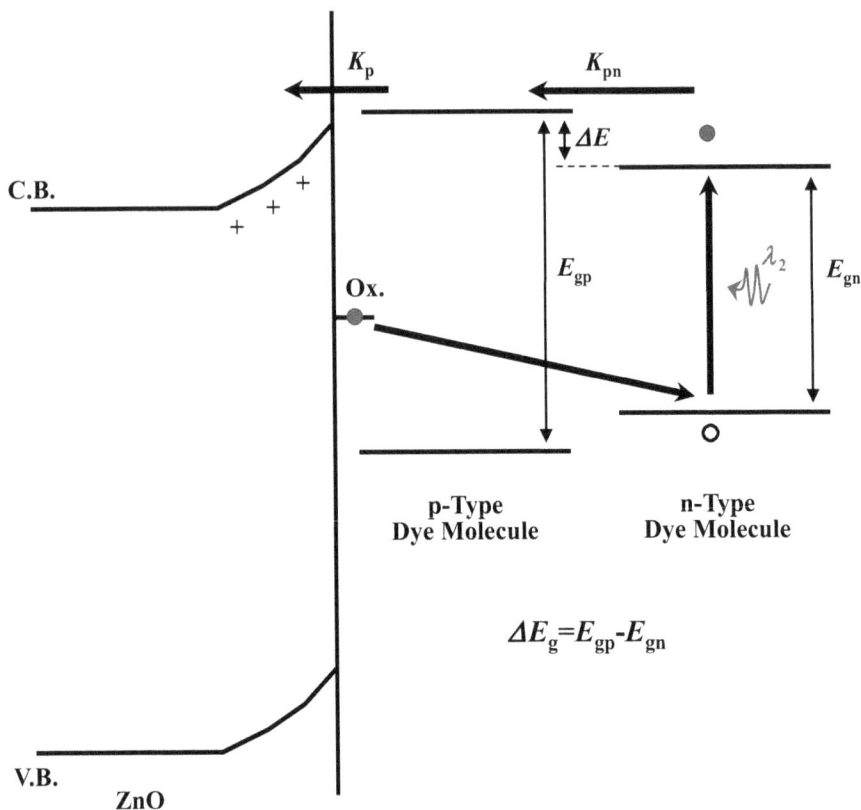

FIGURE 12.12 Mechanism of the two-dye sensitization. (Reprinted with permission from Yoshimura et al. [17]. Copyright (1981) The Physical Society of Japan and The Japan Society of Applied Physics.)

In the case that p-type and n-type dye molecules form an energy structure shown in Figure 12.12, K_{pn} is proportional to $e^{-\Delta E/k_B T}$, where ΔE is the difference in energy between the excited states of the p-type and n-type dye molecules, and k_B is the Boltzmann constant. When K_p is assumed not to depend on dye molecules, the following relationship is obtained:

$$\log I_P \propto -\left(\frac{\log e}{k_B T}\right)\Delta E + a \qquad (12.2)$$

Here, I_p is the photocurrent and a is a constant. At room temperature, $-\log e / k_B T \approx -17$.

Figure 12.13 shows photocurrents generated at the maximum absorption wavelengths of n-type dyes for various combinations of p-type dyes (FL, EO, and RB) and n-type dyes (CV, MG, and BG). The horizontal axis, ΔE_g, represents the difference in the energy gap between p-type and n-type dye molecules, that is, $\Delta E_g = E_{gp} - E_{gn}$. Here, E_{gp} and E_{gn} are, respectively, energy gaps of p-type and n-type dye molecules. It is found that each point tends to be on a line with a slope of −15. This value is close to the slope of −17 in Equation (12.2). Assuming that ΔE approximately equals ΔE_g,

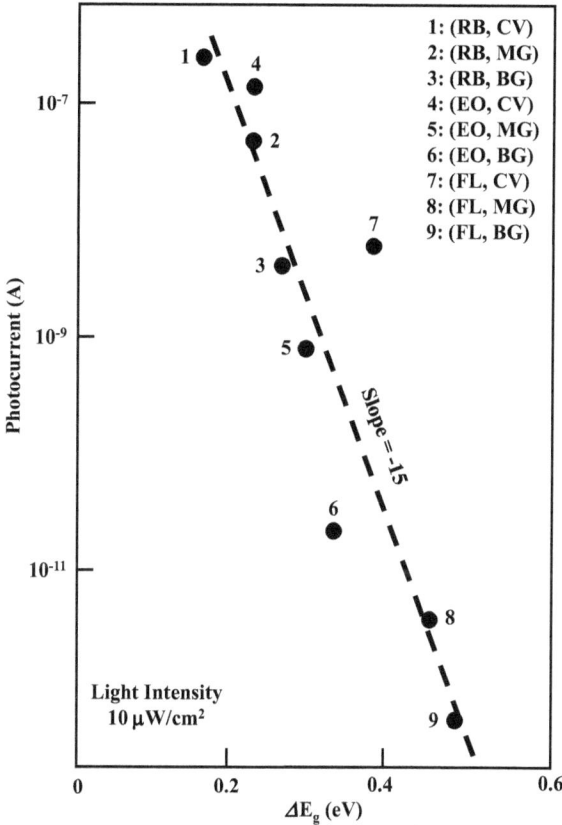

FIGURE 12.13 Photocurrents generated at the maximum absorption wavelengths of n-type dyes for various combinations of p-type dyes and n-type dyes. The photocurrents are plotted as a function of $\Delta E_g = E_{gp} - E_{gn}$. (Reprinted with permission from Yoshimura et al. [17]. Copyright (1981) The Physical Society of Japan and The Japan Society of Applied Physics.)

the coincidence of the line slope with $-\log e / k_B T$ confirms the consistency of the two-dye sensitization model shown in Figure 12.12.

Since K_{pn} increases to unity with decreasing ΔE, the ideal energy structure for the two-dye sensitization can be drawn as Figure 12.14. Excited electrons in p-type and n-type dye molecules are smoothly injected into ZnO. Holes left in the dye molecules are compensated by electrons trapped in adsorbed oxygen on ZnO surface.

These results make the molecular MQD sensitization, in which p/n/p/n/... molecular wire structures are constructed as shown in Figure 12.3b, viable.

12.2.2 MOLECULAR-MQD SENSITIZATION IN GUIDED LIGHT CONFIGURATION UTILIZING LP-MLD

As mentioned in Section 3.5.2, surface potential measurement revealed that p-type and n-type dye molecules are self-assembled into an n/p/n structure of n-type

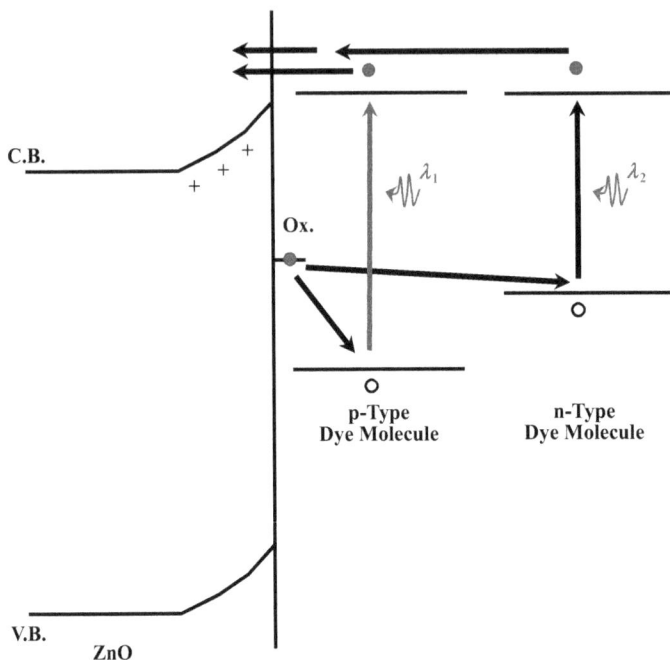

FIGURE 12.14   The ideal energy structure for the two-dye sensitization.

ZnO/p-type dye molecule/n-type dye molecule by the electrostatic force between dye molecules and between a dye molecule and ZnO (see Figure 3.26) [18]. This suggests that molecular wires consisting of p-type and n-type dyes for the molecular-MQD sensitization shown in Figure 12.3b can be grown by LP-MLD utilizing electrostatic force.

Hirotaka Watanabe et al. demonstrated the molecular-MQD sensitization in the guided light configuration utilizing LP-MLD [4]. Figure 12.15 presents LP-MLD with source molecules of RB (p-type) and CV (n-type). When RB molecules are provided to a ZnO thin film (n-type) in a solvent, they are attached to the ZnO surface due to the attractive electrostatic force between RB and ZnO to form a monomolecular layer structure of [ZnO/RB]. Once the surface is covered with RB, the deposition of RB is automatically terminated. After rinsing the surface by a pure solvent, CV molecules are provided to the [ZnO/RB] structure. Then, they are attached to RB molecules due to the attractive electrostatic force between RB and CV to form a [ZnO/RB/CV] structure, which is regarded as a very short molecular MQD on ZnO.

Figure 12.16 shows a ZnO sample for photocurrent measurement. A 600-nm-thick ZnO thin film was deposited on a glass substrate by vacuum evaporation followed by anneal at 400°C for 1 h in air. On the film, a 5000-μm-wide slit-type Al electrode with a gap distance of 60 μm was formed. 8.5 V was applied to the slit-type electrodes. The photocurrent measurement was carried out in two configurations presented in Figure 12.17. In the normally-incident light configuration, a light beam was introduced onto the ZnO surface from an optical fiber. In the photograph, the light beam spot is put on a place apart from the slit-type electrode in order to observe the spot clearly. In the guided light configuration, a light beam was introduced into the ZnO thin film from the edge

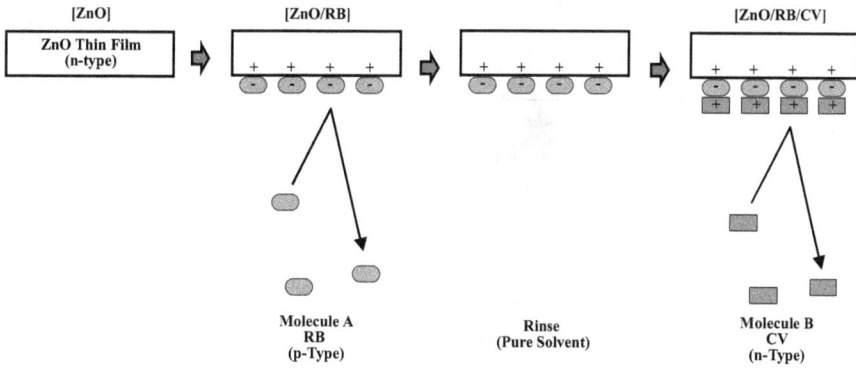

FIGURE 12.15 LP-MLD with source molecules of RB (p-type) and CV (n-type) on ZnO thin film (n-type).

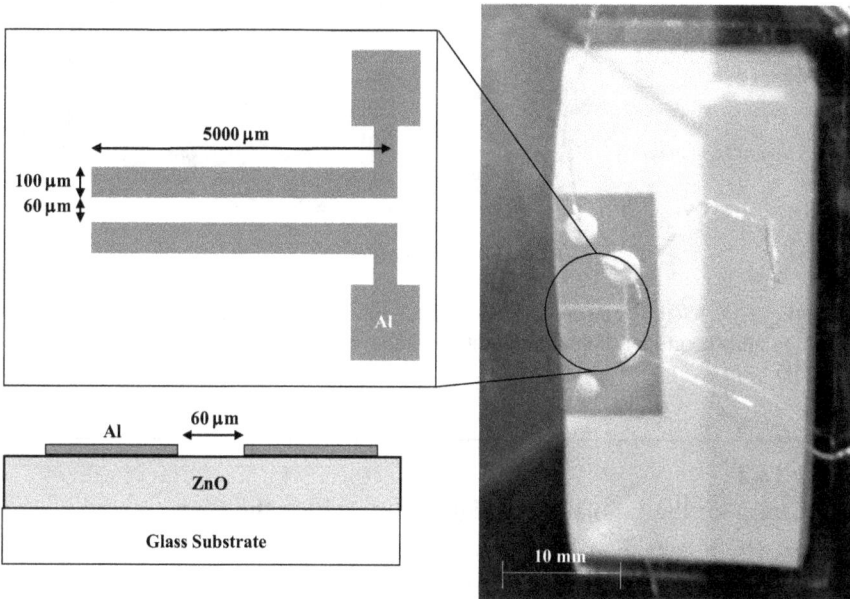

FIGURE 12.16 ZnO sample for photocurrent measurement.

by the butt joint of the optical fiber to propagate in the film. In the photograph, a guided light beam is observed. For the photocurrent measurement, lasers with wavelengths of 405, 532, and 633 nm were used. Since these wavelengths correspond to photon energy below the band gap of ZnO, small light absorption occurs so that the guided light beams propagate over a 5-mm distance.

To estimate photocurrents per unit light power, it was assumed that light beams existing in the gap region of the slit-type electrodes are effective for photocurrent generation. In Table 12.2, parameters for the estimation are listed. The diffraction angle θ in the ZnO thin film is for refractive index of 2. The estimated light power within the gap is summarized in Table 12.3. Here, "Exposed Area S" means the whole area exposed to the light beams, and "Exposed Area within Gap S_G" means the

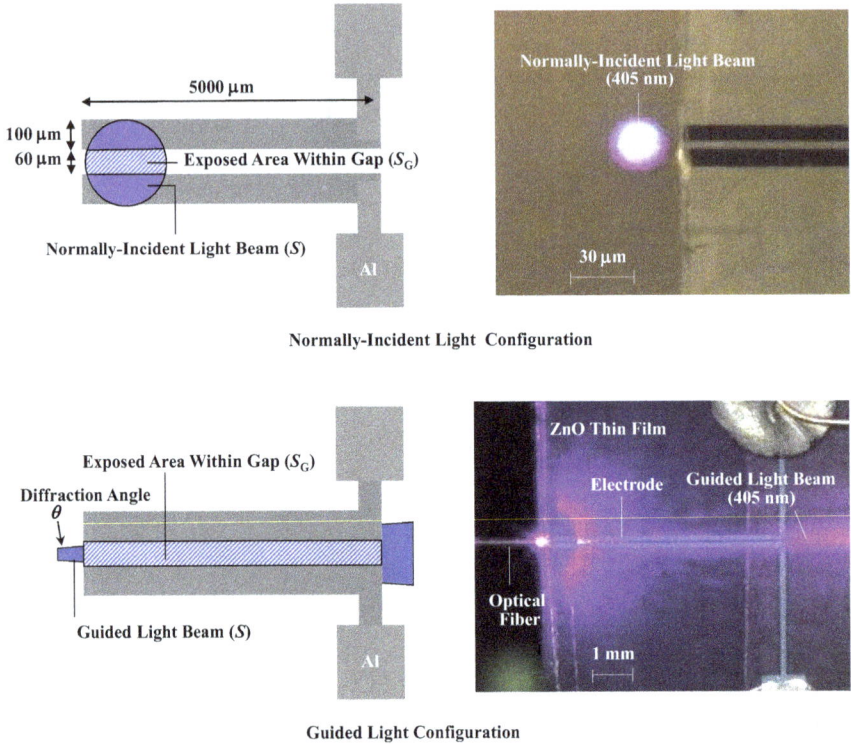

FIGURE 12.17 Setups of photocurrent measurement in the normally-incident light configuration and the guided light configuration. (Reprinted with permission from Yoshimura et al. [4].)

TABLE 12.2

Parameters for the Estimation of Light Power within the Gap

Wavelength (nm)	Output Power from Optical Fiber (mW)	Core Diameter of Optical Fiber (μm)	Diffraction Angle, θ (rad.)
405	1	3	0.021
532	0.35	5	0.017
633	0.23	6	0.017

exposed area within the gap. As the 0th approximation, the light power is assumed to be uniform in the exposed area. For the guided light configuration, the light power within the gap was calculated by assuming that a fraction of the output power from the optical fiber is guided in the ZnO film, where the fraction is given by a ratio of (ZnO thin film thickness: 600 nm)/(core diameter of the optical fiber).

In Figure 12.18a, absorption spectra of RB and CV in alcohol solutions are shown together with an absorption spectrum of ZnO. The spectrum of CV is located in a longer wavelength region comparing with that of RB. Figure 12.18b shows wavelength dependence of photocurrents measured in the guided light configuration for

TABLE 12.3
Estimated Light Power within the Gap

Wavelength (nm)	Exposed Area, S (mm²)	Exposed Area within Gap, S_G (mm²)	S_G/S	Light Power within Gap (mW)
		Normally-Incident Light Configuration		
405	0.053	0.016	0.29	0.294
532	0.053	0.016	0.29	0.103
633	0.053	0.016	0.29	0.068
		Guided Light Configuration		
405	0.54	0.3	0.56	0.112
532	0.42	0.3	0.71	0.030
633	0.42	0.3	0.71	0.016

[ZnO], and for [ZnO/RB], [ZnO/CV], and [ZnO/RB/CV] fabricated by LP-MLD. In [ZnO], large photocurrents are generated at 405 nm, and small photocurrents at 532 and 633 nm. In [ZnO/RB], the photocurrent spectrum extends to 532 nm, and in [ZnO/RB/CV], it further extends to 633 nm. This is attributed to the fact that the absorption band of CV is located in the longer-wavelength region. In [ZnO/CV], due to the low adsorption ability as well as the low electron injection ability of CV, sensitization effect does not appear. By inserting RB between CV and ZnO to form [ZnO/RB/CV], both abilities are raised to enhance the sensitization effect.

As a measure of the photocurrent enhancement induced by the guided light configuration, photocurrent enhancement ratio (PCER) is defined as follows.

$$\text{PCER} = \frac{\left[\text{Photocurrent/Light Power within Gap in Guided Light Configuration}\right]}{\left[\begin{array}{l}\text{Photocurrent/Light Power within Gap in Normally-Incident}\\ \text{Light Configuration}\end{array}\right]}$$

As Figure 12.19 shows, large PCERs of ~5, ~15, and ~8 are, respectively, observed in [ZnO] at 405 nm, in [ZnO/RB] at 532 nm, and in [ZnO/RB/CV] at 633 nm, demonstrating the potentiality of the WTP devices.

12.2.3 POLYMER-MQD SENSITIZATION IN GUIDED LIGHT CONFIGURATION UTILIZING MLD

As mentioned in Section 6.4, by controlling molecular arrangements in poly-azomethine (poly-AM) wires using MLD, polymer MQDs are fabricated [3,11,12], as Figure 6.15 shows. The structure will be used for the polymer-MQD sensitization shown in Figure 12.3c [3,5–12].

In order to clarify the feasibility of the polymer MQD as sensitizer for ZnO thin films, the sensitizing ability of poly-AM wires was investigated. In poly-AM, since

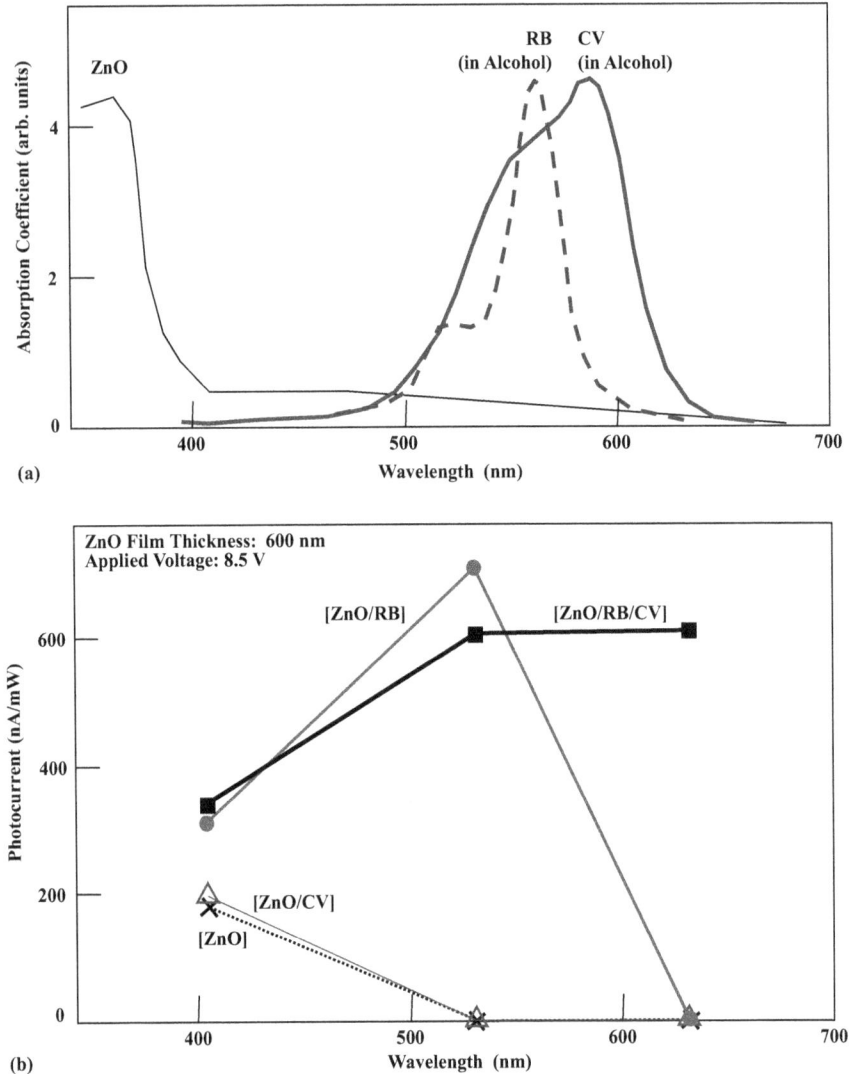

FIGURE 12.18 (a) Absorption spectra of ZnO, RB, and CV, and (b) wavelength dependence of photocurrents measured in the guided light configuration for [ZnO], and for [ZnO/RB], [ZnO/CV], and [ZnO/RB/CV] fabricated by LP-MLD. (Reprinted with permission from Yoshimura et al. [4].)

the absorption band is located in wavelength regions of 350–500 nm as Figure 12.20a shows, sensitization in the blue-green regions is expected. Figure 12.20b presents wavelength dependence of photocurrents measured in the guided light configuration for [ZnO] and [ZnO/poly-AM], which was fabricated by growing a poly-AM thin film on a ZnO thin film utilizing carrier-gas-type MLD. In [ZnO/poly-AM], the photocurrent at 405 nm is enhanced, and the photocurrent spectrum extends to 532 nm, indicating that poly-AM wires have sensitizing ability in the blue-green regions.

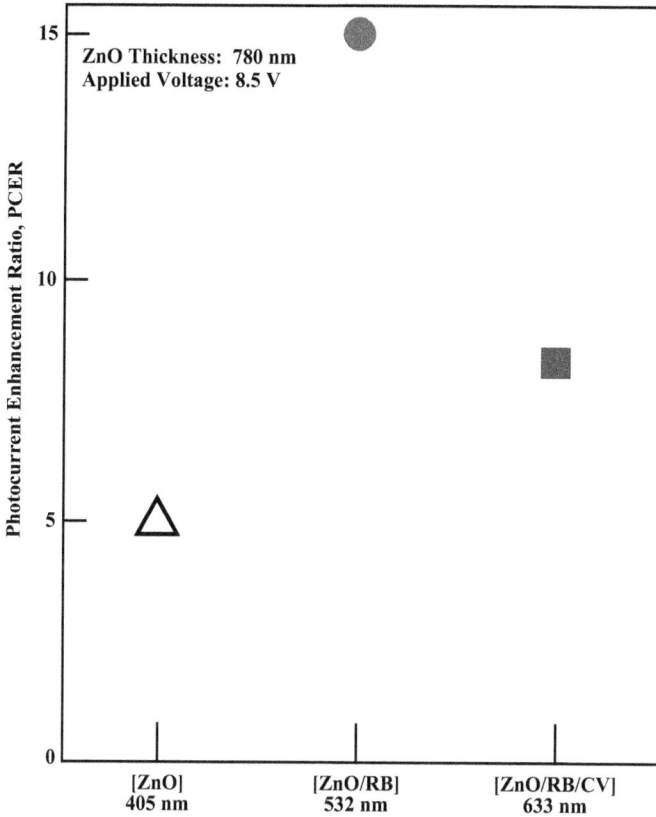

FIGURE 12.19 Photocurrent enhancement ratio (PCER) in [ZnO], [ZnO/RB], and [ZnO/RB/CV]. (Reprinted with permission from Yoshimura et al. [4].)

Figure 12.21 shows wavelength dependence of photocurrents for [ZnO/poly-AM] in the guided light configuration and the normally-incident light configuration. Here, GL and NIL, respectively, represent the guided light configuration and the normally-incident light configuration. PCER of ~12 at 532 nm is larger than PCER of ~3 at 405 nm. The light absorption of poly-AM is smaller at 532 nm than at 405 nm. These results lead us to believe that the photocurrent enhancement in the guided light configuration is more remarkable in wavelength regions with smaller light absorption. This is because in a wavelength region, where the thin films have large light absorption, the thin films can absorb light to some extent even in the normally-incident light configuration to generate photocurrents; consequently, the PCER is reduced.

By utilizing polymer MQDs like 3QD described in Section 6.4.4, spectral sensitization for wide wavelength regions will become possible.

12.2.4 THREE-DYE MOLECULAR-MQD SENSITIZATION UTILIZING LP-MLD

As can be seen in the reflection spectra of dyed ZnO powder layers presented in Figure 12.6, absorption bands of p-type and n-type dyes are, respectively, located in

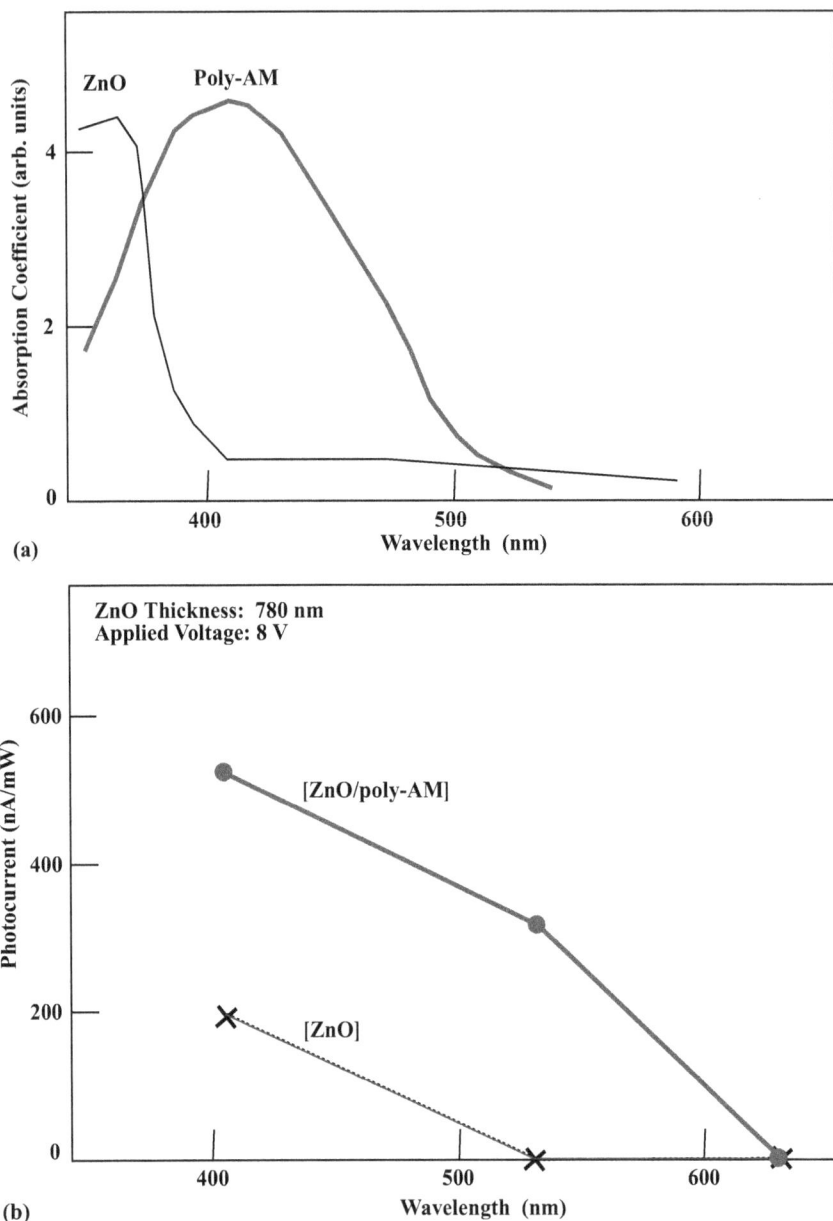

FIGURE 12.20 (a) Absorption spectra of ZnO and poly-AM and (b) wavelength dependence of photocurrents measured in the guided light configuration for [ZnO] and [ZnO/poly-AM].

short and long wavelength regions. By introducing many kinds of p-type and n-type dye molecules into molecular MQDs, spectral sensitization with an extended wavelength range will be realized due to the superposition of the absorption bands. In the present section, sensitization of ZnO by molecular MQDs consisting of three kinds

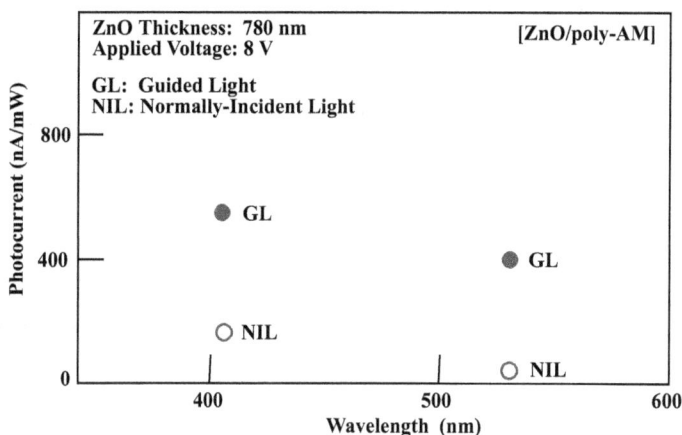

FIGURE 12.21   Wavelength dependence of photocurrents for [ZnO/poly-AM] in the guided light configuration (GL) and the normally-incident light configuration (NIL).

of dye molecules is described. The molecular MQDs are grown by LP-MLD, and the sensitization effect is evaluated by measuring photocurrent spectra in each growth step. Based on energy-level schemes of dye molecules, the mechanism of the three-dye molecular-MQD sensitization is discussed.

The research on the three-dye molecular-MQD sensitization was terminated at a primitive stage due to the retirement of the author. Since there is some uncertainty of the photocurrent magnitude, the discussions are based on the profiles of the spectra.

Three-dye molecular MQDs were grown on 5-μm-thick ZnO powder layers by LP-MLD with source molecules of RB, CV (or BG), and EO. The process is the same as that presented in Figures 3.29 and 3.30. In Steps 1, 2, and 3, [ZnO/RB], [ZnO/RB/CV (or BG)], and [ZnO/RB/CV (or BG)/EO] were formed, respectively. In each MLD step, dye molecules solved in isopropyl alcohol (IPA) with a dye concentration of 1.6×10^{-3} mol/L were provided for 5 min to the ZnO powder layer on a glass substrate with slit-type indium tin oxide (ITO) electrodes (gap distance: 60 μm). Photocurrents were measured by applying 3 V across the electrode gap [13].

Figure 12.22a shows photocurrent spectra in each step for [ZnO/RB/CV/EO] growth. RB is adsorbed on ZnO in Step 1 to generate a photocurrent spectrum with a peak around 570 nm, where the absorption peak of RB is located. In Step 2, CV is connected to RB. Then, photocurrents by CV are superposed to those by RB, raising photocurrents in the longer wavelength region. In Step 3, EO is connected to CV. Photocurrents by EO with a peak around 530 nm that is close to the absorption peak of EO are superposed. Thus, a superposition of photocurrents by RB, CV, and EO is obtained. For [ZnO/RB/BG/EO] growth, as shown in Figure 12.22b, when BG is connected to RB in Step 2, large photocurrents generated by BG appear in the long wavelength region. In Step 3, EO is connected to BG to add photocurrents by EO to that of [ZnO/RB/BG]. As a result, a wide photocurrent spectrum extending from 450 to 700 nm is obtained. These results suggest that the three-dye molecular MQDs act as light-harvesting antennas to enhance the sensitization effects.

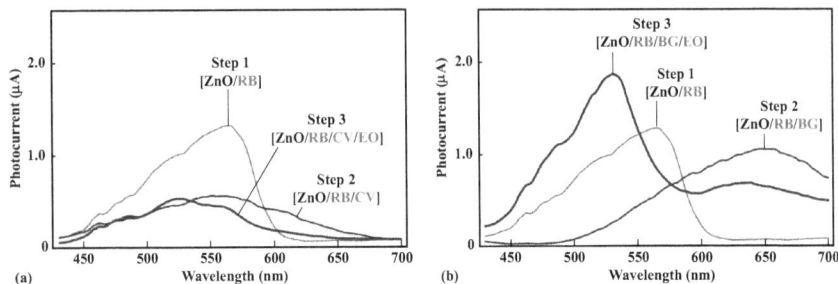

FIGURE 12.22 Photocurrent spectra in each MLD step for (a) [ZnO/RB/CV/EO] growth and (b) [ZnO/RB/BG/EO] growth.

Figure 12.23a shows an experimental setup for *in-situ* photocurrent spectrum measurement during LP-MLD, which was carried out by Bai et al. A sample of a 5-μm-thick ZnO powder layer on a quartz substrate with slit-type ITO electrodes (gap distance: 60 μm) is put in an LP-MLD cell. Photocurrent spectra are measured *in-situ* in each MLD step [13].

The LP-MLD process for the *in-situ* photocurrent spectrum measurement and photographs of the LP-MLD cell in Steps 1, 2, and 3 are shown in Figure 12.23b. Prior to the LP-MLD process, IPA is injected into the LP-MLD cell to fill it with IPA. In Step 1, RB-containing IPA is injected into the cell to fill it for 5 min, achieving the RB adsorption on the ZnO surfaces to grow [ZnO/RB]. After the RB-containing IPA is removed by injection of IPA for rinse, a photocurrent spectrum for Step 1 is measured in IPA without taking out the sample from the cell. After the measurement, in Step 2, CV-containing IPA is injected into the cell to fill it for 5 minutes, growing [ZnO/RB/CV]. After the CV-containing IPA is removed, a photocurrent spectrum for Step 2 is measured. In Step 3, similarly, EO-containing IPA is injected into the cell to grow [ZnO/RB/CV/EO] and a photocurrent spectrum for Step 3 is measured.

Figure 12.24 shows preliminary results. In Step 1, RB generates photocurrents in a range of around 400–580 nm. In Step 2, photocurrents attributed to CV are generated in the longer wavelength region. In Step 3, a wide photocurrent spectrum generated by RB, CV, and EO appears.

To determine energy-level schemes of the three-dye molecular MQDs, ionization potential *IP*, which corresponds to the energy of HOMO of dye molecules, was measured by the photoemission yield spectroscopy in air (PYSA) using the Surface Analyzer [Model AC-3] (RIKEN KEIKI CO., Ltd.) [21]. The measurement was carried out by Dr. Yoshiyuki Nakajima of RIKEN KEIKI. An example of the PYSA measurement is shown in Figure 12.25 for RB on ZnO. The results for five kinds of dye molecules are listed in Table 12.4 with the data for ZnO [20]. Energy gap E_g and electron affinity χ, which correspond to the energy of LUMO of dye molecules, are also listed there. E_g was estimated from the peak wavelengths of the reflection spectra for dye molecules adsorbed on ZnO surfaces (see Figure 12.6), and the electron affinity was determined by $\chi = IP - E_g$. The energy levels are summarized in Figure 12.26 [13].

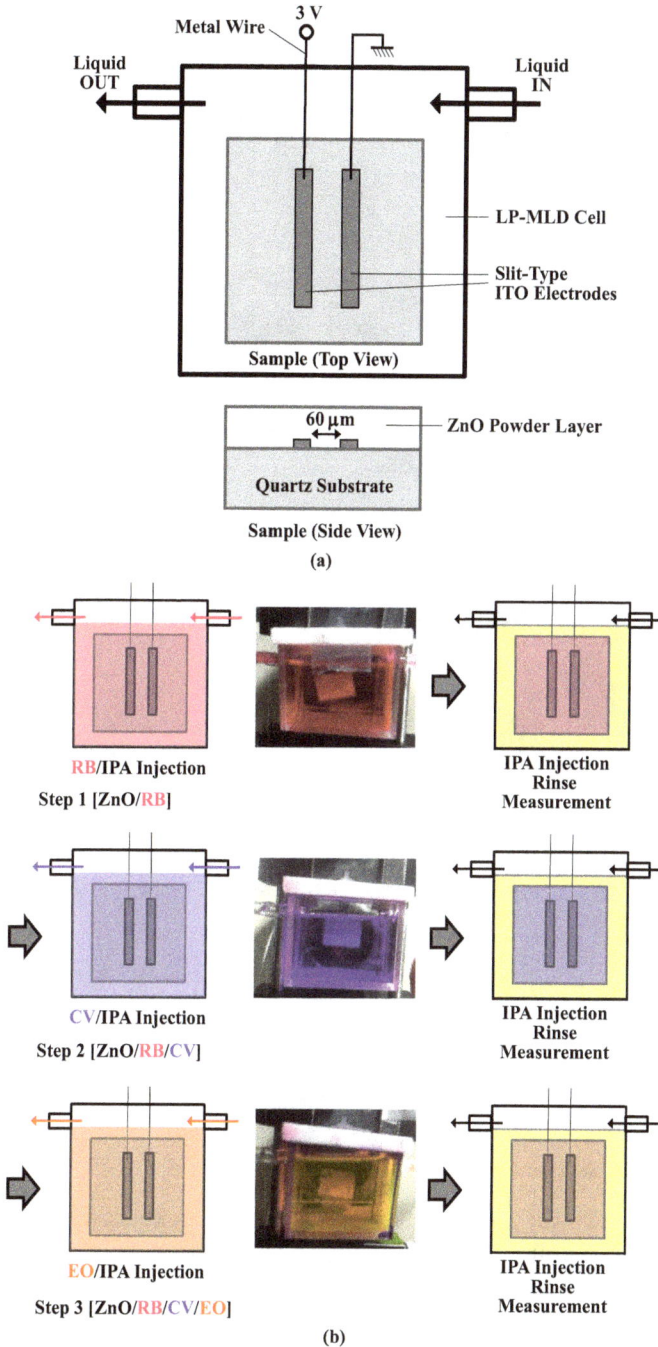

FIGURE 12.23 (a) Experimental setup and (b) LP-MLD process with photographs of LP-MLD cell for *in-situ* photocurrent spectrum measurement during LP-MLD. (Reprinted with permission from Yoshimura et al. [13].)

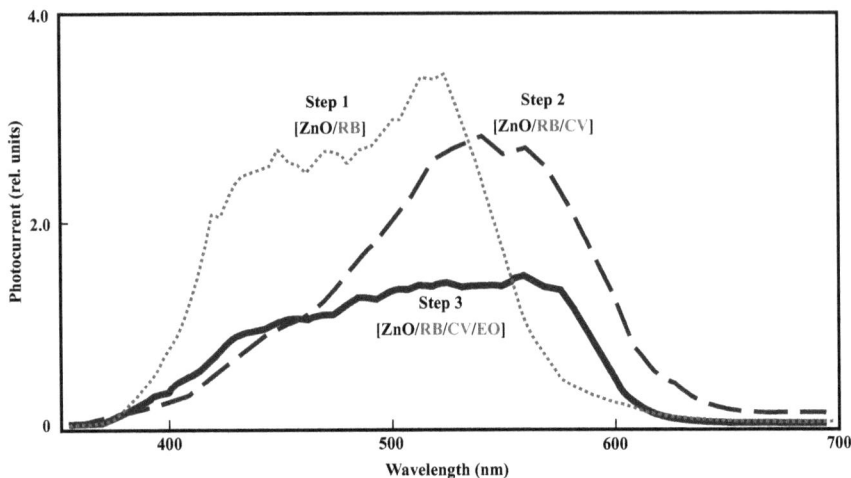

FIGURE 12.24 Preliminary results of *in-situ* photocurrent spectrum measurement for [ZnO/RB/CV/EO].

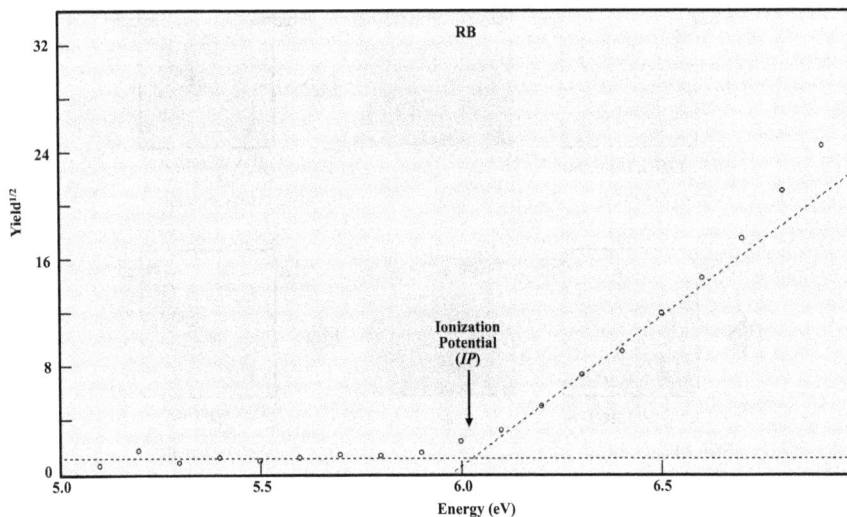

FIGURE 12.25 An example of the PYSA measurement for RB on ZnO.

In Figure 12.27, energy-level schemes of three-dye molecular MQDs are drawn. The upper and lower lines, respectively, represent the energy of LUMO and HOMO. An electron excited in RB is directly injected into ZnO. An electron excited in CV (or BG) is injected into ZnO through RB. An electron excited in EO is injected into ZnO through CV (or BG) and RB. Thus, a wide photocurrent spectrum is generated by the superposed contribution of RB, CV (or BG), and EO.

It should be noted in Figure 12.22b that the photocurrent generated by RB in a wavelength region of 500–580 nm drastically decreases after BG is connected to RB in Step 2. This is an unexpected result. One possible explanation is as follows. Because the LUMO energy of RB and BG are close as can be seen in Figure 12.28, an excited electron in

TABLE 12.4
Energy of HOMOs and LUMOs of Dye Molecules (Unit: eV)

Dye	Ionized Potential, *IP* (HOMO)	Energy Gap, E_g	Electron Affinity, χ (LUMO)
FL	5.97	2.43 (510 nm)	3.54
EO	5.96	2.30 (540 nm)	3.66
RB	6.03	2.18 (570 nm)	3.85
CV	5.78	2.07 (600 nm)	3.71
BG	5.70	1.88 (660 nm)	3.82
ZnO	7.3	3	4.3

Unit: eV.

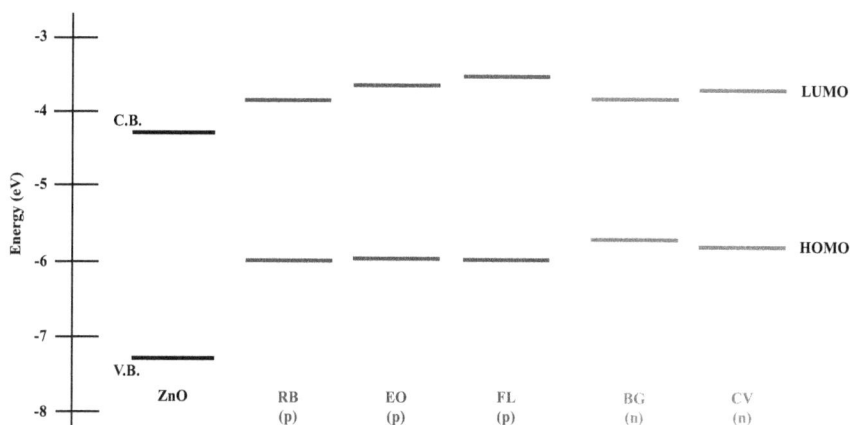

FIGURE 12.26 Energy levels of dye molecules.

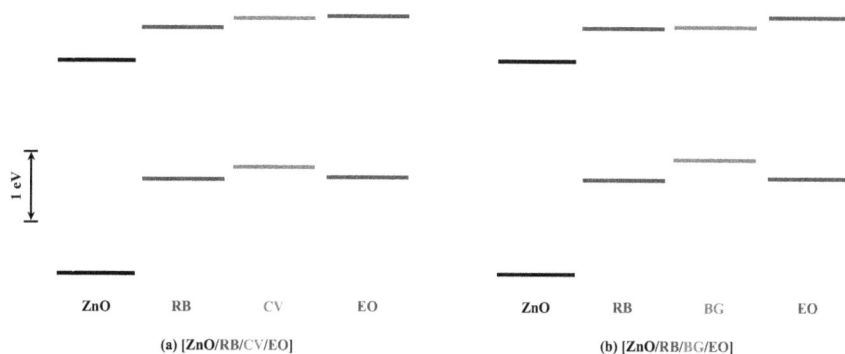

FIGURE 12.27 Energy-level schemes of three-dye molecular MQDs for (a) [ZnO/RB/CV/EO] and (b) [ZnO/RB/BG/EO]. (Reprinted with permission from Yoshimura et al. [13].)

RB transfers to the LUMO of BG, and an electron in the HOMO of BG transfers to the HOMO of RB. This induces a cyclic electron transition of HOMO (RB) → LUMO (RB) → LUMO (BG) → HOMO (BG) → HOMO (RB) as indicated by broken-line arrows in Figure 12.28, resulting in the reduction of the electron injection rate from RB into ZnO.

An example of an energy-level scheme is shown in Figure 12.29 for a light-harvesting antenna of a five-dye molecular MQD. To enhance the light-harvesting performance, it is important to arrange the dye molecules so that an energy-level scheme with a gradual energy slope similar to that of light-harvesting antennas in plants is formed. This facilitates the electron transfer from the light-harvesting antenna to the semiconductor

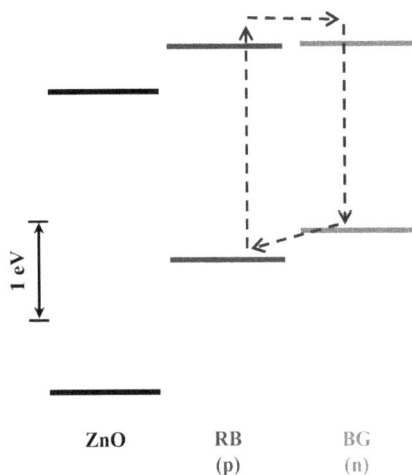

FIGURE 12.28 Tentative mechanism of the decrease in the photocurrent generated by RB after BG deposition on RB in Step 2 in Figure 12.22b. (Reprinted with permission from Yoshimura et al. [13].)

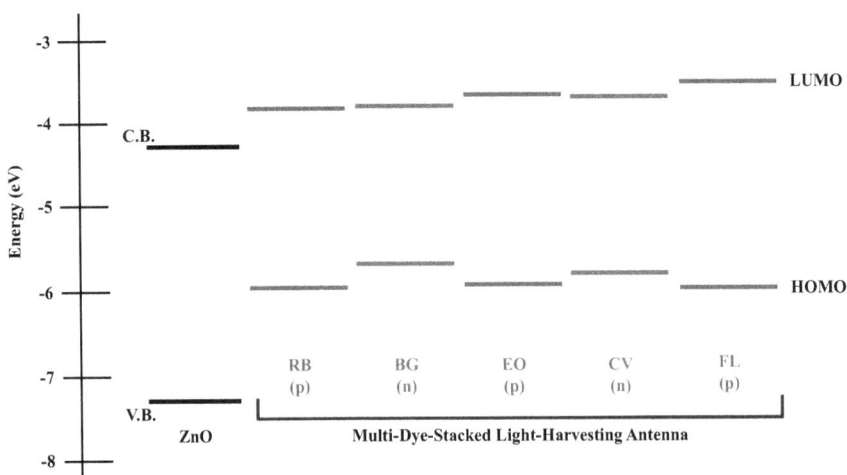

FIGURE 12.29 An example of an energy-level scheme for a light-harvesting antenna of a five-dye molecular MQD. (Reprinted with permission from Yoshimura et al. [13].)

to achieve high-efficiency sensitization. From the view point of the gradual energy slope formation, the energy-level schemes of [ZnO/RB/CV/EO] and [ZnO/RB/BG/EO] shown in Figure 12.27 are preferable.

In addition, to widen the wavelength region of photocurrent spectra, MLD utilizing molecule groups shown in Figure 3.8 is effective. Figure 12.30 shows an energy-level scheme in the case that RB and EO are used for Molecule Group I, and CV and BG for Molecule Group II. Figure 12.31 shows a photocurrent spectrum of [ZnO/EO/CV+BG], which is fabricated by using EO and Molecule Group II of CV+BG. As expected, a broad photocurrent spectrum appears [22].

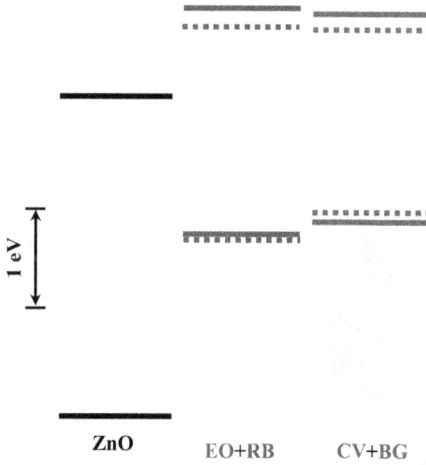

FIGURE 12.30 Energy-level scheme of [ZnO/RB+EO/CV+BG] grown by LP-MLD utilizing Molecule Group I (RB, EO) and Molecule Group II (CV, BG).

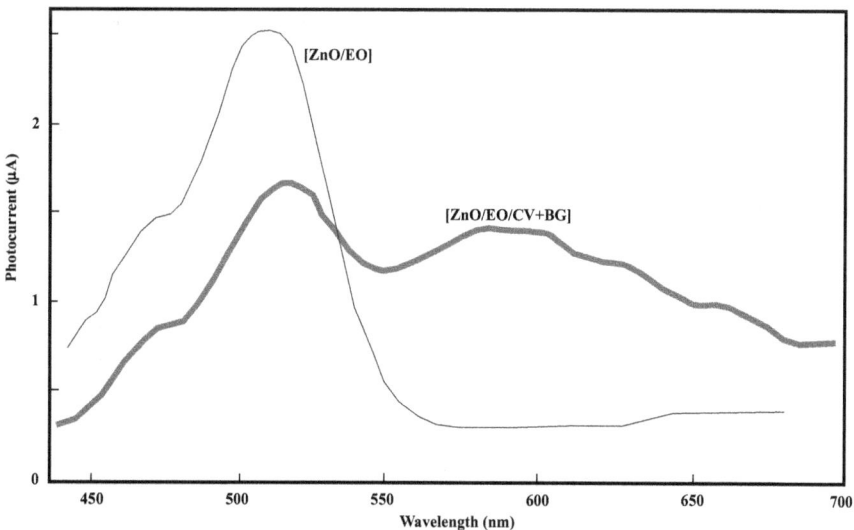

FIGURE 12.31 Photocurrent spectrum of [ZnO/EO/CV+BG].

12.3 FILM-BASED INTEGRATED SOLAR ENERGY CONVERSION SYSTEMS

In conventional solar cell modules, as illustrated in Figure 12.32a, semiconductors such as Si are placed all over the modules, so it is foreseen that the cost of the modules will increase owing to a shortage of semiconductor materials. In modules with Fresnel lens/mirror concentrators, although semiconductor materials are saved by focusing solar beams on small semiconductor cells, the module size becomes large and solar-beam tracking mechanisms are necessary.

In the film-based integrated solar energy conversion system illustrated in Figure 12.32b [2,5–8,10], small/thin semiconductor dies are placed partially in a light beam collecting film. The system is flexible and compact, and it enables the reduction of semiconductor material consumption and wide-angle collection of light beams. Its advantages and possible applications are summarized in Table 12.5.

Figure 12.33 presents various types of film-based integrated solar energy conversion systems. In the distributed-conversion type structure shown in Figure 12.33a,

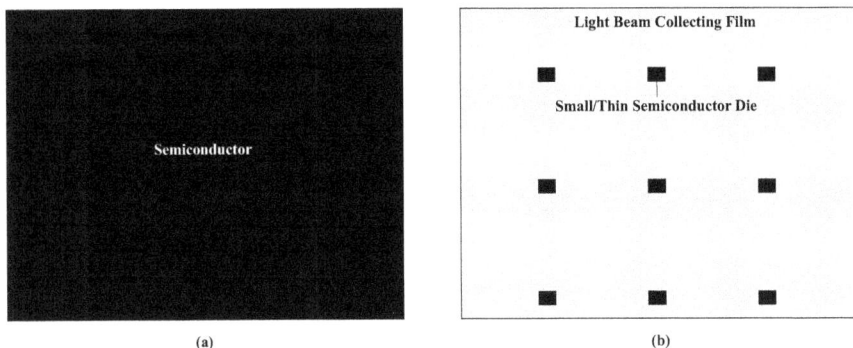

FIGURE 12.32 Concepts of (a) the conventional solar cell module and (b) the film-based integrated solar energy conversion system.

TABLE 12.5

Advantages and Possible Applications of the Film-Based Integrated Solar Energy Conversion Systems

Advantages

- Reduction of semiconductor/ITO consumption, cost, size, and weight
- Flexibility and wide-angle light beam collection capability
- Functions brought by implementing optical circuits (e.g., demultiplexing of light)
- Functions brought by embedding batteries and ICs
- Integration of the WTP Devices

Applications

Cars, Robots, Laptop Computers, Cellular Phones, Hand-Carrying Equipment, Desks, Cloths, Sheets, Umbrella, Buildings, Fences, Walls, Roads, etc.

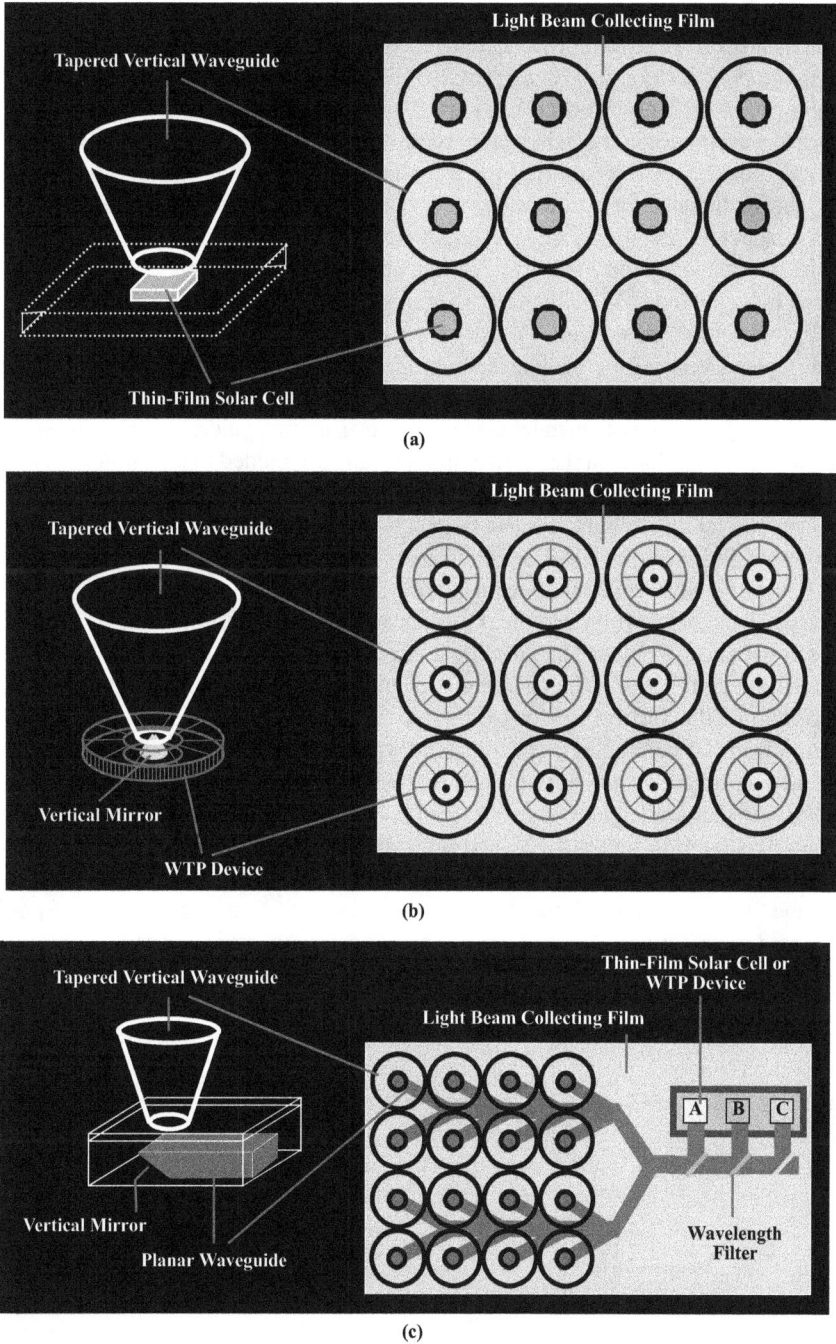

FIGURE 12.33 Various types of film-based integrated solar energy conversion systems. (a) Distributed-conversion type structure with thin-film solar cells, (b) distributed-conversion type structure with WTP devices, and (c) centralized-conversion type structure with thin-film solar cells or WTP devices.

thin-film solar cells made from small/thin semiconductor dies are embedded in a light beam collecting film by the heterogeneous integration process such as the photolithographic packaging with selectively occupied repeated transfer (PL-Pack with SORT) [10,23–26] described in Section 9.8. Incident light beams received by tapered vertical waveguides in the light beam collecting film are guided to the thin-film solar cells. When the aperture ratio of the solar cell to the tapered vertical waveguide is 1:3, the semiconductor consumption is reduced to 1/9 comparing to the case that semiconductors are spread all over the module area. In the distributed-conversion type structure shown in Figure 12.33b, the collected incident light beams are introduced into embedded WTP devices via cone-shaped vertical mirrors.

In the centralized-conversion-type structure shown in Figure 12.33c [27], a film with tapered vertical waveguides and an optical waveguide film with planar waveguides having vertical mirrors are stacked to build a light beam collecting film. Incident light beams collected by the tapered vertical waveguides are introduced into the planar waveguides via the vertical mirrors, and are guided to thin-film solar cells or WTP devices. By embedding wavelength filters into the optical waveguide film for demultiplexing, light beams can branch to cells/devices having matched spectral responses, for example, UV-blue beams to cells/devices with GaN, visible beams to those with GaAs, and infrared beams to those with InP.

Guided beam insertion into the thin-film solar cells is performed via vertical mirrors as depicted in Figure 12.34a while guided beam insertion into the WTP devices is performed via mode-size converters as depicted in Figure 12.34b. When the aperture ratio of the vertical mirror to the tapered vertical waveguide is 1:3 and light beams from 16 tapered vertical waveguides come together into one planar waveguide, the semiconductor consumption is reduced to $(1/9) \times (1/16) = 1/144$ comparing to the case that semiconductors are spread all over the module area.

As illustrated in Figure 12.35, the self-organized lightwave network (SOLNET) might be useful for optical coupling in the centralized-conversion type structure. For the "free spaces to microscale planar waveguides" coupling, an array of luminescent targets (or wavelength filters) is deposited at the vertical mirror sites on the optical waveguide film. After a photo-induced refractive-index increase (PRI) material layer is placed on the film surface, the layer is exposed to blanket write beams through a mask with a window array to fabricate an array of reflective SOLNETs (R-SOLNETs). By etching the surrounding regions of the SOLNETs, an array of tapered vertical waveguides self-aligned to the vertical mirrors is obtained. For the "microscale planar waveguides to nanoscale waveguides in WTP devices" coupling, luminescent targets (or wavelength filters) are deposited on the nanoscale waveguide core edges. After the PRI material is inserted between the waveguides, R-SOLNETs

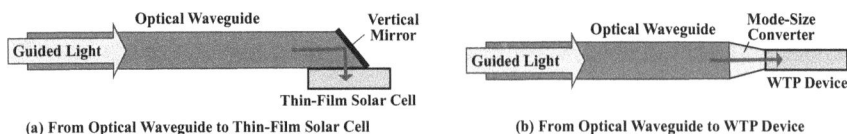

(a) From Optical Waveguide to Thin-Film Solar Cell   (b) From Optical Waveguide to WTP Device

FIGURE 12.34   Guided beam insertion into (a) the thin-film solar cells via vertical mirrors and (b) the WTP devices via mode-size converters.

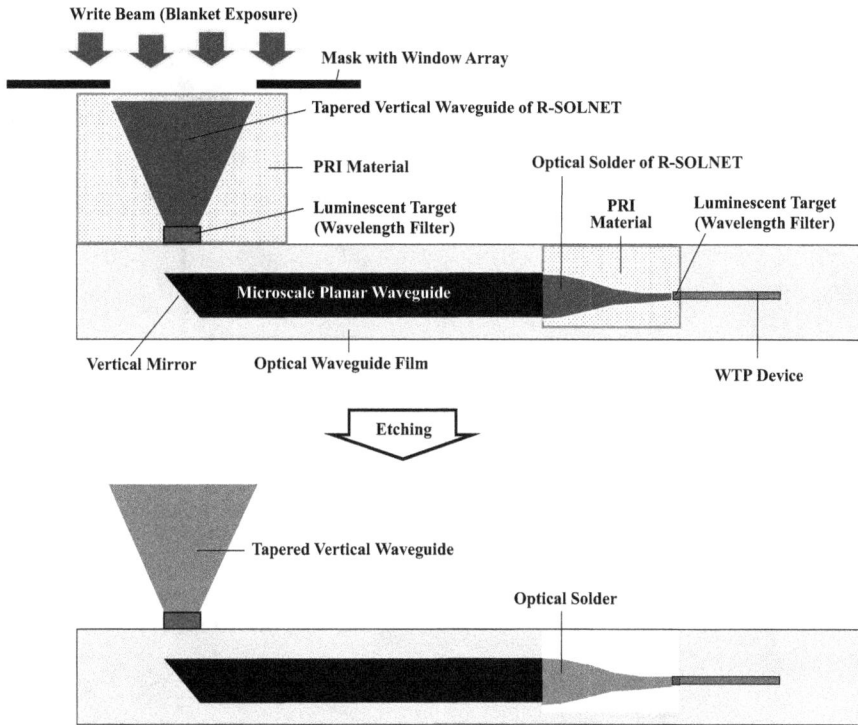

FIGURE 12.35 SOLNETs for optical coupling in the centralized-conversion-type structure.

is formed by introducing write beams from the microscale planar waveguides to the PRI material. By etching the surrounding regions of the SOLNETs, optical solder is obtained. If necessary, the optical solder is covered with low-refractive-index cladding materials.

12.4 THIN-FILM ARTIFICIAL PHOTOSYNTHESIS DEVICES

Artificial photosynthesis of various materials such as hydrogen/oxygen molecules [28] and formate molecules [29] has been reported. Figure 12.36 shows a water-splitting process for the H_2/O_2 photosynthesis developed by Sayama et al. [28]. Particles of semiconductor 1 (Pt-WO$_3$) and semiconductor 2 (Pt-SrTiO$_3$:Cr-Ta) are dispersed in water. An electron in semiconductor 1 is excited by light of λ_1 in wavelength from valence band (V.B.) to conduction band (C.B.) to generate a hole in the V.B. The electron is carried by a shuttle redox mediator to V.B. of semiconductor 2, and is excited by light of λ_2 to C.B. O_2 is generated by holes in the V.B. of semiconductor 1 and H_2 by electrons in the C.B. of semiconductor 2. This process involving the two-step electron excitation is similar to the z-scheme process in the photosynthesis of plants [15].

By replacing the dispersed particles with thin-film semiconductors, the relative position between the semiconductors 1 and 2 can be controlled to optimize the H_2/O_2 generation efficiency [2,5,10,27,30]. Proposed configurations are illustrated in Figure 12.37a for

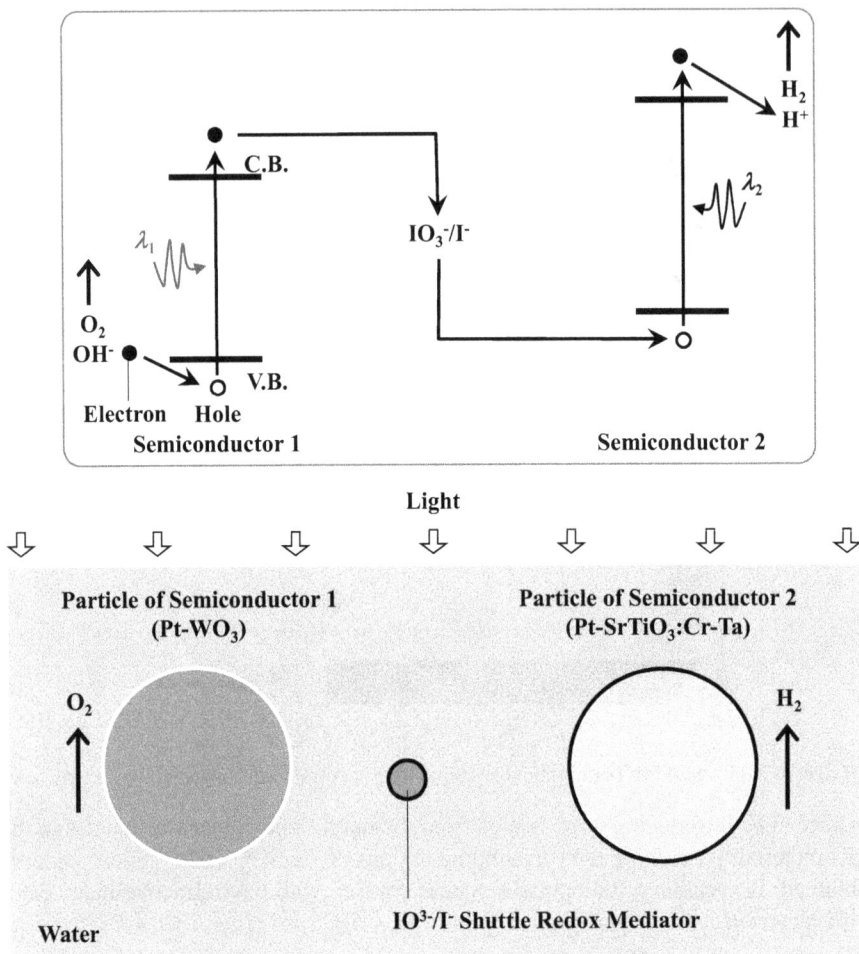

FIGURE 12.36 Water-splitting process for the H_2/O_2 photosynthesis developed by Sayama et al. Particles of semiconductors 1 and 2 are dispersed in water.

the thin-film artificial photosynthesis via the two-step electron excitation. By employing divided thin films or thin films having windows, the ion diffusion is promoted. The thin-film semiconductors are sensitized if necessary.

In the waveguide-type thin-film artificial photosynthesis shown in Figure 12.37b, λ_1-light and λ_2-light are guided in thin-film semiconductors 1 and 2, respectively. The guided light configuration enhances the light absorption of the thin-film semiconductors. Furthermore, the waveguide-based devices, which can increase the light power density, are preferable to improve the conversion efficiency because the H_2O oxidization is induced by the four-electron process.

Figure 12.38a presents the junction-type thin-film artificial photosynthesis with inserted organic MQDs [2,5,10,27,30]. Three-dimensional (3-D) optical circuits with embedded wavelength filters are implemented. Light beams branch into two paths for wavelengths of λ_1 and λ_2, and the two light beams are, respectively, introduced

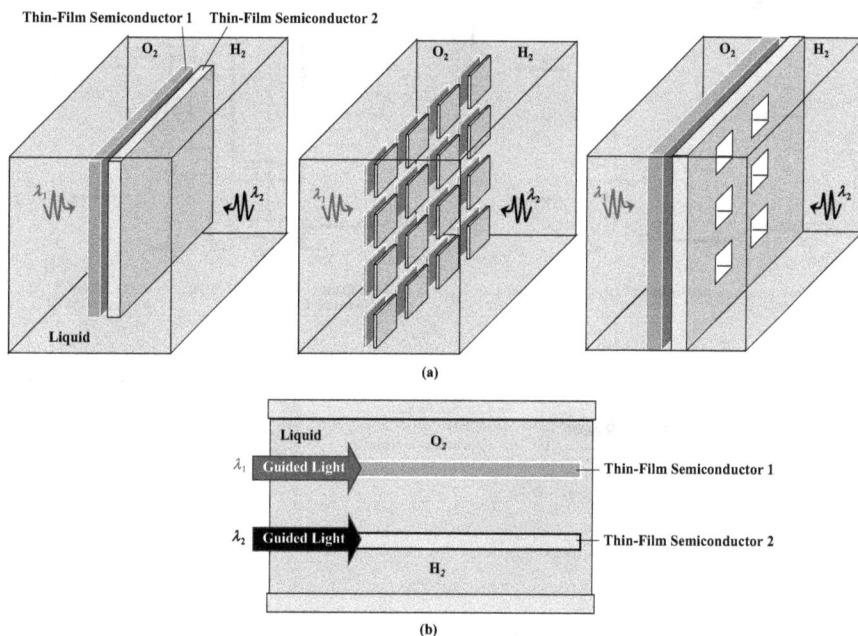

FIGURE 12.37 Proposed configurations for (a) the thin-film artificial photosynthesis and (b) the waveguide-type thin-film artificial photosynthesis.

onto thin-film semiconductors 1 and 2. In the case of the molecular MQD, an electron in Molecule M1 is excited by λ_1-light from HOMO to LUMO to generate a hole in the HOMO. The electron transfers to HOMO of M 2, and is excited by λ_2-light to LUMO. The hole left in the HOMO of M1 transfers to V.B. of thin-film semiconductor 1 to generate O_2, and the electron in the LUMO of M2 transfers to C.B. of thin-film semiconductor 2 to generate H_2. Implementation of light-harvesting functions into the molecular MQDs will further enhance the conversion efficiency. The similar water splitting might be performed by the polymer MQDs. In all these proposals, by employing the guided light configuration, the conversion efficiency would be enhanced. As Figure 12.38b shows, it might also be possible to perform the water splitting by only molecular MQDs or polymer MQDs.

Possible fabrication processes for the thin-film photosynthesis devices are shown in Figure 12.39 [30]. In process I, thin-film semiconductors 1 and 2 are, respectively, grown on growth substrates 1 and 2. The thin films are attached to each other with spacers between them. The growth substrates are removed to obtain stand-alone coupled thin films, which are put into the liquid. In process II, patterned thin-film semiconductor 1 is deposited on a removable layer of a substrate with spacers. Then, another removable layer and spacers are formed on it, and patterned thin-film semiconductor 2 is deposited on them. After one more removable layer with spacers and a cover layer are formed on them, all the removable layers are removed by etching to obtain a structure with thin-film semiconductors located in a designated configuration three-dimensionally. Finally, liquid is injected into the structure. In some cases, different kinds of liquid are injected into a 3-D micro-fluidic circuit, as depicted in Figure 12.40.

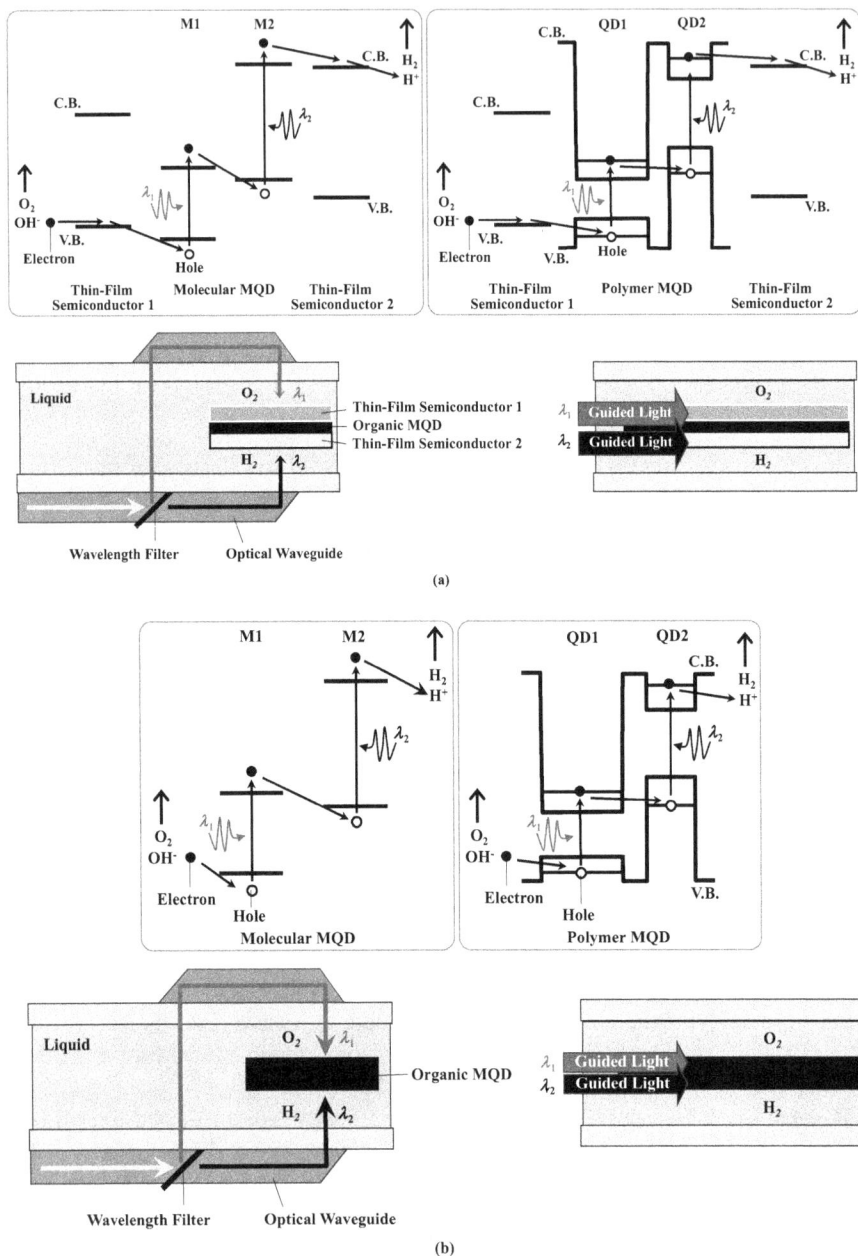

FIGURE 12.38   (a) Junction-type thin-film artificial photosynthesis with inserted organic MQDs and (b) thin-film artificial photosynthesis with organic MQDs.

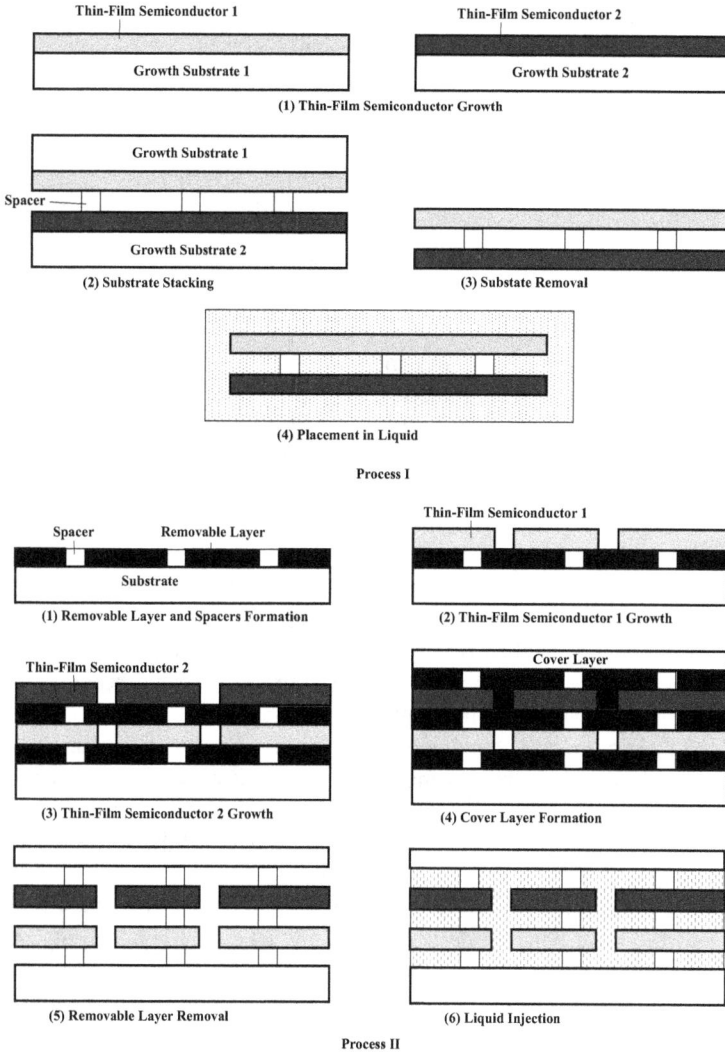

FIGURE 12.39 Possible fabrication processes for thin-film photosynthesis devices.

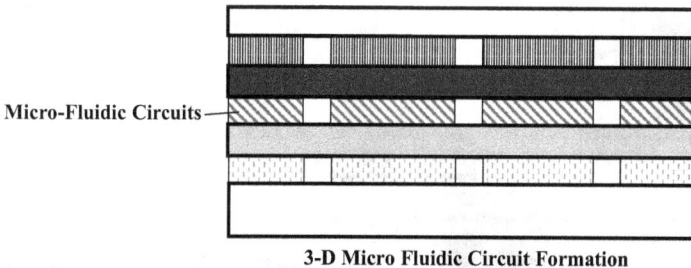

FIGURE 12.40 3-D micro-fluidic circuit containing different kinds of liquid.

12.5 ADDITIONAL PROPOSALS RELATED TO ENERGY AND MOLECULAR SENSING

A concept of the integrated photonic/electronic/chemical system (IPECS) is shown in Figure 12.41. IPECS contains optical circuits and fluidic circuits for supplying liquids such as electrolyte, water, and various solvents to the system. Embedded wavelength filters in IPECS enable efficient artificial photosynthesis as well as photovoltaics. Since the structure of IPECS is similar to that of the three-dimensional-stacked opto-electronic multichip module (3-D-stacked OE MCM) shown in Figure 11.3, similar fabrication processes are applicable to the IPECS construction [2,5,10,27,30].

As described in Section 3.5.2, surface potential of ZnO changes depending on adsorbed molecules. This implies that molecular sensing is possible by detecting the surface potential. As shown in Figures 12.42a, when no adsorbed molecules exist, there is no band bending in ZnO. Here, the effects of surface states are neglected. When oxygen molecules are adsorbed on the ZnO surface as Figure 12.42b shows, band bending to the positive side, that is, surface potential change to the negative side, is induced due to electron transfer from ZnO to the oxygen molecules. When p-type molecules are adsorbed on the ZnO surface, as shown in Figure 12.42c, the band bending further increases to the positive side, which implies that the surface potential becomes more negative, due to electron transfer from ZnO to the p-type molecules. Conversely, when n-type molecules are adsorbed on the ZnO surface, as shown in Figure 12.42d, the band bending decreases, which implies that the surface potential becomes less negative, due to electron transfer from the n-type molecules to ZnO.

Figure 12.43 shows band bending changes induced by light exposure in ZnO. In the initial state, adsorbed oxygen molecules on the ZnO surface extract electrons from ZnO, resulting in low electrical conductivity. When ZnO is exposed to light in vacuum, as shown in Figure 12.43a, holes are generated in V.B. of ZnO by the band-to-band transition. Then, trapped electrons in the oxygen molecules transfer to the V.B, and at the same time, the oxygen molecules are released from the ZnO surface. As a result, the band bending decreases and the electrical conductivity increases. Finally, all the adsorbed oxygen molecules are released from the surface, namely, all the trapped electrons go back to ZnO to eliminate the band bending and raise the conductivity. After the light is turned off, the final state is kept to exhibit high conductivity. When ZnO is exposed to light in air, as shown in Figure 12.43b, the trapped

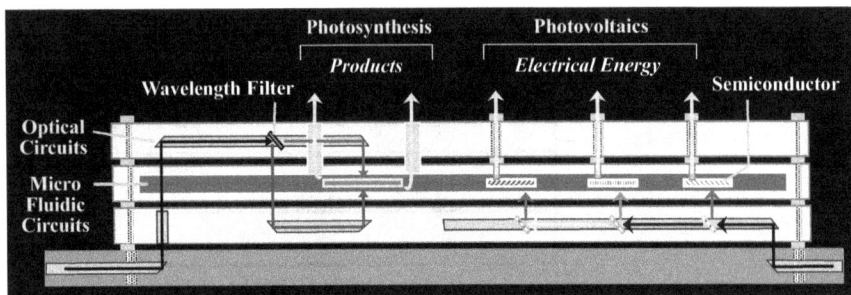

FIGURE 12.41 Concept of the integrated photonic/electronic/chemical system (IPECS).

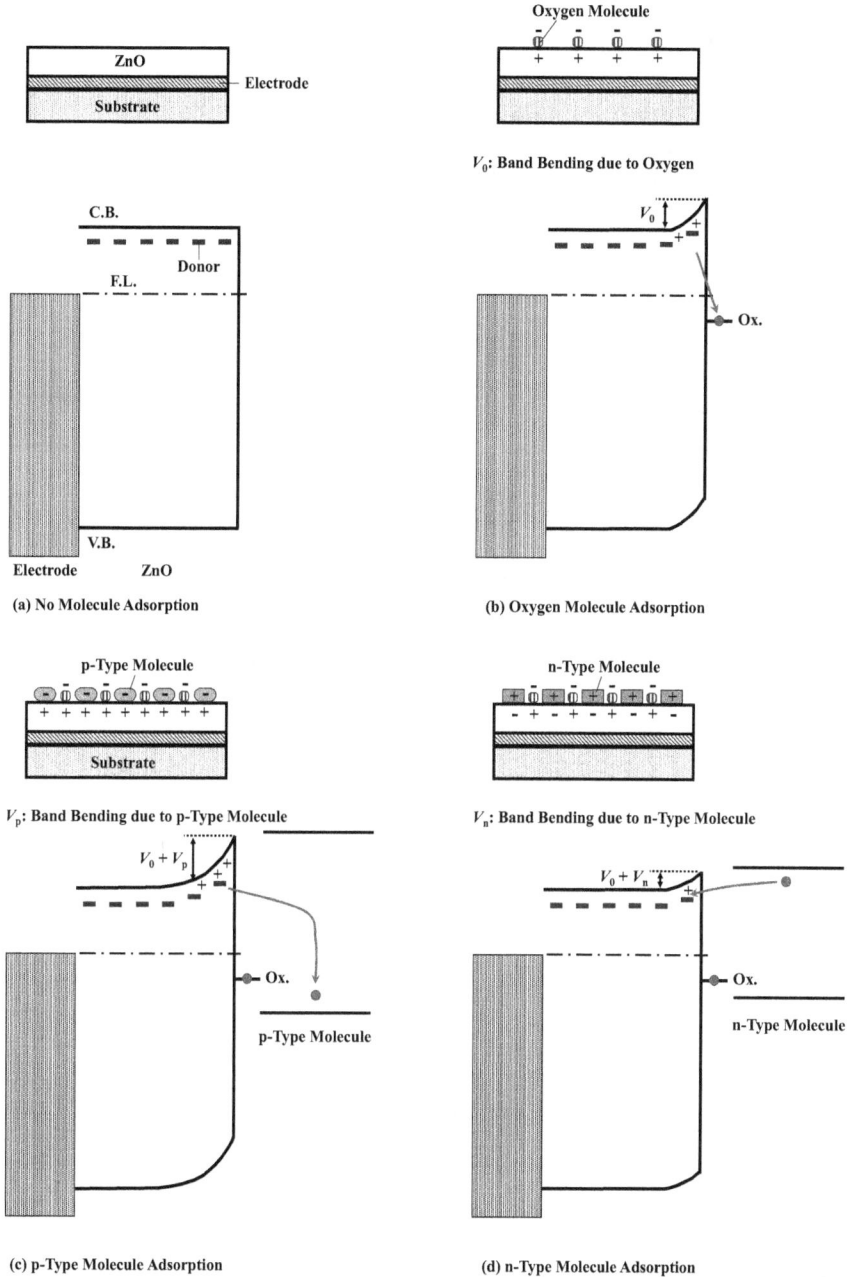

(a) No Molecule Adsorption

(b) Oxygen Molecule Adsorption

(c) p-Type Molecule Adsorption

(d) n-Type Molecule Adsorption

FIGURE 12.42 Band bending induced by adsorption of molecules on ZnO surfaces.

electrons in the oxygen molecules transfer to V.B. of ZnO and the oxygen molecules are released from the ZnO surface similarly to the in-vacuum case. However, surrounding oxygen molecules are adsorbed on the ZnO surface to trap electrons from

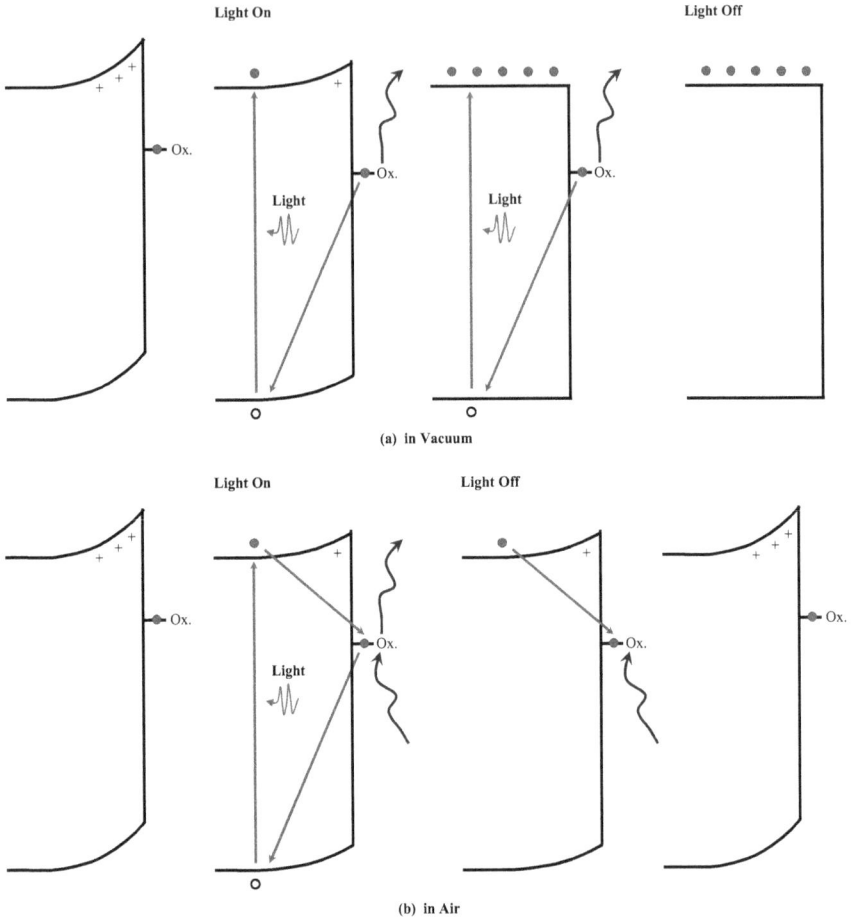

FIGURE 12.43 Band bending changes induced by light exposure in ZnO (a) in vacuum and (b) in air.

ZnO, and the band bending remains to some extent. After the light is turned off, oxygen molecules are adsorbed on the ZnO surface to extract electrons from ZnO, and the band bending returns to the initial state with low conductivity.

Based on the oxygen adsorption/desorption process described above, an oxygen sensing device might be available. Figure 12.44 presents schematic photocurrent vs. time characteristics in ZnO. In vacuum, after light is turned on, photocurrents increase with time to reach a large current level. After the light is turned off, the currents are preserved. In air, on the other hand, after light is turned on, the photocurrents increase with time but are saturated at a small current level. After the light is turned off, the currents decrease to the initial level. Figure 12.45 presents experimental results for the photocurrent vs. time characteristics in a ZnO powder layer. It is found that they are parallel to the schematic current behavior shown in Figure 12.44.

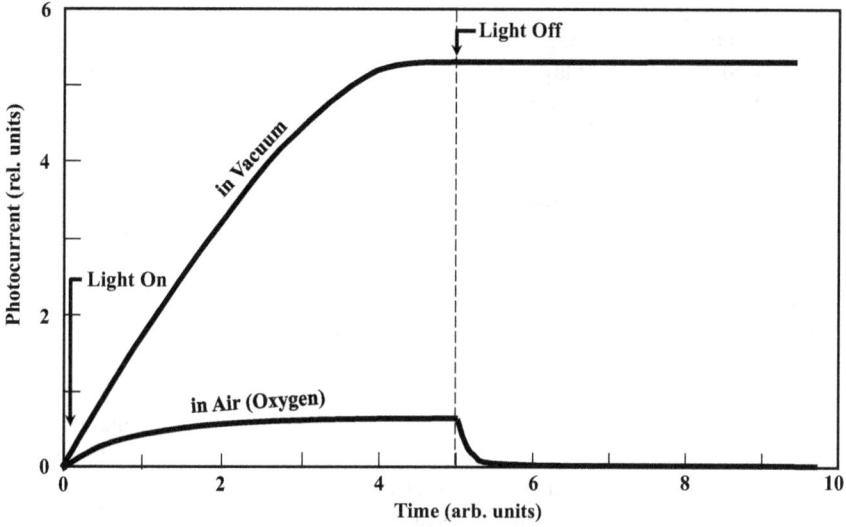

FIGURE 12.44 Schematic photocurrent vs. time characteristics in ZnO.

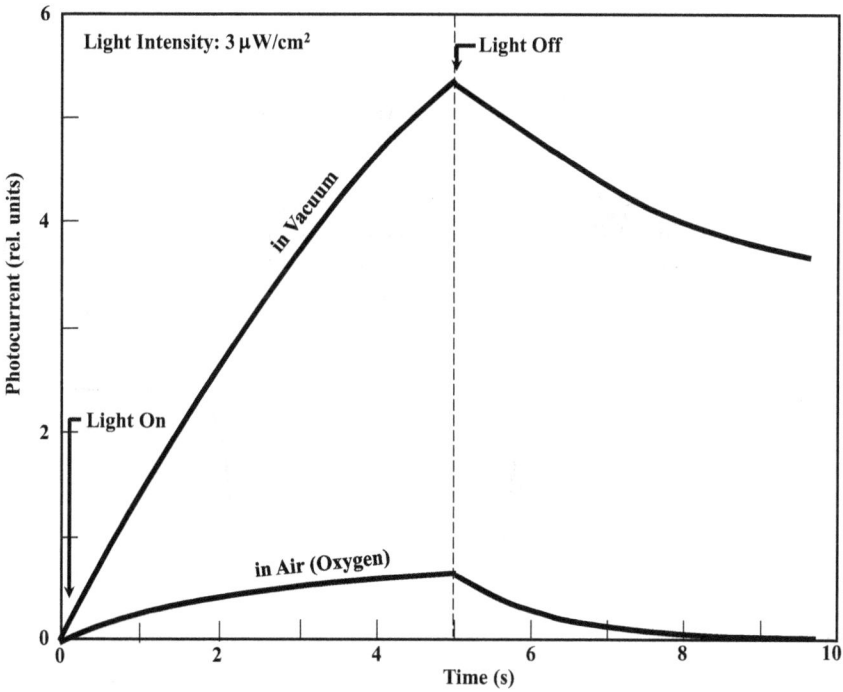

FIGURE 12.45 Experimental results for the photocurrent vs. time characteristics in a ZnO powder layer.

Figure 12.46 presents a proposed molecular sensing device. A ZnO layer is formed on source and drain electrodes, and a voltage is applied between them. When light is introduced onto the ZnO layer in a vacuum chamber, large photocurrents are generated, which correspond to the state shown in Figure 12.43a. By injecting oxygen molecules into the chamber, they extract electrons from the ZnO layer to reduce the photocurrents. This corresponds to the state shown in Figure 12.43b. By evacuating the oxygen molecules from the chamber, the photocurrents increase to the initial level. This device is regarded as a transistor, where molecules have a gate function, enabling the molecular sensing by detecting photocurrents.

Figure 12.47 shows a proposed integrated photoluminescence (PL) analysis chip with 3-D optical circuits. Micro-fluidic channels (MFCs), in which emissive materials are contained, are inserted between two optical waveguide films. When an excitation light beam of λ_{ex} from a vertical-cavity surface-emitting laser (VCSEL) is introduced into an MFC n from the upper optical waveguide, PL is generated. The PL is collected by the lower optical waveguide, and is guided to a photodetector (PD) passing through a wavelength filter m that transmits light beams of λ_m. By using arrayed MFCs, where different materials are individually injected, PL intensity measurements at λ_m can be done for the different materials at one time. By embedding wavelength filters for different wavelengths in series to de-multiplex the PL, spectral analysis of PL can be performed efficiently for many kinds of materials at one time [27].

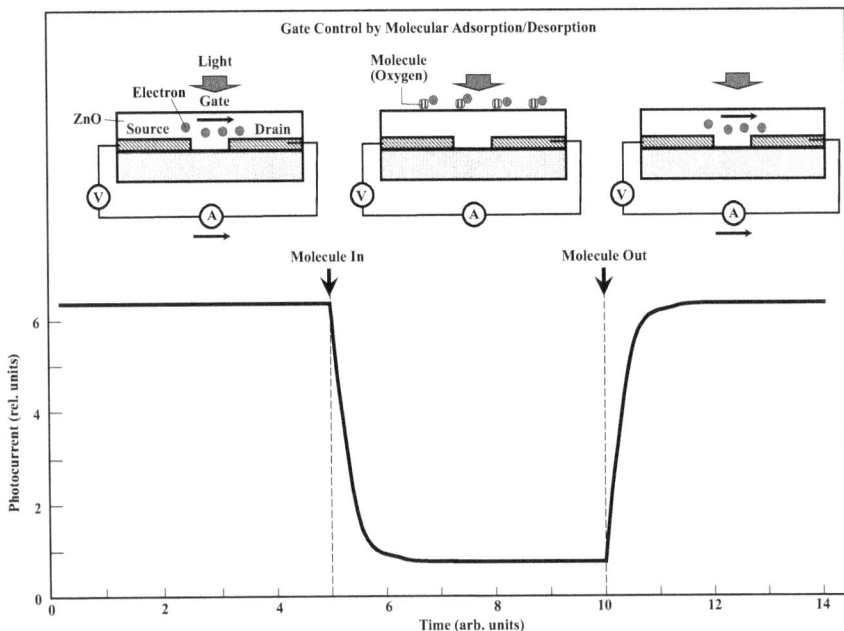

FIGURE 12.46 Proposed molecular sensing device that is regarded as a transistor, where molecules have a gate function.

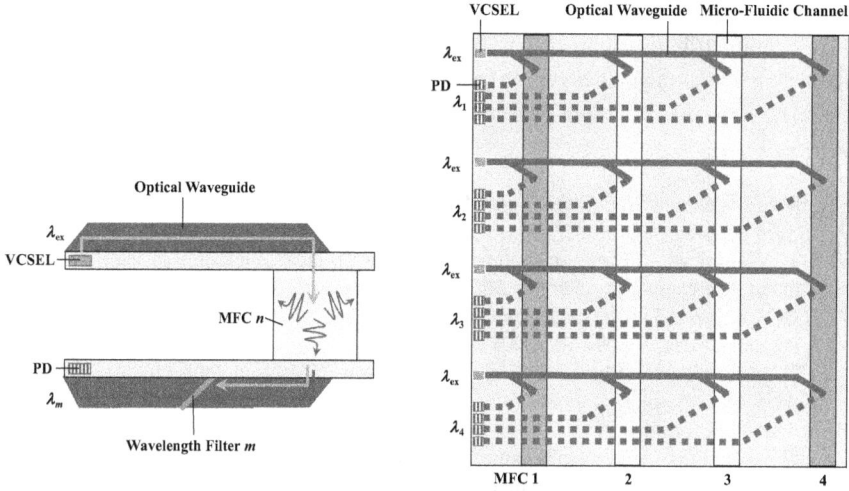

FIGURE 12.47   Proposed integrated PL analysis chip with 3-D optical circuits.

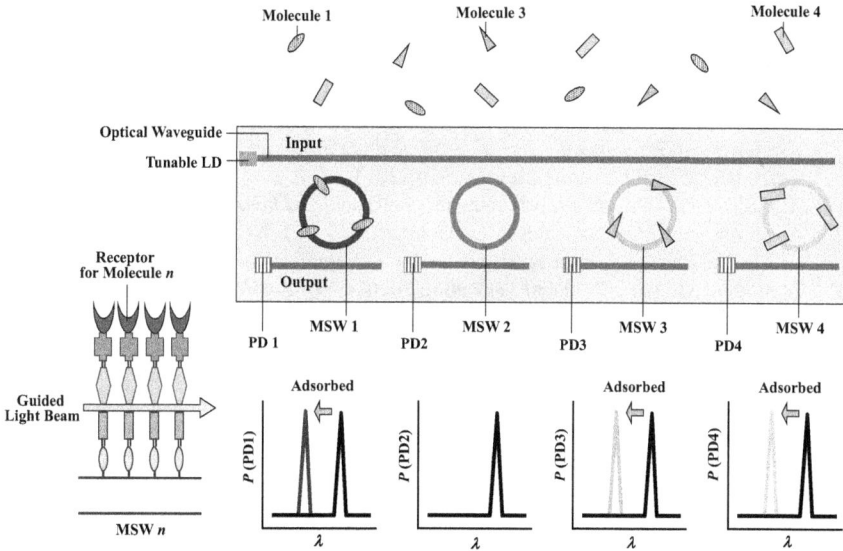

FIGURE 12.48   Proposed molecular recognition chip consisting of molecule-sensitive waveguides (MSWs).

Proposed molecular recognition chips consist of molecule-sensitive waveguides (MSWs), as shown in Figure 12.48. The MSW is, for example, a waveguide ring resonator with receptors for a particular kind of molecules. When the wavelength of light beams in the input waveguide is scanned, optical power transmitted to a PD at the edge of the output waveguide changes with a peak at the resonant wavelength of the MSW. When molecules are adsorbed on the MSW, the refractive index changes, inducing shifts of the resonant wavelength. For example, when Molecules 1, 3, and 4 are contained in the atmosphere, resonant frequencies of MSWs 1, 3, and 4 shift. Thus, constituent molecules can be recognized at one time [27].

REFERENCES

1. B. O'Regan and M. Gratzel, "A low-cost, high-efficiency solar cell based on dye-sensitized colloidal TiO_2 films," *Nature* **353**, 737–740 (1991).
2. T. Yoshimura, "Proposed applications of 3-D optical interconnect technologies to integrated chemical systems," *2007 Digest of the LEOS Summer Topical Meetings [Bio-Inspired Sensors and Application/Imprinting on Photonic Integrated Circuits]*, Portland, Oregon, 129–130 (2007).
3. T. Yoshimura, A. Oshima, D. Kim, and Y. Morita, "Quantum dot formation in polymer wires by three-molecule molecular layer deposition (MLD) and applications to electro-optic/photovoltaic devices," *ECS Trans.* **25**(4), 15 (2009).
4. T. Yoshimura, H. Watanabe, and C. Yoshino, "Liquid-phase molecular layer deposition (LP-MLD): Potential applications to multi-dye sensitization and cancer therapy," *J. Electrochem. Soc.* **158**, 51–55 (2011).
5. T. Yoshimura, *Thin-Film Organic Photonics: Molecular Layer Deposition and Applications*, CRC/Taylor & Francis, Boca Raton, Florida (2011).
6. T. Yoshimura, "Nanostructured materials formed by molecular layer deposition for enhanced solar energy utilization with optical waveguides," in *Energy Efficiency and Renewable Energy through Nanotechnology, Green Energy and Technology* (Ed. L. Zang), 351–390, Springer-Verlag London Limited, London (2011).
7. T. Yoshimura, "Organic quantum dots grown by molecular layer deposition for photovoltaics," in *Polymers for Energy Storage and Conversion* (Ed. V. Mittal), 103–136, Scrivener/Wiley Publishing, Beverly, MA (2013).
8. T. Yoshimura, "Thin-film molecular nanophotonics," in *Photonics, Vol. 2: Nanophotonic Structures and Materials* (Ed. D. L. Andrews), 261–310, Wiley, Hoboken, NJ (2015).
9. T. Yoshimura, "Molecular layer deposition and applications to solar cells and cancer photodynamic therapy," Japanese Patent, Tokukai 2012–45351 (2012) [in Japanese].
10. T. Yoshimura, *Self-Organized 3D Integrated Optical Interconnects: With All-Photolithographic Heterogeneous Integration*, Jenny Stanford Publishing, Singapore (2021).
11. T. Yoshimura, R. Ebihara, and A. Oshima, "Polymer wires with quantum dots grown by molecular layer deposition of three source molecules for sensitized photovoltaics," *J. Vac. Sci. Technol. A* **29**, 051510 (2011).
12. T. Yoshimura and S. Ishii, "Effect of quantum dot length on the degree of electron localization in polymer wires grown by molecular layer deposition," *J. Vac. Sci. Technol. A* **31**, 031501 (2013).
13. T. Yoshimura, S. Bai, H. Tateno, C. Yoshino, "In situ photocurrent spectra measurements during growth of three-dye-stacked structures by the liquid-phase molecular layer deposition," *J. Appl. Phys.* **122**, 015309 (2017).
14. C. Brauchle, U. P. Wild, D. M. Burland, G. C. Bjorkund, and D. C. Alvares, "Two-photon holographic recording with continuous-wave lasers in the 750–1100-nm range," *Opt. Lett.* **7**, 177–179 (1982).
15. V. D. McGinniss and R. E. Schwerzel, "Photopolymerizable composition containing a photosensitive donor and photoinitiating acceptor," U.S. Patent 4,571,377 (1986).
16. K. Kiyota, T. Yoshimura, and M. Tanaka, "Electrophotographic behavior of ZnO sensitized by two dyes," *Photogr. Sci. Eng.* **25**, 76–79 (1981).

17. T. Yoshimura, K. Kiyota, H. Ueda, and M. Tanaka, "Mechanism of spectral sensitization of ZnO coadsorbing p-type and n-type dyes," *Jpn. J. Appl. Phys.* **20**, 1671–1674 (1981).

18. T. Yoshimura, K. Kiyota, H. Ueda, and M. Tanaka, "Contact potential difference of ZnO layer adsorbing p-type dye and n-type dye," *Jpn. J. Appl. Phys.* **18**, 2315–2316 (1979).

19. T. Yoshimura, K. Kiyota, H. Ueda, and M. Tanaka, "Influence of illumination on the surface potential of ZnO adsorbing p-type dye and n-type dye," *Jpn. J. Appl. Phys.* **19**, 1007–1008 (1980).

20. H. J. Meier, "Sensitization of electrical effects in solids," *Phys. Chem.* **69**, 719–729 (1965).

21. Y. Nakajima, D. Yamashita, A. Ishizaki, B. Pellissier, and M. Uda, "Electronic structures of dyes and phthalocyanines estimated with "photo-electron spectroscopy in air (PESA)," *Mater. Res. Symp. Proc.* **1029**, 1029-F04-02 (2008).

22. T. Liu and T. Yoshimura, "Sensitization of ZnO in stacked structures containing multiple dyes grown using liquid phase molecular layer deposition," *ECS Trans.* **69**, 217–222 (2015).

23. T. Yoshimura, M. Ojima, Y. Arai, and K. Asama, "Three-dimensional self-organized micro optoelectronic systems for board-level reconfigurable optical interconnects –performance modeling and simulation–," *IEEE J. Select. Top. Quantum Electron.* **9**, 492–511 (2003).

24. T. Yoshimura, K. Kumai, T. Mikawa, and O. Ibaragi, "Photolithographic packaging with selectively occupied repeated transfer (PL-pack with SORT) for scalable optical link multi-chip-module (S-FOLM)," *IEEE Trans. Electron. Packag. Manufact.* **25**, 19–25 (2002).

25. T. Yoshimura and Y. Arai, "3D optoelectronic micro system," U.S. Patent 2006026 1432A1 (2006).

26. T. Yoshimura and J. Roman, W-C. Wang, M. Inao, and M. T. McCormack, "Device transfer method," U.S. Patent 6,669,801 (2003).

27. T. Yoshimura, "Integrated photonic/electronic/chemical systems, solar energy conversion systems, light collectors, and optical waveguides," Japanese Patent, Tokukai 2009–4717 (2009) [in Japanese].

28. K. Sayama and H. Arakawa, *Dream of Artificial Synthesis*, 102–109, Newton: Next-Generation Technology, Newton Press, Tokyo (2002) [in Japanese].

29. S. Sato, T. Arai, T. Morikawa, K. Uemura, TM. Suzuki, H. Tanaka, and T. Kajino, "Selective CO_2 conversion to formate conjugated with H_2O oxidation utilizing semiconductor/complex hybrid photocatalysts," *J. Am. Chem. Soc.* **133**, 15240–15243 (2011).

30. T. Yoshimura and K. Asama, "Integrated chemical systems and integrated light energy conversion systems," Japanese Patent, Tokukai 2007–107085 (2007) [in Japanese].

13 Expected Applications of LP-MLD to Cancer Therapy

As described in previous chapters, molecular layer deposition (MLD) is promising in the photonics/electronics and energy conversion fields. MLD is also expected to be applied to the biomedical field, especially to cancer therapy. Because a human body is a liquid system, liquid-phase MLD (LP-MLD), where the human body and cancer cells are, respectively, regarded as the MLD chamber and substrates, is used. For cancer therapy, three capabilities of MLD, namely, tailored organic material synthesis, ultrathin/conformal organic material synthesis on arbitrary structures, and selective organic material synthesis, are effective, serving the selective *in-situ* drug synthesis at cancer cell sites. In the present chapter, the molecular targeted drug delivery by the *in-situ* drug synthesis utilizing LP-MLD is proposed, including the low-molecular-weight drug delivery into cancer cells, the antibody drug delivery into cancer cells, and the drug delivery into cancer stem cells. Photo-assisted cancer therapy, namely, the self-organized lightwave network (SOLNET)-assisted laser surgery and photodynamic therapy (PDT), where LP-MLD and SOLNET are employed, is also proposed. Although the author has no biomedical background, some of the proposals would contribute to produce hints on the cancer therapy research.

13.1 MLD IN HUMAN BODIES

Because a human body is a liquid system, LP-MLD is expected to be a powerful process in the biomedical field, particularly for cancer therapy. The human body and cancer cells, respectively, correspond to the MLD chamber and substrates. Figure 13.1 shows a comparison between MLD in the photonics/electronics applications and MLD in the biomedical applications. Sensitization of photovoltaic devices is an example of the former, in which MLD achieves tailored organic material growth at selected sites. Cancer therapy is an example of the latter, in which LP-MLD achieves *in-situ* drug synthesis in cancer cells [1–6].

Three capabilities of MLD are presented in Figure 13.2; tailored organic material synthesis, ultrathin/conformal organic material synthesis on arbitrary structures, including porous, tortuous, deforming, and floating objects, and selective organic material synthesis. These capabilities serve the selective *in-situ* drug synthesis for cancer therapy. The selective synthesis ability of MLD is especially important in applications to the molecular targeted drug delivery.

In the below parts, expected applications of LP-MLD to cancer therapy are proposed.

DOI: 10.1201/9781003094012-13

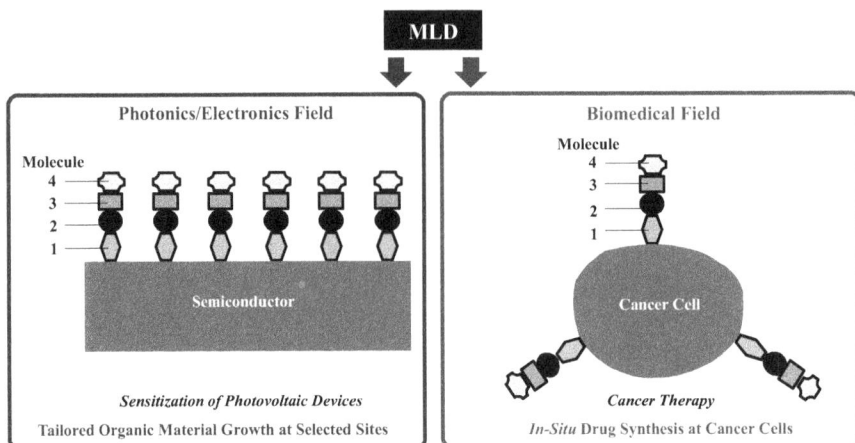

FIGURE 13.1   Comparison between MLD in the photonics/electronics applications and MLD in the biomedical applications.

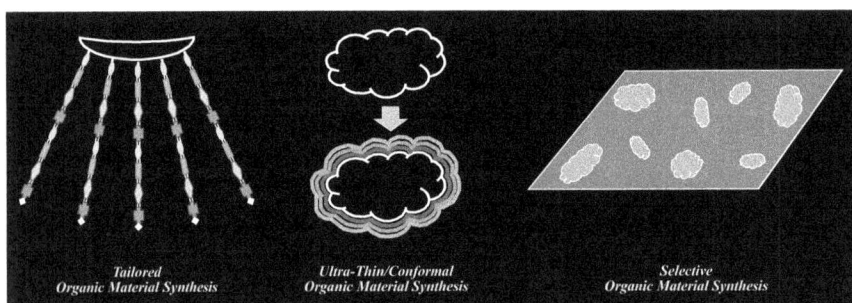

FIGURE 13.2   Three capabilities of MLD, which serve the selective *in-situ* drug synthesis for cancer therapy.

13.2   MOLECULAR TARGETED DRUG DELIVERY BY *IN-SITU* SYNTHESIS IN CANCER CELLS UTILIZING LP-MLD

As depicted in Figure 13.3, in medication for cancer, drugs attack normal cells during attack on cancer cells due to a lack of selectivity, causing side effects. LP-MLD is expected to solve the problem by performing selective attack on cancer cells.

Currently, efficient and safe drug delivery is realized by using antibodies. Antibodies with attached strong cancer killing drugs enable the exact targeting to attack cancer cells with small side effects. Antibodies with attached quantum dots enable the imaging of cancer cell distribution [7], antibodies with attached paramagnetic agents enable the labeling for the magnetic resonance imaging (MRI) system [8], and antibodies with attached radioactive compounds enable the enhanced radiotherapy [9]. One drawback of the antibody drugs is their large molecular weight. This prevents them from attacking the inside of cancer cells. On the contrary, low-molecular-weight drugs can pass through the cell membranes, attacking the inside

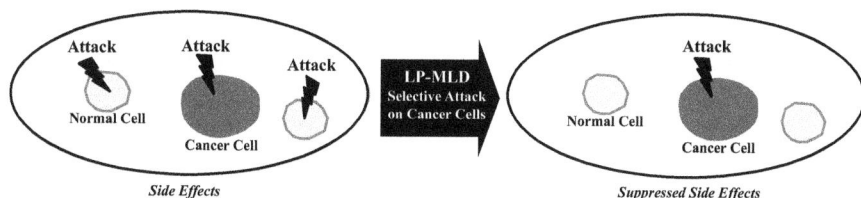

Side Effects   *Suppressed Side Effects*

FIGURE 13.3 Problem in medication for cancer: attack of drugs on normal cells during attack on cancer cells, causing side effects. The solution is selective attack on cancer cells performed by LP-MLD, suppressing the side effects.

of cancer cells while they have a drawback, that is, they cause side effects due to imperfect targeting selectivity. Meanwhile, it was recently revealed that destruction of cancer stem cells is essential to achieve perfect healing [10].

In the present section, to solve the abovementioned problems, concepts of molecular targeted drug delivery, where *in-situ* drug synthesis at cancer cell sites by LP-MLD is involved, are proposed, including selective delivery of low-molecular-weight drugs and antibody drugs to cancer cells, and selective delivery of drugs to cancer stem cells without attacking normal cells.

13.2.1 GENERAL CONCEPT

Figure 13.4 shows an LP-MLD process for the molecular targeted drug delivery by *in-situ* synthesis at cancer cells [1–6]. In Step 1, Molecule A is injected into a human boby. Molecule A is adsorbed by cancer cells selectively as an anchoring molecule for selective synthesis initiation. After extra molecule A is excreted from the body, in Step 2, Molecule B is injected to be connected to Molecule A. Similarly, by injecting Molecules C and D successively, drugs having a structure of A/B/C/D are synthesized selectively at the cancer cells.

In the LP-MLD process, the selective molecular connection is achieved utilizing the selective chemical reaction, electrostatic force, or π–π interaction between source molecules. Molecule A for the anchor is important to enhance the targeting selectivity. For example, since it is known that porphyrin is adsorbed by cancer cells selectively, it could be used for Molcule A as the anchor. When β-carotene is chosen as Molecule B, the selective connection of Molecule B to Molecule A might be possible by the π–π interaction. If Molecules A and B, respectively, have reactive groups of $-NH_2$ and $-CHO$, MLD could be performed in the same manner as the poly-azomethine (poly-AM) growth described in Chapters 3, 4, and 6. In the successive processes for Molecules C and D, the similar scheme could be applied.

Figure 13.5 gives a demonstration of the anchoring effect of dye molecules [4]. When n-type crystal violet (CV) molecules are provided on an n-type ZnO powder layer in solvent, the ZnO surface remains white, indicating that CV molecules are not adsorbed on ZnO. This means that immobilization of CV on ZnO is not possible. When p-type rose bengal (RB) molecules are provided on a ZnO powder layer, the ZnO surface becomes pink, indicating that RB molecules are adsorbed on ZnO. When CV molecules are provided on the RB-adsorbed ZnO powder layer, the ZnO

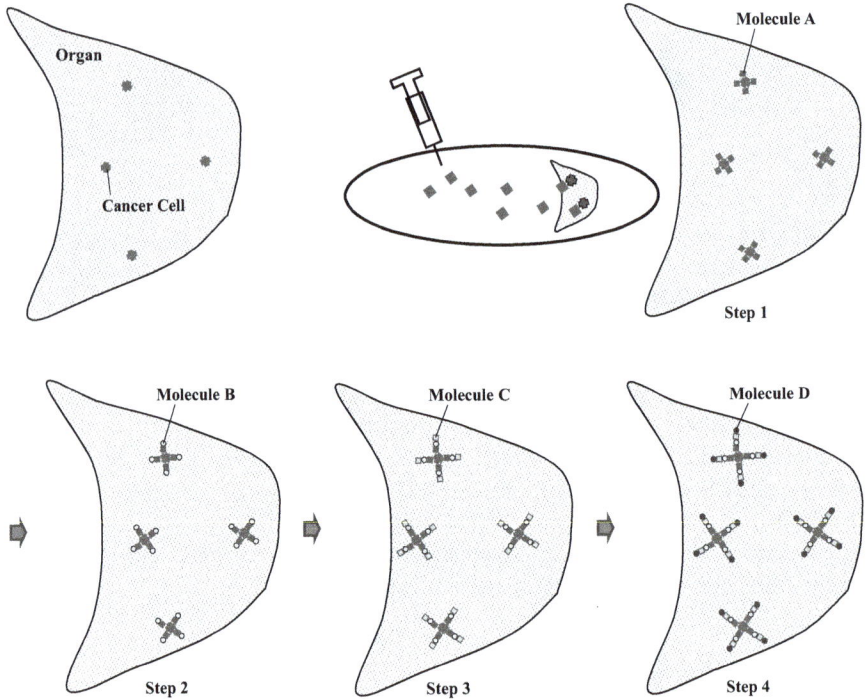

FIGURE 13.4 LP-MLD process for the molecular targeted drug delivery by *in-situ* synthesis at cancer cells. (Reprinted with permission from Yoshimura et al. [2].)

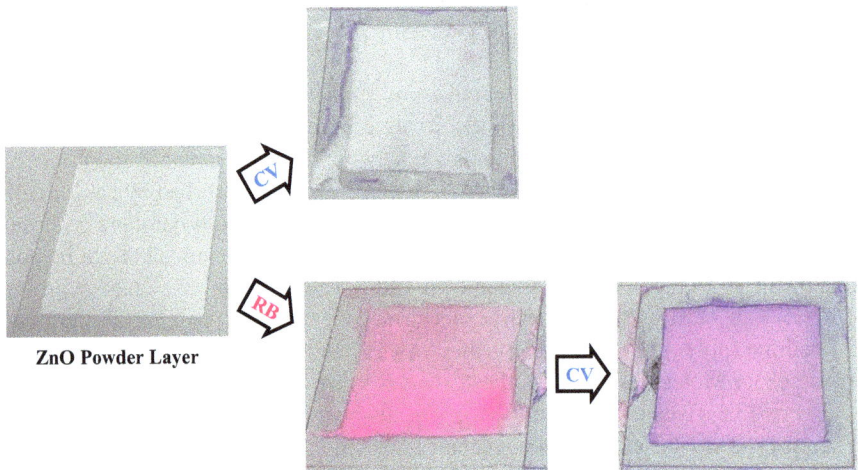

FIGURE 13.5 Demonstration of the anchoring effect of dye molecules: immobilization of CV molecules on ZnO by anchoring molecules of RB.

surface becomes violet, which is the color of CV, indicating that CV molecules are adsorbed on ZnO via RB molecules. These results imply that RB acts as the anchor to immobilize CV on ZnO. The driving force for this immobilization is the electrostatic force as precisely described in Section 3.5.2. Such an anchoring mechanism is expected to be applied to the molecular targeted drug delivery by LP-MLD. In addition, the mechanisms of selective MLD described in Chapter 4, such as area selective growth and selective growth from seed cores, might also be applicable to the anchoring.

An example of multi-functional materials with a structure of A/B/C/D grown by LP-MLD on cancer cells is depicted in Figure 13.6 [3]. Molecule A is the anchoring molecule for the selective synthesis initiation. Molecules B, C, and D could be functional molecules like a luminescent agent, a sensitizer for PDT, a two-photon absorption agent, a near infrared (IR) light absorption agent, a paramagnetic agent for MRI labeling and magnetically-actuated systems, and a radio-enhancement agent. Figure 13.7 presents another example. After Molecule A for the anchor is deposited at cancer cells, Molecules B and C are successively injected to form a structure of A/B/C. Molecules B and C are chosen so that they become a cancer killer when they combine to attack the cancer cells.

The example depicted in Figure 13.8 is for a case that drug molecules are too large to reach deep parts of cancer through narrow channels [3]. To solve the problem, the drug molecules are synthesized from small component molecules at cancer cell sites by LP-MLD. Molecule A is the anchoring molecule. Molecules B, C, and D are component molecules of the drug molecule. Sequential injection of Molecules A, B, C, and D forms the large drug molecules at the cancer cell sites. This method might widen the range of drug choices. When the drug molecule is toxic for some parts in human bodies, drug delivery with less toxic conditions might become possible by the *in-situ* drug synthesis from nontoxic component molecules.

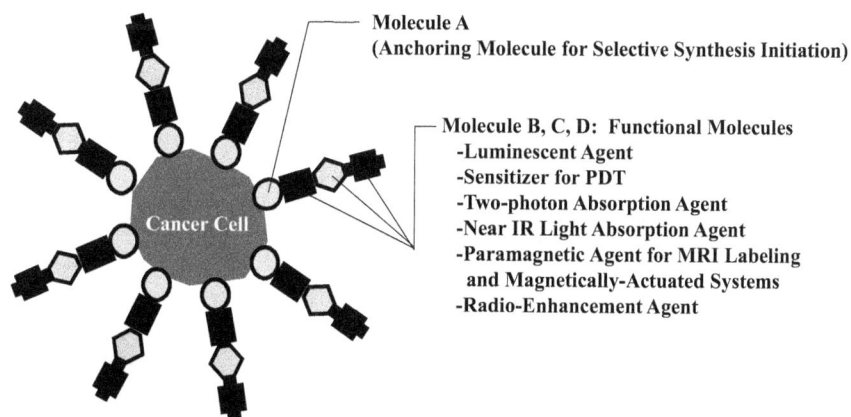

Molecule A
(Anchoring Molecule for Selective Synthesis Initiation)

Molecule B, C, D: Functional Molecules
 -Luminescent Agent
 -Sensitizer for PDT
 -Two-photon Absorption Agent
 -Near IR Light Absorption Agent
 -Paramagnetic Agent for MRI Labeling
   and Magnetically-Actuated Systems
 -Radio-Enhancement Agent

Cancer Cell

FIGURE 13.6 An example of multi-functional materials grown by LP-MLD on cancer cells. (Reprinted with permission from Yoshimura et al. [3].)

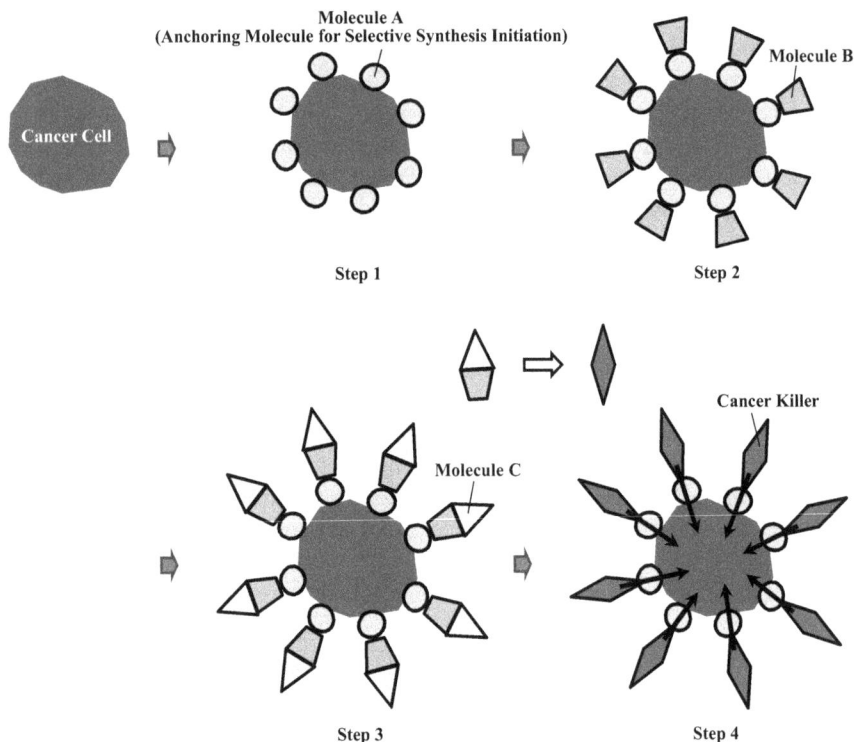

FIGURE 13.7   An example of multi-functional materials grown by LP-MLD on cancer cells to produce cancer killers.

13.2.2   LOW-MOLECULAR-WEIGHT DRUG DELIVERY INTO CANCER CELLS

Although low molecular weight drugs have an advantage that they can attack inside of the cancer cells, they have a drawback of imperfect targeting selectivity [10]. For example, gefitinib, which shuts down the intracellular signal transduction in cancer cells by being attached to the epidermal growth factor receptor (EGFR) to be an effective molecular targeted drug for lung cancer, causes interstitial pneumonia as the side effect.

If *in-situ* synthesis of the toxic drug is done selectively within cancer cells by connecting small nontoxic component molecules using LP-MLD, drug delivery into inside of the targeted cancer cells might be possible without side effects. Figure 13.9 presents a conceptual illustration of an LP-MLD process for the molecular targeted low-molecular-weight drug delivery [4]. Molecule A is an anchor toward the adenosine triphosphate (ATP)-binding site. The toxic drug is divided into nontoxic component molecules; Molecules B, C, D, and E, for LP-MLD. First, Molecule A is injected into a human body to be selectively connected to the ATP-binding sites of the tyrosine kinase domain of EGFR. After extra Molecule A is excreted from the

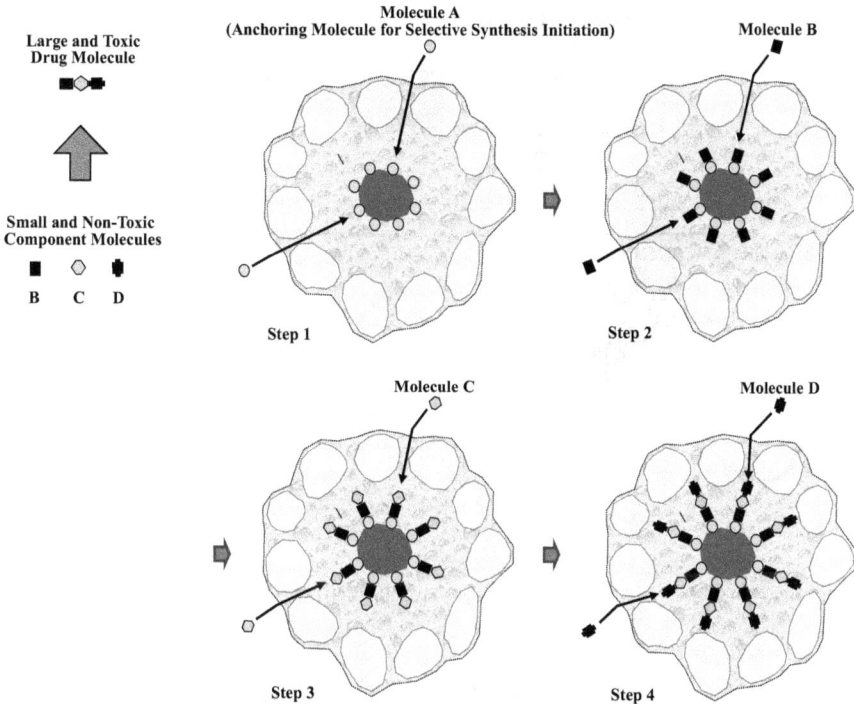

FIGURE 13.8 Delivery of large and toxic drugs by synthesizing the drugs from small and nontoxic component molecules in cancer cell sites by LP-MLD. The drugs reach deep parts of the cancer through narrow channels. (Reprinted with permission from Yoshimura et al. [3].)

human body, Molecule B is injected to be selectively connected to Molecule A. By successive injections of Molecules C, D, and E, the synthesis of the toxic drug is completed. The synthesized drug acts as a tyrosine kinase inhibitor, which disconnects the intracellular signal transduction in cancer cells by protecting ATP to be attached to the ATP-binding sites, enabling cancer growth suppression.

13.2.3 ANTIBODY DRUG DELIVERY TO CANCER CELLS

Figure 13.10 shows an LP-MLD process for the antibody drug delivery to grow multifunctional materials in cancer cells one by one in designated arrangements [4]. First, Molecule A of the antibody is injected into a human body to be selectively attached on cancer cells as an anchor for synthesis initiation, and extra molecule A is excreted from the human body. Next, Molecule B, which is the sensitizer for PDT, is injected to be connected to Molecule A. Similarly, by successively injecting Molecule C, which is the luminescent agent for imaging, and Molecule D, which is the radio-enhancement agent, multi-functional materials having a structure of A/B/C/D can be constructed in the cancer cells.

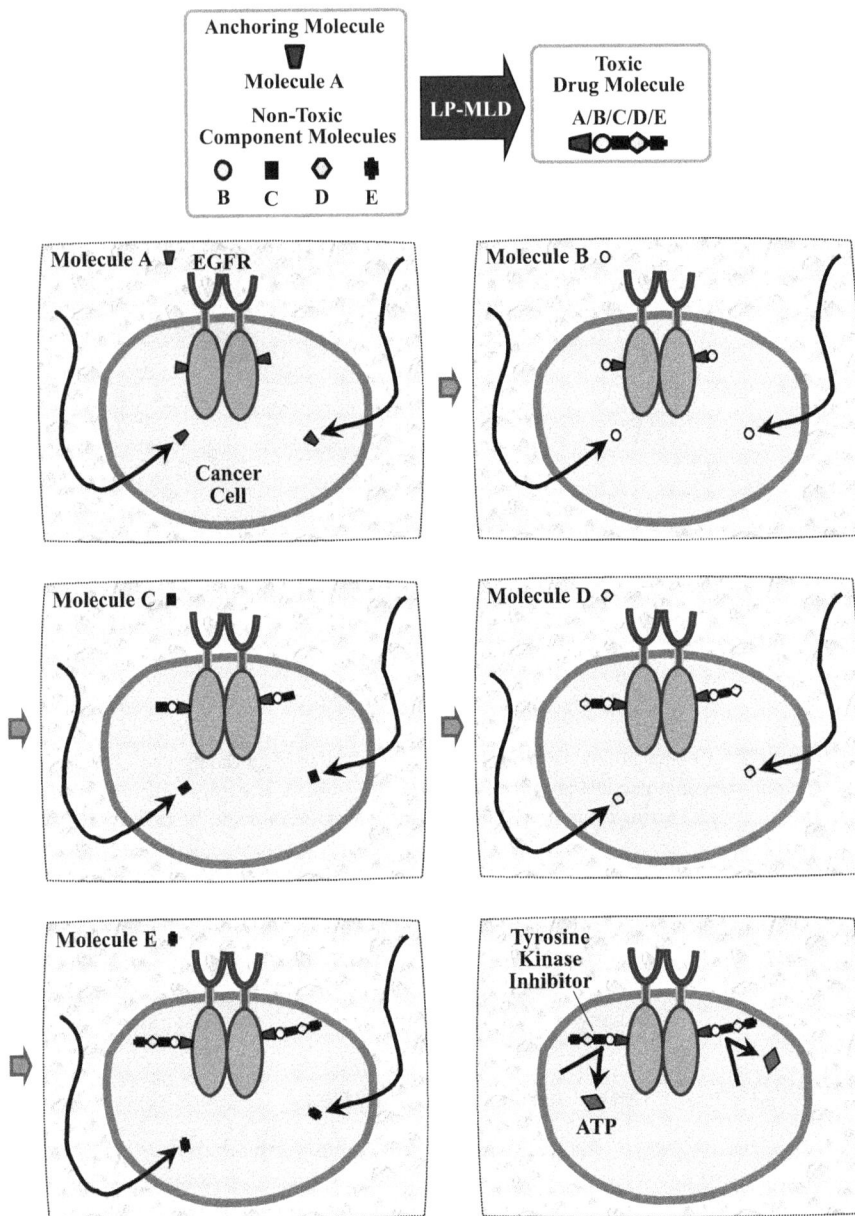

FIGURE 13.9 Conceptual illustration of an LP-MLD process for the molecular targeted low-molecular-weight drug delivery into inside of the cancer cells.

13.2.4 DRUG DELIVERY INTO CANCER STEM CELLS

It is known that destruction of cancer stem cells is important for perfect healing [10]. The difficulty in the cancer stem cell destruction comes from several reasons such as excretion of drugs by the ATP-binding cassette (ABC) transporters, cell protecting by the cell cycle arrest, and reducing systems for the oxidative stress. In the present

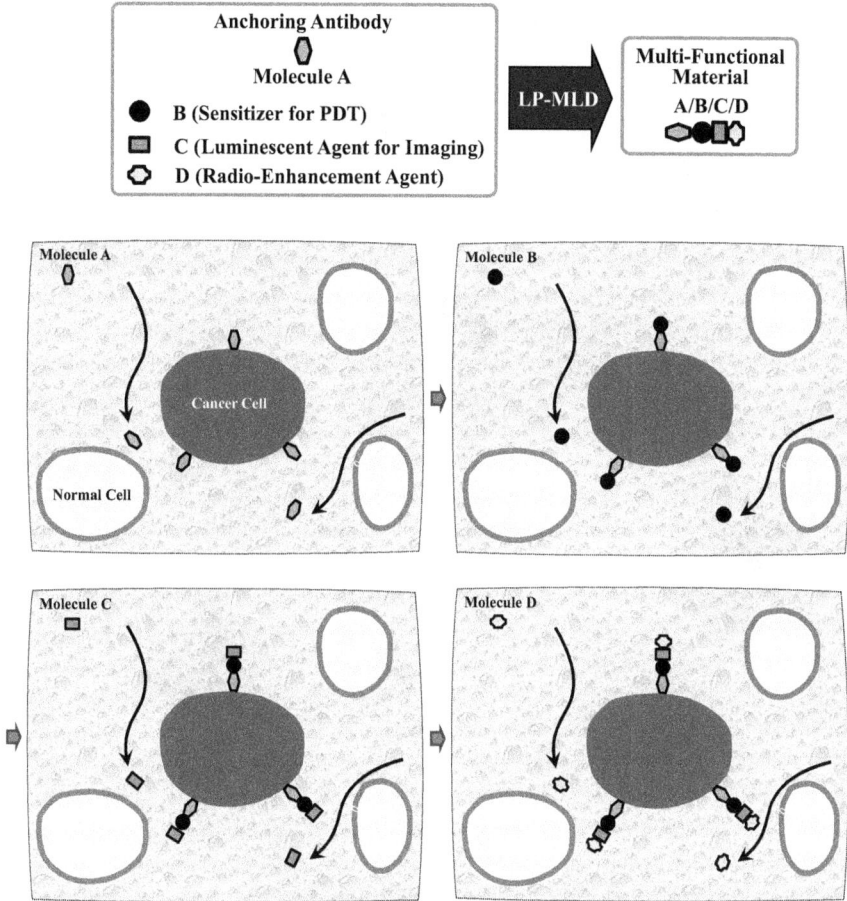

FIGURE 13.10 Conceptual illustration of an LP-MLD process for the antibody drug delivery to grow multi-functional materials in cancer cells one by one in designated arrangements.

proposal, LP-MLD is applied to suppress the drug excretion by the ABC transporters [3].

Figure 13.11 shows a model of the drug excretion by the ABC transporters. Rapamycin, which is an inhibitor of mammalian target of rapamycin (mTOR), is known as a molecular targeted drug toward leukemic stem cells. The drug is carried away from the cancer stem cell sites by the ABC transporters distributed in the surrounding region. This reduces the effect of the drug to attack the cancer stem cells.

LP-MLD is expected to solve this problem as conceptually illustrated in Figure 13.12. The drug is divided into several component molecules, say, five component molecules of Molecules A, B, C, D, and E, that can pass by the ABC transporters and reach the cancer stem cells. First, Molecule A acting as the anchor toward the mTOR is injected into a human body to be connected to mTOR within the cancer stem cells selectively. After extra Molecule A is excreted from the human body, by successive injections of Molecules B, C, D, and E, the drug is synthesized within the cancer stem cells. It acts as an mTOR inhibitor, enabling cancer growth suppression.

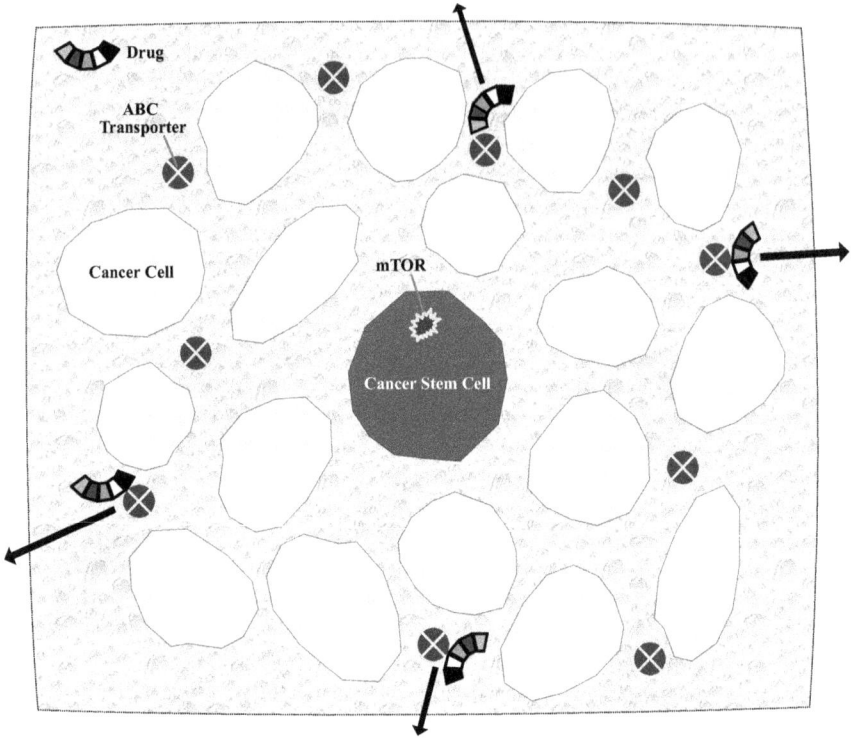

FIGURE 13.11   A model of the drug excretion by the ABC transporters in the drug delivery into cancer stem cells.

13.3   PHOTO-ASSISTED CANCER THERAPY

Photonics technologies have been introduced into the cancer therapy more and more, such as the PDT [11,12], the optical diagnostics, and the light-harvesting antenas utilizing the fluorescence resonance energy transfer (FRET) [13]. Photonics and molecular nanotechnologies are the main streams for cancer therapy. In the present section, expected applications of LP-MLD and SOLNET to the laser surgery and PDT are proposed.

13.3.1   SOLNET-ASSISTED LASER SURGERY

Figure 13.13 shows the concept of the SOLNET-assisted laser surgery [1–4,6]. Firstly, multi-functional materials containing luminescent molecules are grown in cancer cells by LP-MLD. In the LP-MLD process, after anchoring molecules that have a capability to be adsorbed by cancer cells selectively are injected, luminescent molecules are successively injected to form luminescent targets. Next, an optical waveguide array and a photo-induced refractive-index increase (PRI) material are placed in a region surrounding the cancer cells, and write beams are introduced from the optical waveguides. The luminescent targets absorb the write beams and

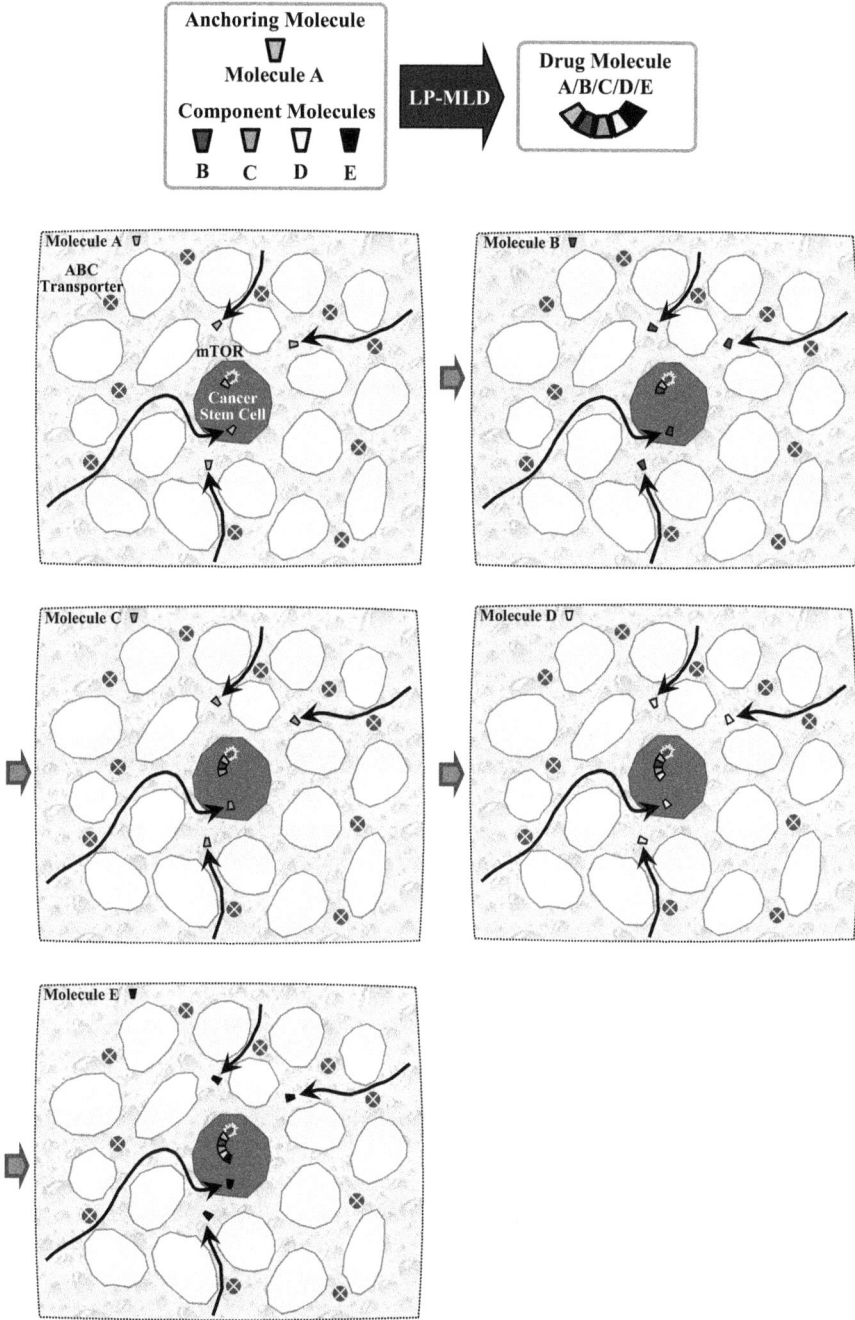

FIGURE 13.12 Conceptual illustration of an LP-MLD process for the *in-situ* synthesis within cancer stem cells in the molecular targeted drug delivery.

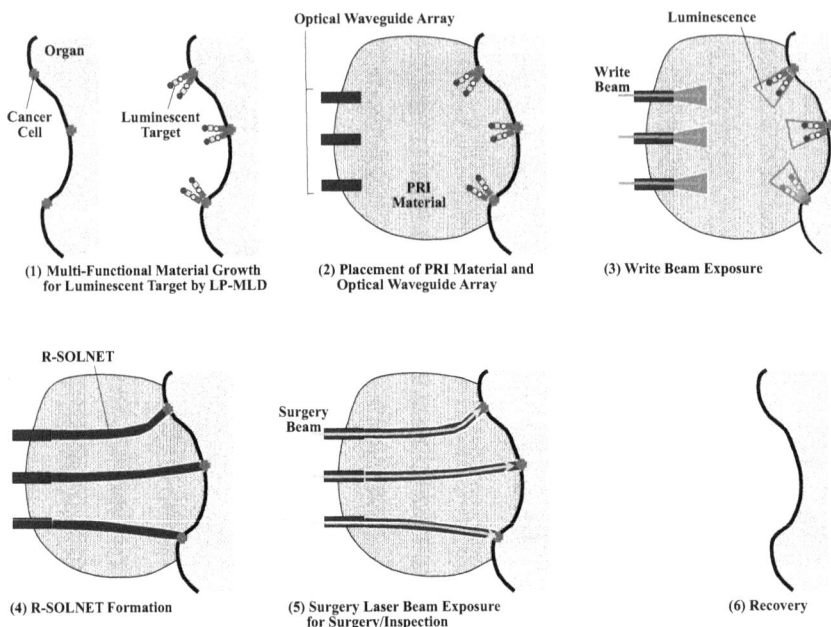

(1) Multi-Functional Material Growth for Luminescent Target by LP-MLD

(2) Placement of PRI Material and Optical Waveguide Array

(3) Write Beam Exposure

(4) R-SOLNET Formation

(5) Surgery Laser Beam Exposure for Surgery/Inspection

(6) Recovery

FIGURE 13.13   Concept of the SOLNET-assisted laser surgery.

generate luminescence to form reflective-SOLNET (R-SOLNET) that connects the optical waveguides to the cancer cells. The detail of R-SOLNET is described in Chapter 10. By introducing surgery laser beams into the R-SOLNETs via the optical waveguides, cancer cells are destroyed selectively. By detecting the backward luminescence emitted from the luminescent targets, monitoring of the cancer destruction might be possible. Wavelengths longer than ~650 nm are preferable for the write beams and the surgery beams because hemoglobin has a high-transmittance window in these wavelengths.

As described in Section 13.3.2, the SOLNET-assisted method is applicable to PDT by adding molecules of sensitizers for PDT as source molecules in LP-MLD.

Figure 13.14 shows performance of the SOLNET-assisted laser surgery simulated by the two-dimensional finite-difference time-domain (2-D FDTD) method. When write beams are emitted from the waveguide cores on the left, luminescence is emitted from the luminescent targets to form R-SOLNETs between the waveguide cores and the nearest targets. Surgery beams emitted from the waveguide cores are guided to the nearest targets. The SOLNET-assisted method might be effective for the removal of scattered small cancer cells.

An experimental demonstration of R-SOLNET formation between an optical fiber with a core diameter of 50-μm and a luminescent target, which corresponds to a cancer cell site, is presented in Figure 13.15 [4]. The luminescent target is made of tris(8-hydroxyquinolinato) aluminum (Alq3) dispersed in polyvinyl alcohol (PVA). The PRI material, which is a mixture of Norland Optical Adhesive NOA65 ($n = 1.52$) and NOA81 ($n = 1.56$), is sensitized by CV. When a 405-nm write beam is emitted

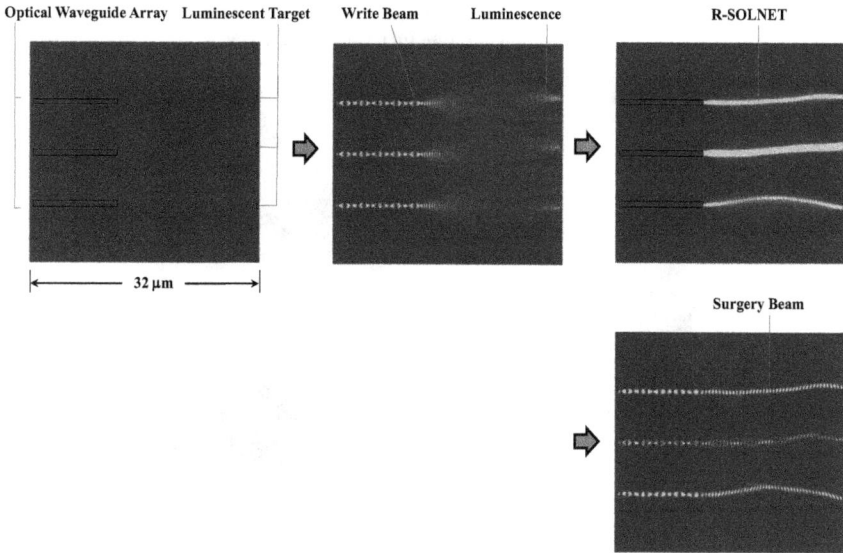

FIGURE 13.14 Performance of the SOLNET-assisted laser surgery simulated by the 2-D FDTD method.

from the optical fiber, green/blue luminescence is generated from the luminescent target. At the same time, red luminescence from CV doped in the PRI material is observed, which enables us to trace R-SOLNET formation. With writing time t_W, SOLNET is formed toward the luminescent target, and finally the optical fiber and the target are connected by R-SOLNET.

13.3.2 TWO-PHOTON AND SOLNET-ASSISTED PHOTODYNAMIC THERAPY (PDT)

In conventional PDT, sensitizers like talaporfin sodium [12] are selectively adsorbed on cancer cells. By exposing the sensitizers to excitation light, they generate singlet oxygen to damage the cancer cells. One concern in the PDT is that the excitation light reaches only a limited depth of objects containing cancer cells.

The two-photon PDT will solve the problem, as shown in Figure 13.16 [1,4,6,14]. Multi-functional materials for the two-photon PDT are formed at cancer sites by successively injecting anchoring molecules and molecules of two-photon PDT sensitizers, which exhibit the two-photon photochemistry, into a human body. The two-photon PDT is carried out as follows. A photon with a wavelength of λ_1 excites an electron from S_0 state to S_n state in the two-photon PDT sensitizer, and the excited electron transfers to the T_1 state. Then, a photon with a wavelength of λ_2 further excites the electron to the T_n state to induce chemical reactions for cancer destruction. This two-step excitation process enables us to use longer wavelength light, which can penetrate into deep parts of the objects, comparing to the one-step excitation case. As depicted in Figure 13.17, the two-step excitation also enables three-dimensional attack on cancer cells since the two-photon photochemical reaction occurs only in

Write Beam Wavelength: 405 nm
Concentration of CV in PRI Material: 0.1wt%

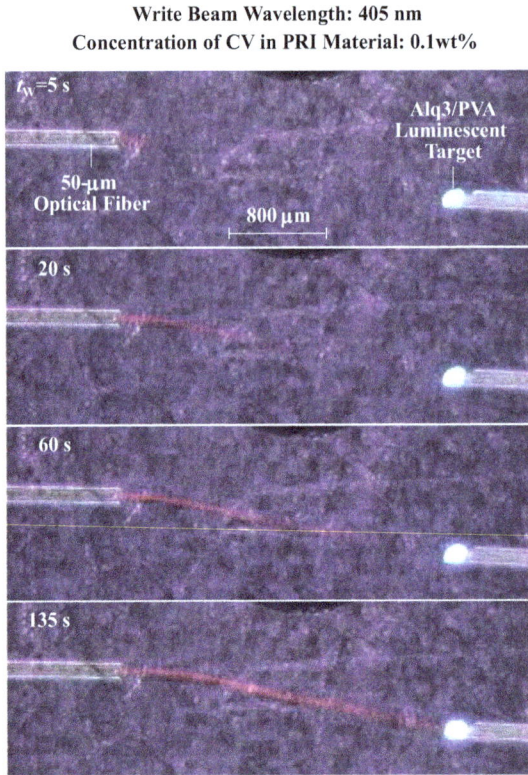

FIGURE 13.15 Experimental demonstration of R-SOLNET formation between an optical fiber with a core diameter of 50-mm and an Alq3/PVA luminescent target, which corresponds to a cancer cell site.

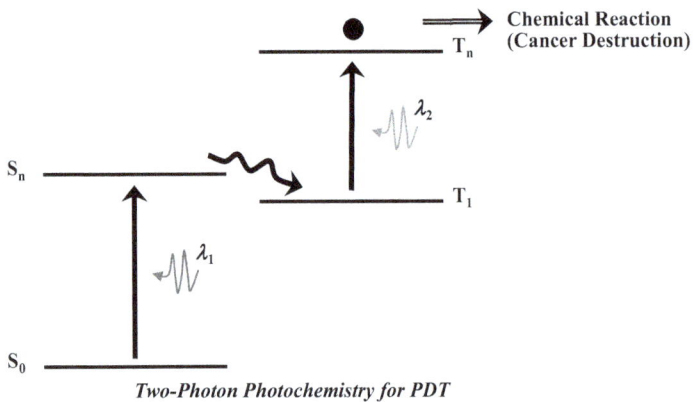

FIGURE 13.16 Concept of two-photon PDT for cancer.

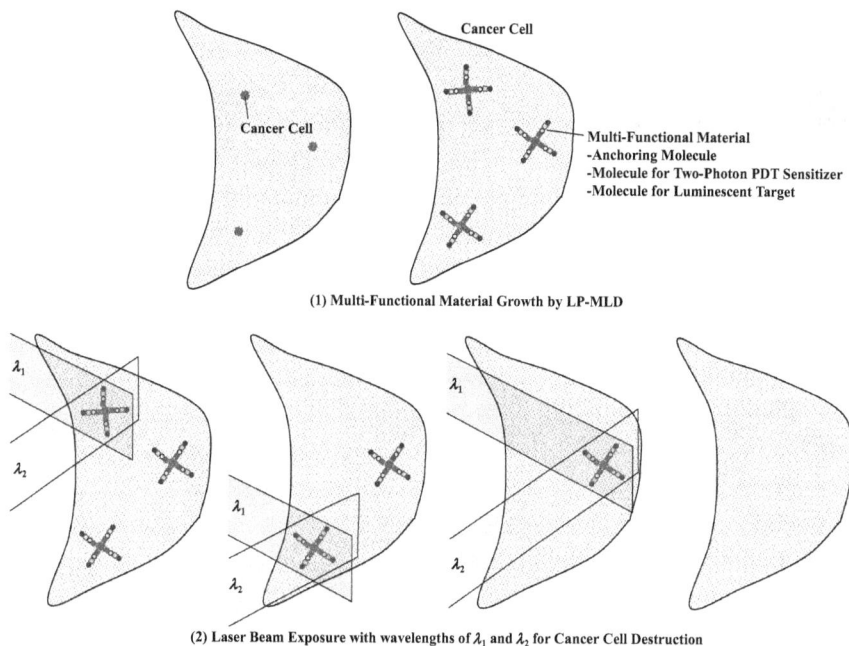

(1) Multi-Functional Material Growth by LP-MLD

(2) Laser Beam Exposure with wavelengths of λ_1 and λ_2 for Cancer Cell Destruction

FIGURE 13.17 Three-dimensional attack on cancer cells by two-photon PDT.

regions, where photons with λ_2 and photons with λ_2 coexist. The regions, where cancer killing is performed, can be selected by controlling the positions of the light beams with the two different wavelengths.

By combining the SOLNET-assisted method presented in Figure 13.13 with the PDT, the SOLNET-assisted PDT shown in Figure 13.18 might be available [14]. In this case, anchoring molecules, molecules of two-photon PDT sensitizers, and molecules for luminescent targets are sequentially provided to grow multi-functional materials on cancer cells. After forming R-SOLNETs that connect optical waveguides to the cancer cells, λ_1-beams and λ_2-beams are guided toward the cancer cells through the R-SOLNETs to destroy them located at deep parts. The SOLNET-assisted PDT can also work in the conventional one-photon PDT. In this case, one-photon PDT sensitizers and guided beams with a single wavelength are employed.

Although many molecules with the two-photon photochemistry function are known, such as porphyrin, biacetyl, camphorquinone, benzyl, etc. [15,16], for applications to the human body, advanced molecules that are solved in blood safely and exhibit light absorption in a wavelength range of the light transmission window of hemoglobin are required. Organic multiple quantum dots (MQDs) could be the candidates of the two-photon sensitizers (see Figure 12.5).

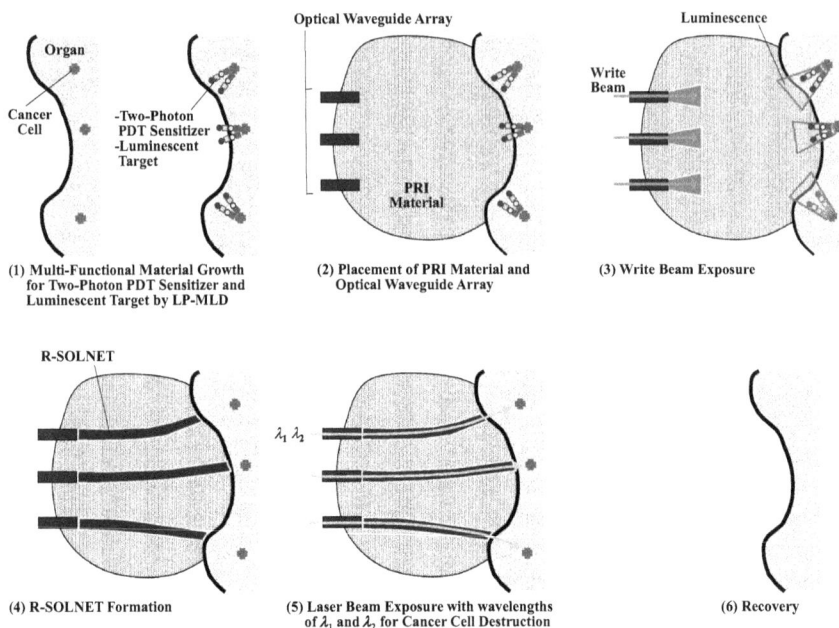

FIGURE 13.18   Concept of SOLNET-assisted PDT for cancer.

REFERENCES

1. T. Yoshimura, "Molecular layer deposition and applications to solar cells and cancer photodynamic therapy," Japanese Patent, Tokukai 2012–45351 (2012) [in Japanese].
2. T. Yoshimura, H. Watanabe, and C. Yoshino, "Liquid-phase molecular layer deposition (LP-MLD): Potential applications to multi-dye sensitization and cancer therapy," *J. Electrochem. Soc.* **158**, 51–55 (2011).
3. T. Yoshimura, C. Yoshino, K. Sasaki, T. Sato, and M. Seki, "Cancer therapy utilizing molecular layer deposition (MLD) and self-organized lightwave network (SOLNET) -proposal and theoretical prediction-," *IEEE J. Select. Top. Quantum Electron* **18**, 1192–1199 (2012).
4. T. Yoshimura, "In-situ drug synthesis at cancer cells for molecular targeted therapy by molecular layer deposition -conceptual proposal-," in *Nano Based Drug Delivery* (Ed. J. Naik), 429–458, IAPC Publishing, Zagreb (2015).
5. T. Yoshimura, "Proposed applications of 3-D optical interconnect technologies to integrated chemical systems," *2007 Digest of the LEOS Summer Topical Meetings [Bio-Inspired Sensors and Application/Imprinting on Photonic Integrated Circuits]*, Portland, Oregon, 129–130 (2007).
6. T. Yoshimura, *Thin-Film Organic Photonics: Molecular Layer Deposition and Applications*, CRC/Taylor & Francis, Boca Raton, Florida (2011).
7. E. M. Rivera, C. T. Provencio, A. Steinbrueck, P. Rastogi, A. Dennis, J. Hollingsworth, and E. Serrano, "Imaging heterostructured quantum dots in cultured cells with epifluorescence and transmission electron microscopy," *Proc. SPIE* **7909**, 79090N (2011).
8. S. A. Boppart, "Magnetomotive molecular probes for targeted contrast enhancement and therapy," *Proc. SPIE* **7910**, 791004 (2011).

9. N. J. Withers, N. N. Glazener, J. B. Plumley, B. A. Akins, A. C. Rivera, N. C. Cook, G. A. Smolyakov, G. S. Timmins, and M. Osinski, "Locally increased mortality of gamma-irradiated cells in presence of lanthanide-halide nanoparticles," *Proc. SPIE* **7909**, 79090L, 2011.

10. J. Yoshida and H. Saya, *Experimental Medicine* 29 (2011) 3378 [in Japanese].

11. H. Kobayashi and P. L. Choyke, "Near-infrared photoimmunotherapy of cancer," *Acc. Chem. Res.* **52**, 2332–2339 (2019).

12. J. Akimoto and K. Aizawa, "Intraoperative photodynamic diagnosis and therapy for malignant glioma using novel photosensitizer Talaporfin," in *Meeting Digest of IPDA*, 10–18, May 28, 2010, IEICE.

13. K. Boeneman, D. E. Prasuhn, J. S. Melinger, M. Ancona, M. H. Stewart, K. Susumu, A. Huston, I. L. Medintz J. B. Blanco-Canosa, and P. E. Dawson, "Quantum dots as a FRET donor and nanoscaffold for multivalent DNA photonic wires," *Proc. SPIE* **7909**, 79090R (2011).

14. T. Yoshimura, *Self-Organized Lightwave Networks: Self-Aligned Coupling Optical Waveguides*, CRC/Taylor & Francis, Boca Raton, Florida (2018).

15. C. Brauchle, U. P. Wild, D. M. Burland, G. C. Bjorkund, and D. C. Alvares, "Two-photon holographic recording with continuous-wave lasers in the 750–1100-nm range," *Opt. Lett.* **7**, 177–179 (1982).

16. V. D. McGinniss and R. E. Schwerzel, "Photopolymerizable composition containing a photosensitive donor and photoinitiating acceptor," U.S. Patent 4,571,377 (1986).

14 Review of Forefront Research on MLD

In Chapters 3–13, after the fundamentals of molecular layer deposition (MLD) are explained, along the author's interests, the organic multiple quantum dot (MQD), which is a typical tailored organic thin-film material grown by MLD, is described, and expected applications of MLD to photonics/electronics, energy conversion, and biomedical/cancer therapy fields are proposed. Meanwhile, MLD has been applied to many other fields. In the present chapter, accomplishments reported by forefront research groups in the world are reviewed, covering the basic research on growth mechanisms, applications to microelectronics such as resists for lithography, diffusion barriers, and flexible insulators/semiconductors/conductors, and applications to molecular sieving/catalysis such as batteries, gas separation/water purification, dry reforming, and photocatalysis. The MLD applications are categorized into five groups: (1) tailored materials grown by MLD, (2) tailored materials grown by MLD combined with atomic layer deposition (ALD), (3) pore-size-controlled organic–inorganic hybrid materials grown by MLD, (4) pore-size-controlled metal oxide materials prepared utilizing MLD, and (5) others. Because there are a lot of variations of MLD processes, they are classified in the initial part. Considerable parts of the present review were prepared by referring to previously-published review papers [1–10]. To make the explanation clear, the author prepared new, modified, or simplified figures based on the original articles.

14.1 MATERIALS GROWN BY MLD

Figure 14.1 summarizes precursor combinations in ALD and MLD. In ALD, there are two types of combinations: "Inorganic Molecule + Inorganic Molecule" and "Metalorganic Molecule + Inorganic Molecule." Both combinations produce inorganic thin films. In MLD, there are three types of combinations, namely, "Organic Molecule + Organic Molecule," "Metalorganic Molecule + Organic Molecule," and "Inorganic Molecule + Organic Molecule." The first combination produces pure organic thin films while the second and third combinations produce organic–inorganic hybrid thin films.

MLD appeared in 1991 [11]. At the initial stage, it was applied to the growth of pure organic thin films such as polyamic acid/polyimide, poly-azomethine (poly-AM), and polyamide [11–14]. Around the middle of the 2000s, MLD began to be applied to the growth of metal organic frameworks (MOFs) and organic–inorganic hybrid thin films, in which metals are used as connections between organic linkers [15–18]. The MOFs and organic–inorganic hybrid thin films have dramatically widened the application fields of MLD.

DOI: 10.1201/9781003094012-14

ALD

MLD

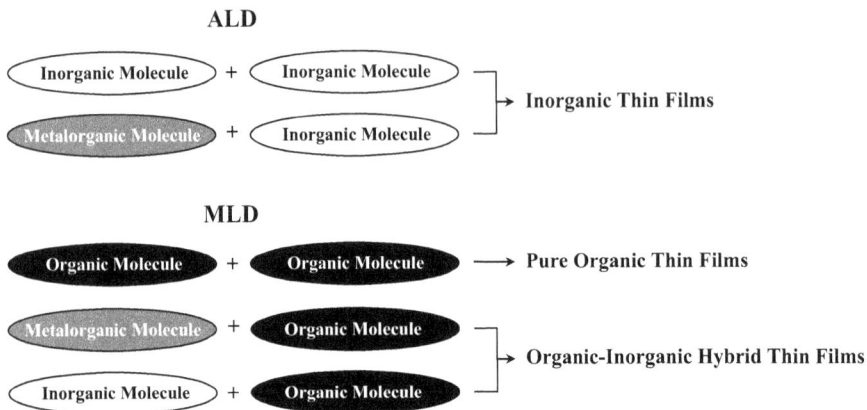

FIGURE 14.1   Precursor combinations in ALD to produce inorganic thin films, and those in MLD to produce pure organic thin films and organic–inorganic hybrid thin films.

Typical MLD processes are presented in Figure 14.2. Figure 14.2a shows pure organic thin film growth by the combination of "Organic Molecule + Organic Molecule." The figure depicts an example of MLD for poly-AM thin film growth with a combination of "terephthalaldehyde (TPA) + p-phenylenediamine (PPDA)" [12]. Bitzer and Richardson achieved growth of polyimide chains standing upright on a clean Si surface by curing an amic intermediate formed using MLD with a combination of "PPDA + pyromellitic dianhydride (PMDA)" [13]. Putkonen et al. succeeded in MLD of polyimide thin films at a substrate temperature of 160°C with a combination of "PMDA + ethylene diamine (ED)" [19].

Figure 14.2b shows organic–inorganic hybrid thin film growth by the combination of "Metalorganic Molecule + Organic Molecule." Alucone thin films are grown with a combination of "trimethyl aluminum (TMA) + ethylene glycol (EG)" [18]. Figure 14.2c shows organic–inorganic hybrid thin film growth by the combination of "Inorganic Molecule + Organic Molecule." Titanicone thin films are grown with a combination of "$TiCl_4$ + 2,4-hexadiyn-1,6-diol (HDD)" [20]. In general, metal alkoxides are called metalcones. They are named referring to contained metals as follows: alucone (contained metal: Al), indicone (contained metal: In), titanicone (contained metal: Ti), zincone (contained metal: Zn), zircone (contained metal: Zr), and so on. Organic–inorganic hybrid materials containing Fe, Mn, Co, Er, or Hf have also been grown by MLD [21–24].

14.2   CLASSIFICATION OF MLD

Although MLD was originally carried out by basic processes shown in Figures 3.5 and 3.6, various advanced processes have been developed to date. These MLD processes are classified in the present section.

Reactivity between reactive groups for MLD precursors is summarized in Figure 14.3. For the basic reactive groups, reactivity between different kinds of groups is large while reactivity between the same kind of groups is small. In addition, in the following reactive group combinations, that is, [small circle, square, or triangle] and

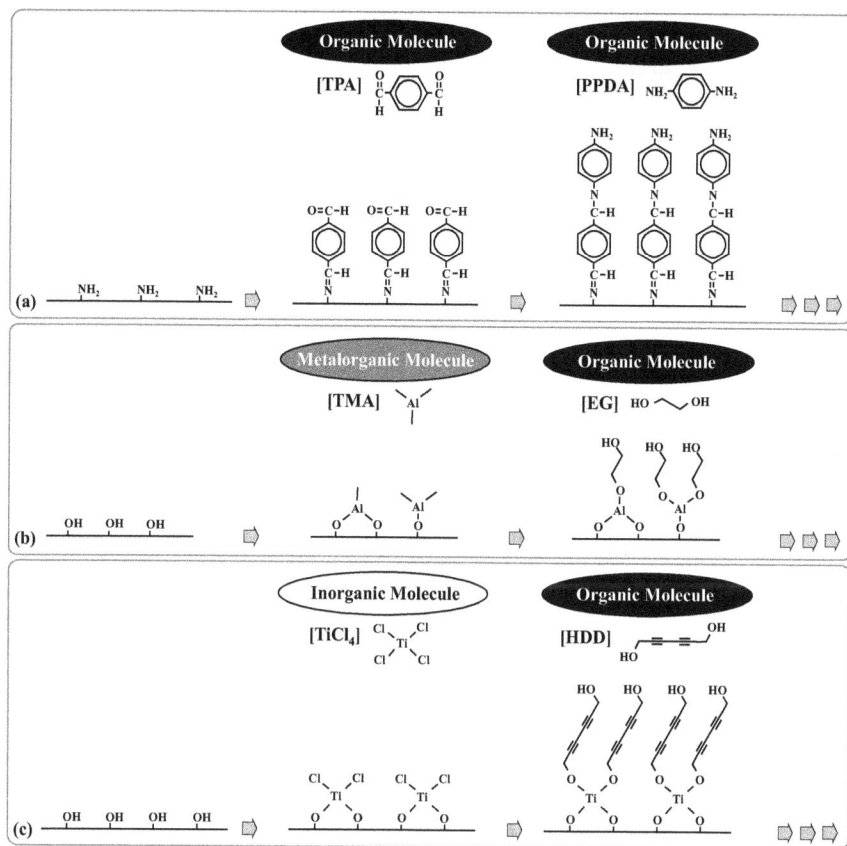

FIGURE 14.2   Typical MLD processes. (a) Pure organic thin film growth by the combination of "Organic Molecule + Organic Molecule," (b) organic-inorganic hybrid thin film growth by the combination of "Metalorganic Molecule + Organic Molecule," and (c) organic-inorganic hybrid thin film growth by the combination of "Inorganic Molecule + Organic Molecule." For (b): (Adapted with permission from [18] A. A. Dameron et al., *Chem. Mater.* **20**, 3315–3326 (2008). Copyright (2008) American Chemical Society.) For (c): (Adapted with permission from [20] K. Yoon et al., *Nanoscale Res. Lett.* **7**, 7–71 (2012).)

[large circle or square], reactivity is small. In the case of MLD utilizing electrostatic force, "Large Reactivity" and "Small Reactivity" are, respectively, replaced with "Attraction" and "Repulsion." For the excitation-assisted reactive groups, they exhibit large reactivity only when they are exposed to external excitation like photons and chemical beams. For the excitation-activated reactive groups, they are normally inert but are activated to exhibit large reactivity once they are exposed to external excitation. In the figure, crosses represent inactivation of reactive groups.

MLD processes are classified into five types, that is, Type-A, B, C, D, and E. These are schematically shown in Figures 14.4–14.8. Type-A MLD presented in Figure 14.4 is the basic MLD with precursors of bifunctional molecules: homo-bifunctional or hetero-bifunctional molecules. These precursors have the basic reactive groups

Basic Reactive Groups

Excitation-Assisted Reactive Groups

Excitation-Activated Reactive Groups

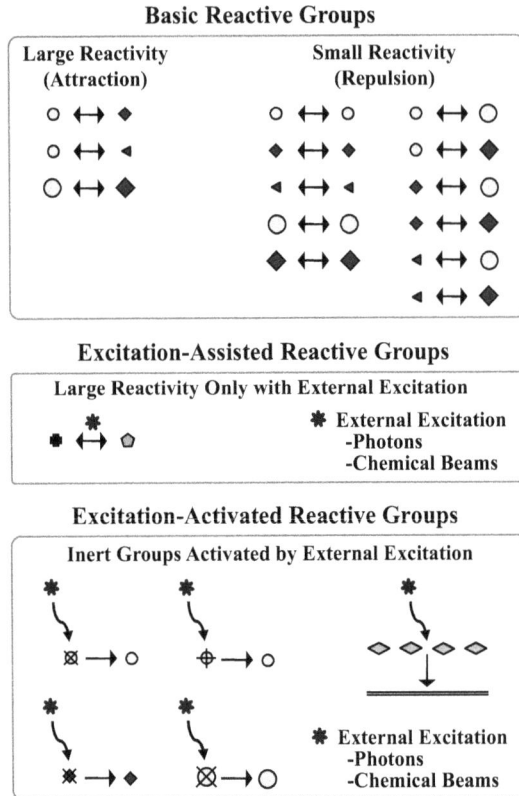

FIGURE 14.3 Summary of reactivity between reactive groups for MLD precursors.

shown in Figure 14.3. In some cases, the MLD process involves steps for attaching side molecules to main wires. Type-B MLD presented in Figure 14.5 uses precursors of multifunctional molecules: homo-multifunctional or hetero-multifunctional molecules, having the basic reactive groups. Type-A MLD and Type-B MLD are carried out by sequentially providing precursors with designated sequences to a substrate, whose surface is terminated by reactive groups exhibiting large reactivity with reactive groups in Molecule A, to form tailored organic thin films with designated molecular arrangements.

Figure 14.6 shows Type-C MLD, in which the chemical reaction is assisted by simultaneous external excitation during the film growth. Precursors have the excitation-assisted reactive groups shown in 14.3. The external excitation could be photons and chemical beams. In the case of photons, Type-C MLD is the photoactivated MLD (*p*MLD), where the film growth is performed under light exposure.

Figure 14.7 shows Type-D MLD, in which processes to activate inactivated reactive groups in precursors by external excitation are involved. Precursors have the excitation-activated reactive groups shown in Figure 14.3. In Type-D MLD <Molecules with Homo-Reactive Groups>, firstly, Molecule A is supplied to a substrate to form a monomolecular layer A. Here, Molecule A has same reactive groups, and one of them

Type-A MLD: Utilizing Bifunctional Molecules

<Homo-Bifunctional Molecules>

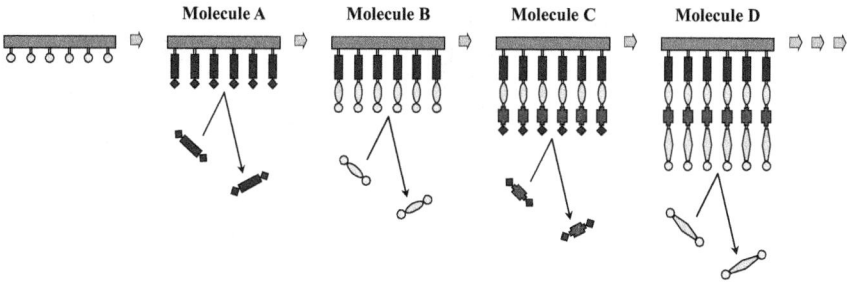

<Hetero-Bifunctional Molecules>

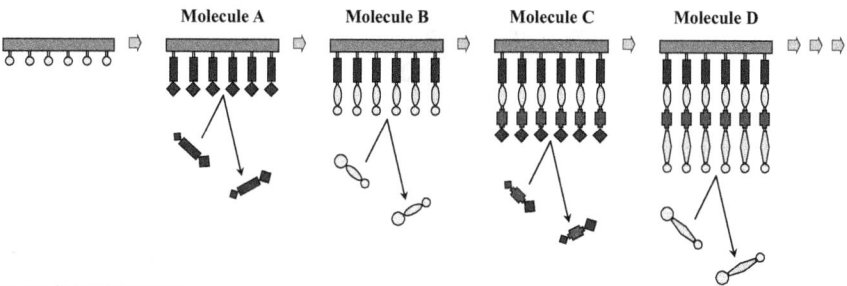

<with Side Molecules>

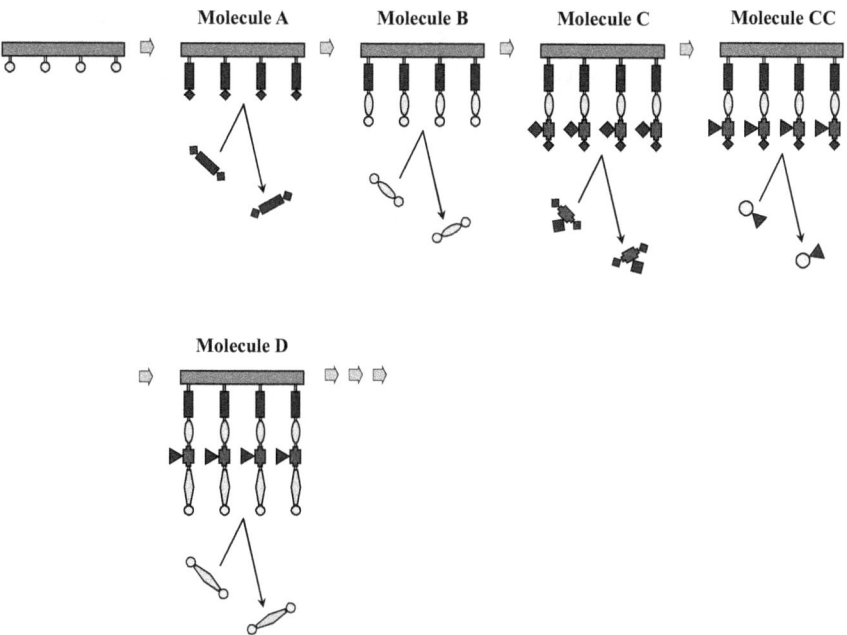

FIGURE 14.4 Type-A MLD, which is the basic MLD with precursors of homo-bifunctional molecules or hetero-bifunctional molecules. In some cases, the MLD process involves steps for attaching side molecules to main wires.

Type-B MLD: Utilizing Multifunctional Molecules

<Homo-Multifunctional Molecules>

<Hetero-Multifunctional Molecules>

FIGURE 14.5 Type-B MLD, which uses precursors of homo-multifunctional molecules or hetero-multifunctional molecules.

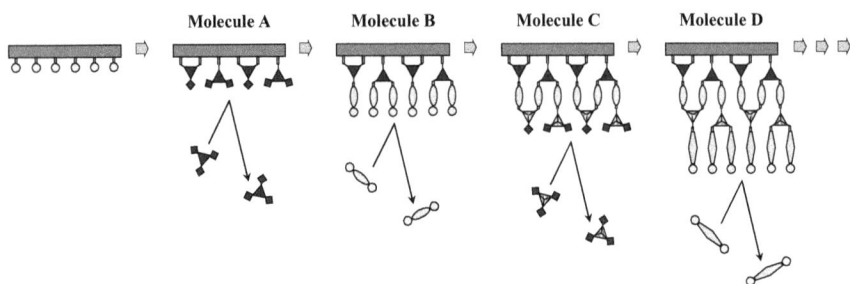

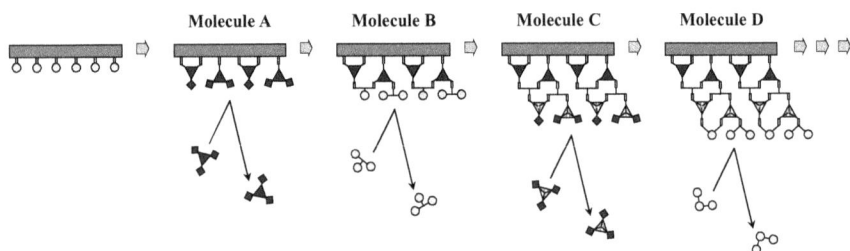

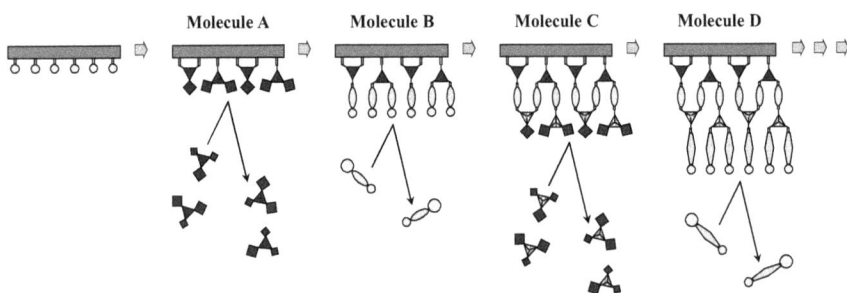

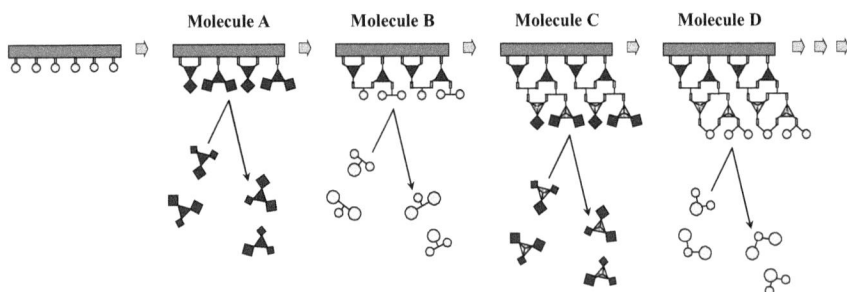

is inactivated to be the excitation-activated reactive group. Consequently, the inactivated reactive groups appear at the surface. By introducing external excitation 1, the inactivated reactive groups at the surface are activated. Next, Molecule B is supplied to the surface to form a monomolecular layer B on A. Molecule B has the same

Type-C MLD: Assisted by *External Excitation*

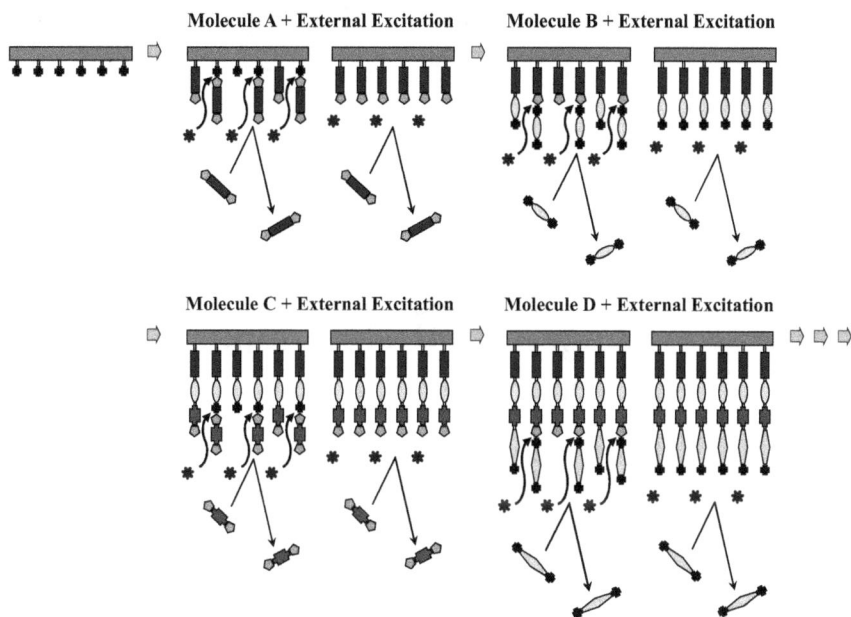

Molecule A + External Excitation

Molecule B + External Excitation

Molecule C + External Excitation

Molecule D + External Excitation

FIGURE 14.6 Type-C MLD, in which the chemical reaction is assisted by simultaneous external excitation during the film growth. Molecules having excitation-assisted reactive groups shown in Figure 14.3 are used for the precursors.

reactive groups, which are different from those of Molecule A, and one of them is inactivated. By introducing external excitation 2, the inactivated reactive groups at the surface are activated. By repeating the similar process, organic tailored thin films are formed. In Type-D MLD <Molecules with Hetero-Reactive Groups 1>, Molecule A, which has different reactive groups and one of them is inactivated, is supplied to form a monomolecular layer A. By introducing external excitation 1, the inactivated reactive groups appearing at the surface are activated. Next, Molecule B is supplied to form a monomolecular layer B on A. Molecule B has different reactive groups, which are different from those of Molecule A, and one of them is inactivated. By introducing external excitation 2, the inactivated reactive groups at the surface are activated. The similar process is repeated. Type-D MLD <Molecules with Hetero-Reactive Groups 2> is almost the same as Type-D MLD <Molecules with Hetero-Reactive Groups 1> except that Molecules A, B, C, and D have common inactivated reactive groups. The inactivated reactive groups are represented by small crossed circles. In this case, common reactive groups represented by small circles always appear at the surface after each external excitation step. Type-D MLD <Molecules with Groups for Horizontal Connection> is also carried out by using precursors with the excitation-activated reactive groups. In the example depicted in Figure 14.7, Molecules B and D have the excitation-activated reactive groups. By introducing external excitation sequentially, horizontal polymerization is induced to form horizontal connections.

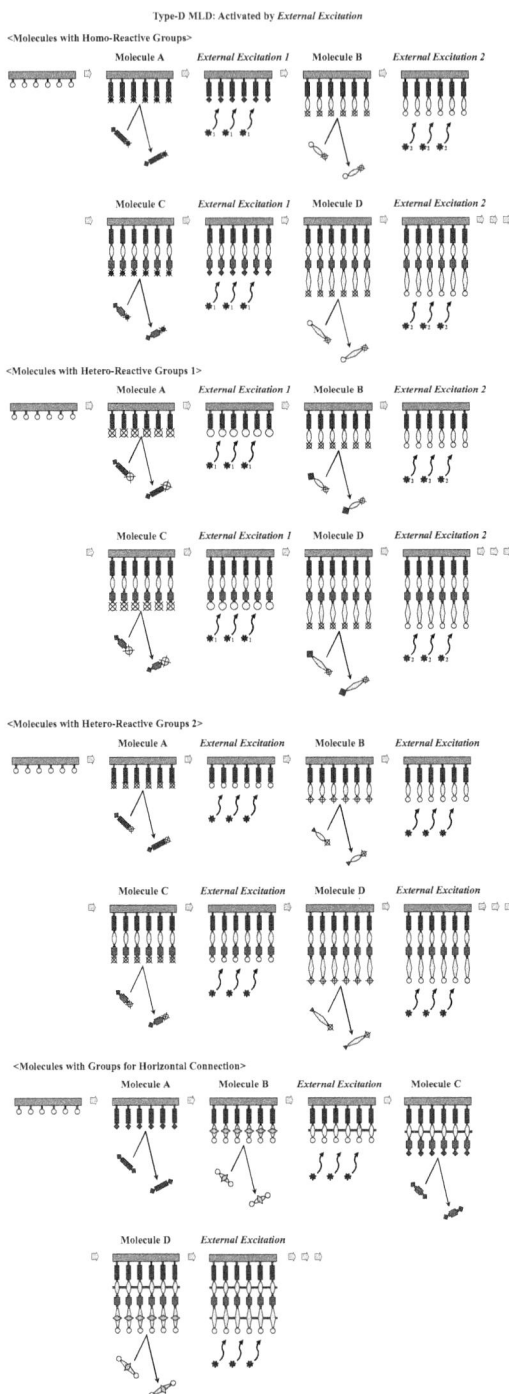

FIGURE 14.7 Type-D MLD, in which sequential external excitation is involved to activate inactivated reactive groups in precursors. Molecules having excitation-activated reactive groups shown in Figure 14.3 are used for the precursors.

Type-E MLD: Utilizing Ring-Opening Molecules

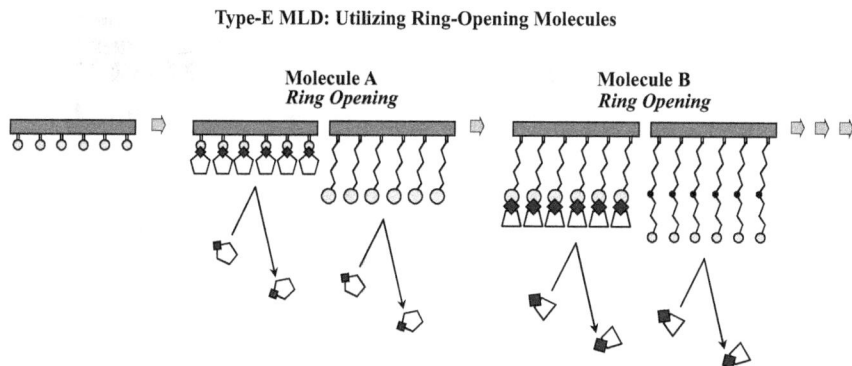

FIGURE 14.8 Type-E MLD utilizing ring-opening molecules, which is regarded as a special case of Type D MLD.

Figure 14.8 shows Type-E MLD utilizing precursors of ring-opening molecules. When the precursors react with reactive groups at the surface, the rings of the precursors open to produce new reactive groups at the surface. This process can be regarded as a special case of Type-D MLD with sequential external excitation. Namely, the reactive groups at the surface contribute to connecting molecules, and at the same time, they act as the external excitation.

14.3 CONCERNS AND SOLUTIONS IN MLD

According to the growth mechanism of MLD described in Section 3.4, the growth rate of MLD thin films is determined by molecular sizes and molecular orientations, as schematically illustrated in Figure 3.14. However, the actual growth rate sometimes deviates from the ideal rate. It has been revealed that the deviation is caused by two factors: the double reaction and the uptake of molecules [1,18,25]. Another concern in the MLD process is the limited choice of precursors [26]. In the present section, these two concerns and the solutions of them are reviewed.

14.3.1 DOUBLE REACTION AND UPTAKE OF MOLECULES

In the ideal MLD model shown in Figure 3.5, when Molecule A is provided on a substrate surface, a monomolecular layer A is formed and reactive groups in Molecule A appear at the surface as reactive sites. When Molecule B is provided on the monomolecular layer A surface, a monomolecular layer B is formed and reactive groups in Molecule B appear at the surface as new reactive sites. The similar process is repeated for Molecules C, D, and so on to allow the MLD process to proceed normally.

The double reaction breaks the normal MLD process, as schematically illustrated in Figure 14.9a [1,18,25]. When Molecule B is provided on a monomolecular layer A, both reactive groups in the molecule are sometimes connected to two reactive sites at the surface, causing the double reaction. The double reaction prevents creation of new reactive sites, decreasing the number of reactive sites at the monomolecular layer B surface. With the progress of the MLD step, the number of reactive sites decreases more and more to suppress the growth rate. Once the reactive site density reaches a

(a) Decrease in Reactive Sites by Double Reaction

(b) Increase in Reactive Sites by Uptake of Molecules

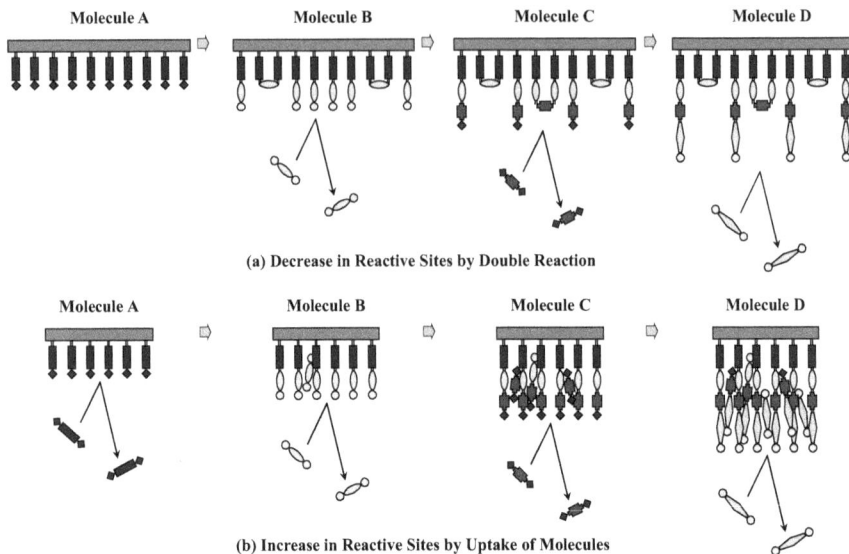

FIGURE 14.9   Concerns in MLD. (a) A decrease in reactive sites by the double reaction and (b) an increase in reactive sites by the uptake of excess molecules by an already-deposited film.

value low enough not to cause further double reaction, the growth rate becomes constant. This situation appears after the step for Molecule D in Figure 14.9a.

Another factor that breaks the normal MLD process is the uptake of excess molecules by the already-deposited film [1,18,25]. As schematically illustrated in Figure 14.9b, when incident molecules are adsorbed by the film, they create excess reactive sites. The sites initiate excess growth of polymer wires to increase the growth rate, breaking the normal MLD process. Because the desorption of the excess molecules is promoted by raising the substrate temperature, the growth rate decreases with increasing the temperature.

Bergsman et al. established a model, in which the film growth termination caused by the double reaction is compensated by the uptake of incident molecules to continue MLD film growth [27]. They set up a rate equation, which involves the number of reactive sites and the rate constants for the double reaction and the uptake of molecules, and derived the following results for certain parameters:

- Three percent of the chains experience termination events during each MLD cycle.
- Any given chain has ~50% chance to be terminated before reaching 22 MLD cycles.

Figure 14.10 schematically depicts film thickness vs. number of MLD cycles, n_{MLD}, characteristics. Reactive site termination by the double reaction causes a growth rate decrease while reactive site creation by the uptake of molecules causes a growth rate increase. When the double reaction and the uptake of molecules coexist, the growth rate reaches a steady state, where the growth rate is constant. The rate is determined by the balance of the termination rate and the molecule uptake rate. Similar consideration based on the picosecond acoustics analysis was reported [28].

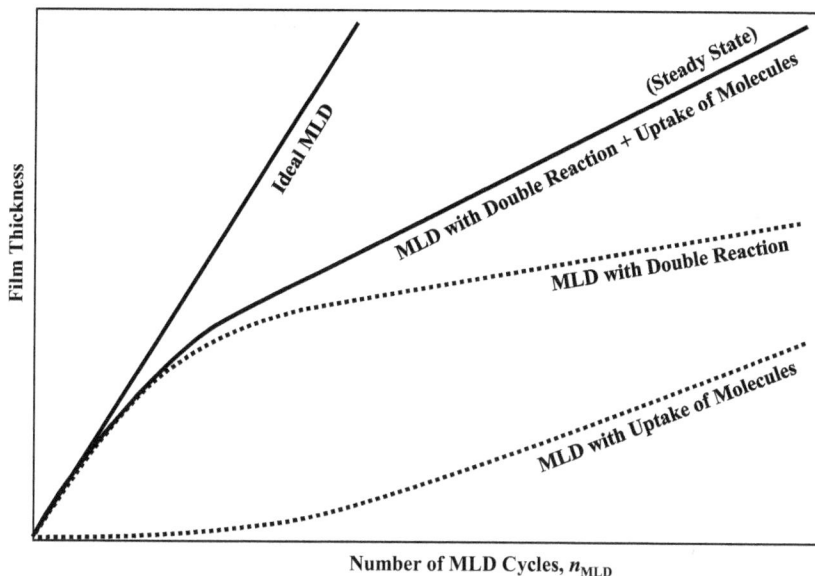

FIGURE 14.10 Schematic illustrations of film thickness vs. number of MLD cycles characteristics. (Adapted with permission from [27] D. S. Bergsman et al., *Chem. Mater.* **30**, 5087–5097 (2018). Copyright (2018) American Chemical Society.)

For the ideal MLD, the double reaction and the uptake of incident molecules should be prevented. Ban and George reported that the double reaction can be avoided by introducing aromatic molecules with rigid structures [29]. Bergsman et al. grew polyurea MLD thin films using *p*-phenylene diisocyanate (PDIC) or 1,4-diisocyana-tobutane (DICB) for Molecule A and PD or 1,4-diaminobutane (BD) for Molecule B [30]. The molecular structures are drawn in Figure 14.11. They found that the film growth rate decreases in the order of polyurea [PDIC, PD], [PDIC, BD], and [DICB, BD], indicating that the growth rate decreases with increasing the backbone flexibility. These results suggest that the double reaction can be suppressed by using rigid molecules for precursors that induce steric hindrance. For the uptake of molecules, it is expected to be solved by optimizing the substrate temperature.

Novel MLD processes, which are effective to solve the double reaction problem, have been developed. Figure 14.12 shows an MLD process developed by Lee et al. [16]. It is classified into Type-D MLD <Molecules with Hetero-Reactive Groups 2>. On a surface terminated by -OH, 7-octenyltrichlorosilane (7OTS) and H_2O are provided as Molecule A to form a monomolecular layer A. The monomolecular layer is exposed to external excitation 1 of O_3 to activate the groups at the surface. After titanium isopropoxide (TIP) is provided as Molecule B to form a monomolecular layer B, the layer is exposed to external excitation 2 of H_2O to activate the groups at the surface. By repeating these processes, MLD of an organic–inorganic hybrid thin film is accomplished with suppressed double reaction because one of the reactive groups in each incident molecule is inactive until it is exposed to the external excitation.

Figure 14.13 shows an MLD process developed by Adamczyk et al. [25]. The process is classified into Type-E MLD. On a surface terminated by -OH, azasilane

Molecule A

[PDIC] [DICB]

Molecule B

[PD] [BD]

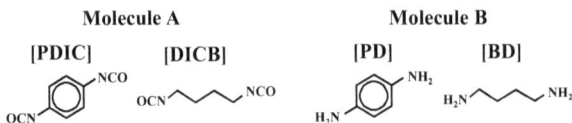

FIGURE 14.11 Precursors for growth of polyurea MLD thin films. The growth rate decreases in the order of polyurea [PDIC,PD], [PDIC,BD], and [DICB,BD], indicating that the growth rate decreases by increasing the backbone flexibility [30].

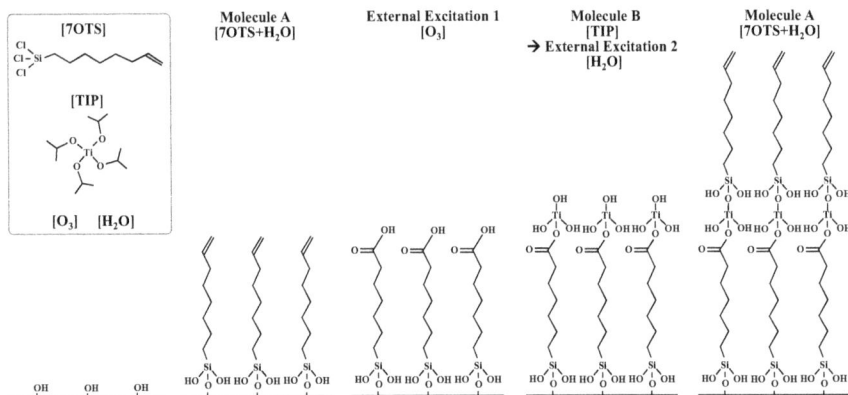

[7OTS]

[TIP]

[O₃] [H₂O]

Molecule A
[7OTS+H₂O]

External Excitation 1
[O₃]

Molecule B
[TIP]
→ External Excitation 2
[H₂O]

Molecule A
[7OTS+H₂O]

FIGURE 14.12 Type-D MLD <Molecules with Hetero-Reactive Groups 2>, which is effective to suppress the double reaction. (Adapted with permission from [16] B. H. Lee et al., *J. Am. Chem. Soc.* **129**, 16034–16041 (2007). Copyright (2007) American Chemical Society.)

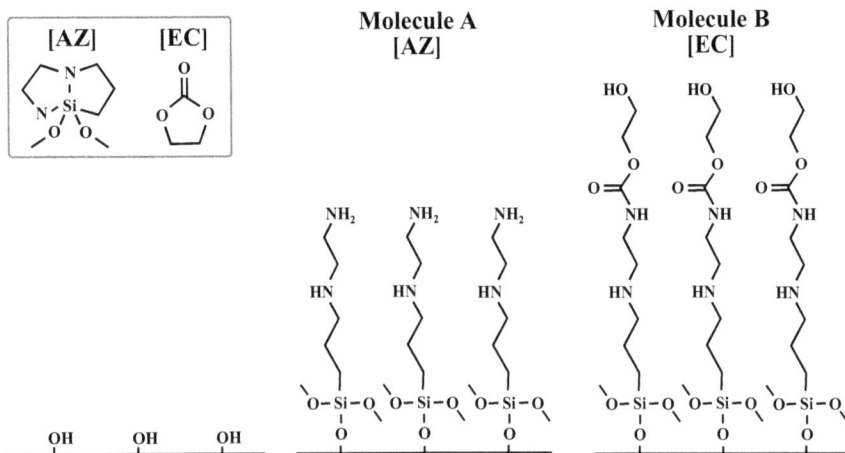

[AZ] [EC]

Molecule A
[AZ]

Molecule B
[EC]

FIGURE 14.13 Type-E MLD, which is effective to suppress the double reaction. (Adapted with permission from [25] N. M. Adamczyk et al., *Langmuir* **24**, 2081–2089 (2008). Copyright (2008) American Chemical Society.)

(AZ) is provided as Molecule A. When AZ reacts with -OH at the surface, the AZ molecule opens to generate a reactive group of -NH$_2$ at the surface of the monomolecular layer A. Next, ethylene carbonate (EC) is provided as Molecule B. When EC reacts with -NH$_2$ at the surface, the EC molecule opens to generate a reactive group of -OH at the surface of the monomolecular layer B. By repeating these processes, an MLD thin film can be grown without the double reaction because each incident molecule has only one reactive group, and the reactive group cannot react with the newly-generated reactive groups at the surface.

Yoon et al. developed an MLD process called ABC Cycle MLD [31] that combines Type-E MLD with Type-A MLD <Hetero-Bifunctional Molecules>. The concept is shown in Figure 14.14. On a monomolecular layer A, hetero-bifunctional Molecule B is provided. In this step, no double reaction occurs because one of the two different reactive groups in Molecule B cannot be connected to reactive sites on the surface. When Molecule C is provided on the monomolecular layer B, Molecule C opens to generate a new reactive group at the surface of the monomolecular layer C. By repeating these processes, an MLD thin film can be grown without the double reaction. Figure 14.15 shows a case where TMA, ethanolamine (EA), and maleic anhydride (MA) are used for Molecule A, B, and C, respectively.

Type-E MLD + Type-A MLD <Hetero-Bifunctional Molecules>

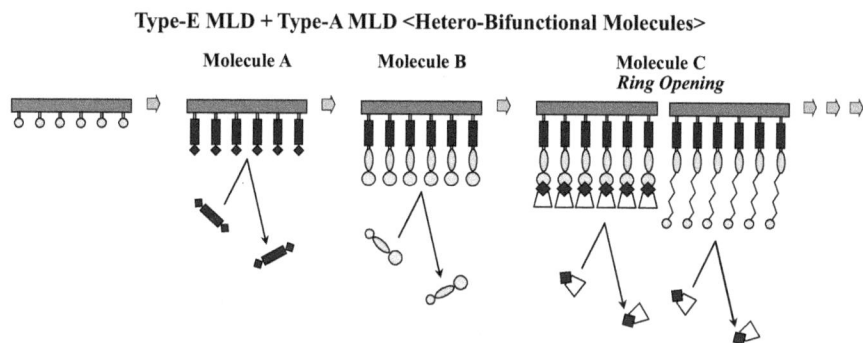

FIGURE 14.14 Concept of ABC Cycle MLD that combines Type-E MLD with Type-A MLD <Hetero-Bifunctional Molecules>.

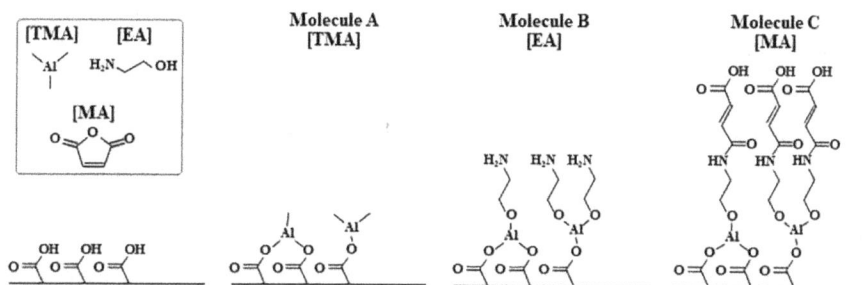

FIGURE 14.15 ABC Cycle MLD, which is effective to suppress the double reaction. (Adapted with permission from [31] B. Yoon et al., *Chem. Mater.* **21**, 5365–5374 (2009). Copyright (2009) American Chemical Society.)

14.3.2 Limited Choice of Precursors

In most of MLD processes, reactions between an electrophile (acid anhydride, chloride, aldehyde, isocyanate, isothiocyanate) and a nucleophile (amine, alcohol) are utilized to form polymers, and consequently, the choice of reactive molecules for precursors is limited. Photo-assisted MLD, which is classified into Type-C MLD with the external excitation of photons, extends the choice of available reactions for precursors and resultant products.

Li et al. demonstrated MLD utilizing the thiol-ene photoinduced coupling [32]. Figure 14.16 shows MLD utilizing octane-diene (C8-DE) for Molecule A and 1,4-benzenedimethanethiole (BDT) for Molecule B. Thiol-ene photoinduced coupling occurs via a free-radical chain mechanism. Upon ultra-violet (UV) irradiation at 254 nm, heterolytic cleavage of the thiol group produces thiyl radical, which causes the chain reaction. Lillethorup et al. succeeded in pMLD with the radical step-growth polymerization by using 1,3-diiodopropane (DIP) and ethylene glycol dimethacrylate (EGM) [26]. pMLD films were grown by the photoactivated iodo-ene process shown in Figure 14.17. Photoactivated homolysis of the C-I bond induced

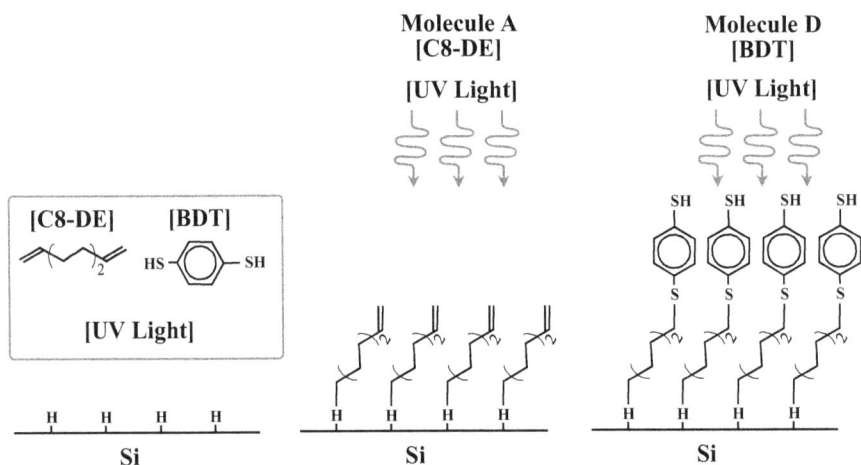

FIGURE 14.16 Type-C MLD with the external excitation of photons utilizing the thiol-ene photoinduced coupling. (Adapted with permission from [32] Y.-H Li et al., *Langmuir* **26**, 1232–1238 (2010). Copyright (2010) American Chemical Society.)

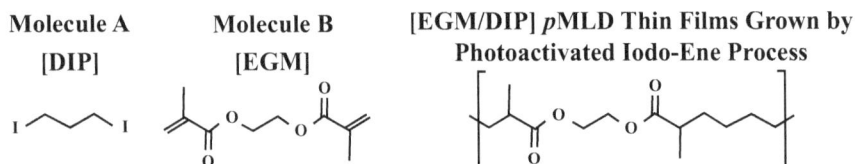

FIGURE 14.17 Photoactivated iodo-ene process in pMLD. (Adapted with permission from [26] M. Lillethorup et al., *Chem. Mater.* **29**, 9897–9906 (2017). Copyright (2017) American Chemical Society.)

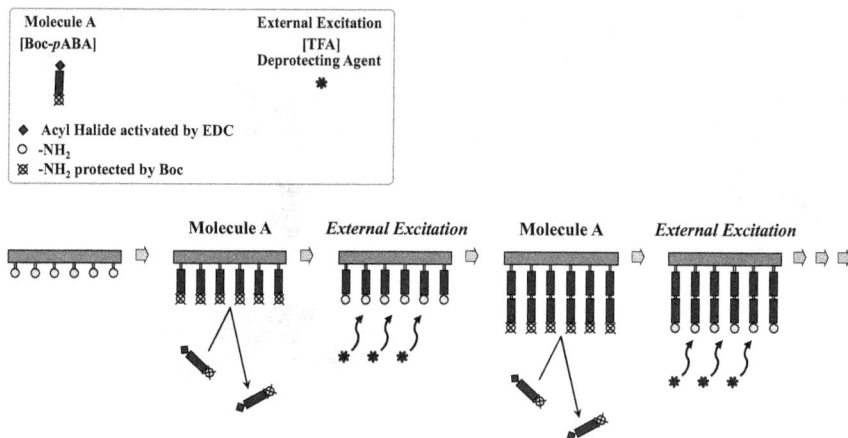

FIGURE 14.18   Concept of Type-D MLD <Molecules with Hetero-Reactive Groups 2> performed in liquid phase. This LP-MLD process is useful to extend the choice of reactive molecules for precursors, and at the same time, to prevent the double reaction.

in DIP by 250 nm UV light is involved in the *p*MLD process, enabling C–C bond creation without catalysts.

Liquid-phase MLD (LP-MLD) also extends the choice of reactive molecules for precursors because it can use precursors that are hard to be evaporated. Fichtner et al. performed LP-MLD, as schematically illustrated in Figure 14.18 [33]. The process is classified into Type-D MLD <Molecules with Hetero-Reactive Groups 2>. Molecule A is Boc-*p*ABA. Here, *p*ABA is *p*-aminobenzoic acid and Boc is tert-butyloxycarbonyl. Boc-*p*ABA has two reactive groups. One is -NH$_2$ protected by Boc, and the other is acyl halide activated by 1-ethyl-3-(3-dimethylaminopropyl)carbodiimide (EDC). By introducing Molecule A onto a substrate surface terminated by -NH$_2$ in liquid, a monomolecular layer A is formed and the -NH$_2$ groups protected by Boc appear at the surface. When external excitation of trifluoroacetic acid (TFA), which is a deprotecting agent, is introduced onto the surface, the -NH$_2$ groups are deprotected. Molecule A is introduced onto the surface to form the second monomolecular layer A. By repeating these processes, polyamide thin films are grown. The LP-MLD process is useful to widen the precursor choice, and at the same time to prevent the double reaction.

14.4   FEATURES AND APPLICATION FIELDS OF MLD

In Figure 14.19, features and application fields of MLD are summarized. As mentioned in Chapter 5, MLD has (1) "tailored organic material growth" capability, which enables optical/electronic, chemical/biomedical, and mechanical/thermal property control. MLD also has (2) "ultrathin/conformal organic material growth" capability and (3) "selective organic material growth" capability. These three features are preferable in applications to photonics/electronics, energy conversion, biomedical/cancer therapy, and microelectronics fields. In addition, MLD has (4) "pore-size-controlled material growth" capability, which enables molecular/atomic sieving and surface area increase.

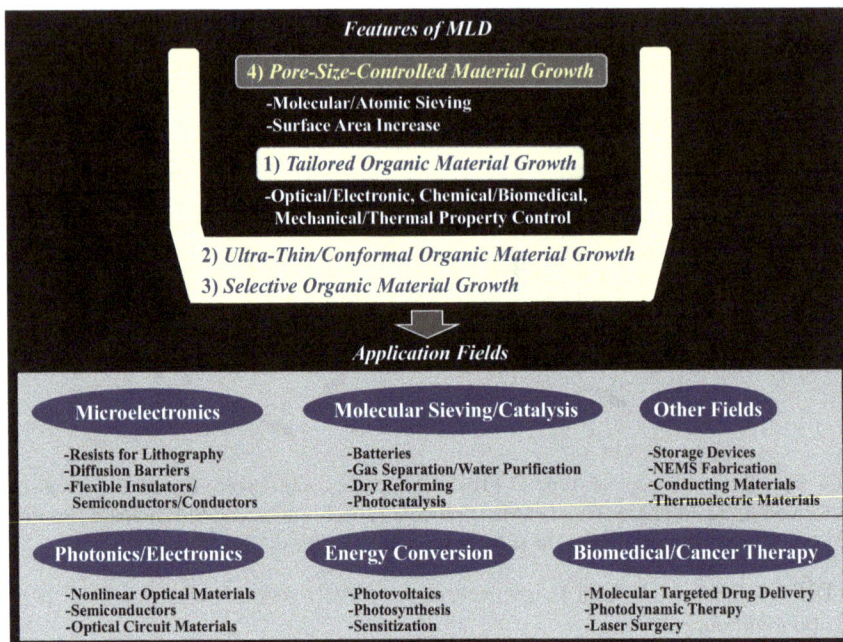

FIGURE 14.19 Features and application fields of MLD

For the applications of MLD to the photonics/electronics field (nonlinear optical materials, semiconductors, optical circuit materials), the energy conversion field (photovoltaics, photosynthesis, sensitization), and the biomedical/cancer therapy field (molecular targeted drug delivery, photodynamic therapy, laser surgery), detailed description is presented in Chapters 3–13 based on the study performed along the author's research strategy. Applications to other fields, however, are not presented in the previous chapters, including the microelectronics field (resists for lithography, diffusion barriers, flexible insulators/semiconductors/conductors), the molecular sieving/catalysis field (batteries, gas separation/water purification, dry reforming, photocatalysis), and other fields (storage devices, nanoelectro-mechanical systems (NEMS) fabrication, conducting materials, thermoelectric materials). These are reviewed in the present chapter based on the achievements reported by forefront research groups.

The applications are categorized as follows:

- Tailored materials grown by MLD.
- Tailored materials grown by MLD combined with ALD.
- Pore-size-controlled organic–inorganic hybrid materials grown by MLD.
- Pore-size-controlled metal oxide materials formed utilizing MLD.
- Others.

In most of the applications described in the present chapter, the feature of "ultrathin/conformal organic material growth" capability is commonly important. As a typical example, nanoparticle surface coating reported by Liang and Weimer is presented schematically in Figure 14.20 [34]. Quenching of the photocatalytic activity of TiO_2

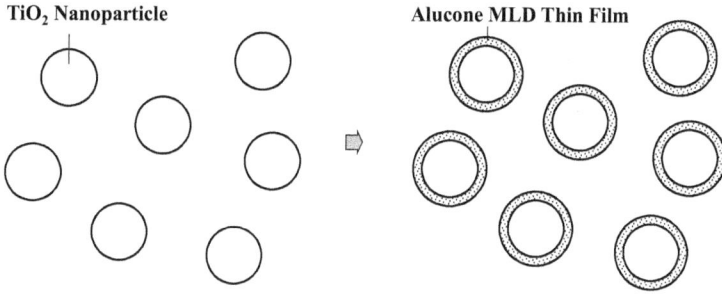

FIGURE 14.20   Schematic illustration for coating of nanoparticles with ultrathin/conformal organic films.

nanoparticles was achieved by forming 20-nm-thick conformal alucone MLD thin films on the nanoparticle surfaces.

14.5   TAILORED MATERIALS GROWN BY MLD

The feature of "tailored organic material growth" capability of MLD is essential in the microelectronics field as in the photonics/electronics, energy conversion, and bio-medical/cancer therapy fields. In the present section, applications of tailored materials grown by MLD to microelectronics are reviewed.

14.5.1   RESISTS FOR LITHOGRAPHY

Zhou and Bent developed a novel way to produce molecularly-designed polyurea-based chemically-amplified (CA) resists for lithography. In the resist thin films, which are highly-conformal and homogeneous, the composition and functionality are precisely controlled with a monomolecular level [35–38].

Figure 14.21 shows the concept of a photoresist thin film with tunable acid-cleavage sites grown by MLD with Molecules A, B, and D. Here, Molecule D is an acid-labile molecule that is cleaved by acid treatment. When no acid-labile molecules are used in the MLD process, the MLD thin film stays in the initial state after the acid treatment. When Molecule D of the acid-labile molecule is used at a certain MLD step, the MLD thin film breaks into two portions at the position, where the acid-labile molecules are inserted, and the upper portion is removed away after acid treatment. The breaking position can be controlled with a monomolecular level.

An MLD process for photoresist thin film growth is shown in Figure 14.22. This process is Type-A MLD <Homo-Bifunctional Molecules>. For Molecules A and B, PDIC and ED are used, respectively. For Molecule D, 2,2'-(propane-2,2-diylbis(oxy)) diethanamine (PDDE) is used. PDDE is the acid-labile molecule. By the acid treatment, PDDE is cleaved and the upper portion of the resist thin film is removed.

In order to use a photoresist thin film in photolithography processes, photo-acid-generators (PAGs) such as triphenylsulfonium triflate (TPSOTf) are diffused into the film by, for example, soaking, as shown in Figure 14.23. In the parts exposed to UV light, photo-generated acid from the PAGs diffuses to acid-labile molecules

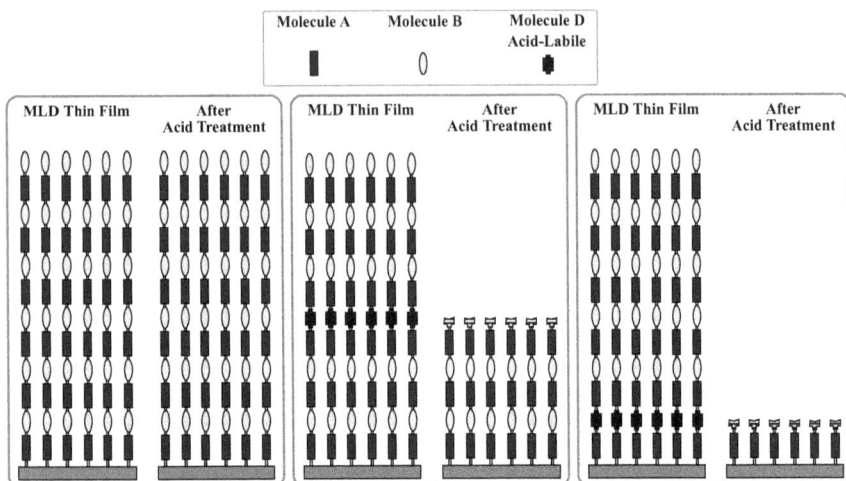

FIGURE 14.21 Concept of a photoresist thin film with tunable acid-cleavage sites grown by MLD.

to break them. By development with the acid treatment, films of UV-exposed areas are removed. In this case, however, a problem was found, that is, the photoresist thin films are not completely removed due to limited lengths of PAG diffusion in the soaking process. Well-packed MLD thin films exhibit high surface-protecting ability while they do not allow dopants to diffuse efficiently into the films. Namely, a trade-off relationship exists between the surface-protecting ability and the PAG diffusion ability.

Zhou et al. overcame the trade-off problem by embedding the acid-labile molecules and the PAG molecules at designated positions in the polymer wire backbones. The concept is presented in Figure 14.24. Molecules A, B, D, and FF are used for MLD precursors. Here, Molecule FF is the PAG molecule. By switching source molecules at designated MLD steps, acid-labile molecules and PAG molecules can be inserted at intended positions. This solves the problem of the trade-off relationship between the surface-protecting ability and the PAG diffusion ability to realize ideal resist thin films.

An MLD process to grow a resist thin film with acid-labile molecules and PAG molecules embedded at designated positions in polymer backbones is shown in Figure 14.25. This process is Type-A MLD <with Side Molecules>. Molecules A and B are PDIC and ED, respectively. Molecule D of the acid-labile molecule is PDDE. Molecule F is the receptor to attach Molecule FF for PAG formation to the polymer wire as a side molecule. Molecules F and FF are, respectively, 2,2′-thiobis(ethylamine) (TBEA) and methyl triflate (MeOTf). For Steps 1–3, the process is the same as that shown in Figure 14.22. In Steps 4 and 5, Molecule F and Molecule FF are successively introduced to produce trialkylsulfonium triflate (TAST), which is the PAG molecule attached to the main polymer wires. The MLD

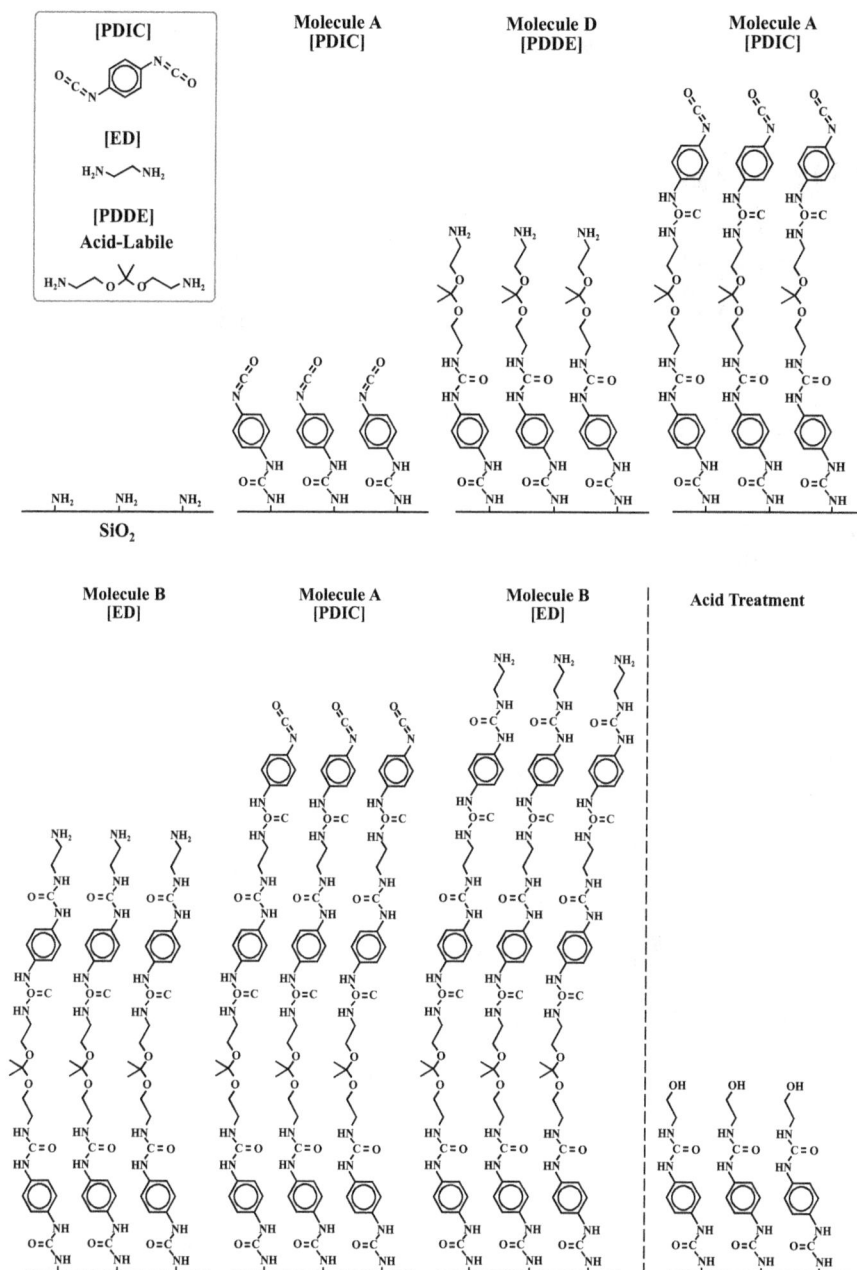

FIGURE 14.22 An MLD process for photoresist thin film growth. (Adapted with permission from [35] H. Zhou et al., *ACS Appl. Mater. Interfaces* **3**, 505–511 (2011). Copyright (2011) American Chemical Society.)

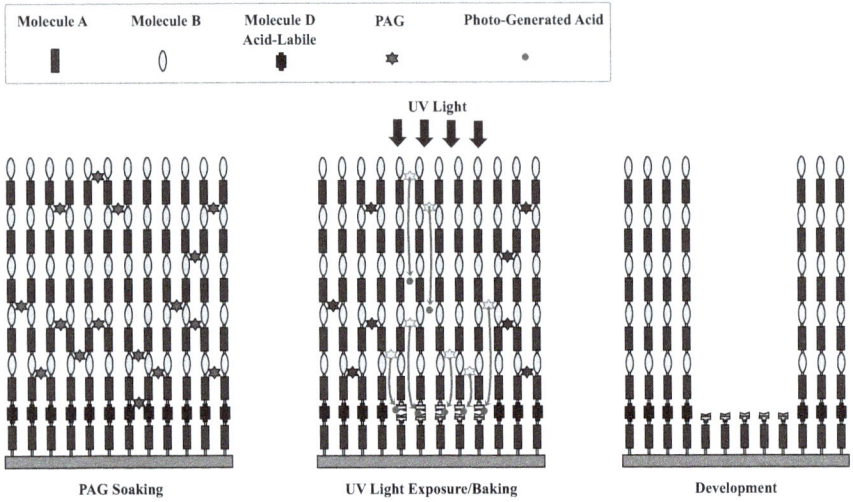

FIGURE 14.23 Concept of a photoresist thin film with diffused PAGs for photolithography processes. (Drawn by the author based on the description in [35].)

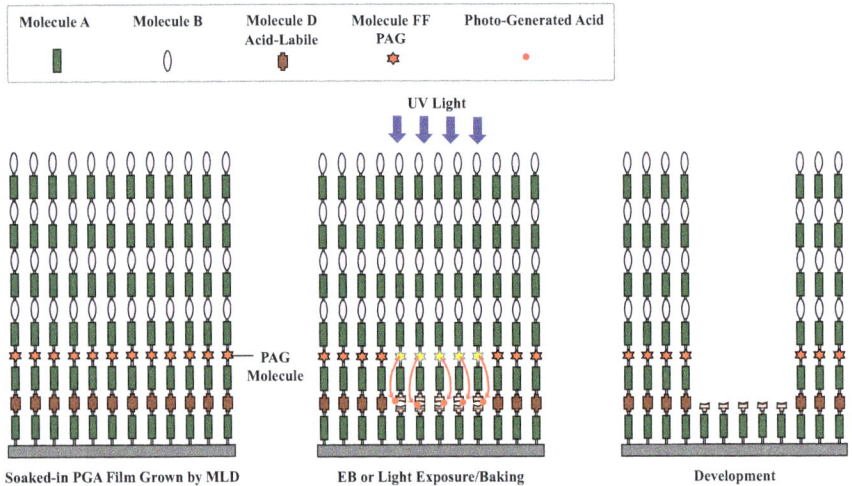

FIGURE 14.24 Concept of a resist thin film with acid-labile molecules and PAG molecules embedded at designated positions in polymer backbones. (Drawn by the author based on the description in [36,37].)

process continues until the film thickness reaches an intended value. Thus, a resist thin film with polyurea backbones containing the acid-labile molecules and the PAG molecules at designated positions is grown. By UV light exposure and development, photolithography is completed. Using the polyurea-based resist thin films, the electron beam (EB) lithography was also achieved.

This study on the resist thin films is regarded as an essential role model for the MLD applications.

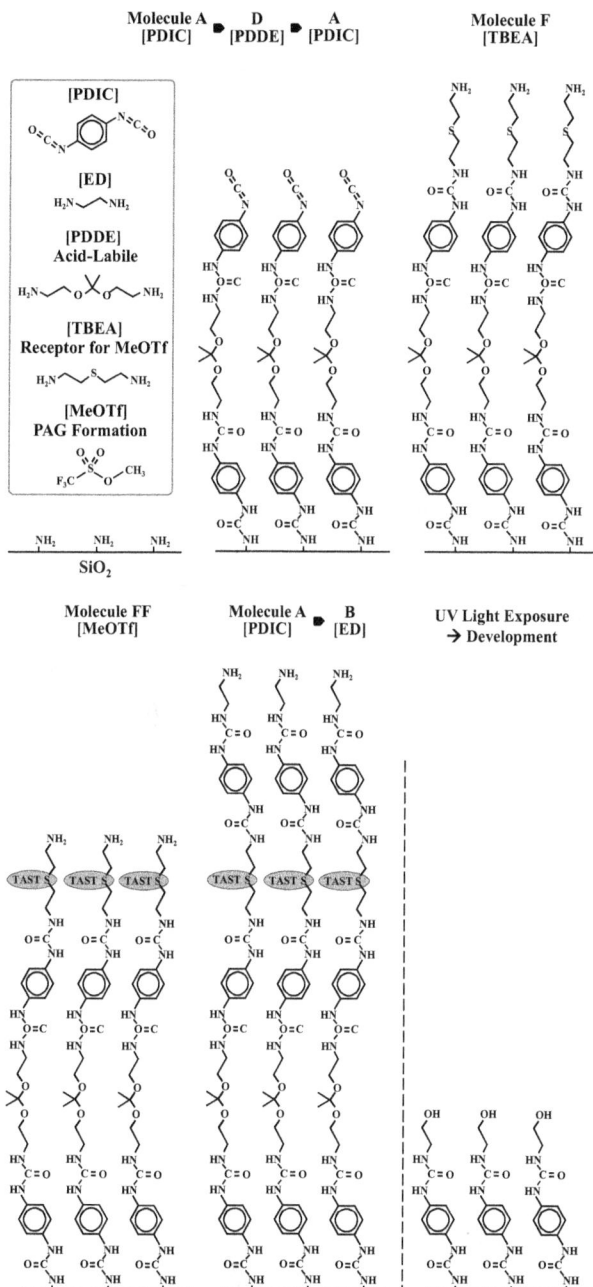

FIGURE 14.25 An MLD process to grow a resist thin film with acid-labile molecules and PAG molecules embedded at designated positions in polymer backbones. (Drawn by the author based on the description in [36,37].)

14.5.2 Diffusion Barriers

Loscutoff et al. grew polyurea and polythiourea thin films by Type-A MLD <Homo-Bifunctional Molecules>. Oxygen-free polythiourea thin films are useful for applications that are sensitive to oxidation. They can be applied to copper diffusion barriers because sulfur forms strong bonds with copper. They are also applied to nanoglue utilizing a sulfur property that it promotes adhesion between dissimilar materials. Figure 14.26 shows an MLD process for forming a layered structure of polyurea/polythiourea/polyurea. The polyurea block is grown by PDIC and ED, and the polythiourea block by PDIC and TBEA. Thicknesses of the blocks are controlled by the number of MLD cycles. By using 1,4-phenylene diisothiocyanate (PDTIC) and ED, oxygen-free polythiourea thin films are made available [39,40].

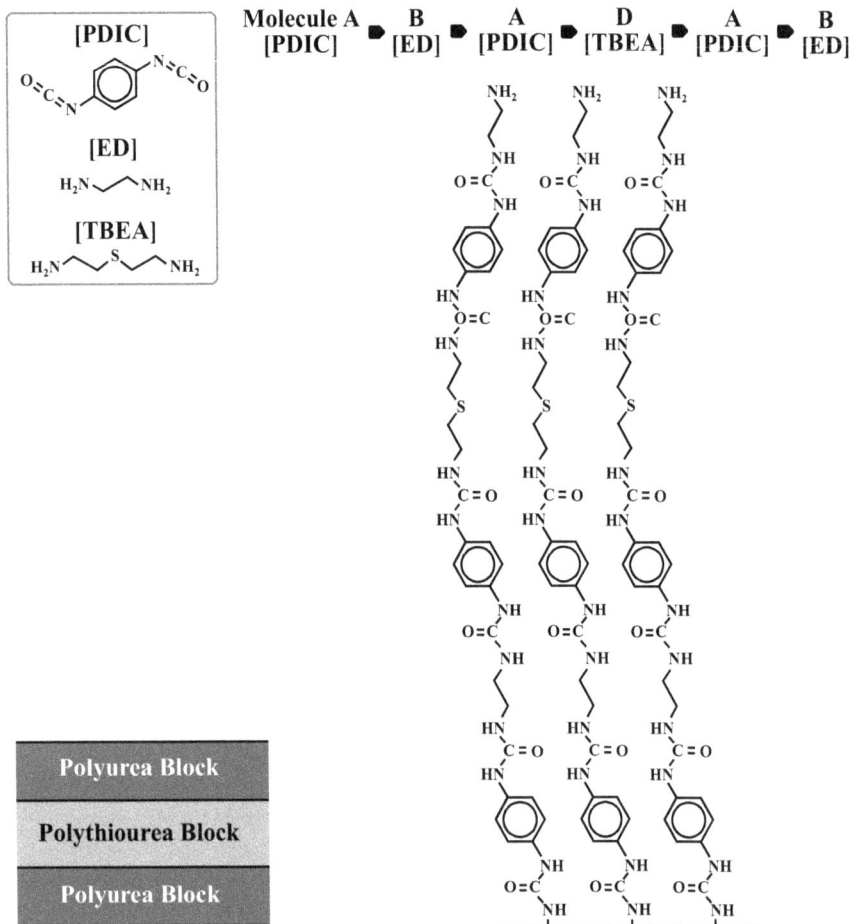

FIGURE 14.26 An MLD process for forming a layered structure of polyurea/polythiourea/polyurea. (Drawn by the author based on the description in [39,40].)

14.5.3 INSULATORS

Lee et al. grew organic–inorganic hybrid thin films consisting of self-assembled organic layers (SAOLs) by Type-D MLD <Molecules with Hetero-Reactive Groups 2>, and applied them to a gate insulator of organic pentacene thin-film transistors (TFTs). The process is similar to that drawn in Figure 14.12. In the present case, TMA is used instead of TIP for Molecule B. The gate insulator films exhibit high dielectric strength of 4 MV/cm, hole mobility of 0.57 cm²/Vs operating at −4 V, and an on/off current ratio of ~10^3 [41].

Abdulagatovet et al. fabricated cross-linked titanicone MLD thin films by Type-B MLD <Homo-Multifunctional Molecules> shown in Figure 14.27 [42]. Molecule A is $TiCl_4$ and Molecule B is EG or glycerol (GL). Because EG and GL are, respectively, bifunctional and trifunctional, the titanicone film grown using GL is expected to contain more cross-links between the polymer wires than that grown using EG. It was revealed that the latter exhibits much higher stability, elastic modulus, and hardness than the former. Similar cross-linked alucone thin films were fabricated using TMA and GL.

Zhou et al. fabricated cross-linked polyurea thin films by Type-B MLD <Homo-Multifunctional Molecules> shown in Figure 14.28 using 1,4-diisocyanatobutane (DICB) and diethylenetriamine (DETA) [43]. Compared to the noncross-linked polyurea MLD thin films, the cross-linked polyurea MLD thin films exhibit a decrease in molecular volume of each polymeric repeat units, an increase in film density by ~50%, and improvement of thermal stability. The decomposition/desorption temperature is increased from ~260°C to ~290°C.

Zhou et al. also succeeded in the growth of carbosiloxane thin films by MLD using 1,2-bis- (dimethylamino)dimethylsilyl]ethane (DDSE) and O_3, as Figure 14.29

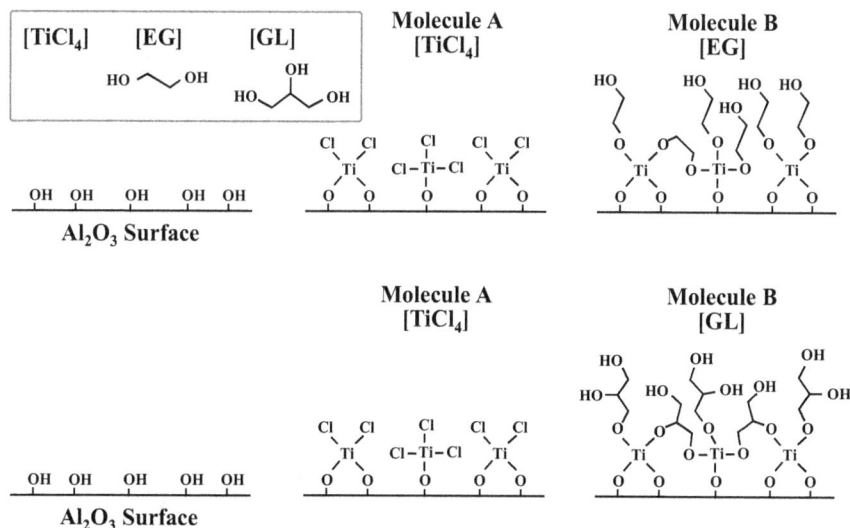

FIGURE 14.27 MLD processes to grow cross-linked titanicone thin films. (Adapted with permission from [42] A. I. Abdulagatov et al., *Chem. Mater.* **24**, 2854–2863 (2012). Copyright (2012) American Chemical Society.)

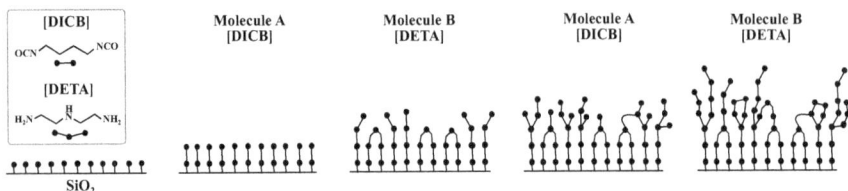

FIGURE 14.28 An MLD process to grow a cross-linked polyurea thin film. (Adapted with permission from [43] H. Zhou et al., *Macromolecules* **46**, 5638–5643 (2013). Copyright (2013) American Chemical Society.)

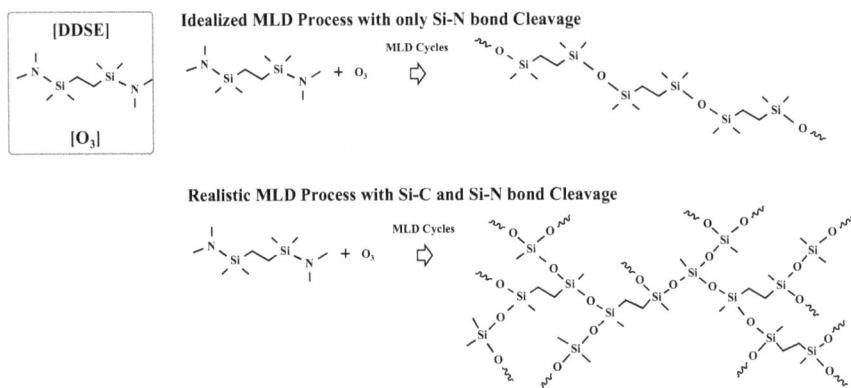

FIGURE 14.29 Growth types of carbosiloxane thin films by MLD. (Adapted with permission from [44] H. Zhou and S. F. Bent, *J. Phys. Chem.* C **117**, 19967–19973 (2013). Copyright (2013) American Chemical Society.)

shows [44]. Two types of MLD models are proposed: an idealized process with only Si-N bond cleavage and a realistic process with Si-C and Si-N bond cleavage. The thin films are stable to acid, base, organic solvents, and also thermally stable at 400°C. The carbosiloxane thin films are expected as pore-sealing dielectric materials for low-k materials in integrated circuits, separation membranes with molecular sieving functions, etc.

14.5.4 SEMICONDUCTORS

Conjugated polymers are useful as semiconductors for organic electronics. In conjugated polymer thin films, however, the constituent one-dimensional polymer wires sometimes have distorted structures like twists and bends to divide electron wavefunctions in the wires, preventing electrons from being transported. To solve the problem, Cho et al. incorporated inorganic cross-linkers of -O-Zn-O- between the conjugated polymer wires to form two-dimensional polydiacetylene (PDA) with organic–inorganic hybrid structures using MLD. The inorganic linkers enable control of the wire alignment, stability, and electronic properties [45].

Cho et al. applied the zinc oxide cross-linked PDA (ZnOPDA) to an n-type semiconductor thin film of an organic TFT. A ZnOPDA thin film was grown on a SiO_2

gate insulator by Type-D MLD <Molecules with Groups for Horizontal Connection>. Molecule A, Molecule B, and external excitation were, respectively, diethylzinc (DEZ), HDD, and photons of UV light. The growth process is similar to that for MLD of two-dimensional PDA with titanium oxide linkers drawn in Figure 14.35, where $TiCl_2$ is used instead of DEZ for Molecule A. After Molecule B is connected to Molecule A, the external excitation is introduced to form horizontal PDA wires, constructing a two-dimensional system consisting of PDA and -O-Zn-O- linkers. The ZnOPDA TFT exhibited a high field effect electron mobility of $>1.3 \, cm^2/Vs$ while a TFT using pure PDA exhibited low mobility of $<10^{-8} cm^2/Vs$. The ZnOPDA also enhanced thermal/mechanical stability.

Huang et al. fabricated an organic TFT, in which an organic–inorganic hybrid thin film is used as an n-type semiconductor layer. The layer is grown by Type-A MLD <Homo-Bifunctional Molecules> with DEZ and hydroquinone (HQ) as Figure 14.30 shows. The TFT exhibits a high field effect electron mobility of ~$5.7 \, cm^2/Vs$ with an on/off current ratio of $>10^3$ after post-annealing at 200°C [46].

The conduction mechanism is schematically illustrated in Figure 14.30. Electrons are provided from oxygen vacancies denoted by Vo. The electrons are transported through π–π stacks of aromatic rings with strong intermolecular interactions. This mechanism wherein the electron-donating portion and the electron transporting channel portion are separated is similar to the mechanism of the high-electron-mobility transistor (HEMT), as explained in Section 2.3 [47].

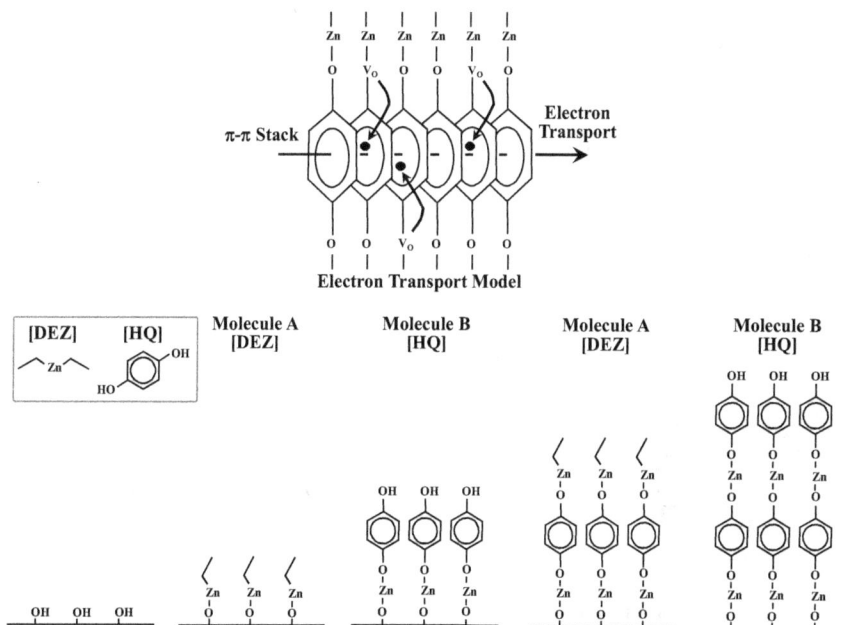

FIGURE 14.30 An MLD process for forming an organic–inorganic hybrid n-type semiconductor thin film of an organic pentacene TFT, and the conduction mechanism in the film. (Drawn by the author based on the description in [46].)

14.6   TAILORED MATERIALS GROWN BY MLD COMBINED WITH ALD

By combining MLD with ALD, the structural and compositional freedom in produced materials increases to strengthen the feature of "tailored organic material growth" capability of MLD. MLD and ALD can be performed in the same chamber seamlessly by switching precursors.

Figure 14.31 shows typical structures grown by MLD combined with ALD: nanolaminates, superlattices, and alloys [6]. Although the author does not completely understand the difference between nanolaminates and superlattices, probably the layers in superlattices are relatively thinner comparing to those in nanolaminates. MLD layers are organic or organic–inorganic hybrid thin films, and ALD layers are inorganic films, typically oxides. The layer thickness is controlled by the number of cycles; n_{MLD} for MLD and n_{ALD} for ALD. These multilayer structures have the following advantages.

- The optical, electrical, and mechanical properties, such as refractive index, dielectric constant, electrical conductivity, density, and elastic modulus, can widely be tuned between the organic material property and the inorganic material property by adjusting the cycle ratio of n_{MLD}: n_{ALD}.
- Mechanical toughness can be enhanced by inserting compliant MLD films between brittle ALD films.

In the present section, applications of tailored materials grown by MLD combined with ALD to the microelectronics fields are reviewed.

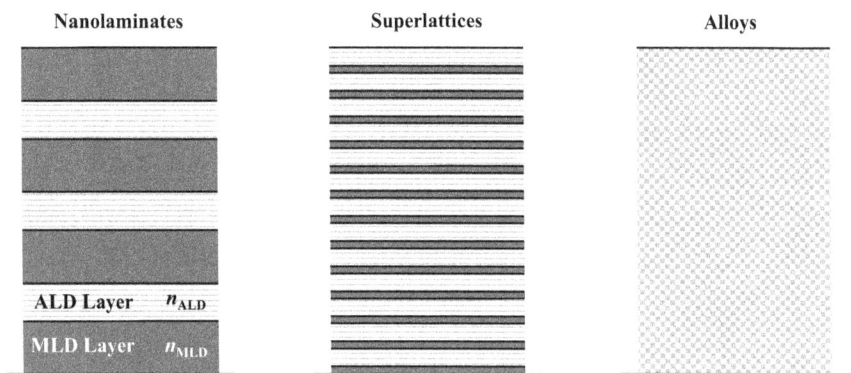

MLD Layer: Organic Films, Organic-Inorganic Hybrid Films
ALD Layer: Inorganic Films (Typically Oxides)

FIGURE 14.31   Typical structures grown by MLD combined with ALD; nanolaminates, superlattices, and alloys. (Drawn by the author based on the description in [6].)

14.6.1 DIFFUSION BARRIERS

Moisture and oxygen barrier coatings are important to prevent organic electronic/optoelectronic devices from being degraded. Lee applied nanolaminates to the moisture barrier coating. Although a single Al_2O_3 ALD film of 25 nm thickness is an effective moisture barrier coating, it exhibits small critical strain, generating cracks. By introducing a nanolaminate structure consisting of Al_2O_3 ALD layers and alucone MLD layers, the required Al_2O_3 thickness is reduced to 5–2 nm, resulting in an increase in critical strain. The nanolaminate was applied to thin-film encapsulation (TFE) for organic light-emitting diodes (OLEDs) with polymer substrates. It is also expected to be a moisture/gas barrier for packaging of the NEMS/micro-electro-mechanical system (MEMS) [48].

Park et al. previously investigated the TFE technology to prevent water and oxygen permeation into flexible OLEDs. They reported TFE with a multilayer structure consisting of Al_2O_3 ALD films and alucone MLD films [49,50].

Recently, Park et al. developed a superlattice encapsulation film consisting of Al_2O_3 ALD layers and 4-mercaptophenol (4MP) MLD layers for OLEDs. The schematic illustration is depicted in Figure 14.32. The 4MP MLD layer, which is a monomolecular layer, namely, $n_{MLD} = 1$, was grown by Type-A MLD <Hetero-bifunctional Molecules>. The Al_2O_3 ALD layer was grown using TMA and H_2O. It was found that a 100-nm-thick superlattice encapsulation film exhibits much higher barrier performance than a 100-nm-thick single Al_2O_3 ALD film, improving the stability. The total 4MP layer count of 20 is sufficient for satisfactory barrier properties. Bending tests (3-mm radius) revealed that n_{ALD} of 8 is the best, corresponding to an organic/inorganic thickness ratio of 1/1.5.

As Figure 14.32 shows, for a single Al_2O_3 ALD film, the pinhole/defect channels propagate within the entire Al_2O_3 ALD film, and consequently, moisture is transported in the channels to reach the underlying substrate directly with short transit time. For a superlattice, on the other hand, the 4MP monolayers decouple the

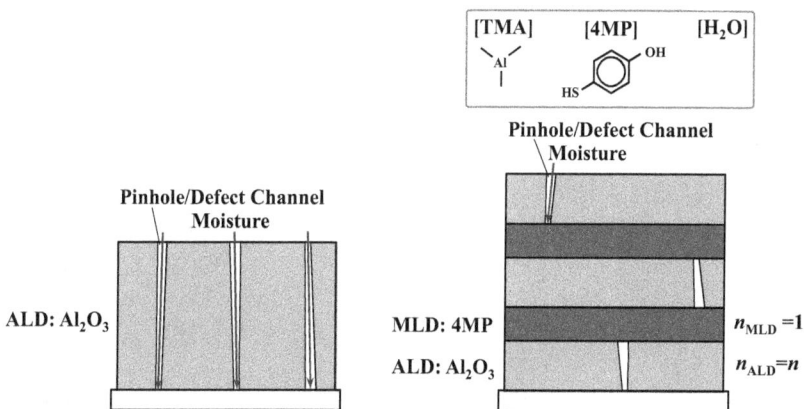

FIGURE 14.32 Schematic illustration of a superlattice encapsulation film consisting of 4MP MLD layers and Al_2O_3 ALD layers for OLEDs. (Drawn by the author based on the description in [50].)

pinhole/defect channels in Al_2O_3 layers, blocking the propagation of the channels. The 4MP monolayers act as buffers for the pinholes/defects to interrupt the direct transportation of moisture to the substrate, delaying moisture from reaching the substrate. The 4MP monolayers also give flexibility to the encapsulation films, reducing the film stress.

14.6.2 INSULATORS

Salmi et al. demonstrated that optical, electrical, and mechanical properties of nanolaminates can be widely tuned by controlling the layer thicknesses [51]. The nanolaminates consist of five MLD/ALD bilayers. The MLD layer is polyimide grown by Type-A MLD <Homo-Bifunctional Molecules> with PMDA and diaminohexane (DAH), and the ALD layer is Ta_2O_5 grown using $Ta(OEt)_5$ and H_2O. They investigated the properties of the nanolaminates for three sets of layer thickness combinations: 5-nm polyimide/15-nm Ta_2O_5, 10-nm polyimide/10-nm Ta_2O_5, and 15-nm polyimide/5-nm Ta_2O_5. $n_{MLD}/(n_{MLD} + n_{ALD})$ increases in this order. The results are depicted in Figure 14.33. The graph is drawn qualitatively by the author for simplicity. The refractive index, dielectric constant, density, and elastic modulus decrease with increasing $n_{MLD}/(n_{MLD} + n_{ALD})$, namely, by increasing the polyimide content. Furthermore, they found that the nanolaminates exhibit smaller leakage currents and less C-V hysteresis than a single polyimide film and a single Ta_2O_5 film.

Lee et al. reported the decrease in refractive index, elastic modulus, and hardness accompanied with an increase in the organic content in structures consisting of alucone and Al_2O_3 [52].

Lee et al. fabricated superlattices consisting of SAOLs and ZrO_2 layers, and applied them to gate insulators of organic pentacene TFTs. As shown in Figure 14.34, SAOLs are grown by Type-D MLD <Molecules with Hetero-Reactive Groups 2> with Molecule A of 7OTS+H_2O and external excitation of O_3. ZrO_2 layers are grown by ALD with Molecule C of zirconium tetra-tert-butoxide (ZTB) and Molecule D

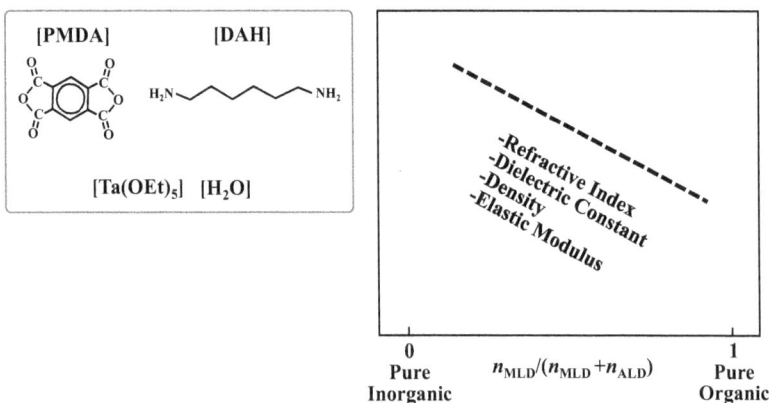

FIGURE 14.33 Qualitatively-drawn dependence of optical, electrical, and mechanical properties on $n_{MLD}/(n_{MLD} + n_{ALD})$ in nanolaminates consisting of polyimide MLD layers and Ta_2O_5 ALD layers. (Drawn by the author based on the description in [51].)

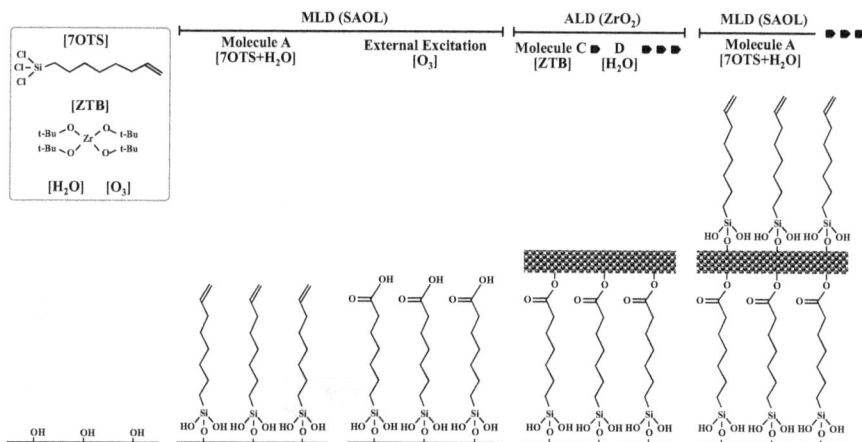

FIGURE 14.34 A process for forming a superlattice consisting of SAOLs grown by MLD and ZrO$_2$ ALD layers. The superlattice is applied to a gate insulator of an organic pentacene TFT. (Adapted with permission from [53] B. H. Lee et al., *Thin Solid Films* **517**, 4056–4060 (2009). Copyright (2009) Elsevier.)

of H$_2$O. 2.2-nm SAOL and 3-nm ZrO$_2$ layer were sequentially stacked as to form a 23-nm superlattice for a gate insulator film. The film exhibited high dielectric constant of ~16 at 1 MHz and high dielectric strength. The TFT exhibited hole mobility of 0.63 cm^2/Vs, operation voltage of −1 V, and on/off current ratio of ~10^3. Lee et al. also fabricated superlattices by the similar process, in which ZTB is replaced with TIP, and applied them to gate insulators of organic pentacene TFTs [53].

Yoon et al. fabricated highly-stable insulators of superlattices consisting of 0.6-nm titanium oxide cross-linked polydiacetylene (TiOPDA) and 2.3-nm TiO$_2$. As schematically depicted in Figure 14.35, TiOPDA is grown by Type-B MLD <Homo-Multifunctional Molecules> combined with Type-D MLD <Molecules with Groups for Horizontal Connection>. Molecule A is TiCl$_4$, Molecule B is HDD, and the external excitation is photons of UV light. TiO$_2$ is grown by ALD with TiCl$_4$ and H$_2$O. The superlattice film exhibits superior thermal and mechanical stability at 400°C in air [20].

14.6.3 SEMICONDUCTORS

Park et al. developed SAOL/ZnO superlattices, which work as n-type semiconductors. The SAOLs offer structural flexibility while ZnO layers provide semiconducting properties. The superlattices exhibit superior thermal/mechanical stability and flexibility, and are transparent in the visible region [54].

They fabricated an organic TFT depicted in Figure 14.36 using the SAOL/ZnO superlattice as a semiconductor layer. On a polymer substrate of polyethersulfone with a Cr gate electrode, a gate insulator of an SAOL/Al$_2$O$_3$ superlattice is formed, followed by the growth of an SAOL/ZnO superlattice semiconductor. Source and drain electrodes are deposited on the top.

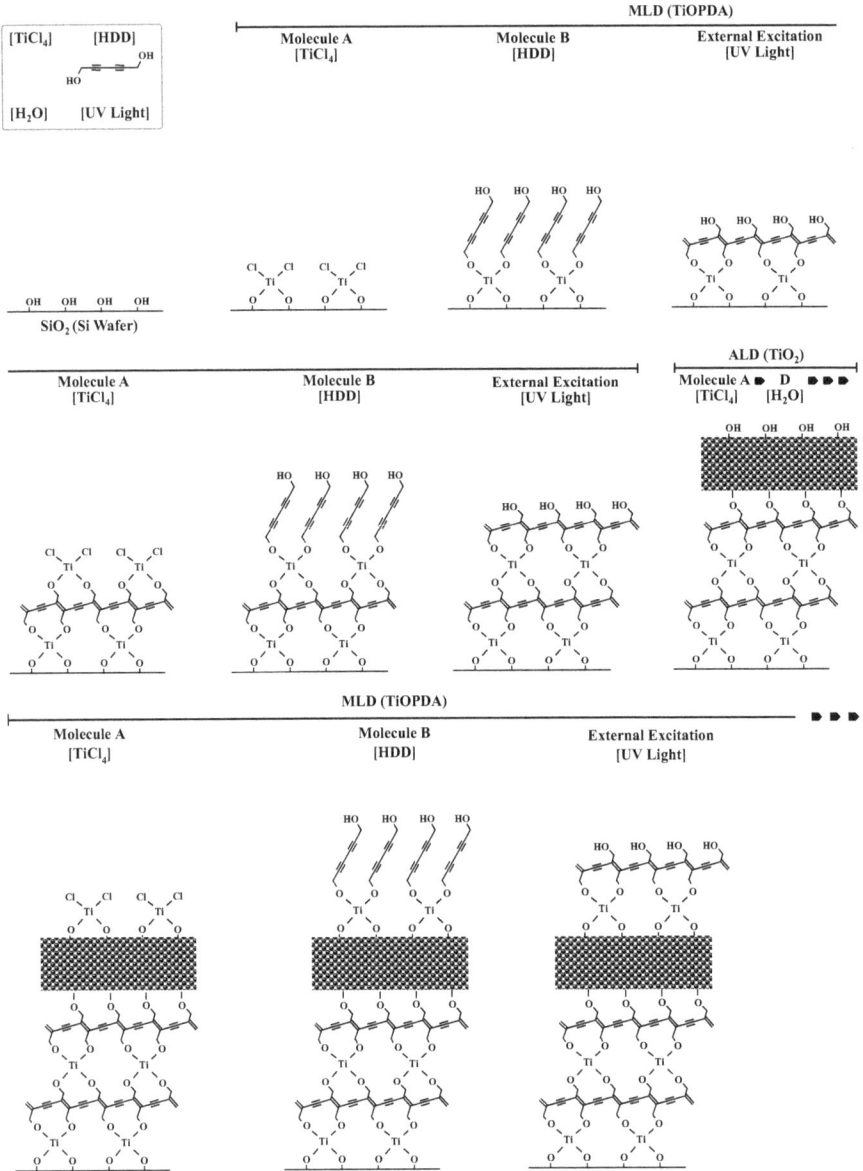

FIGURE 14.35 A process for forming a superlattice consisting of TiOPDA MLD layers and TiO$_2$ ALD layers. The superlattice is applied to a gate insulator of an organic pentacene TFT. (Adapted with permission from [20] K. Yoon et al., *Nanoscale Res. Lett.* **7**, 7–71 (2012).)

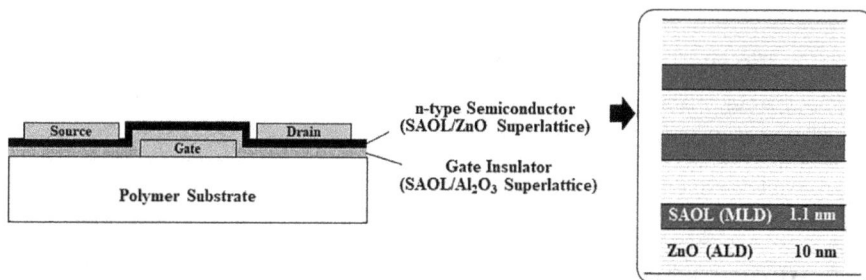

FIGURE 14.36 Organic TFT using the SAOL/ZnO superlattice as an n-type semiconductor layer and the SAOL/Al$_2$O$_3$ superlattice as a gate insulator. (Adapted with permission from [54] Y. Park et al., *Org. Electron.* **12**, 348–352 (2011). Copyright (2011) Elsevier.)

A schematic illustration of the growth process for the SAOL/ZnO superlattice is shown in Figure 14.37. SAOLs are grown by Type-D MLD <Molecules with Hetero-Reactive Groups 2> with Molecules A of 7OTS+H$_2$O and external excitation of O$_3$. ZnO layers are grown by ALD with Molecule C of DEZ and Molecule D of H$_2$O. The thicknesses of the SAOL and ZnO layer are 1.1 and 10 nm, respectively.

The SAOL/ZnO TFT exhibits field effect mobility of >7 cm^2/Vs under operation voltage of $-1\sim3$ V, an on/off current ratio of 10^6, and a threshold voltage of 0.6 V. While the mobility of the SAOL/ZnO TFT is lower than that of a pure ZnO TFT (14 cm^2/Vs), the operating voltage, on/off current ratio, and threshold voltage are comparable. The SAOL/ZnO TFTs is transparent in the visible region, and exhibits no significant degradation in performance when the devices are bent to a radius as small as 6 mm, indicating that the SAOL/ZnO TFT is potentially useful for flexible and invisible electronics. In addition, a vertical p-n junction was fabricated by stacking a p-type pentacene film and an n-type SAOL/ZnO superlattice, and diode performance was observed.

Han et al. developed nanolaminates consisting of zinc oxide-crosslinked hydroxybenzene (ZnOHB) and ZnO, and applied the nanolaminates to organic TFTs illustrated in Figure 14.38. The nanolaminates are n-type semiconductors. The ZnOHB layers were grown by Type-B MLD <Homo-Multifunctional Molecules> with Molecule A of DEZ and Molecule B of 1,2,4-trihydroxybenzene (THB). ZnO layers were grown by ALD with Molecule A of DEZ and Molecule D of H$_2$O. An example of the molecular sequence for the nanolaminate growth is depicted in Figure 14.39. The ZnOHB/ZnO TFT, in which the thicknesses of the ZnOHB and ZnO layers were, respectively, 2.1 and 2.9 nm, exhibited field effect mobility of >0.6 cm^2/Vs under gate bias of 50 V and drain voltage of 50 V, an on/off current ratio of 10^6, and a threshold voltage of 21 V [55].

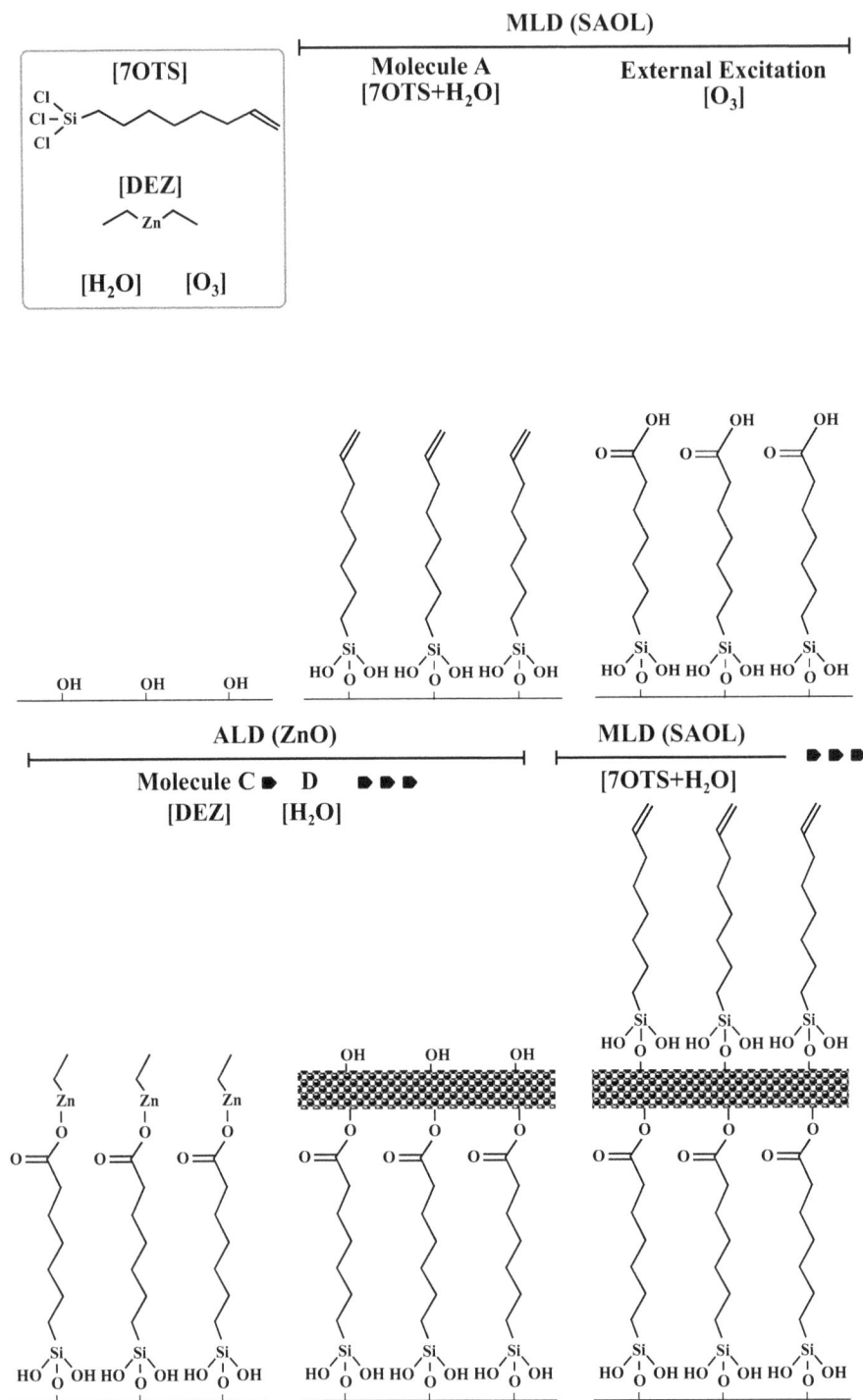

FIGURE 14.37 A process for forming an SAOL/ZnO superlattice. (Drawn by the author based on the description in [53,54].)

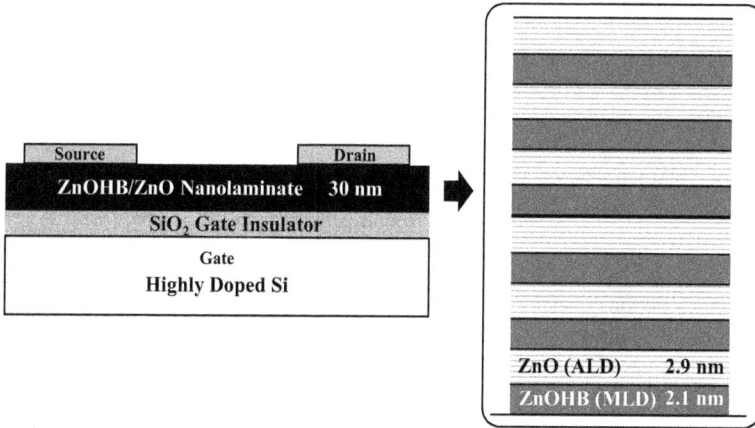

FIGURE 14.38 Organic TFT using the ZnOHB/ZnO nanolaminate as an n-type semiconductor layer. (Drawn by the author based on the description in [55].)

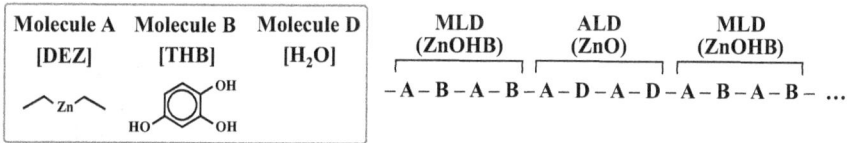

FIGURE 14.39 An example of the molecular sequence for ZnOHB/ZnO nanolaminate growth. (Drawn by the author based on the description in [55].)

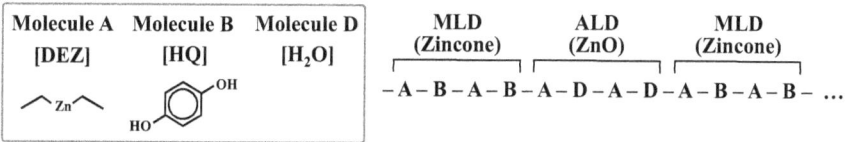

FIGURE 14.40 An example of the molecular sequence for zincone/ZnO alloy film growth. (Drawn by the author based on the description in [56].)

14.6.4 CONDUCTORS

Yoon et al. developed transparent conductors of zincone/ZnO alloy films. The zincone was grown by Type-A MLD <Homo-Bifunctional Molecules> with Molecule A of DEZ and Molecule B of HQ, and ZnO by ALD with Molecule A of DEZ and Molecule D of H_2O. An example of the molecular sequence for the alloy film growth is depicted in Figure 14.40. An alloy film grown with an MLD/ALD cycle ratio of $n_{MLD} : n_{ALD} = 2:2$ at substrate temperature of $T_s = 150°C$ exhibited high electrical conductivity of 170 S/cm, which is approximately one order of magnitude higher than that of the pure ZnO ALD film grown at $T_s = 150°C$. The alloy film was transparent in whole of the visible region, and exhibited smaller elastic modulus compared to a pure ZnO ALD film. The mechanism of the high electrical conductivity in the zincone/ZnO alloy films is explained as follows. Many oxygen vacancies (electron donors) in ZnO ALD layers or zincone layers release electrons to increase the carrier concentration in the alloy, and the electrons are transported through the electron coupling between the π-conjugated phenylene rings and the ZnO parts with high electron mobility [56].

Lee et al. developed transparent conductors of indicone/In$_2$O$_3$ superlattices. The indicone is grown by Type-B MLD <Homo-Multifunctional Molecules> with diethyl (1,1,1-trimethyl-N-(trimethylsilyl)silanaminato)-indium) (INCA-1) and HQ, and In$_2$O$_3$ by ALD with INCA-1 and H$_2$O$_2$. The indicone consists of indium and aromatic linkers, which are chosen because of their several advantages over aliphatic molecules: less frequent double reaction, participation of the delocalized π-electrons in charge transfer (CT), etc. As qualitatively drawn in Figure 14.41, there is an MLD window in the growth rate vs. substrate temperature characteristics, corresponding to Case II in Figure 3.16 [57].

The indicone/In$_2$O$_3$ superlattices were fabricated with various cycle ratios of $n_{MLD}:n_{ALD}$, and the film properties were measured. The results are summarized schematically in Figure 14.42. The graphs are drawn qualitatively by the author for simplicity. By increasing the MLD fraction, the refractive index decreases and the transmittance increases, which is attributed to a decrease in the film density and crystallinity; the band gap increases, which is attributed to the quantum confinement of electrons; and the resistivity R increases according to a decrease in the carrier concentration. R of the superlattice for $n_{MLD}:n_{ALD} = 1:99$ is 1.6 times larger than R of pure In$_2$O$_3$, which is attributed to the carrier mobility decrease possibly arising from a scattering effect increase. In the superlattices, In$_2$O$_3$ contributes to superior electrical properties and indicone to large elasticity to produce flexible transparent conductor. The superlattice with a cycle ratio of $n_{MLD}:n_{ALD} = 1:99$ exhibits superior electrical/optical properties, which are preserved even after repeated mechanical bending.

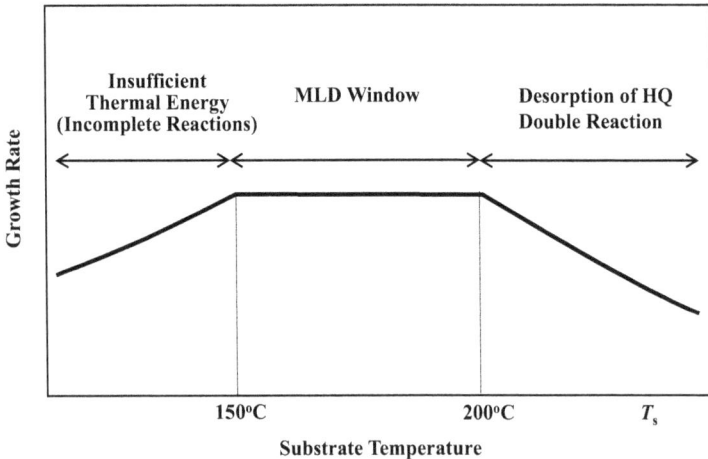

FIGURE 14.41 MLD window in the growth rate vs. substrate temperature characteristics for indicone. (Adapted with permission from [57] S. Lee et al., *Appl. Surf. Sci.* **525**, 146383 (2020). Copyright (2020) Elsevier.)

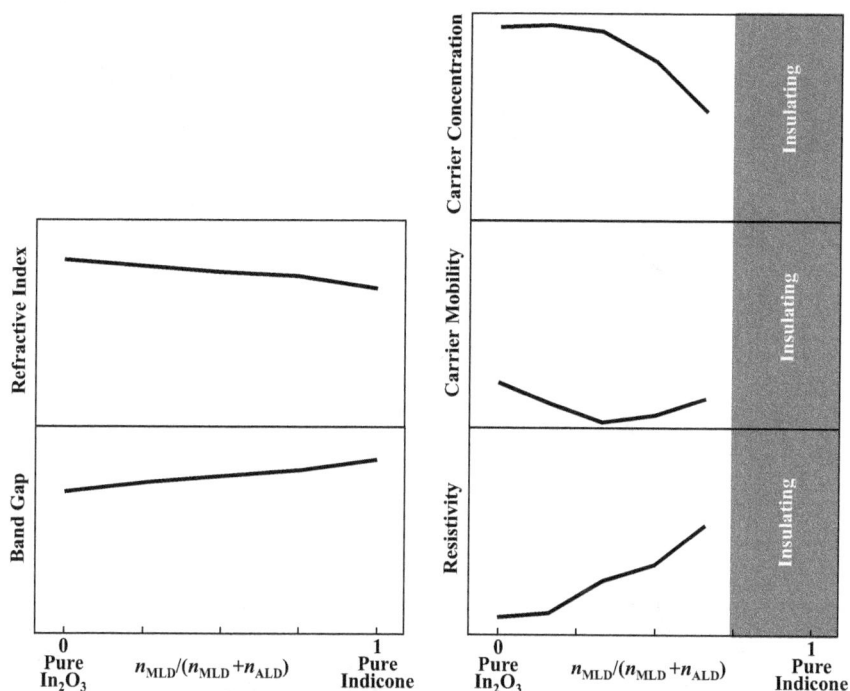

FIGURE 14.42 Qualitatively-drawn dependence of film properties on $n_{MLD}/(n_{MLD} + n_{ALD})$ in indicone/In$_2$O$_3$ superlattices. (Adapted with permission from [57] S. Lee et al., *Appl. Surf. Sci.* **525**, 146383 (2020). Copyright (2020) Elsevier.)

14.7 PORE-SIZE-CONTROLLED ORGANIC–INORGANIC HYBRID MATERIALS GROWN BY MLD

Organic–inorganic hybrid materials, which consist of linkers of organic molecules connected by metal atoms, provide a molecular sieving function due to their porous structures, as shown in Figure 14.43. For example, atoms and ions pass through an organic–inorganic hybrid thin film while large molecules are rejected by the thin film. The fourth MLD feature of "pore-size-controlled material growth" capability enables us to adjust the molecular sieving effect by controlling the thin film structures.

In the applications to the molecular sieving, thin film formation on porous/tortuous surfaces is sometimes required. The second MLD feature of "ultrathin/conformal organic material growth" capability enables it. As shown in Figure 14.44, source molecules are provided to surfaces of inner parts of porous/tortuous structures as well as the outer surfaces. Therefore, the whole of the porous/tortuous structure surfaces can be covered with uniform thin films formed with monomolecular growth steps by MLD.

In the present section, applications of MLD to the molecular sieving are reviewed.

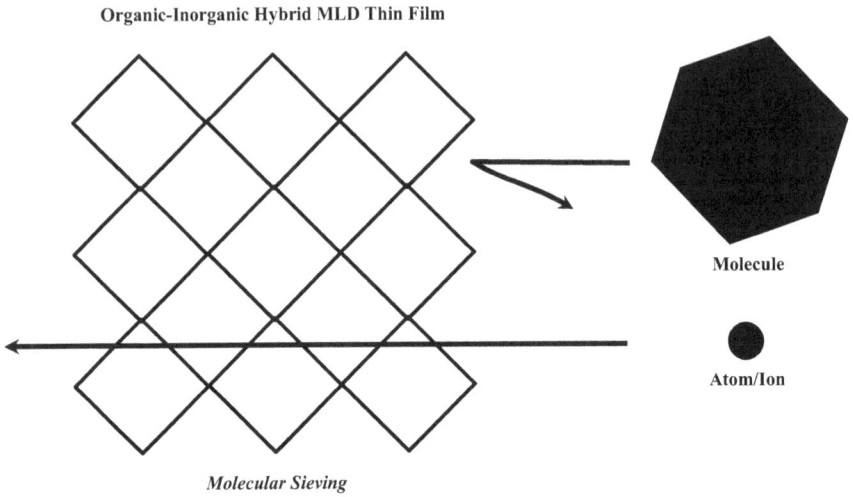

FIGURE 14.43 Molecular sieving function of organic–inorganic hybrid thin films.

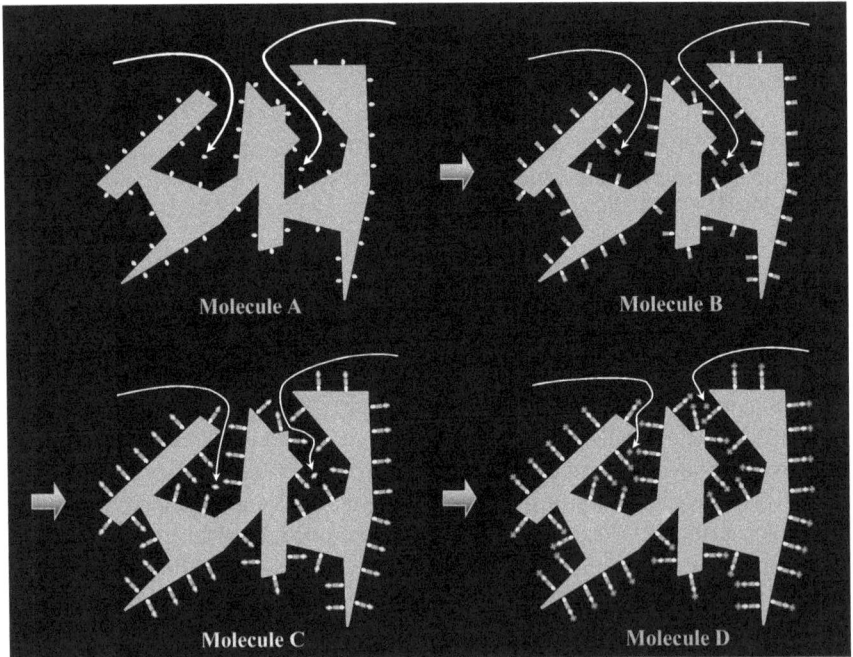

FIGURE 14.44 MLD for porous/tortuous structures.

14.7.1 BATTERIES

In the Si nanocomposite anode, which is promising for the Li-ion batteries due to its high-capacity characteristics, there are issues of rapid capacity fate, poor rate capability, and low coulombic efficiency (CE). These are caused by the following two problems [29,58]:

- Large volume change (~300%) of Si nanoparticles occurs during lithiation/delithiation cycling. This results in mechanical integrity in the composite electrodes and disconnection between the Si nanoparticles and conductive binders, which breaks the conductive network to degrade the electrochemical performance.
- Si nanoparticles are continuously exposed to liquid electrolytes, resulting in parasitic side reaction between liquid electrolytes and surfaces of electrodes.

To solve these problems, surface coating of Si nanoparticles is essential. ALD is a popular coating technique for particles. However, it cannot be applied to the Si nanoparticle anodes because the volume change is too large for inorganic thin films grown by ALD. The "spring-like" nature of elastic flexible organic thin films formed by MLD contributes to relaxation of the stress caused by the volume expansion of Si nanoparticles.

Ban and Piper et al. succeeded in improvement of the Si nanocomposite anode performance by coating the Si nanoparticles with alucone MLD thin films, as schematically illustrated in Figure 14.45. Using TMA and GL for precursors, Si nanoparticle surfaces are coated with AlGL-alucone MLD thin films. The precursors penetrate into porous/tortuous structures to achieve the thin film coating on all of the Si nanoparticle surfaces. The covalent bonding enhances adhesion between the thin films and the nanoparticles as well as between the nanoparticles and the binder to maintain the conductive network. The GL molecule is trifunctional, so cross-linked three-dimensional (3-D) structures are constructed, improving fracture toughness [29,58].

The alucone MLD thin films also improve the electrochemical cycling behavior. Li ions are trapped in the MLD thin films. Then, the films become electrically conductive to enhance lithium diffusion, enabling the rapid lithiation process.

The lithium-sulfur (Li-S) battery is a high-energy storage entity. The main problems are as follows [59]:

- Parasitic side reaction between liquid electrolytes and electrodes.
- Dissolution and the shuttle effect of intermediate lithium polysulfides (Li_2S_n) in liquid electrolytes. The lithium polysulfides are highly soluble in liquid electrolytes, and they shuttle freely between the cathode and anode.
- Formation of lithium dendrites that short the cell.

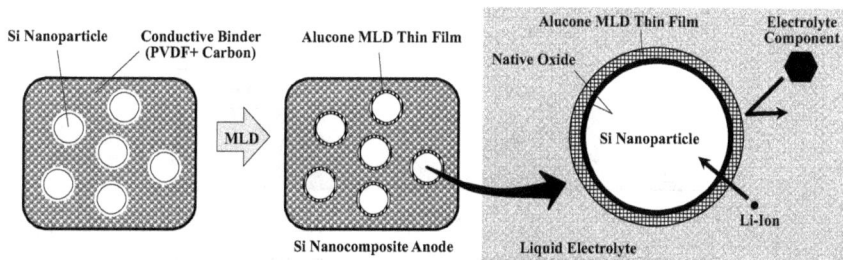

FIGURE 14.45   Process for coating Si nanoparticles with alucone MLD thin films in a Si nanocomposite anode. (Drawn by the author based on the description in [29,58].)

Li et al. solved the problems by coating the C-S composite cathode and Li metal anode with alucone MLD thin films, as shown in Figure 14.46. Li-ions pass through the alucone MLD thin films while the electrolyte components are blocked by the films, suppressing the side reaction, the dissolution and shuttle effect of lithium polysulfides, and lithium dendrite formation to realize long cycle life at high temperatures in carbonate-based electrolytes.

Lithium terephthalate (LiTP) is expected to be a Li-ion battery anode with small volume change of ~6% during lithiation/delithiation. LiTP, however, has two problems: (1) instability/dissolution in liquid electrolytes and (2) negligible electric conductivity. Nisula and Karppinen succeeded in fabricating LiTP thin-film electrodes by MLD [60].

Figure 14.47 schematically shows a Li-ion battery, in which a LiTP thin film is used for an organic anode. The LiTP thin film was grown by MLD with Li(thd), where thd represents 2,2,6,6-tetramethyl-3,5-heptanedionate, and terephthalic acid (TPA). On the anode, a solid-state electrolyte of a lithium phosphorus oxynitride

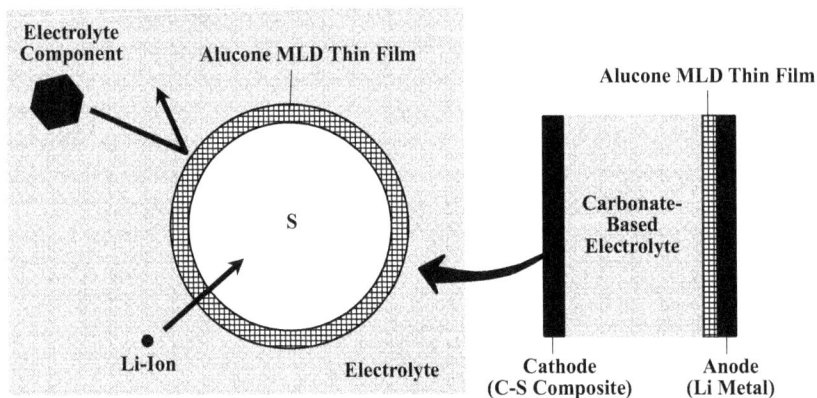

FIGURE 14.46 Coating of the C-S composite cathode and the Li metal anode with alucone MLD thin films in Li-S batteries. (Drawn by the author based on the description in [59].)

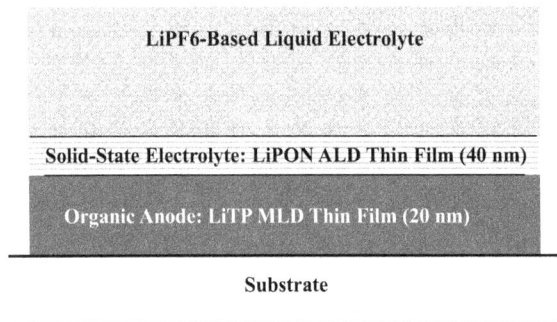

FIGURE 14.47 Li-ion battery, in which a LiTP thin film is used for an organic anode. (Drawn by the author based on the description in [60].)

(LiPON) ALD thin film is formed. The LiPON thin film acts as a protective layer, which solves the problem (1). It was demonstrated that the LiTP thin film is electrochemically active and the LiPON thin film stabilizes the LiTP electrode, resulting in excellent cycle life in LiPF6-based liquid electrolyte. By solving the problem (2), for example, by reducing the LiTP film thickness, the LiTP is expected to be used as a high-rate anode material.

Li metal batteries have higher energy density compared to Li-ion batteries. However, there are problems such as insufficient electrochemical performance and robustness during the plating/stripping process, and generation of mossy/dendritic Li and electrochemically inert Li layers.

Zhao et al. overcame the problems by introducing a novel anode structure with dual protective layers depicted in Figure 14.48 [61]. On a Li metal anode, an inner Al_2O_3 ALD thin film and an outer alucone MLD thin film are successively formed. The structure was inspired by the natural solid electrolyte interphase (SEI) structures. Generally, naturally-formed SEI on Li metal consists of two layers, that is, the inner inorganic layer (such as Li_2O, Li_3N, LiF, LiOH, and Li_2CO_3) and the outer organic layer with higher oxidation states (such as $ROCO_2Li$, ROLi, and $RCOO_2Li$).

They coated Li metal with various types of protective layers, and found that the electrochemical stability is improved in the following order. Here, the numbers represent the MLD and ALD cycles.

Li (with no protective layers) < 50 Al_2O_3 ALD/Li < 50 alucone MLD/Li
< 50 Al_2O_3 ALD/50 alucone MLD/Li < 50 alucone MLD/50 Al_2O_3 ALD/Li

This indicates that the dual protective layer structure depicted in Figure 14.48 is the optimum. The flexible and porous outer alucone MLD thin film provides the channel for electrolyte diffusion and relieves the volume change of Li metal during cycling. The dense inner Al_2O_3 ALD thin film transports Li ions while it blocks the pathway of electrons and suppresses the reaction between Li metal and electrolytes, improving performance of Li metal batteries.

Meng et al. coated Li metal anodes with lithium-containing crosslinked LiGL lithicone MLD thin films using lithium tert-butoxide (LTB) and GL as precursors. The anodes realized a superior Li-stripping/plating cycling stability with little formation of Li dendrites and solid electrolyte interphase [62].

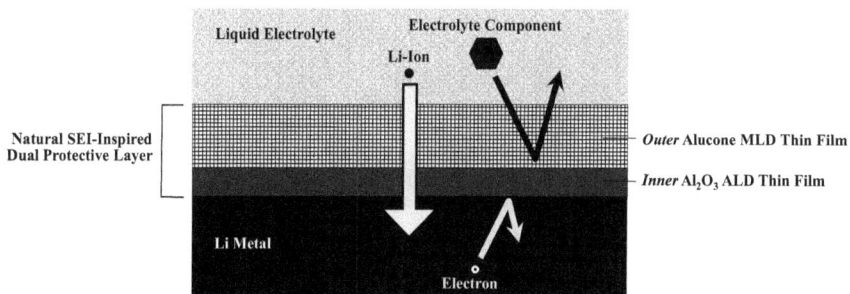

FIGURE 14.48 Novel Li anode structure with natural SEI-inspired dual-protective layers. (Drawn by the author based on the description in [61].)

Pearse et al. applied ALD/MLD coating to all-solid-state batteries. A structure of the all-solid-state battery with ALD/MLD protective coating layers is schematically shown in Figure 14.49a. Typically, the solid-state electrolyte is a sulfide-based material like $Li_{10}SnP_2S_{12}$. The ALD/MLD protective coating layers located between the electrodes and the solid-state electrolyte can improve the lifetime of the anode (Li or Na metal) and suppress the mossy metal dendrites. Furthermore, the method is promising to fabricate 3-D micro solid-state batteries, as shown in Figure 14.49b [63].

In Na-ion batteries, similarly to the Li-ion batteries, alucone MLD thin films can be used to construct safe and durable high-voltage cathodes for sodium as clarified by Kaliyappan et al. [64].

14.7.2 GAS SEPARATION/WATER PURIFICATION

MOFs have porous structures with a pore size typically in the range of ~1 nm, and the pore size can be controlled by varying organic linkers. MOF thin films grown by MLD are promising in applications to membranes for separation and desalination, sensors for chemical species, and antibacterial photodynamic therapy due to their porous, luminescent, and antibacterial properties.

Lausund et al. attempted to control the pore size using appropriate organic linkers [65]. Figure 14.50 shows precursors for MLD. Molecule A is $ZrCl_4$. For Molecule B,

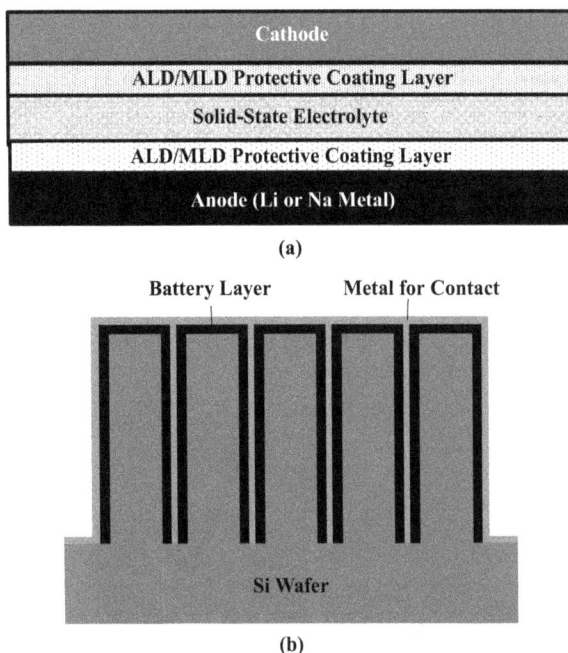

FIGURE 14.49 (a) All-solid-state battery with ALD/MLD protective coating layers. (b) 3-D micro solid-state batteries. (Adapted with permission from [63] A. Pearse et al., *ACS Nano* **12**, 4286–4294 (2018). Copyright (2018) American Chemical Society.)

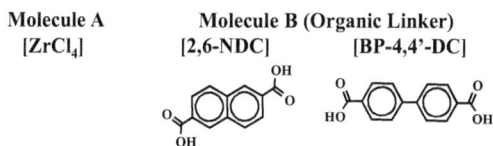

Molecule A [ZrCl₄]   Molecule B (Organic Linker) [2,6-NDC]   [BP-4,4'-DC]

FIGURE 14.50 Precursors for MLD to produce MOFs having porous structures. (Drawn by the author based on the description in [65].)

2,6-naphthalenedicarboxylate (2,6-NDC) or biphenyl-4,4'-dicarboxylate (BP-4,4'-DC) is used to investigate the influence of organic linker lengths on pore sizes. BP-4,4'-DC molecule is longer than 2,6-NDC. It was found that the amount of water adsorption is larger for BP-4,4'-DC than for 2,6-NDC, indicating that the pore size increases with the organic linker length.

14.8 PORE-SIZE-CONTROLLED METAL OXIDE MATERIALS PREPARED UTILIZING MLD

Pore-size controlled metal oxide materials, which are produced from organic–inorganic hybrid thin films grown by MLD, provide a molecular sieving function and enhanced catalysis. The fabrication process is shown in Figure 14.51.

Step 1: Organic–inorganic hybrid thin films are grown on substrate surfaces by MLD.
Step 2: Organic parts that construct templates of the thin films are removed to prepare porous metal oxide thin films with well-defined porous structures and precisely-controlled thicknesses down to several angstroms.

By adjusting sizes of organic precursors, pore sizes of porous metal oxide thin films can be adjusted, enabling the sieving function control. When large organic precursors are used, the pore size becomes large, blocking large-size molecules and allowing transport of small-/middle-size molecules through the thin films. When small organic precursors are used, the pore size becomes small, allowing transport of only small-size molecules.

In the present section, applications of MLD to the molecular sieving/catalysis are reviewed.

14.8.1 CATALYSIS STABILIZATION, GAS SEPARATION/WATER PURIFICATION, AND DRY REFORMING

Liang et al. demonstrated deposition of porous Al_2O_3 thin films on SiO_2 nanoparticles by using MLD in the fluidized bed reactor, which enables thin film coating on surfaces of fine particles without significant particle aggregation, as follows (Figure 14.52) [7,66]:

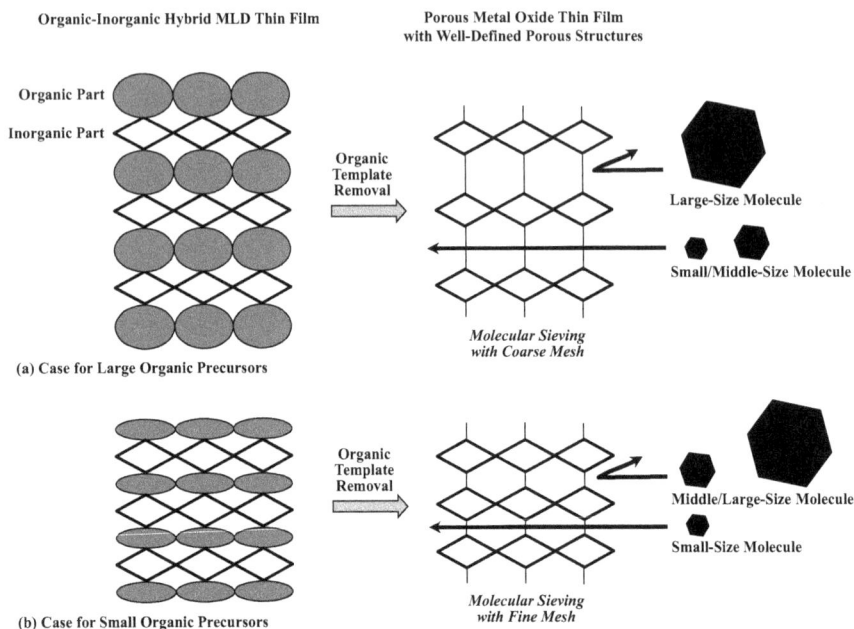

FIGURE 14.51 Fabrication process for porous metal oxide thin films for (a) large organic precursors and (b) small organic precursors.

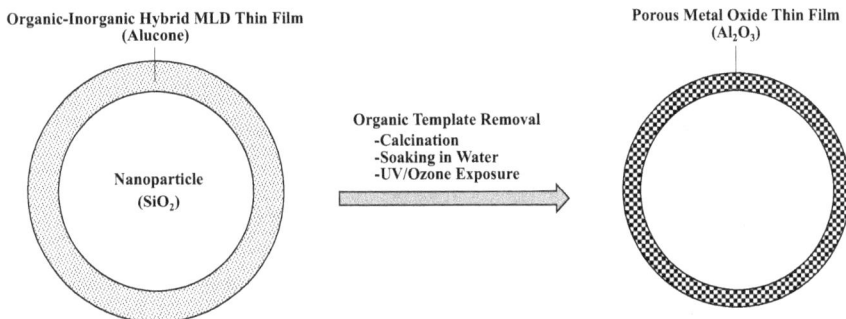

FIGURE 14.52 Deposition of porous Al_2O_3 thin films on SiO_2 nanoparticles by organic template removal in alucone MLD thin films. (Drawn by the author based on the description in [7,66].)

1. 25-nm-thick alucone thin films are deposited on 250-nm-diameter SiO_2 nanoparticles by MLD in the fluidized bed reactor using TMA and EG.
2. Organic templates (sacrificial components) in the alucone thin films are removed by calcination, soaking in water, or UV/ozone exposure to form 8-nm-thick porous Al_2O_3 thin films.

In the calcination, the temperature is precisely controlled with a raising rate of 1°C/min to form porous Al_2O_3 films having ~0.6-nm micropores and ~3.8-nm mesopores.

In the soaking in water, the alucone thin films are gently soaked in water vapor for 1–3 days at room temperature to form porous Al_2O_3 films having ~0.6-nm micropores and ~4.3-nm mesopores.

To avoid sintering of metal catalysts at high temperature, the core-shell structure, in which a metal catalyst particle (core) is covered with a porous oxide thin film (shell), is preferable. The structure isolates the metal catalyst particles, and is effective in catalytic combustion, streaming reforming, automobile-exhaust control, etc. [7,66].

Liang et al. fabricated Pt nanoparticle catalysts encapsulated with porous Al_2O_3 thin films. As Figure 14.53 shows, 1.7-nm-size Pt particles on 30–75 nm-diameter support particles of SiO_2 are coated with 4-nm-thick alucone MLD thin films, and the thin films are converted to porous Al_2O_3 thin films by annealing at 400°C in air to produce thermally stable and highly-dispersed Pt nanoparticle catalysts.

Liang et al. also demonstrated that the porous oxide shells realize reaction selectivity through reactant molecular discrimination, namely, molecular sieving (Figure 14.53). In the case of liquid-phase hydrogen generation of olefins in ethyl acetate solution, the conversion efficiency was, respectively, 6.9% and 9.1% for cis-cyclooctene and for n-hexane when the porous Al_2O_3 thin-film shells were not formed. When the shells were formed, the conversion efficiency was, respectively, <0.02% and 4.5% for cis-cyclooctene and for n-hexane, exhibiting a large molecular sieving effect.

Dry reforming of methane (DRM) is an efficient way to convert greenhouse gas (CO_2 and CH_4) to synthesis gas. However, the catalysts for DRM suffer from

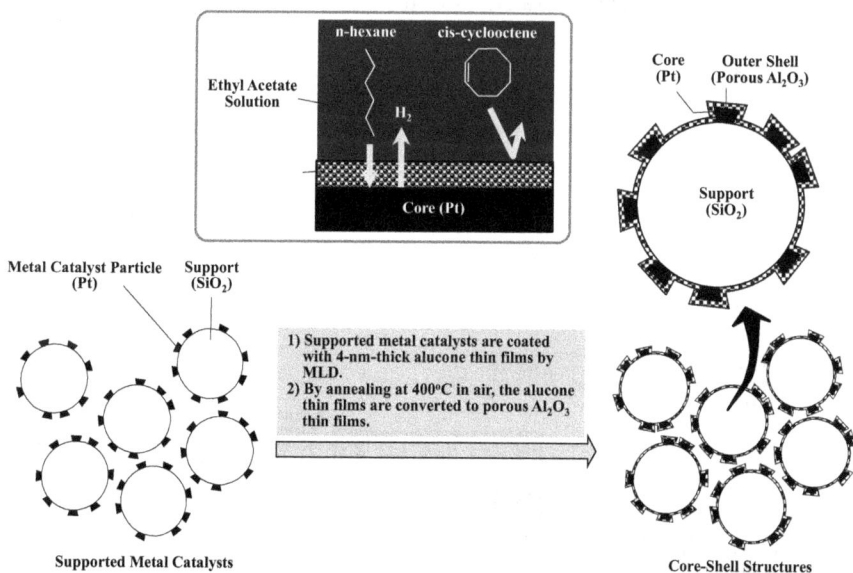

FIGURE 14.53 Fabrication process of thermally stable and highly-dispersed Pt nanoparticle catalysts utilizing MLD and calcination, and reaction selectivity through the molecular sieving. (Drawn by the author based on the description in [7,66].)

deactivation during the reforming reaction. It is caused by sintering of catalytically active nanoparticles and formation of filamentous carbon [7,66].

Ingale et al. solved the problems by applying MLD for the stabilization of the nanoparticles [67]. NiO/SiO_2 structures, where mesoporous SiO_2 is used as catalyst supports due to their high surface area and internal porosity, are coated with alucone MLD thin films using TMA and EG. The structures are reduced at 500°C in 5% H_2 stream, at the same time, the organic moieties in the alucone thin films are decomposed to form porous Al_2O_3 protective films for the Ni/SiO_2 catalyst. Three MLD cycles are enough for the stabilization of the catalyst at DRM operation temperature. Deactivation of the Ni/SiO_2 catalyst is prevented more efficiently by the porous Al_2O_3 protective films than by Al_2O_3 ALD thin films.

To improve the gas separation selectivity in SAPO-34 zeolite, inter-crystalline-pathway defects should be reduced. Yu et al. succeeded in it as follows. With the similar process described in Figure 14.53, as depicted in Figure 14.54, zeolite membranes are coated with microporous Al_2O_3 by MLD and calcination. The membranes have pore sizes of 0.36 nm. They are defect-free and have negligible flow in larger pores/defects, exhibiting very high H_2 separation selectivity: 1040 for H_2/N_2 and 20 for H_2/CO_2 [68].

Water purification enables removal of viruses, hardness, dissolved organic matter, and salts. Song et al. demonstrated a highly controllable method to prepare TiO_2 nanofiltration (NF) membranes with ~1-nm pores for water purification by using MLD. The fabrication process is shown in Figure 14.55. Titanicone MLD thin films are formed on surfaces of an anodic aluminum oxide (AAO) membrane support using $TiCl_4$ and EG. The films are converted to porous TiO_2 thin films by calcination to form the microporous TiO_2 membranes. Before the calcination, the membranes are dense and impermeable to water. After the calcination, the membranes become

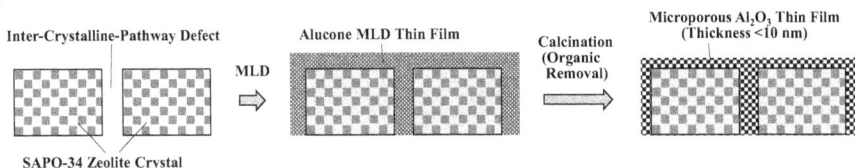

FIGURE 14.54 Fabrication process of defect-free inorganic membranes for H_2 separation utilizing MLD and calcination. (Adapted with permission from [68] M. Yu et al., *J. Am. Chem. Soc.* **133**, 1748–1750 (2011). Copyright (2011) American Chemical Society.)

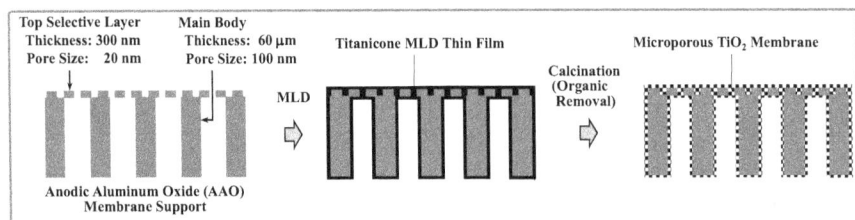

FIGURE 14.55 Fabrication process of TiO_2 nanofiltration membranes for water purification utilizing MLD and calcination. (Adapted with permission from [69] Z. Song et al., *J. Membr. Sci.* **510**, 72–78 (2016). Copyright (2011) American Chemical Society.)

permeable to water while they reject methylene blue (MB) and salts. The TiO_2 NF improves antifouling performance and recovery capability [69].

Molecular-scale membrane separation is used in many applications like sea water desalination, waste water treatment, and gas generation. MLD can control the separation layers of NF and tight ultrafiltration (TUF) membranes with monomolecular levels, and adjust the pore size by choosing organic constituents with certain chain lengths [70].

Wu et al. demonstrated a highly controllable method to prepare TiO_2 nanofiltration membranes for water purification [70]. Titanicone MLD thin films were formed on surfaces of plate ceramic ultrafiltration (UF) membranes composed of sintered TiO_2 nanoparticles. The thin films were converted to microporous TiO_2 thin films by calcination. Low-temperature calcination produced an amorphous TiO_2 thin film that worked as an NF membrane. High-temperature calcination produced a crystalline TiO_2 film that worked as a TUF membrane.

14.8.2 PHOTOCATALYSIS

Chen et al. fabricated nanoporous N-doped TiO_2 thin films on carbon nanocoils (CNCs). First, titanicone thin films were grown on CNCs by MLD with a four-step ABCB reaction sequence shown in Figure 14.56. This MLD process is regarded as a combination of Type-A MLD and Type-B MLD with $TiCl_4$ for Molecule A, EA for Molecule B, and malonyl chloride (MC) for Molecule C. Then, the titanicone thin films were annealed in H_2/Ar atmosphere at 600°C to produce nanoporous N-doped TiO_2 nanotubes. Amine groups in the titanicone MLD thin films allow nitrogen doping. The distance between adjacent Ti atoms can be tuned through the organic linker length [71].

It was revealed by evaluation using MB that the nanoporous TiO_2 thin films exhibit photocatalytic activity under visible light irradiation, and that MLD cycle of 200 is the optimum to enhance the photocatalytic activity. By replacing $TiCl_4$ with TMA, the method described above is applicable for preparing nanoporous Al_2O_3 thin films.

Tong et al. coated a vertically-oriented TiO_2 nanotube array (TNTA) conformally with ultrathin N-doped carbon films via the carbonization of polyimide MLD thin films, and simultaneously hydrogenated the thin films to create core/shell nanostructures with precisely-controlled shell thickness, as illustrated in Figure 14.57. The core/shell nanostructures provide larger heterojunction interfaces to substantially reduce the recombination of photogenerated electron–hole pairs, and the hydrogenation enhances solar absorption of the TNTA. In addition, the N-doped carbon thin films act as high catalytic active surfaces for oxygen evolution reaction, as well as protective films to prevent oxidation of hydrogen-treated TiO_2 nanotubes by electrolytes

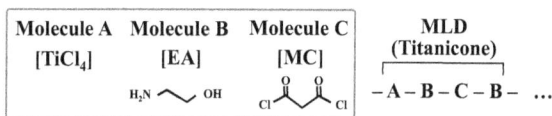

FIGURE 14.56 Four-step ABCB reaction sequence to form titanicone MLD thin films, which are converted to nanoporous N-doped TiO_2 nanotubes. (Drawn by the author based on the description in [71].)

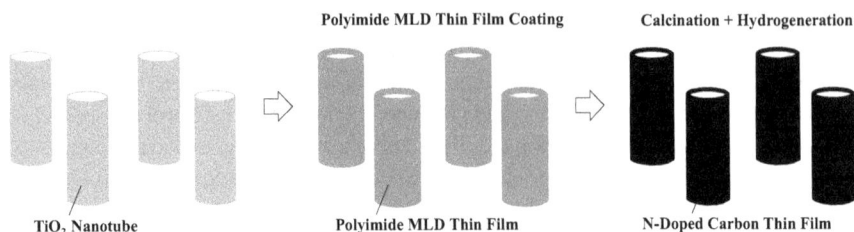

FIGURE 14.57 Deposition process of ultrathin N-doped carbon film via the carbonization of polyimide MLD thin films on surfaces of a vertically oriented TiO$_2$ nanotube array to create core/shell nanostructures with precisely-controlled shell thicknesses. (Drawn by the author based on the description in [72].)

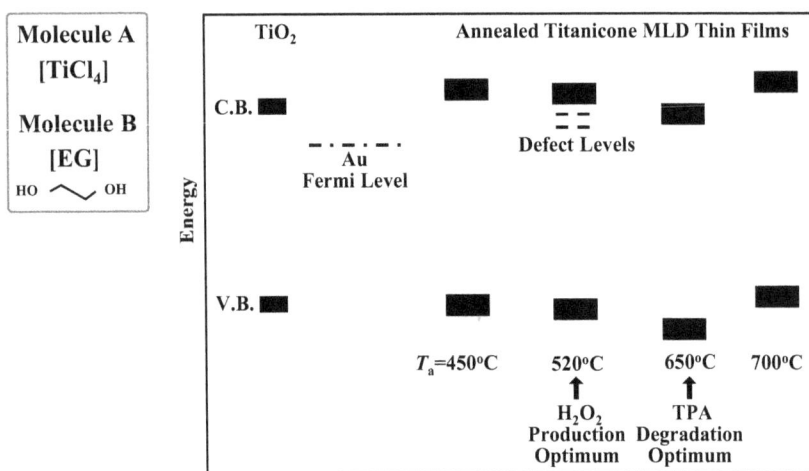

FIGURE 14.58 Energy-level schemes of annealed titanicone MLD thin films. (Adapted with permission from [73] D. Sarkar et al., *J. Phys. Chem.* C **120**, 3853–3862 (2016). Copyright (2016) American Chemical Society.)

or air. As a result, the TNTA coated with ~1-nm-thick N-doped carbon thin films displayed remarkably-improved photocurrent and photostability [72].

In titanium oxide thin films, by controlling oxygen vacancies in the film, the positions of the conduction/valence bands and defects can be tuned to enhance the photocatalytic activity. The oxygen vacancies also act as adsorption centers for Lewis acids and bases [73].

Sarkar et al. achieved the oxygen vacancy control by adjusting annealing conditions for titanicone MLD thin films to produce nonstoichiometric titanium oxide TiO$_{2-\delta}$ [73]. The titanicone MLD thin films were annealed at various temperatures (350°C–850°C) to obtain porous oxygen-deficient titanium oxide. Annealing temperatures T_a of 650°C and 520°C, respectively, led to the optimal performance for the photodegradation of TPA molecules and the direct photocatalytic production of H$_2$O$_2$. These are not typically attainable by stoichiometric titanium oxide. The mechanisms of the enhancement were explained as follows, referring to Figure 14.58.

- Photodegradation of TPA: $TiO_{2-\delta}$ films obtained by annealing at 650°C exhibit downshift of the valence band, which enhance the photodegradation activity.
- Photocatalytic production of H_2O_2: In the gold-decorated hybrid catalyst based on annealed titanicone films, the direct photocatalytic production is owing to the introduction of defect states. The defect levels are located between the conduction band and the Fermi level of gold in $TiO_{2-\delta}$ films obtained by annealing at 520°C. They mediate the charge separation and improve the injection of the excited electrons from the oxide to the gold nanoclusters, where molecular oxygen is adsorbed and reduced.

14.9 OTHER APPLICATIONS

Yoshida et al. succeeded in conductivity switching in polyimide thin films grown by Type-A MLD <Homo-Bifunctional Molecules> using Molecule A of PMDA and Molecule B of ODA. By using a scanning prober microscope (SPM), the electrical conductivity of the polyimide films was switched between a high conductive state and a low conductive state. The "High"/"Low" current ratio was about 500 with a recording voltage of 5 V for 500 ms. The high conductive dot pattern was defined with a pitch of ~50 nm [74,75].

The mechanism of the conductivity switching is the CT in PMDA-ODA, as illustrated in Figure 14.59. ODA acts as an electron donor, and PMDA as an electron acceptor. By applying a high electric field, CT occurs between PMDA and ODA (or carrier injection from the tip to the polymer chain occurs). The high conductive state becomes a low conductive state after several minutes due to the instability of the CT state. The retention time can be controlled by the donor/acceptor strength and molecular structures. The functional polyimide MLD thin films are expected to be applied to nonvolatile and rewritable recording media for high-density data storage systems.

Lushington et al. controlled conductivity of Al_2O_3/carbon composites by employing the ABCB reaction sequence shown in Figure 14.60 to form alucone MLD thin films. This MLD process is regarded as a combination of Type-A MLD <Homo-Bifunctional Molecules> and Type-B MLD <Homo-Multifunctional Molecules> with TMA for Molecule A, EG for Molecule B, and terephthaloyl chloride (TC) for

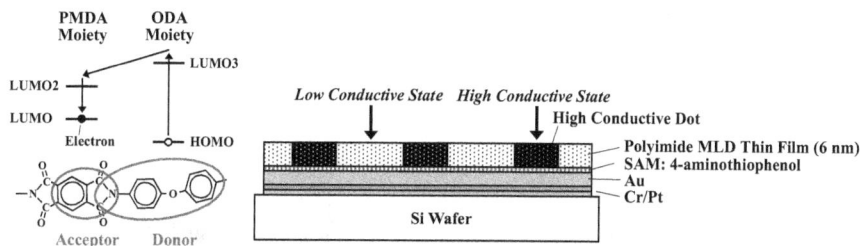

FIGURE 14.59 Mechanism of the conductivity switching in polyimide MLD thin films. (Drawn by the author based on the description in [75].)

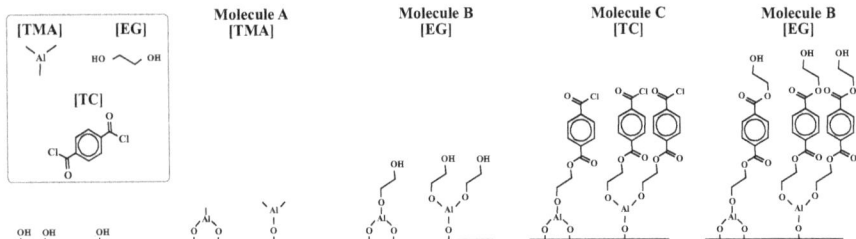

FIGURE 14.60 ABCB reaction sequence to form alucone MLD thin films, which are converted to conductive Al_2O_3/carbon composites. (Adapted with permission from [76] A. Lushington et al., *Appl. Surf. Sci.* **357**, 1319–1324 (2015). Copyright (2015) Elsevier.)

Molecule C. They grew alucone thin films with MLD sequences of [A(BC)$_n$A(BC) $_n$A…], and pyrolyzed the films under a reducing atmosphere to obtain a conductive Al_2O_3/carbon composite. The carbon content in the alucone MLD thin film is tuned by the number of the BC cycles, namely, the carbon content increases with increasing n. The low sheet resistance of the pyrolyzed alucone MLD thin films is attributed to nanocrystalline graphite generated in the film. By increasing the carbon content in the alucone MLD thin film, the degree of the sheet resistance reduction becomes large. This approach also enables us to fabricate porous Al_2O_3 with various pore sizes, indicating a possible application to batteries [76].

A problem in carbonaceous materials for microwave absorption is the microwave reflection on the materials due to the electromagnetic mismatch. Yan et al. solved the problem by fabricating microwave absorption materials consisting of the tailored wire-in-tube structure schematically shown in Figure 14.61. After polyimide MLD thin films are grown on ZnO nanowire surfaces using PMDA and ED, the wire-in-tube ZnO at carbon nanostructure is formed by calcination. The nanostructure has a core of ZnO nanowire, a carbon shell, and a void between them. The void mediates the impedance mismatch between air and the material. The void also contributes to an increase in the microwave transmission path by creating multi-reflection and scattering sites. The highly-conductive carbon shell contributes to formation of a 3-D conductive network. The heteroatoms, such as nitrogen, oxygen, or defects remaining in the carbon shell, act as polarization centers that contribute to the microwave absorption [77].

Seghete et al. applied alucone MLD thin films to sacrificial layers in an NEMS fabrication process depicted in Figure 14.62. On a Si wafer with a thermal oxide layer and alignment markers for EB lithography, an alucone MLD layer, an Al_2O_3 ALD adhesive layer, and a W ALD layer are successively formed. On the Al_2O_3/W ALD layer, a Ni mask for reactive ion etching (RIE) is formed utilizing EB lithography, and the Al_2O_3/W ALD layer is patterned by RIE. After Au electrode formation, the alucone MLD layer and the Ni mask are removed by wet etching to complete an NEMS structure [78].

Surface modification of polydimethylsiloxane (PDMS) is important for microfluidic devices and microcontact printing. Gong et al. succeeded in surface modification of PDMS utilizing MLD as follows. PDMS is exposed to TMA for 1 hour and H_2O for 30 min, allowing the vapors to infiltrate and react to create a mechanical and diffusion buffer layer in the PDMS surface region. 100-cycle ALD with TMA and

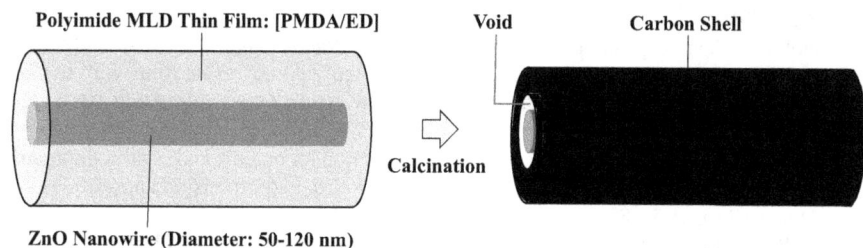

FIGURE 14.61 Microwave absorption materials consisting of the tailored wire-in-tube structure, which is formed from polyimide MLD thin films via calcination. (Adapted with permission from [77] L. Yan et al., *Chem. Eng. J.* **382**, 122860 (2020). Copyright (2020) Elsevier.)

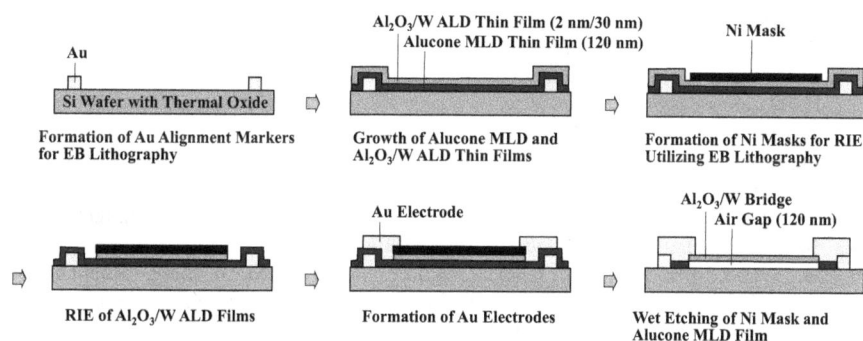

FIGURE 14.62 NEMS fabrication process, in which alucone MLD thin films are used as sacrificial layers. (Adapted with permission from [78] D. Seghete et al., *Sens. Actuators A* **155**, 8–15 (2009). Copyright (2009) Elsevier.)

H_2O, and 100-cycle Type-B MLD <Hetero-Multifunctional Molecules> with TMA and glycidol having epoxy group and -OH group produce stable hydrophilic surface on PDMS [79].

REFERENCES

1. S. M. George, B. Yoon, and A. A. Dameron, "Surface chemistry for molecular layer deposition of organic and hybrid organic-inorganic polymers," *Acc. Chem Res.* **42**, 498–508 (2009).
2. G. N. Parsons, S. M. George, and M. Knez, "Progress and future directions for atomic layer deposition and ALD-based chemistry," *MRS Bull.* **36**, 865–871 (2011).
3. M. Leskelä, M. Ritala, and O. Nilsen, "Novel materials by atomic layer deposition and molecular layer deposition," *MRS Bull.* **36**, 877–884 (2011).
4. H. Zhou and S. F. Bent, "Fabrication of organic interfacial layers by molecular layer deposition: Present status and future opportunities," *J. Vac. Sci. Technol. A* **31**, 040801 (2013).
5. D. C. Cameron and T. V. Ivanova, "Molecular layer deposition," *ECS Trans.* **58**, 263–275 (2013).

6. P. Sundberg and M. Karppinen, "Organic and inorganic–organic thin film structures by molecular layer deposition: A review," *Beilstein J. Nanotechnol.* **5**, 1104–1136 (2014).

7. X. Liang and A. W. Weimer, "An overview of highly porous oxide films with tunable thickness prepared by molecular layer deposition," *Curr. Opin. Solid State Mater. Sci.* **19**, 115–125 (2015).

8. X. Meng, "An overview of molecular layer deposition for organic and organic-inorganic hybrid materials: Mechanism, growth characteristics, and promising applications," *J. Mater. Chem. A* **5**, 18326–18378 (2017).

9. A. W. Weimer, "Particle atomic layer deposition," *J. Nanopart. Res.* **21**, 9–42 (2019).

10. X. Meng, "Atomic and molecular layer deposition in pursuing better batteries," *J. Mater. Res.* **36**, 2–25 (2021).

11. T. Yoshimura, S. Tatsuura, and W. Sotoyama, "Polymer films formed with monolayer growth steps by molecular layer deposition," *Appl. Phys. Lett.* **59**, 482–484 (1991).

12. T. Yoshimura, S. Tatsuura, W. Sotoyama, A. Matsuura, and T. Hayano, "Quantum wire and dot formation by chemical vapor deposition and molecular layer deposition of one-dimensional conjugated polymer," *Appl. Phys. Lett.* **60**, 268–270 (1992).

13. T. Bitzer and N. V. Richardson, "Demonstration of an imide coupling reaction on a Si(100)-2×1 surface by molecular layer deposition," *Appl. Phys. Lett.* **71**, 662–664 (1997).

14. A. Kubono, N. Yuasa, H.-L. Shao, S. Umemoto, and N. Okui, "In-situ study on alternating vapor deposition polymerization of alkyl polyamide with normal molecular orientation," *Thin Solid Films* **289**, 107–111 (1996).

15. O. Nilsen and H. Fjellvag, "Thin films prepared with gas phase deposition technique," International Patent WO 2006/071126 A1 (2006).

16. B. H. Lee, M. K. Ryu, S.-Y. Choi, K.-H. Lee, S. Im, and M. M. Sung, "Rapid vapor-phase fabrication of organic–inorganic hybrid superlattices with monolayer precision," *J. Am. Chem. Soc.* **129**, 16034–16041 (2007).

17. O. Nilsen and H. Fjellvag, "Synthesis of molecular metalorganic compounds," International Patent WO 2008/111850 A2 (2008).

18. A. A. Dameron, D. Seghete, B. B. Burton, S. D. Davidson, A. S. Cavanagh, J. A. Bertrand, and S. M. George, "Molecular layer deposition of alucone polymer films using trimethylaluminum and ethylene glycol," *Chem. Mater.* **20**, 3315–3326 (2008).

19. M. Putkonen, J. Harjuoja, T. Sajavaara, and L. Niinisto, "Atomic layer deposition of polyimide thin films," *J. Mater. Chem.* **17**, 664–669 (2007).

20. K. H. Yoon, K. S. Han, and M. M. Sung, "Fabrication of a new type of organic-inorganic hybrid superlattice films combined with titanium oxide and polydiacetylene," *Nanoscale Res. Lett.* **7**, 7–71 (2012).

21. A. Tanskanena and M. Karppinen, "Iron-based inorganic-organic hybrid and superlattice thin films by ALD/MLD," *Dalton Trans.* **44**, 19194–19199 (2015).

22. A. Tanskanen and M. Karppinen, "Iron-terephthalate coordination network thin films through in-situ atomic/molecular layer deposition," *Scientific Reports* **8**, 8976 (2018).

23. E. Ahvenniemi and M. Karppinen, "ALD/MLD processes for Mn and Co based hybrid thin films," *Dalton Trans.* **45**, 10730–10735 (2016).

24. L. Mai, Z. Giedraityte, M. Schmidt, D. Rogalla, S. Scholz, A. D. Wieck, A. Devi, and M. Karppinen, "Atomic/molecular layer deposition of hybrid inorganic–organic thin films from erbium guanidinate precursor," *J. Mater. Sci.* **52**, 6216–6224 (2017).

25. N. M. Adamczyk, A. A. Dameron, and S. M. George, "Molecular layer deposition of poly(p-phenylene terephthalamide) films using terephthaloyl chloride and p-phenylene-diamine," *Langmuir* **24**, 2081–2089 (2008).

26. M. Lillethorup, D. S. Bergsman, T. E. Sandoval, and S. F. Bent, "Photoactivated molecular layer deposition through iodo-ene coupling chemistry," *Chem. Mater.* **29**, 9897–9906 (2017).

27. D. S. Bergsman, R. G. Closser, and S. F. Bent, "Mechanistic studies of chain termination and monomer absorption in molecular layer deposition," *Chem. Mater.* **30** 5087–5097 (2018).

28. R. A. Nye, A. P. Kelliher, J. T. Gaskins, P. E. Hopkins, and G. N. Parsons, "Understanding molecular layer deposition growth mechanisms in polyurea via picosecond acoustics analysis," *Chem. Mater.* **32**, 1553–1563 (2020).

29. C. Ban and S. M. George, "Molecular layer deposition for surface modification of lithium-ion battery electrodes (review)," *Adv. Mater. Interfaces* **3**, 1600762 (2016).

30. D. S. Bergsman, R. G. Closser, C. J. Tassone, B. M. Clemens, D. Nordlund, and S. F. Bent, "Effect of backbone chemistry on the structure of polyurea films deposited by molecular layer deposition," *Chem. Mater.* **29**, 1192–1203 (2017).

31. B. Yoon, D. Seghete, A. S. Cavanagh, and S. M. George, "Molecular layer deposition of hybrid organic–inorganic alucone polymer films using a three-step ABC reaction sequence," *Chem. Mater.* **21**, 5365–5374 (2009).

32. Y.-H Li, D. Wang, and J. M. Buriak, "Molecular layer deposition of thiol–ene multilayers on semiconductor surfaces," *Langmuir* **26**, 1232–1238 (2010).

33. J. Fichtner, Y. Wu, J. Hitzenberger, T. Drewello, and J. Bachmannz, "Molecular layer deposition from dissolved precursors," *ECS J. Solid State Sci. Technol.* **6**, N171–N175 (2017).

34. X. Liang and A. W. Weimer, "Photoactivity passivation of TiO2 nanoparticles using molecular layer deposited (MLD) polymer films," *J. Nanopart. Res.* **12**, 135–142 (2010).

35. H. Zhou and S. F. Bent, "Molecular layer deposition of functional thin films for advanced lithographic patterning," *ACS Appl. Mater. Interfaces* **3**, 505–511 (2011).

36. H. Zhou and S. F. Bent, "Novel photoresist thin films with in-situ photoacid generator by molecular layer deposition," *Proc. SPIE* **8682**, 86820U (2013).

37. D. Bergsman, H. Zhou, and S. F. Bent, "Molecular layer deposition of nanoscale organic films for nanoelectronics applications," *ECS Trans.* **64**, 87–96 (2014).

38. H. Zhou, J. M. Blackwell, H.-B.-R. Lee, and S. F. Bent, "Highly sensitive, patternable organic films at the nanoscale made by bottom-up assembly," *ACS Appl. Mater. Interfaces* **5**, 3691–3696 (2013).

39. P. W. Loscutoff, H.-B.-R. Lee, and S. F. Bent, "Deposition of ultrathin polythiourea films by molecular layer deposition", *Chem. Mater.* **22**, 5563–5569 (2010).

40. P. W. Loscutoff, H. Zhou, S. B. Clendenning, and S. F. Bent, "Formation of organic nanoscale laminates and blends by molecular layer deposition," *ACS Nano* **4**, 331–341 (2010).

41. B. H. Lee, K. H. Lee, S. Im, and M. M. Sun, "Vapor-phase molecular layer deposition of self-assembled multilayers for organic thin-film transistor," *J. Nanosci. Nanotechnol.* **9**, 6962–6967 (2009).

42. A. I. Abdulagatov, R. A. Hall, J. L. Sutherland, B. H. Lee, A. S. Cavanagh, and S. M. George, "Molecular layer deposition of titanicone films using TiCl$_4$ and ethylene glycol or glycerol: Growth and properties," *Chem. Mater.* **24**, 2854–2863 (2012).

43. H. Zhou, M. F. Toney, and S. F. Bent, "Cross-linked ultrathin polyurea films via molecular layer deposition," *Macromolecules* **46**, 5638–5643 (2013).

44. H. Zhou and S. F. Bent, "Highly stable ultrathin carbosiloxane films by molecular layer deposition," *J. Phys. Chem. C* **117**, 19967–19973 (2013).

45. S. Cho, G. Han, K. Kim, and M. M. Sung, "High-performance two-dimensional polydiacetylene with a hybrid inorganic–organic structure," *Angew. Chem. Int. Ed.* **50**, 2742–2746 (2011).

46. J. Huang, H. Zhang, A. Lucero, L. Cheng, K. C. Santosh, J. Wang, J. Hsu, K. Choa, and J. Kim, "Organic–inorganic hybrid semiconductor thin films deposited using molecular-atomic layer deposition (MALD)," *J. Mater. Chem. C* **4**, 2382–2389 (2016).

47. T. Mimura, S. Hiyamizu, T. Fujii, and K. Nanbu, "A new field-effect transistor with selectively doped GaAs/n-Al$_x$Ga$_{1-x}$As heterojunctions," *Jpn. J. Appl. Phys.* **19**, L225–L227 (1980).

48. Y. C. Lee, "Atomic layer deposition/molecular layer deposition for packaging and interconnect of N/MEMS," *Proc. SPIE* **7928**, 792802 (2011).

49. J.-S. Park, H. Chae, H. K. Chung, and S. I. Lee, "Thin film encapsulation for flexible AM-OLED: A review," *Semicond. Sci. Technol.* **26**, 034001 (2011).

50. J. Park, J. Seth, S. Chob, and M. M. Sung, "Hybrid multilayered films comprising organic monolayers and inorganic nanolayers for excellent flexible encapsulation films," *Appl. Surf. Sci.* **502**, 144109 (2020).

51. L. D. Salmi, E. Puukilainen, M. Vehkamäki, M. Heikkilä, and M. Ritala, "Atomic layer deposition of Ta$_2$O$_5$/polyimide nanolaminates," *Chem. Vap. Depos.* **15**, 221–226 (2009).

52. B. H. Lee, B. Yoon, V. R. Anderson, S. M. George, "Alucone Alloys with tunable properties using alucone molecular layer deposition and Al$_2$O$_3$ atomic layer deposition," *J. Phys. Chem. C* **116**, 3250–3257 (2012).

53. B. H. Lee, K. K. Im., K. H. Lee, S. Im, and M. M. Sung, "Molecular layer deposition of ZrO$_2$-based organic–inorganic nanohybrid thin films for organic thin film transistors," *Thin Solid Films* **517**, 4056–4060 (2009).

54. Y. Park, K. S. Han, B. H. Lee, S. Cho, and M. M. Sung, "High performance n-type organic–inorganic nanohybrid semiconductors for flexible electronic devices," *Org. Electron.* **12**, 348–352 (2011).

55. K. S. Han and M. M. Sung, "Molecular layer deposition of organic-inorganic hybrid films using diethylzinc and trihydroxybenzene," *J. Nanosci. Nanotechnol.* **14**, 6137–6142 (2014).

56. B. Yoon, B. H. Lee, and S. M. George, "Highly conductive and transparent hybrid organic–inorganic zincone thin films using atomic and molecular layer deposition," *J. Phys. Chem. C* **116**, 24784–24791 (2012).

57. S. Lee, G. Baek, J.-H. Lee, T. T. N. Van, A. S. Ansari, B. Shong, and J.-S. Park, "Molecular layer deposition of indicone and organic-inorganic hybrid thin films as flexible transparent conductor," *Appl. Surf. Sci.* **525**, 146383 (2020).

58. D. M. Piper, J. J. Travis, M. Young, S.-B. Son, S. C. Kim, K. H. Oh, S. M. George, C. Ban, and S.-H. Lee, "Reversible high-capacity Si nanocomposite anodes for lithium-ion batteries enabled by molecular layer deposition," *Adv. Mater.* **26**, 1596–1601 (2014).

59. X. Li, A. Lushington, Q. Sun, W. Xiao, J. Liu, B. Wang, Y. Ye, K. Nie, Y. Hu, Q. Xiao, R. Li, J. Guo, T.-K. Sham, and X. Sun, "Safe and durable high-temperature lithium–sulfur batteries via molecular layer deposited coating," *Nano Lett.* **16**, 3545–3549 (2016).

60. M. Nisula and M. Karppinen, "Atomic/molecular layer deposition of lithium terephthalate thin films as high rate capability Li-ion battery anodes," *Nano Lett.* **16**, 1276–1281 (2016).

61. Y. Zhao, M. Amirmaleki, Q. Sun, C. Zhao, A. Codirenzi, L. V. Goncharova, C. Wang, K. Adair, X. Li, X. Yang, F. Zhao, R. Li, T. Filleter, M. Cai, and X. Sun, "Natural SEI-inspired dual-protective layers via atomic/molecular layer deposition for long-life metallic lithium anode," *Matter* **1**, 1215–1231 (2019).

62. X. Meng, K. C. Lau, H. Zhou, S. K. Ghosh, M. Benamara, and M. Zou, "Molecular layer deposition of crosslinked polymeric lithicone for superior lithium metal anodes," *Energy Mater. Adv.* 2021, 9786201 (2021).

63. A. Pearse, T. Schmitt, E. Sahadeo, D. M. Stewart, A. Kozen, K. Gerasopoulos, A. A. Talin, S. B. Lee, G. W. Rubloff, and K. E. Gregorczyk, "Three-dimensional solid-state lithium-ion batteries fabricated by conformal vapor-phase chemistry, *ACS Nano* **12**, 4286–4294 (2018).

64. K. Kaliyappan, T. Or, Y.-P. Deng, Y. Hu, Z. Bai, and Z. Chen, "Constructing safe and durable high-voltage P2 layered cathodes for sodium ion batteries enabled by molecular layer deposition of alucone," *Adv. Funct. Mater.* **30**, 1910251 (2020).

65. K. B. Lausund, M. S. Olsen, P.-A. Hansen, H. Valen, and O. Nilsen, "MOF thin films with bi-aromatic linkers grown by molecular layer deposition," *J. Mater. Chem. A* **8**, 2539–2548 (2020).

66. X. H. Liang, D. M. King, P. Li, S. M. George, and A. W. Weimer, "Nanocoating hybrid polymer films on large quantities of cohesive nanoparticles by molecular layer deposition," *AIChE J.* **55**, 1030–1039 (2009).

67. P. Ingale, C. Guan, R. Kraehnert, R. N. d'Alnoncourt, A. Thomas, and F. Rosowski, "Design of an active and stable catalyst for dry reforming of methane via molecular layer deposition," *Catal. Today* **362**, 47–54 (2021).

68. M. Yu, H. H. Funke, R. D. Noble, and J. L. Falconer, "H_2 separation using defect-free, inorganic composite membranes," *J. Am. Chem. Soc.* **133**, 1748–1750 (2011).

69. Z. Song, M. Fathizadeh, Y. Huang, K. H. Chu, Y. Yoon, L. Wang, W. L. Xu, and M. Yu, "TiO_2 nanofiltration membranes prepared by molecular layer deposition for water purification," *J. Membr. Sci.* **510**, 72–78 (2016).

70. S. Wu, Z. Wang, S. Xiong, and Y. Wang, "Tailoring TiO_2 membranes for nanofiltration and tight ultrafiltration by leveraging molecular layer deposition and crystallization," *J. Membr. Sci.* **578**, 149–155 (2019).

71. C. Chen, P. Li, G. Wang, Y. Yu, F. Duan, C. Chen, W. Song, Y. Qin, and M. Knez, "Nanoporaous nitrogen-doped titanium dioxide with excellent photocatalytic activity under visible light irradiation produced by molecular layer deposition," *Angew. Chem. Int. Ed.* **52**, 9196–9200 (2013).

72. X. Tong, Peng Yang, Y. Wang, Y Qin, and X. Guo, "Enhanced photoelectrochemical water splitting performance of TiO_2 nanotube arrays coated with an ultrathin nitrogen-doped carbon film by molecular layer deposition," *Nanoscale* **6**, 6692–6700 (2014).

73. D. Sarkar, S. Ishchuk, D. H. Taffa, N. Kaynan, B. A. Berke, T. Bendikov, and R. Yerushalmi, "Oxygen-deficient titania with adjustable band positions and defects; molecular layer deposition of hybrid organic-inorganic thin films as precursors for enhanced photocatalysis," *J. Phys. Chem. C* **120**, 3853–3862 (2016).

74. S. Yoshida, T. Ono, and M. Esashi, "Deposition of conductivity-switching polyimide film by molecular layer deposition and electrical modification using scanning probe microscope," *Micro Nano Lett.* **5**, 321–323 (2010).

75. S. Yoshida, T. Ono, and M. Esashi, "Local electrical modification of a conductivity-switching polyimide film formed by molecular layer deposition," *Nanotechnology* **22**, 335302 (2011).

76. A. Lushington, J. Liu, M. N. Bannis, B. Xiao, S. Lawes, R. Li, and X. Sun, "A novel approach in controlling the conductivity of thin films using molecular layer deposition," *Appl. Surf. Sci.* **357**, 1319–1324 (2015).

77. L. Yan, M. Zhang, S. Zhao, T. Sun, B. Zhang, M. Cao, and Y. Qin, "Wire-in-tube ZnO@carbon by molecular layer deposition: Accurately tunable electromagnetic parameters and remarkable microwave absorption," *Chem. Eng. J.* **382**, 122860 (2020).

78. D. Seghete, B. D. Davidson, R. A. Hall, Y. J. Chang, V. M. Bright, and S. M. George, "Sacrificial layers for air gaps in NEMS using alucone molecular layer deposition," *Sens. Actuators A* **155**, 8–15 (2009).

79. B. Gong, J. C. Spagnola, and G. N. Parsons, "Hydrophilic mechanical buffer layers and stable hydrophilic finishes on polydimethylsiloxane using combined sequential vapor infiltration and atomic/molecular layer deposition," *J. Vac. Sci. Technol A* **30**, 01A156 (2012).

Appendix I
Additional MLD Processes

Some disclosures in patents are presented for expected variants of molecular layer deposition (MLD) and polymer wire/sheet networks containing inorganic components [1,2].

AI.1 EXPECTED VARIANTS OF MLD

AI.1.1 MLD UTILIZING STERIC HINDRANCE

Figure AI.1 illustrates the bonding mechanism for Molecule 1 with added groups R and Molecule 2 with added groups R'. The same kind of molecules are not bonded due to the steric hindrance between the added groups while different kinds of molecules are bonded because shapes of R and R' are matched to cause no steric hindrance. This bonding mechanism resembles "whether or not a key is fit to a key hole."

Conceptual structures of source molecules for MLD utilizing steric hindrance are summarized in Figure AI.2. Molecules K and H, respectively, have R groups and R' groups, implying that Molecules K and K and Molecules H and H cannot be combined while Molecules K and H can be combined. In Molecules K_D and H_D, donor groups are added, and in Molecules K_A and H_A, acceptor groups are added. Figure AI.3 shows an example of an MLD process, where source molecules are sequentially provided in the order of Molecules K_A, H_A, K, H, K_D, and H_D.

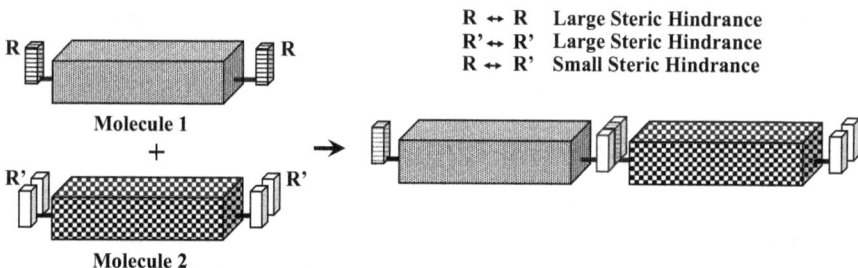

FIGURE AI.1 The bonding mechanism for Molecule 1 with added groups R and Molecule 2 with added groups R'.

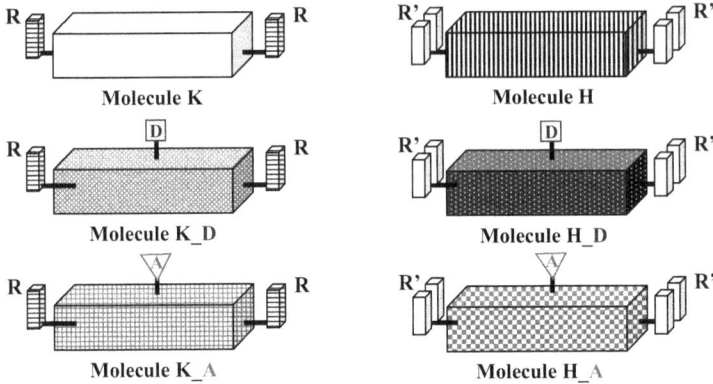

FIGURE AI.2   Conceptual structures of source molecules for MLD utilizing steric hindrance.

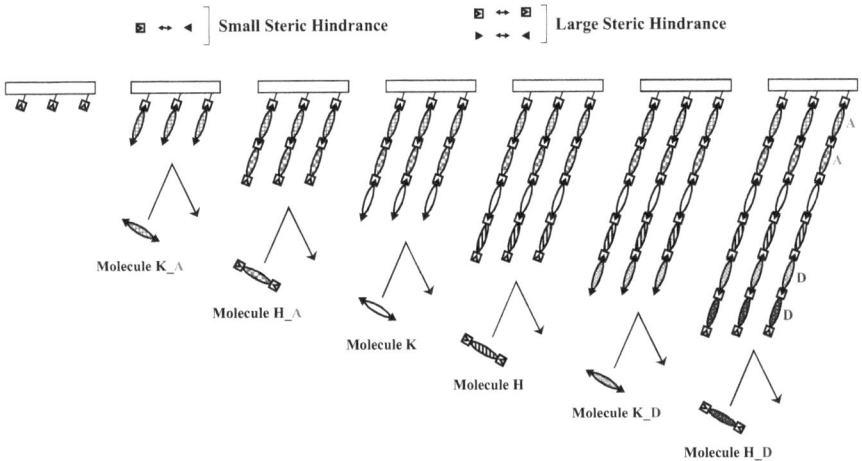

FIGURE AI.3   An example of an MLD process, where source molecules are sequentially provided in the order of Molecules K_A, H_A, K, H, K_D, and H_D.

AI.1.2   MLD Utilizing Adsorptive Force

Conceptual structures of source molecules for MLD utilizing adsorptive force are summarized in Figure AI.4. Molecules n and p, respectively, have n-type groups and p-type groups. Repulsive force works between p-type groups and between n-type groups while attractive force works between p-type and n-type groups. In Molecules n_D and p_D, donor groups are added, and in Molecules n_A and p_A, acceptor groups are added. Figure AI.5 shows an example of an MLD process, where source molecules are sequentially provided in the order of Molecules n_A, p_A, n, p, n_D, and p_D.

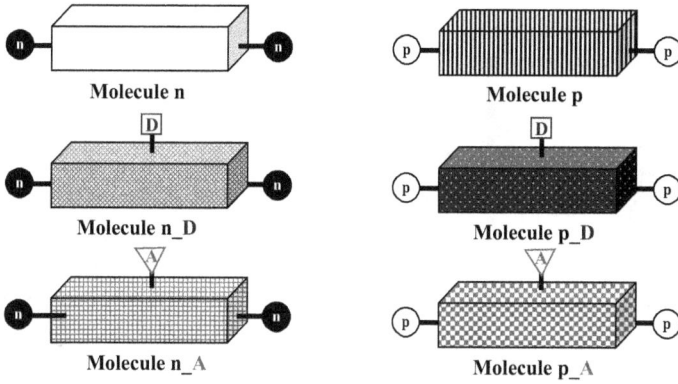

FIGURE AI.4   Conceptual structures of source molecules for MLD utilizing adsorptive force.

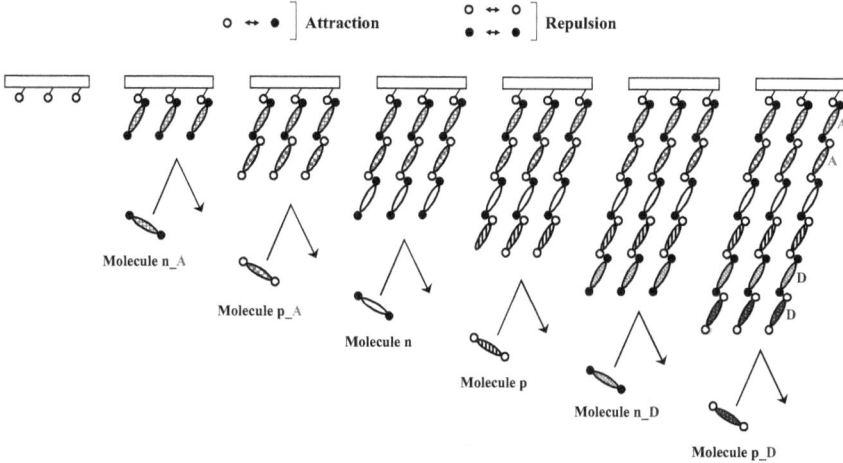

FIGURE AI.5   An example of an MLD process, where source molecules are sequentially provided in the order of Molecules n_A, p_A, n, p, n_D, and p_D.

AI.1.3   MLD EQUIPMENT

Figure AI.6 illustrates an example of full-specification MLD equipment. It possesses a gas shower system connected to molecular cells with mass flow controllers and heating/cooling devices. The substrate holder has housing heaters and circulation pipes for cooling medium like liquid nitrogen. Heating/cooling devices for the periphery of the chamber body are also placed. The equipment has capabilities of light irradiation from windows, plasma generation, and electron beam (EB) shower application to activate chemical reaction, as well as a capability of voltage application to substrate surfaces using grid electrodes. This equipment can perform the photo-assisted MLD with photoactive source molecules and the electric-field-assisted MLD.

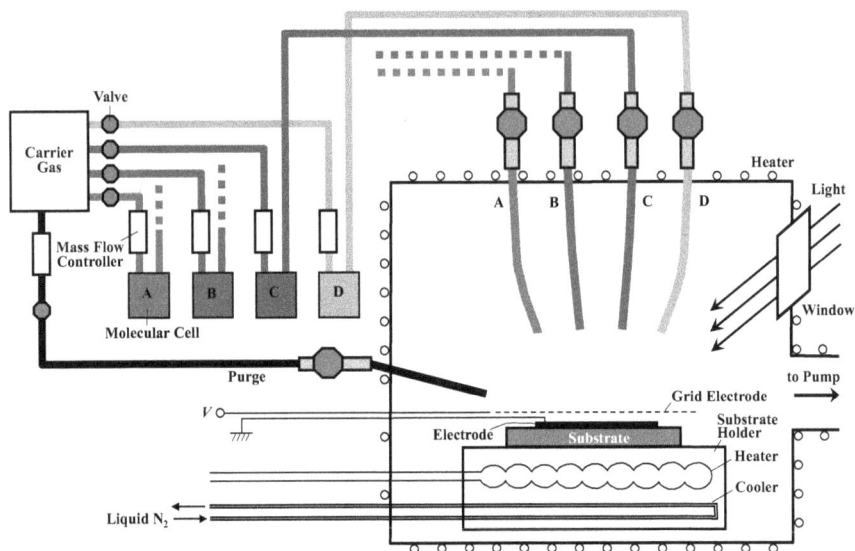

FIGURE AI.6 An example of full-specification MLD equipment, possessing a light irradiation system, and grid electrodes for plasma generation and EB shower/voltage application.

AI.2 POLYMER WIRE/SHEET NETWORKS CONTAINING INORGANIC COMPONENTS

AI.2.1 GENERAL DESCRIPTION OF MLD PROCESSES FOR POLYMER WIRE/SHEET GROWTH

In Figure AI.7, an MLD process for polymer wire growth is schematically illustrated. Here, W, X, Y, and Z represent atoms. R_A and R_B are combined while R_A and R_A, and R_B and R_B are not combined. By providing source molecules sequentially in the order of Molecules W, X, Y, and Z, a polymer wire of W-X-Y-Z grows.

In Figure AI.8, an MLD process for polymer sheet growth is schematically illustrated. Here, R_A and R_B are combined while R_A and R_A, and R_B and R_B are not combined. R_α and R_β are combined while R_α and R_α, and R_β and R_β are not combined. Furthermore, R_A and R_B are not combined with R_α and R_β. Firstly, source molecules are sequentially provided in the order of Molecules W1, W2, W1, W2, ..., to grow a polymer wire of -W-W-W-W-. Then, by providing Molecules X1 and X2 simultaneously, a two-dimensional network grows. By providing Y1/Y2 and X1/X2 repeatedly, the two-dimensional network extends.

	Large Reactivity	Small Reactivity
	$R_A \leftrightarrow R_B$	$R_A \leftrightarrow R_A$
		$R_B \leftrightarrow R_B$

$$
\begin{array}{cccc}
R_1 & R_3 & R_5 & R_7 \\
| & | & | & | \\
R_A-W-R_A & R_B-X-R_B & R_A-Y-R_A & R_B-Z-R_B \\
| & | & | & | \\
R_2 & R_4 & R_6 & R_8 \\
\text{Molecule W} & X & Y & Z
\end{array}
$$

$$
\begin{array}{cccc}
\text{Molecule W} & X & Y & Z \\
R_1 & R_1\ R_3 & R_1\ R_3\ R_5 & R_1\ R_3\ R_5\ R_7 \\
\bigl\rangle\!-W-R_A & \bigl\rangle\!-W-X-R_B & \bigl\rangle\!-W-X-Y-R_A & \bigl\rangle\!-W-X-Y-Z-R_B \\
R_2 & R_2\ R_4 & R_2\ R_4\ R_6 & R_2\ R_4\ R_6\ R_8
\end{array}
$$

MLD Process for Polymer Wire Growth

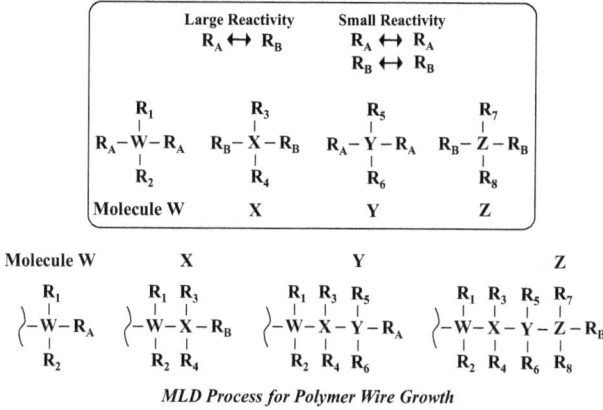

FIGURE AI.7 An MLD process for polymer wire growth.

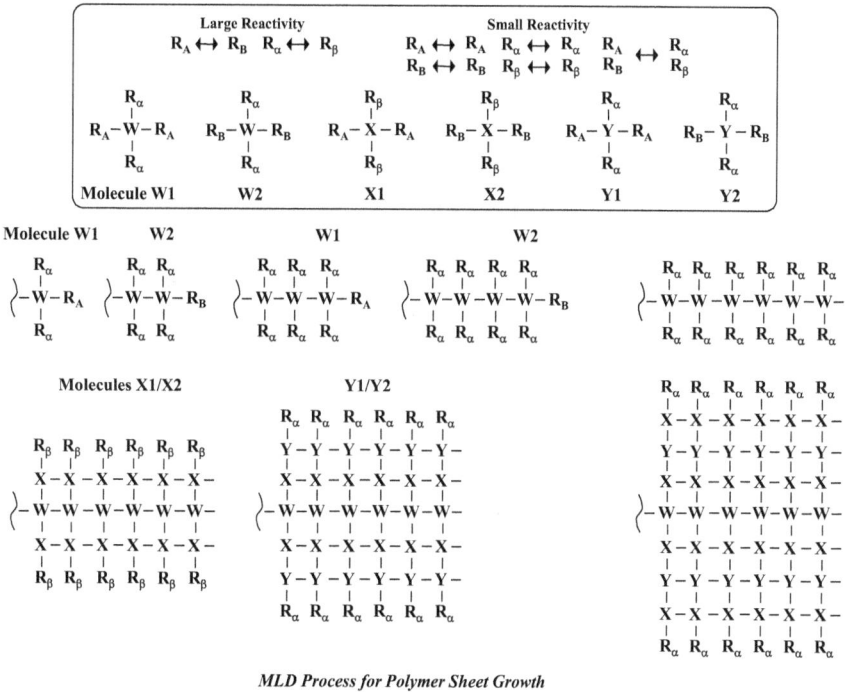

	Large Reactivity		Small Reactivity		
	$R_A \leftrightarrow R_B$ $R_\alpha \leftrightarrow R_\beta$		$R_A \leftrightarrow R_A$ $R_\alpha \leftrightarrow R_\alpha$	$R_A \leftrightarrow R_\alpha$	
			$R_B \leftrightarrow R_B$ $R_\beta \leftrightarrow R_\beta$	$R_B \leftrightarrow R_\beta$	

$$
\begin{array}{cccccc}
R_\alpha & R_\alpha & R_\beta & R_\beta & R_\alpha & R_\alpha \\
| & | & | & | & | & | \\
R_A-W-R_A & R_B-W-R_B & R_A-X-R_A & R_B-X-R_B & R_A-Y-R_A & R_B-Y-R_B \\
| & | & | & | & | & | \\
R_\alpha & R_\alpha & R_\beta & R_\beta & R_\alpha & R_\alpha \\
\text{Molecule W1} & W2 & X1 & X2 & Y1 & Y2
\end{array}
$$

Molecule W1 W2   W1   W2

$$
\begin{array}{ccccc}
R_\alpha & R_\alpha\ R_\alpha & R_\alpha\ R_\alpha\ R_\alpha & R_\alpha\ R_\alpha\ R_\alpha\ R_\alpha & R_\alpha\ R_\alpha\ R_\alpha\ R_\alpha\ R_\alpha\ R_\alpha \\
\bigl\rangle\!-W-R_A & \bigl\rangle\!-W-W-R_B & \bigl\rangle\!-W-W-W-R_A & \bigl\rangle\!-W-W-W-W-R_B & \bigl\rangle\!-W-W-W-W-W-W- \\
R_\alpha & R_\alpha\ R_\alpha & R_\alpha\ R_\alpha\ R_\alpha & R_\alpha\ R_\alpha\ R_\alpha\ R_\alpha & R_\alpha\ R_\alpha\ R_\alpha\ R_\alpha\ R_\alpha\ R_\alpha
\end{array}
$$

Molecules X1/X2   Y1/Y2

$$
\begin{array}{ccc}
R_\beta\ R_\beta\ R_\beta\ R_\beta\ R_\beta\ R_\beta & R_\alpha\ R_\alpha\ R_\alpha\ R_\alpha\ R_\alpha\ R_\alpha & R_\alpha\ R_\alpha\ R_\alpha\ R_\alpha\ R_\alpha\ R_\alpha \\
X-X-X-X-X-X- & Y-Y-Y-Y-Y-Y- & X-X-X-X-X-X- \\
\bigl\rangle\!-W-W-W-W-W-W- & X-X-X-X-X-X- & Y-Y-Y-Y-Y-Y- \\
X-X-X-X-X-X- & \bigl\rangle\!-W-W-W-W-W-W- & X-X-X-X-X-X- \\
R_\beta\ R_\beta\ R_\beta\ R_\beta\ R_\beta\ R_\beta & X-X-X-X-X-X- & \bigl\rangle\!-W-W-W-W-W-W- \\
 & Y-Y-Y-Y-Y-Y- & X-X-X-X-X-X- \\
 & R_\alpha\ R_\alpha\ R_\alpha\ R_\alpha\ R_\alpha\ R_\alpha & Y-Y-Y-Y-Y-Y- \\
 & & X-X-X-X-X-X- \\
 & & R_\alpha\ R_\alpha\ R_\alpha\ R_\alpha\ R_\alpha\ R_\alpha
\end{array}
$$

MLD Process for Polymer Sheet Growth

FIGURE AI.8 An MLD process for polymer sheet growth.

AI.2.2 POLYMER WIRES CONTAINING SI, GE, AND C

In the example shown in Figure AI.9, R_A and R_B in Figure AI.7 are, respectively, -Cl and -H. By providing source molecules sequentially in the order of Molecules F, A, B, G, H, G, D, C, and F, a polymer wire of -Si-O-(Si-)$_9$-O-Si-, in which donor groups (-NMe$_2$) and acceptor groups (-NO$_2$) are distributed in designated

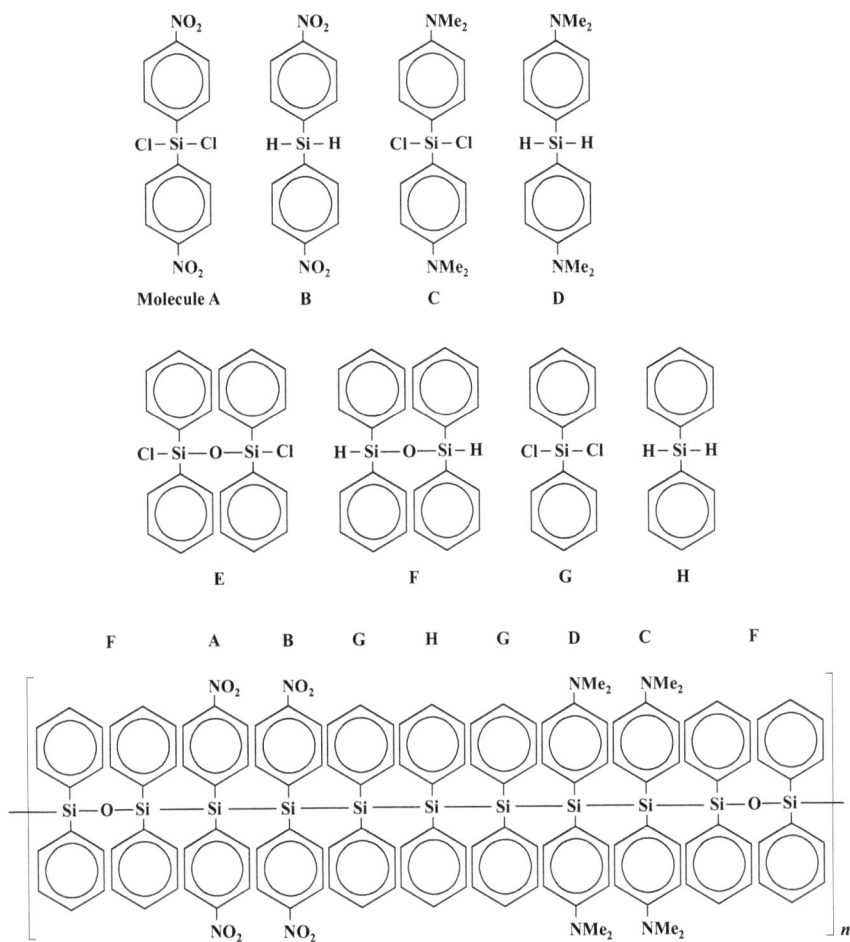

FIGURE AI.9 An example of an MLD process, where R_A and R_B in Figure AI.7 are, respectively, -Cl and -H. Source molecules are sequentially provided in the order of Molecules F, A, B, G, H, G, D, C, and F to grow a polymer wire of -Si-O-(Si-)$_9$-O-Si-, in which donor groups (-NMe$_2$) and acceptor groups (-NO$_2$) are distributed in designated arrangements.

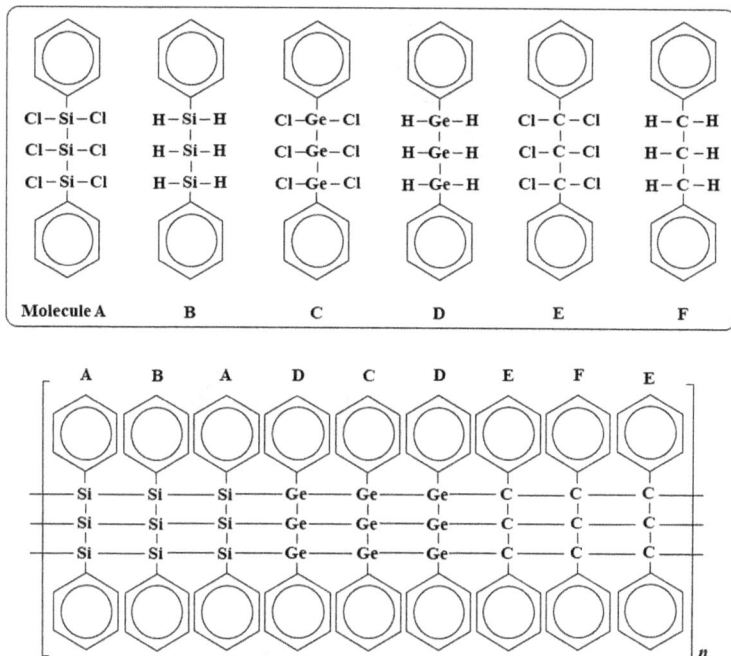

FIGURE AI.10 An example of an MLD process to grow a polymer wire with three-atom width. Source molecules are sequentially provided in the order of Molecules A, B, A, D, C, D, E, F, and E. Si, Ge, and C are distributed in the wire with designated arrangements.

arrangements, grows. The parts of Si wires are quantum dots or wires, and the parts of O are barriers. In the example shown in Figure AI.10, by providing source molecules in the order of Molecules A, B, A, D, C, D, E, F, and E, a polymer wire with three-atom width grows. Si, Ge, and C are distributed in the wire with designated arrangements.

In the illustrations, although the bonds between atoms of Si, O, Ge, and C are drawn with long lines, benzene rings might rotate to reduce the distance between atoms actually.

AI.2.3 Polymer Sheets

In the example shown in Figure AI.11, R_A and R_B are, respectively, -Cl and -H, and R_α and R_β are, respectively, -CHO and -NH_2 in Figure AI.8. By providing source molecules sequentially in the order of Molecules A/B, C/D, and E/F, a polymer sheet grows. Si, Ge, and C are distributed in the wire with designated arrangements. The π electrons are widely spread in the sheet. In the example shown in Figure AI.12, O atoms are distributed to form barriers in the two-dimensional system.

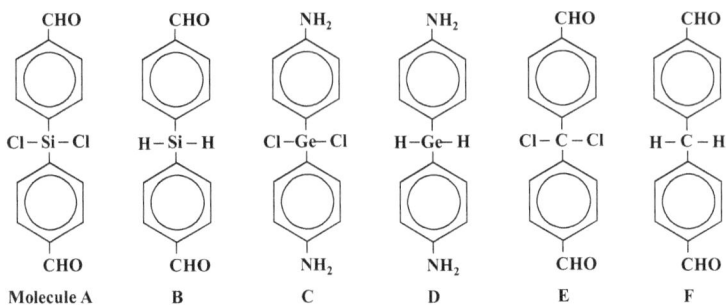

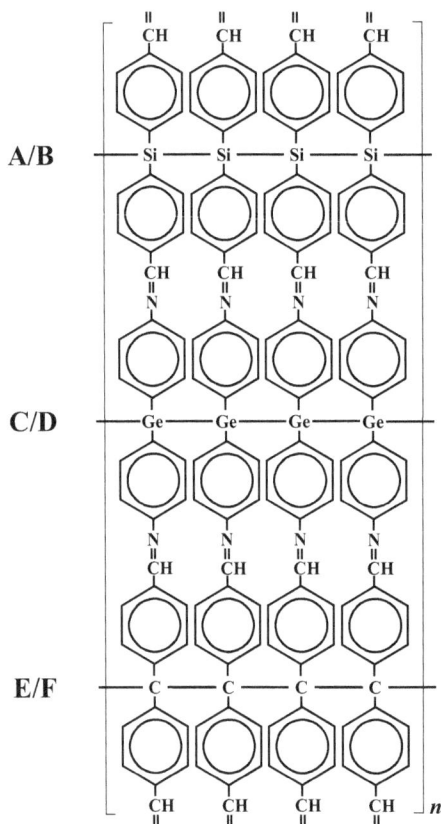

FIGURE AI.11 An example of an MLD process, where R_A, R_B, R_a, and R_b in Figure AI.8 are, respectively, -Cl, -H, -CHO, and -NH_2. Source molecules are sequentially provided in the order of Molecules A/B, C/D, and E/F to grow a polymer sheet.

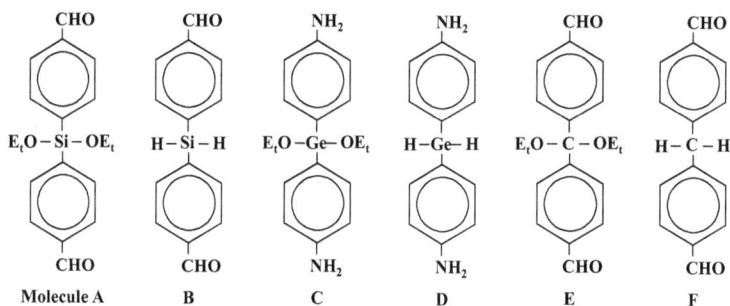

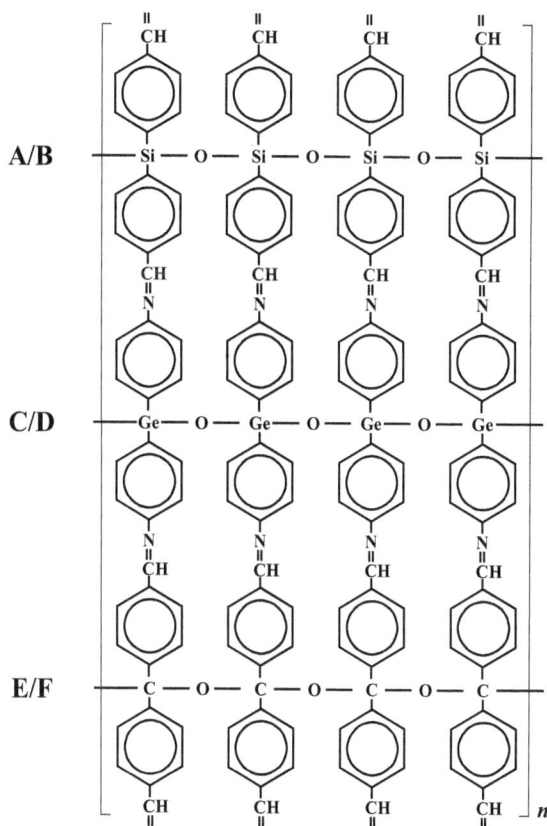

FIGURE AI.12 An example of an MLD process to grow a polymer sheet with barriers in the two-dimensional system.

REFERENCES

1. T. Yoshimura, E. Yano, S. Tatsuura, and W. Sotoyama, Organic functional optical thin film, fabrication and use thereof, U.S. Patent 5,444,811 (1995).
2. T. Yoshimura, Organic nonlinear optical material, U.S. Patent 5,194,548 (1993).

Appendix II
Analogy between Photons and Electrons

To understand the behavior of electrons and photons, it is convenient to consider the analogy between them. The electron wavefunction ψ and the electric field of a light-wave $\mathbf{E}$ are determined by the following wave equations:

$$\text{Schrödinger Equation} \quad \left(-\frac{\hbar^2}{2m}\nabla^2 + V(x)\right)\psi(x) = E\psi(x) \qquad \text{(AII.1)}$$

$$\text{Helmholtz Equation} \quad \left(\nabla^2 + \left(n(x)k_0\right)^2\right)\mathbf{E}(x) = 0 \qquad \text{(AII.2)}$$

Here, m, V, and E are, respectively, the mass, potential energy, and total energy of an electron. $\hbar$ is the Plank constant divided by 2π. n is the refractive index and k_0 is the wavenumber of a light wave in vacuum. Equations (AII.1) and (AII.2) are rewritten as follows:

$$\left.\begin{array}{l} \left(-\dfrac{\hbar^2}{2m}\nabla^2 - \left(E - V(x)\right)\right)\psi(x) = 0 \\[3mm] \left(-\dfrac{\hbar^2}{2m}\nabla^2 - \dfrac{\hbar^2 k_0^{\,2}}{2m}n^2(x)\right)E(x) = 0 \end{array}\right\} \qquad \text{(AII.3)}$$

By comparing these two, the following correspondence is found:

$$\psi(x) \leftrightarrow E(x) \qquad \text{(AII.4)}$$

$$E - V(x) \leftrightarrow \frac{\hbar^2 k_0^{\,2}}{2m}n^2(x) \qquad \text{(AII.5)}$$

$$[\psi(x) \leftrightarrow \mathbf{E}(x)]$$

$\psi(x)$ represents the probability amplitude for an electron. $\left|\psi(x)\right|^2$ is proportional to the probability to find an electron at x. The relationship of (AII.4) leads us to believe that $\mathbf{E}$ represents the probability amplitude for a photon. $\left|\mathbf{E}(x)\right|^2$ is proportional to the probability to find a photon at x.

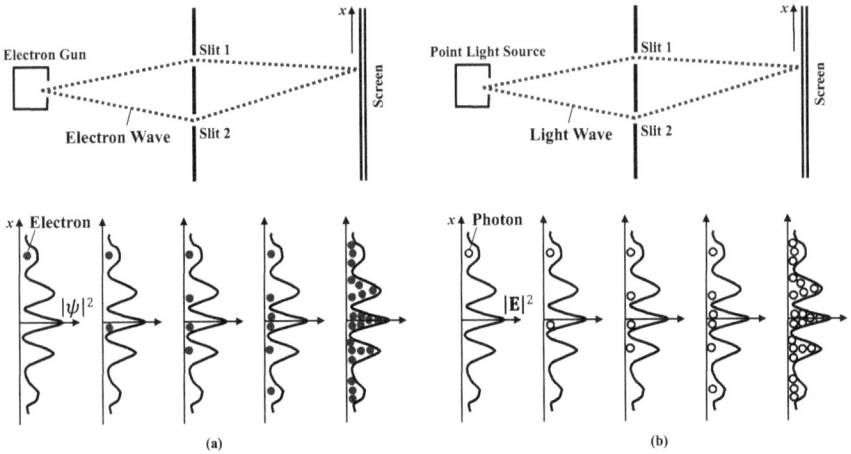

FIGURE AII.1   (a) Electron wave interference and (b) light wave interference.

It is known that an electron wave passing through two slits makes an interference pattern of probability $\propto |\psi(x)|^2$ that an electron arrives at x, as schematically illustrated in Figure AII.1a. Many electrons come to places, where $|\psi(x)|^2$ is large, making bright stripes consisting of many spots on the screen. Similarly, a lightwave passing through two slits makes an interference pattern of probability $\propto |E(x)|^2$ that a photon arrives at x, as illustrated in Figure AII.1b. Many photons come to places, where $|E(x)|^2$ is large, making bright stripes consisting of many spots on the screen.

Figure AII.2a depicts a Mach–Zehnder interferometer for electron waves [1]. An electron wave emitted from an emitter is divided into arm 1 and arm 2, and then converges into an output waveguide to reach a collector. When magnetic fields exist in the area between the two arms, a phase difference proportional to the magnetic field strength is induced between the wave propagating in arm 1 and the wave propagating in arm 2 via the Aharonov–Bohm effect. Then, the collector current oscillates as a function of the magnetic field due to the electron wave interference. This phenomenon is similar to that observed in the Mach–Zehnder light modulator, which operates via the light wave interference controlled using the Pockels effect, as depicted in Figure AII.2b.

$$[-V(x) \leftrightarrow n^2(x)]$$

An electron is confined in a region with low potential energy $V(x)$, namely, in a quantum well, as depicted in Figure AII.3a. The relationship of Equation (AII.5) indicates that "$-V(x)$" corresponds to "$n^2(x)$." This implies that a photon is confined in a region with a high refractive index. Therefore, by constructing a line-shaped region with a high refractive index in a planar substrate, a photon is confined there, as depicted in Figure AII.3b. The line-shaped region is the core of an optical waveguide, and the surrounding area is the cladding.

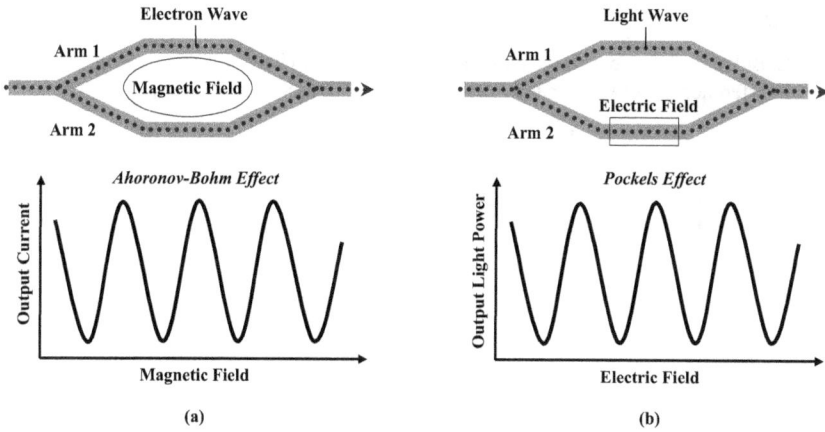

FIGURE AII.2   Mach–Zehnder interferometer for (a) electron waves and (b) light waves.

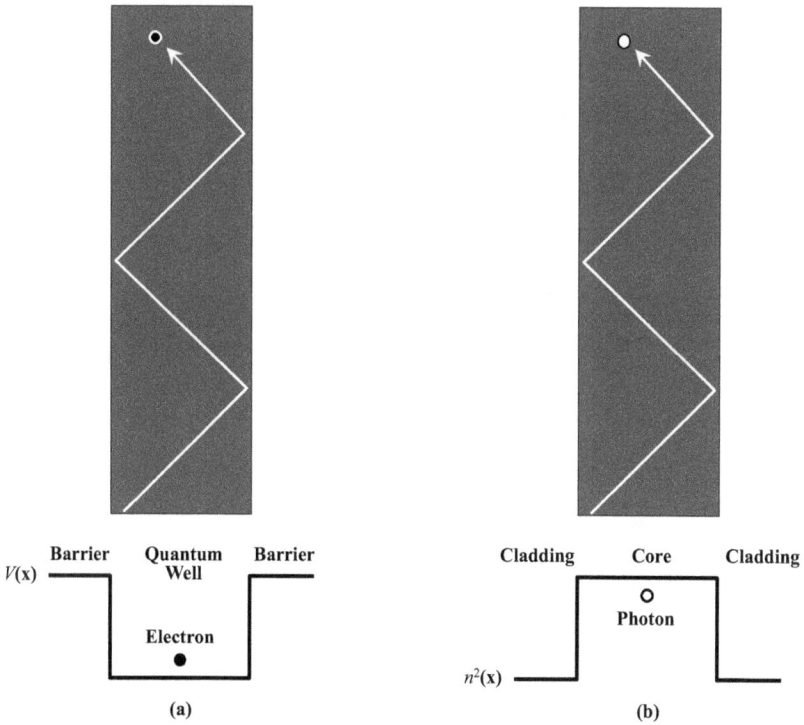

FIGURE AII.3   (a) Confinement of an electron in a quantum well and (b) confinement of a photon in an optical waveguide core.

REFERENCE

1. E. Buks, R. Schuster, M. Heiblum, D. Mahalu, and V. Umansky, "Dephasing in electron interference by a 'which-path' detector," *Nature* **391**, 871–874 (1998).

Epilogue

In January 2009, I had two offers to write books on different subjects. One of them was from CRC Press/Taylor & Francis for a book entitled *Thin-Film Organic Photonics: Molecular Layer Deposition and Applications*. I was afraid that it might be very hard to complete two books at the same time. So, I asked my wife, Yoriko, "How do you think about writing two books at the same time in English?" In reply, she said "I know that you have accomplished everything you once told me." Her answer made me accept the two offers.

During the preparation of these books, Yoriko passed away, on September 27, 2009. She continued to receive encouragement from her friends of Joshigakuin High School and Musashino Academia Musicae, mama-friends in Morimura Gakuen, students of her piano school, and her relatives and family for 5 months before September 27. I deeply thank Yoriko for her sincere support and encouragement since 1977, and have placed her photographs in the epilogue as a memorial to her.

It should be noted that students in my laboratory promoted the research with their excellent efforts and passions. Their faces came to my mind when I wrote their individual accomplishments.

By the way, our daughter, Naoko Yoshimura, designed the front cover of this book. Yoriko and I hope that MLD will contribute to various fields such as high-performance computing, energy, and cancer therapy in future.

Tetsuzo Yoshimura

July 9, 2022

Yoriko Yoshimura Playing Chaconne (Bach/Busoni), Tokyo, 1980s

From the left in the back row: Mari Tadokoro and Yoriko Yoshimura

From the left in the front row: Naoko and Chikako Yoshimura at GROVE, Tokyo, March 2009

Index

For Product Safety Concerns and Information please contact our EU
representative GPSR@taylorandfrancis.com
Taylor & Francis Verlag GmbH, Kaufingerstraße 24, 80331 München, Germany